COMPUTER ORGANIZATION

Pankaj Sharma
(Sr. Astt. Professor)
Department of Information Technology
ABES, Ghaziabad

S.K. KATARIA & SONS®
Publisher of Engineering Books
4885/109, Prakash Mahal, Dr. Subhash Bhargav Lane,
Opposite Delhi Medical Association, Daryaganj, New Delhi–110002 (INDIA)
Phone: +91-11-43551243; *Mobile:* +91-9871775858, 7678425209
e-mail: katariabook@gmail.com, katariabooks@yahoo.com
Website: www.skkatariaandsons.com
(e-books, subscriptions and bulk buying also available on our website)

Published by:

S.K. Kataria & Sons®

4885/109, Prakash Mahal, Dr. Subhash Bhargav Lane,
Opposite Delhi Medical Association, Daryaganj, New Delhi-110002 (INDIA)
Phone: +91-11-43551243; *Mobile:* +91-9871775858, 7678425209
e-mail: katariabook@gmail.com, katariabooks@yahoo.com
Website: www.skkatariaandsons.com

First Edition : January, 2011, Reprint: 2025

Laser Typesetting by : Plus Computers, Delhi

Printed at : New A.S. Printing Press, Delhi

✍ Preface to the First Edition.....

Computer Organization is a core subject for undergraduate students and Postgradutate students in Computer Science Engineering (CSE), Information Technology Streams and Master of Computer Application (MCA). This book is primarily intended to serve as a textbook in accordance with the syllabus of Introduction to Web Technology offered by various universities in India as well as abroad.

In this book, a significant effort has been made to find simple ways to develop theoretical as well as Practical concepts of Web Technology. A very strong emphasis has been given on Practical examples. Neat and Clean diagrams have been used for explanation.

Every effort has been made to make the concepts simple and comprehensive. This book is divided into twelve chapters :

The organisation of these chapters are :

Chapter 1 Introduces the Number Representation and Floating Point Representation.

Chapter 2 Introduces the Digital Computer Generation, Computer Types and Classification.

Chapter 3 Introduces the Register Transfer and Register Language.

Chapter 4 Introduces the Central Processing Unit.

Chapter 5 Introduces the Processor Organization.

Chapter 6 Introduces the Working of Control Unit

Chapter 7 Introduces the Memory and Memory Organization

Chapter 8 Introduces various Peripheral Devices.

The author hopes that the book will fulfill the need of readers and would welcome any suggestions towards the improvement of the book.

Pankaj Sharma
ABES, Ghaziabad

Acknowledgement.....

I get this opportunity to express my gratitude and thanks to the respected Director (administration) of ABES Engg. College, **Dr. S.K. Kansal** for his valuable technical suggestion and constant encouragement, without which this book would not have to come into existence.

I express kind gratitude of **Prof. R.G. Mendiratta**, Director (Academic) ABES Engg. College, Ghaziabad.

I wish to thank **Prof. C.L. Mittal** (Dean UG) ABES Engg. College, Gzb. I also like to remember the motivation initiated by **Shri M.D Punia** my father-in-Law, (Lead District Manager, PNB, Karnal).

I am grateful to **Prof. V.K. Srivastava,** Director of ABES, IT, Ghaziabad, for helping me through his insightful suggestions and construction inputs.

I wish to thank **Prof. Pradeep Sharma**, Head of Department of Information Technology, Department of ABES Engg. College, Ghaziabad, for their constant co-operation, directly inspiration and encourangement.

I like to remember the motivation initiated by **Prof. D. Pandey** (HOD CSE) and Prof. Sunil Sharma (AHODCSE), ABES, Engg. College. I am also thankful to Prof. K.P. Jayant (HODMCA) for his inspiration.

I thank to my elder brother **Mr. Sanjay Sharma** (KIET), Ghaziabad for inspiration and encourangement.

I thank to **Mr. Aswin Perti,** Lecturer in IT Department for his technical suggestions.

I am grateful to my younger Brother **Sandeep** and daughter **Prisha** for inspiring me for this project.

I deeply express my heartful thanks to the **Mr. Sanjeev Kataria** for **S.K. Kataria & Sons** for publishing this book in such a beautiful getup and well in time.

Lastly I am grateful to **Mr. Manoj Tyagi** and his staff from **Plus Computer's** for thier hard work.

Pankaj Sharma

Dedicated This book to

My Parents, Wife Priyanka, Daughter Prisha, Pihu

and

Readers of this book

Contents.....

Chapter-3

Chapter-4

Chapter-5

Chapter

1

Computer Organization, Basics Number System and Codes

Chapter Outline

- Introduction
- Computer Organization
- Number Representation
- Error Detection and Correction
- Hamming Code
- Decimal Number System
- Binary Number System
- Octal Number System
- Hexadecimal Number System
- Conversion of Decimal Number to Binary Number
- Conversion of Real Number
- Binary Addition
- Binary Subtraction
- Compliments

1.0 INTRODUCTION

A digital computer is a programmable device that processes information by manipulating symbols according to logical rules. Digital computers come in a wide variety of types, ranging from tiny, special-purpose devices embedded in cars and other devices to the familiar desktop computer, the minicomputer, the mainframe, and the supercomputer. The fastest supercomputer, as of early 2003, can execute up to 36 trillion instructions (elementary computational operations) per second; this record is certain to be broken. The impact of the digital computer on society has been tremendous. It is used to run everything from spacecraft to factories, healthcare systems to telecommunications, banks to household budgets. Since its invention during World War II, the electronic digital computer has become essential to the economies of the developed world.

Digital computer systems consist of three distinct units. These units are as follows: Input Unit, Central Processing Unit, Output Unit. These units are interconnected by electrical cables to permit communication between them. This allows the computer to function as a system.

1.1 COMPUTER ORGANIZATION

Computer Organization is divided into:

Hardware: CPU, Memory, Bus, Storage device.

Software: OS, Application, Utilities.

Hardware:

1. CPU

The Brain of Machine

Performs ALU Operations

Measurement: Speed in megahertz (Millions of tick per sec.)

Example: Intel Pentium

— AMD K6

— Motorola Power PC

— Sun SPARC

2. RAM

Main memory which is fast, but volatile. Analogous to person's short term memory. Many "On-Off" switches and on is represent by 1, "off" by 0.

Each switch is called binary bit or digit

8 bits = 1 byte

2^{10} bytes = 1024 bytes

2^{20} is called a kilo bytes (1 K) bytes is called a mega bytes (1 M).

3. DISK

Table storage using magnetic or optical media.

Slower to access than RAM

Examples:

— Floopy disk

— Hard disk

— CD Rom

4. THE BUS

Connects CPU to other hardware devices. Speed measured in megahertz (like CPU) but much slower than CPU.

Cache

While accessing RAM is faster than accessing secondary memory it is still quite slow elative to rate at which the CPU runs.

To getrid if this problem host systems add a fast cache memory to the CPU to store recently used instructions and data.

Hardware Summary

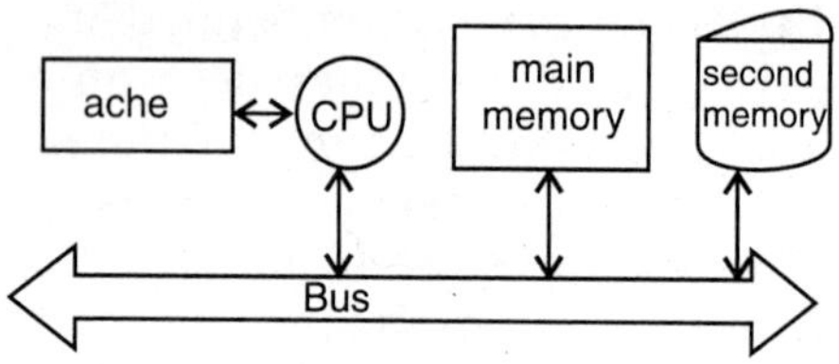

Software: OS

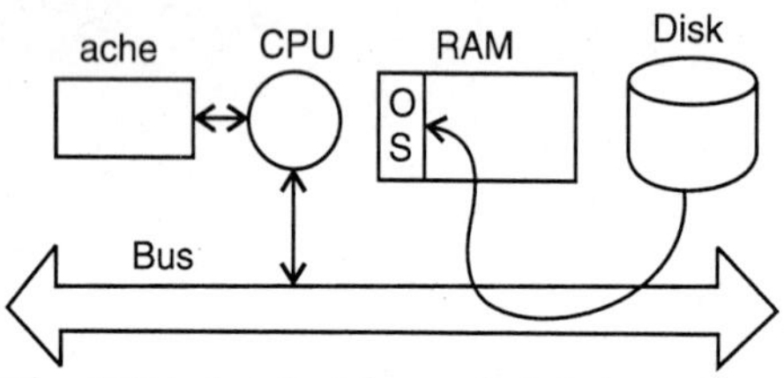

The OS is loaded from disk into main memory when the computer is turned on and remains in memory untill the computer is turned off.

OS acts as the "manager" of the system, making sure that each hardware device interacts smoothly with the others.

It also provides the interface by which the user interacts with computer.

Software : Applications

Applications are non OS programs that perform some useful task, including: word processors spread sheets, data bases, web browsers, C++ compilers.

Example C++ Compliers/environment :

— Code warrior (Mac OS, Win 95, Win NT, Solaris, ...)

— GNU C++ (Unix, Linux)

— Turbo/Borland C++ (Win NT, Win 95)

— VC++ (Win 95, Win NT)

Software : Programs

Programs that are neither OS programs nor application programs are called use programs.

User programs are written by user like C/C++ etc.

Example: Mac OS, Win 95, Win NT, UNIX, LINUX, Solaris, ...	

Putting it all together

Programs and applications that are not running are stores on the disk.

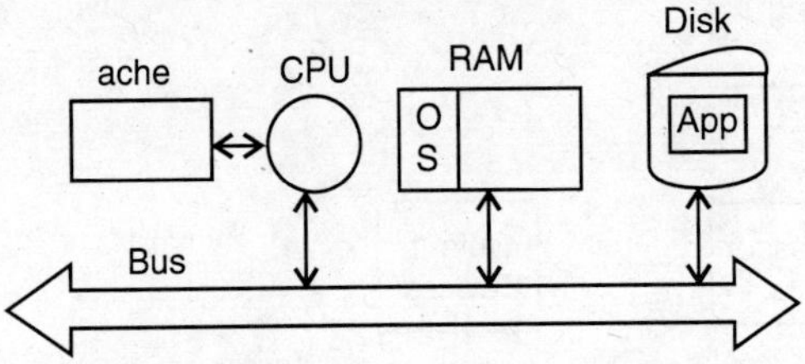

When you launch a program, the OS controls the CPU and leads the program from disk to RAM.

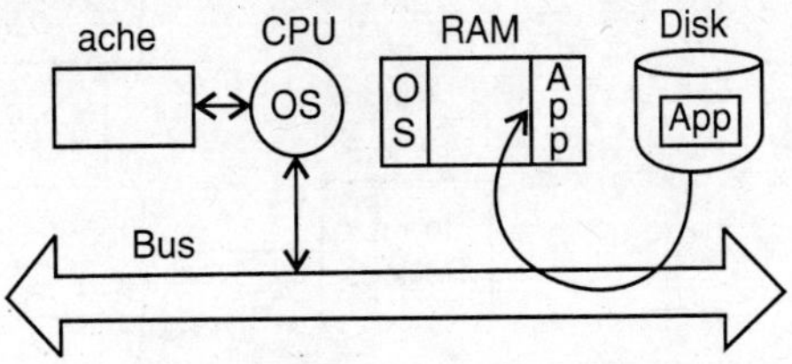

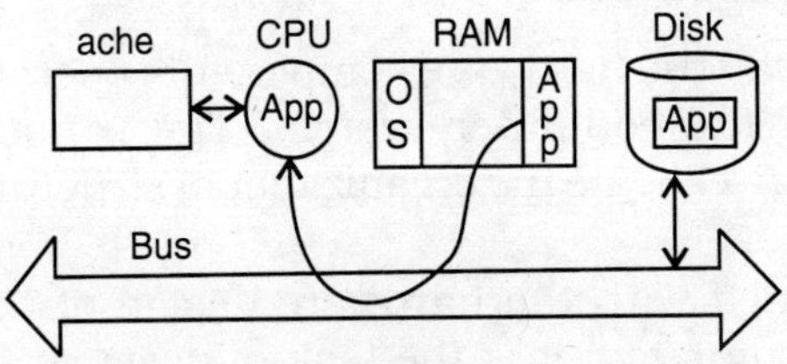

The OS then relinquishes the CPU to the program, which begins to run.

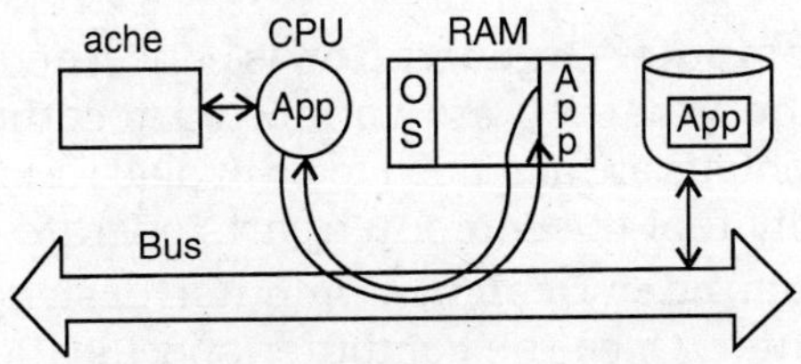

As the program runs, it repeatedly fetches the next instruction (from memory/cache) executes it and stores any results back to memory.

That is all a computer does fetch-decode-execute-store millions of times each second.

Computer System: A computer system is sometimes divided into two functional entities: hardware and software.

The Hardware of the computer consists of all the electronic components and electro-mechanical devices.

The computer software consists of the instructions and data that the computer manipulates to perform various data-processing tasks. A sequence of instructions for the computer is called a **program.**

The data that are **manipulated** by the **program** constitute the **database.**

The hardware of the computer is usually divided into three major parts. The Central Processing Unit (CPU) contains an arithmetic and logic unit for manipulating data a number of registers for storing data, and control circuits for fetching and executing instructions.

The memory of a computer contains storage for instructions and data. It is called a Random Access Memory (RAM) because CPU can access any location in memory at random and retrieve the binary information within a fixed interval of time.

The Inupt Output Processor (IOP) contains electronic circuits for communicating and controlling the transfer of information between the computer and outside world. The input and output devices connected to the computer include keywords, printers, terminals, magnetic disk drives and other communication devices.

Figure 1.1 shows the block diagram of digital computer.

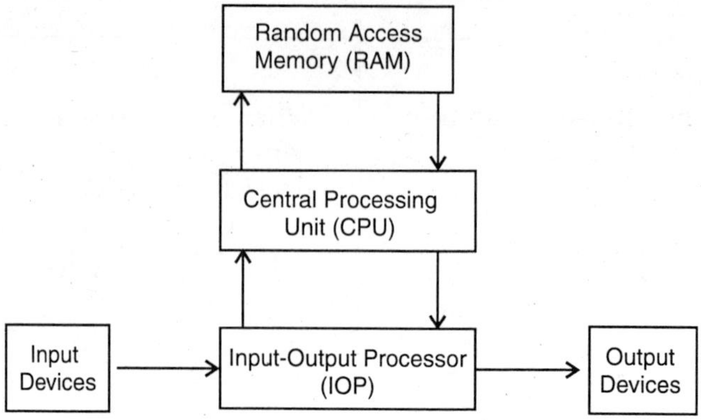

Fig. 1.1. Block diagram of digital computer

Computer Organization is concerned with the way the hardware components operate and the way they are connected together to form the computer system. The various components are assumed to be in place and task is to investigate the organizational structure to verify that the computer parts operate as intended.

Computer Design: Computer design is concerned with the hardware design of the computer. Once the computer specifications are formulated, it is the task of designer to develop hardware for the system. Computer design is concerned with the determination of what *h/w* should be used and how the parts should be connected. This aspect of computer hardware is sometimes referred to as computer implementation.

Computer Architecture: Computer architecture is conerned with the structure and behaviour of the computer as seen by the user. It includes the information formats, the instruction set and techniques for addressing memory. The architectural design of a computer system is concerned with the specifications of the various functional modules such as processors and memories, and structuring them together into a computer system.

Register Transfer Languages: The symbolic notation used to describe the micro-operation transfer among registers is called a register transfer language. The term "register transfer" implies the availability of *h/w* logic circuits that can perform a stated micro-operation and transfer the result of operation to the same or another register. The word "language" is borrowed from programmers, who apply this term to programming languages.

A register transfer language is a system for expressing in symbolic fom the micro-operation sequences among the registers of the digital module.

Register Transfer: Computer registers are designated by capital letters to denote the function of the register. For example, the register that holds an address for the memory unit is usually called a memory address register and is designated by the name MAR.

Other designations for registers are PC (for Program Counter), IR (for Instruction Register) and R_1 (for Processor Register).

The individual flip-flops in a *n*-bit register are numbered in sequence from 0 through *n*-1 starting from 0 in the right most position and increasing the numbers toward the left. Following figure shows the representation.

Information transfer from one register to another is designated in symbolic form by means of replacement operator. The statement

$$R_2 \longleftarrow R_1$$

denotes a transfer of the content of register R_1 into register R_2. It designates the replacement of the content of R_2 by the content of R_1. By the content of the source register R_1 definition does not change after the transfer. Following Fig. 1.2 shows the Register Bits.

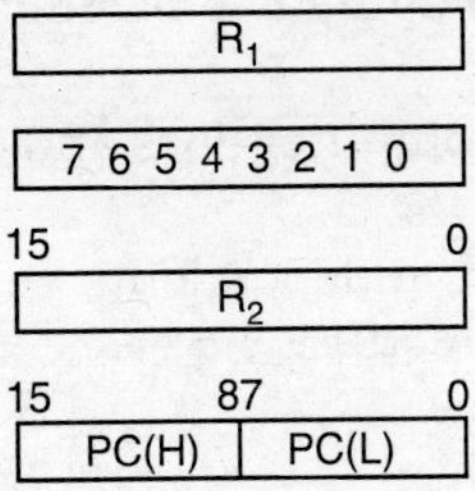

Fig. 1.2. Register

Normally, we want the transfer to occur only under a predetermined control condition. This can be shown by means of an if then statement.

If (P = 1) then ($R_2 \leftarrow R_1$) where P is a control signal generated in the control section. It is sometimes convenient to separate the control variables from the register transfer operation by specifying a control function. A control function is a boolean variable that is equal to/or 0. The control function is included in the statement as follows:

P: $R_2 \leftarrow R_1$

The control condition is terminated with colon.

Basic Symbols for Register Transfers:

Symbol	*Description*	*Examples*
Letters (and numbers)	Denotes a register	MAR, R_2
Parenthesis ( )	Denotes a part of register	R_2 (0 – 7), R_2 (L)
Arrow ←	Denotes transfer of information	$R_2 \leftarrow R_1$
Comma,	Separates two micro-operations	$R_2 \leftarrow R_1$, $R_1 \leftarrow R_2$

Three State Bus Buffers: A bus system can be constructed with three state gates instead of multiplexers. A three state gate is a digital circuit that exhibits three states. Two of the states are signals equivalent to logic 1 and 0 as in a conventional gate. The third state is a high-impedance state. The high impedance state behaves like an open circuit which means that the output is disconnected and does not have a logic significance. Three state gates may perform any conventional logic, such as AND or NAND. However, the one most commonly used in the design of a bus system is buffer gate.

The graphic symbol and bus line with three state buffer by having both normal input and control input. The control input determines the output state. When the control input is equal to '1', the output is enabled and gate behaves like any conventional buffer, with the output equal to the normal input. To when the control input is equal to '0' the output is disabled and gate goes to the high impedance state.

To construct a common bus for four registers of n bits each using three state buffers, we need n circuits with four buffers in each as shown in figure.

Organization: Dynamic interaction of these parts and their management (design).

Computer organization is how operational attributes are linked together and contribute to realize the architectural specifications. Computer architecture is the architectural atrributes like physical address memory, CPU and how they should be made and made to coordinate with each other keeping the future demands and goals in mind. **Computer architecture comes before computer organization. It is like building the design and architecture of house takes maximum time and then organization is building house of bricks or by latest technology keeping the basic layout and architecture of house in mind.**

Architecture describes what the computer does and organization describes how it does it.

Architecture and organization are independents you can change the organization of computer without changing its architecture.

1.2 NUMBER REPRESENTATION

1.2.1 Number System

There are basically two types of number system:

(*a*) **Non-positional Number System:** In the beginning human beings counted on fingers, when ten fingers were not adequate, stones, sticks were used to indicate values.

In this system, we have symbols such as I for 1, II to 2, III for 3 and so on. Each symbol represents the same value regardless of its position in the number and the symbols are simply added to find out the value of a particular number, since it is very difficult to perform arithmetic work.

(*b*) **Positional Number System:** In this system there are only a few symbols called digits and these symbols represent different values depending on the position they occupy in the number. The value of each digit in such a number is determined by three considerations:

1. The digit itself.
2. The position of the digit in the number.
3. The base of the number system.

There are four types of positional number system:

1. Binary Number System (2)
2. Decimal Number System (10)
3. Octal Number System (8)
4. Hexa Decimal Number System (16).

Rules for Representation of Number System:

Rule 1: Total number of digits available (Base of No).

Rule 2: The first digit is always zero (0).

Rule 3: Last digit = Base – 1.

1. Binary Number System:

Base = 2

I^{st} digit = 0

Last digit = Base – 1 = 2 – 1 = 1

Final digits = 0, 1.

2. Decimal Number System:

Base = 10

I^{st} digit = 0

Last digit = Base – 1 = 10 – 1 = 9

Final digits = 0, 1, 2, 3, 4, 5, 6, 7, 8, 9.

3. Octal Number System:

Base = 8

I[st] digit = 0

Last digit = Base – 1 = 8 – 1 = 7

Final digits = 0, 1, 2, 3, 4, 5, 6, 7.

4 Hexa Decimal Number System:

Base = 16

I[st] digit = 0

Last digit = Base – 1 = 16 – 1 = 15

Final digits = 0, 1, 2, 3, 4, 5, 6, 7, 8, 9, *A, B, C, D, E, F*

The meaning of

$A \equiv 10$

$B \equiv 11$

$C \equiv 12$

$D \equiv 13$

$E \equiv 14$

$F \equiv 15.$

1.2.2 Conversion from One Number System to Another

There are two types of converting from one base to another base:

(A) Decimal to any Base
- — Decimal to Binary
- — Decimal to Octal
- — Decimal to Hexa Decimal

(B) Any Base to Decimal
- — Binary to Decimal
- — Octal to Decimal
- — Hexadecimal to Decimal

1.2.2.1 Decimal to Any Base

(*a*) **Decimal to Binary:** To convert the decimal number to binary number the following rules are followed:

Rule 1: First of all write the decimal number.

Rule 2: Divide the number by required base (*i.e.* 2) until the integer position is '0'.

Rule 3: Record the remainder from bottom to top indicated by arrow.

Example:

$(34)_{10} = (?)_2$

2	34	
2	17	= 0 ↑
2	8	= 1
2	4	= 0
2	2	= 0
2	1	= 0
	0	= 1

i.e. $(34)_{10} = (0100010)_2$

For Decimal Fraction to Binary Fraction: To convert the decimal fraction to binary fraction, the repeated multiplication by 2 process is used until the fractional product is '0' or same. The carry generated by each multiplication forms the binary number.

The following rules are to be followed:

Rule 1: Write decimal fractional number.

Rule 2: These numbers are multiplied by 2 and carry binary number be noted down.

Rule 3: If decimal fraction becomes same or approximate nearly then stop process.

Rule 4: The required binary nos. are downward direction.

Example: $(.4)_{10} = (?)_2$

$$0.4 \times 2 = 0.8 = 0$$
$$0.8 \times 2 = 1.6 = 1$$
$$0.6 \times 2 = 1.2 = 1$$
$$0.2 \times 2 = 0.4 = 0$$

Hence, $(.4)_{10} = (.0110)_2$

Example: $(34.4)_{10} = (?)_2 = (0100010.0110)_2$.

Decimal to Octal: To convert the decimal number to octal number, the following rules are followed:

Rule 1: First of all write the decimal number.

Rule 2: Divide the number by required base (*i.e.* 8) until the integer position is '0'.

Rule 3: Record the remainder from bottom to top indicated by arrow.

Example: $(34)_{10} = (?)_8$

$(34)_{10} = (?)_{16}$

```
16 | 34
16 |  2   ⇒2 ↑
   |  0   ⇒4 |
```

i.e. $(34)_{10} = (42)_8$

For Decimal Fraction to Octal Fraction: To convert the decimal fraction to octal fraction, the repeated multiplication by 8 process is used until the fractional product is '0' or same. The carry generated by each multiplication forms the octal number.

The following Rules are to be followed:

Rule 1: Write decimal fractional number.

Rule 2: These numbers are multiplied by 8 and carry octal number be noted down.

Rule 3: If decimal fraction becomes same or approximate nearly, then stop process.

Rule 4: The required octal numbers are downward direction.

Example: $(.4)_{10} = (?)_8$

$$0.4 \times 8 = 3.2 \Rightarrow 3$$
$$0.2 \times 8 = 1.6 \Rightarrow 1$$
$$0.6 \times 8 = 4.8 \Rightarrow 4$$
$$0.8 \times 8 = 6.4 \Rightarrow 6$$

i.e. $(.4)_{10} = (.3146)$

Example: $(34.4)_{10} = (?)_8$

$\Rightarrow (42.3146)_8$

(c) **Decimal to Hexadecimal:** To convert the decimal number to hexadecimal number, the following rules are followed:

Rule 1: First of all write the decimal number.

Rule 2: Divide the number by required base (*i.e.* 16) until the integer position is '0'.

Rule 3: Record the remainder from bottom to top indicated by arrow.

Example:

$$(34)_{10} = (?)_{16}$$

```
16 | 34
16 |  2    ⇒ 2 ↑
   |  0    ⇒ 4 |
```

i.e. $(34)_{10} = (22)_{16}$

For Decimal Fraction to Hexadecimal Fraction: To convert the decimal fraction to hexadecimal fraction, the repeated multiplication by 16 process is used until the fractional product is '0' or same. The carry generated by each multiplication forms the hexadecimal number.

The following Rules are to be followed:

Rule 1: Write decimal fractional number.

Rule 2: These numbers are multiplied by 16 and carry hexadecimal number be noted down.

Rule 3: If the decimal fraction becomes same or approximate nearly then stop process.

Rule 4: The required octal numbers are downward direction.

Example: $(.4)_{10} = (?)_{16}$

$0.4 \times 16 = 6.4 \Rightarrow 6$

$0.4 \times 16 =$ same (stop)

i.e. $(.4)_{10} = (.6)$

Example: $(34.4)_{10} = (?)_{16} = (22.6)_{16}$

1.2.2.3 Any Base to Decimal

(*a*) **Binary to Decimal:** To convert a binary number to decimal number, the following rules are involved for this conversion:

Rule 1: Write the given binary number.

Rule 2: Write the weights of digits from right to left starting from 0 to upwards.

Rule 3: Multiply of binary digits with corresponding weight and added it (*i.e.* this is required decimal nos.)

Example: $(1011)_2 = (?)_{10} = 1011 = 1 \times 2^3 + 0 \times 2^2 + 1 \times 2^1 + 1 \times 2^0$

$= 1 \times 8 + 0 + 2 + 1 = 8 + 2 + 1 \Rightarrow (11)_{10}$

For Binary Fraction to Decimal Fraction: To convert binary fraction to the decimal fraction following rules are to be followed:

Rule 1: Write the binary number.

Rule 2: Assign the position of each bit from right to left before decimal with positive sign and after decimal with negative sign. as follows:

```
                3  2  1  0     -1 -2
e.g. 1011.11 ⇒  1  0  1  1  .   1  1
```

Rule 3: Multiply the each digit of given number with the power of position of the given base.

Example: $(110.11)_2 = (\ )_{10}$

```
  1 1 0 . 1 1
  2 1 0 -1 -2
⇒ 1 1 0 . 1 1
```

$\Rightarrow \quad 1 \times 2 + 1 \times 2^1 + 0 \times 2^0 + 1 \times 2^{-1} + 1 \times 2^{-2}$

$\Rightarrow \quad 4 + 2 + 0 + \frac{1}{2} + \frac{1}{4}$

$\Rightarrow \quad 6 + \frac{1}{2} + \frac{1}{4}$

$\Rightarrow \quad 6 + .50 + .25$

$\Rightarrow \quad 6 + .75$

i.e. $(110.11)_2 = (6.75)_{10}$

(*b*) **Octal to Decimal:** Same as binary to decimal except of base 2, base 8 is should be used.

(*c*) **Hexadecimal to Decimal:** Same as above, base 16 is used.

Binary Arithmetic:

— Addition

— Subtraction

— Division

— Multiplication

Rules for Addition:

$0 + 0 = 0$

$0 + 1 = 1$

$1 + 0 = 1$

$1 + 1 = 10$ (zero with one carry)

Carry

Example:

$A = 1101$

$B = 1101$

1010

Carry

Rules for Subtraction:

$0 - 0 = 0$

$1 - 0 = 1$

$1 - 1 = 0$

$0 - 1 = 1$ (with one borrow)

bbb

e.g. $A = 1101$

$B = 1111$

$(1110)_2$

i.e. b indicates that borrow has been taken.

Rules for Division:

$0/0 \Rightarrow 1$

$0/1 \Rightarrow 0$

$1/0 \Rightarrow 1$

$1/1 \Rightarrow 1$

Note: If divisor is greater than the dividend then the quotient will be (zero) and if divisor is less than or equal to dividend then, quotient will be '1'.

e.g.

```
11 ) 110 ( 10
     11
     0
```

Ans. $(10)_2$.

Multiplication Rules:

$0 \times 0 = 0$

$0 \times 1 = 0$

$1 \times 0 = 0$

$1 \times 1 = 1$

e.g.

```
     1101
      101
  -------
     1101
    0000 ×
   1101 ×
  -------
  1000001
```

Note: If both inputs are one then the output will be copied of multiplication either input is zero then the all output will also be zero.

1.2.3 Complements

Complements are the operations employed for the simplification of the subtraction and logical manipulation of well as for negative number representations. In a base (r) system two types of complements are represented.

(*i*) r's complement

(*ii*) $(r - 1)$'s complement.

(*i*) *r*'s complements: The r's complements of the positive number N is defined as

$$r^n - N$$

where, r is the base of number

N is the digit value

n is the number of digit.

e.g. Find the r's complement of the $(23)_{10}$

Sol.

$r = 10$

$N = 23$

$n = 2$

According to formula

$$r^n - N$$

$\Rightarrow 10^2 - 23$

$\Rightarrow 100 - 23$

$\Rightarrow (77)_{10}$

e.g. Find the 2's complement of the $(10101)_2$

Sol.

$r = 2$

$n = 5$

$N = 10101$

According to formula $(r^n - N)$

$\Rightarrow 2^5 - 10101$
$\Rightarrow 32 - 10101$
$\Rightarrow 100000 - 10101$
$\Rightarrow (01010)_2$

Note: Similarly for octal and hexadecimal for octal $r = 8$ and hexadecimal $r = 16$.

(ii) (r – 1)'s complement: $(r - 1)$'s complement of a positive number N is defined as

$$r^n - N - 1$$

where, r is the base of number
n is the number of digit
N is the digit value

e.g. Find the *q's* complement of the $(13)_{10}$.

Sol.
$r = 10$
$N = 13$
$n = 2$

According to the formula

$$r^n - N - 1$$

$\Rightarrow 10^2 - 13 - 1$
$\Rightarrow 100 - 14$
$\Rightarrow (86)_{10}$

e.g. Find the *f's* complement of the $(10110)_2$.

Sol.
$r = 2$
$n = 5$
$N = 10110$

According to formula

$$r^n - N - 1$$

$\Rightarrow 2^5 - 10110 - 1$
$\Rightarrow 32 - 10110 - 1$
$\Rightarrow 100000 - 10110 - 1$
$\Rightarrow (01001)$

Note: Similarly for the octal and hexadecimal numbers for octal $r - 1 = 7$ and for hexadecimal number $r - 1 = 15$.

1.2.4 Subtraction using *r's* Complement

To find the subtraction using *r's* complement following rules are to be followed:

Rule 1: If we have the numbers in the form $S - T$ (*i.e.* we want to find the subtraction of '*S*' and '*T*'). In this case find the *r's* complement of '*T*' and add it into '*S*'.

Rule 2: If end carry occurs discard it, otherwise find again the *r's* complement of the result which is computed in step-1, and place negative sign before it.

e.g. Subtract $(53)_{10} - (44)_{10}$.

Sol. Both the numbers are decimal numbers hence

$S = 53$
$T = 44$

According to **Rule 1** find 1's complement of T (44) as follows:

$\Rightarrow 10^2 - 44$ $\quad [\because r^n - N]$
$\Rightarrow 100 - 44$

$\Rightarrow (56)_{10}$

Add it into S (53) as follows:

$$\begin{array}{r} 53 \\ 56 \\ \hline 109 \end{array}$$

End carry occurs

According to **Rule 2** discard it hence $(9)_{10}$ **Ans.**

Subtraction using (*r* – 1)'*s* complement: To perform the subtraction using $(r-1)$'s complement following rules are to be followed

Rule 1. Find the $(r-1)$'s complement of T and add it into S.

Rule 2. If end carry occurs, then add it into least significant bit of the computed value. If end carry does not occur, then again find $(r-1)$'s complement of the computed value in **Step 1** and place negative sign before it.

e.g. Find $(r-1)$'s complement of $(44)_{10} - (53)_{10}$.

Sol. Here, $S = 44$

$T = 53$

Find $(r-1)$'s (a's) complement of T and add it as follows:

$\Rightarrow 10^2 - 53 - 1 \quad [r^n - N - 1]$

$\Rightarrow 100 - 54$

$\Rightarrow (46)_{10}$

Add it to S as

$$\begin{array}{r} 44 \\ 46 \\ \hline 90 \end{array}$$

After addition 90

No end carry occurs hence find again 9's complement of 90 as

$\Rightarrow -(10^2 - 90 - 1)$

$\Rightarrow -(100 - 91)$

$\Rightarrow -(9)_{10}$

Note: For fractional number we use the formula

(*i*) $r^n - r^{-m} - N$ For r's complement

(*ii*) $r^n - r^{-m} - N - 1$ for $(r-1)$'s complement.

1.2.5 Integer and Floating Point Representation

Representation of Integers: Representation of integers means the representation of number in the registers. In case of representing integer we use two types of representation.

(*i*) **Positive Integer Representation:** To represent positive integers first of all convert the integer into binary form and then use the representation method as **only signed magnitude method** the bit '0' is used for positive sign and the bit '*f*' is used for representing negative sign at the most significant bit position.

e.g. If we want to represent +5 in a 8–bit register, then we will do as follows.

Sol. The binary conversion of +5 is 101.

∵ We have 8–bit register so in 8-bit

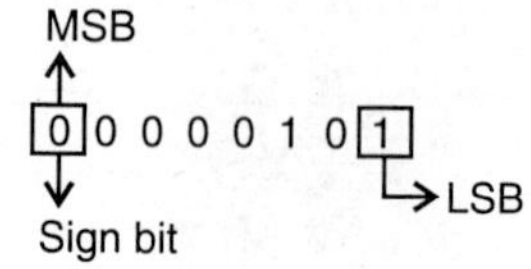

The representation is as follows:

0	0	0	0	0	1	0	1

Sign bit

(*ii*) **Negative Integer Representation:** To represent negative number we use three methods:

(*a*) Using signed magnitude method.

(*b*) Using 1's complement method.

(*c*) Using 2's complement method.

(*a*) Using signed magnitude method: Similar to the positive integer representation except that it will contain the MSB equal to 1 in place of 0 as it was in positive integer representation this is as follows:

e.g. If we want to represent –5 then it will be as

Sol. The binary conversion of +5 is 101 in 8-bit it will be 00000101 because we want to represent –5 then it will be as 10000101.

Sign bit →

1	0	0	0	0	1	0	1

Remainig tow methods will be on pose No. 17

1.2.6 Floating Point Representation

A number which has both an integer part as well as fractional part is called real number or floating-point number. A real number may be either positive or negative.

Scientists have used a convenient notation in which a number is expressed as a combination of a mantissa and an exponent

Thus $m \times r^e$

where m = mantissa

r = radix

e = exponent

The mantissa may be a fraction or an integer. The position of the radis '*r*' and the value of radix are not indicated in the register.

e.g. Assume a fraction representation and a radix 1. The decimal number 537.25 is represented in a register with m = 53725 and e = 3 and is interpreted to represent the floating point number.

$$.53725 \times 10^3$$

A floating point number is said to be normalized if the most significant digit of the mantissa is non-zero.

We cannot normalize a zero because it does not have a nono-zero digit. It is represented in floating point by all 0's in the mantissa and exponent.

e.g. $.00038695 \times 10^5$ is not normalized

But $.38695 \times 10^2$ is normalized.

This number may be written as:

Sign	*Mantissa*	*Sign*	*Exponent*
0	38695	0	2

Example: If we want to represent the floating point number in a 4-byte (32 bits) word then we will use first 24 bits for mantissa and remaining 8 bits for exponent.

Among 24 bits of mantissa 23 bits are used for magnitude of mantissa and 1-bit is used for representing sign bit of mantissa.

Similarly, among 8-bits of exponent 7 bits are used for the magnitude of exponent and 1-bit is used for sign bit of exponent.

The implementation will be as follows:

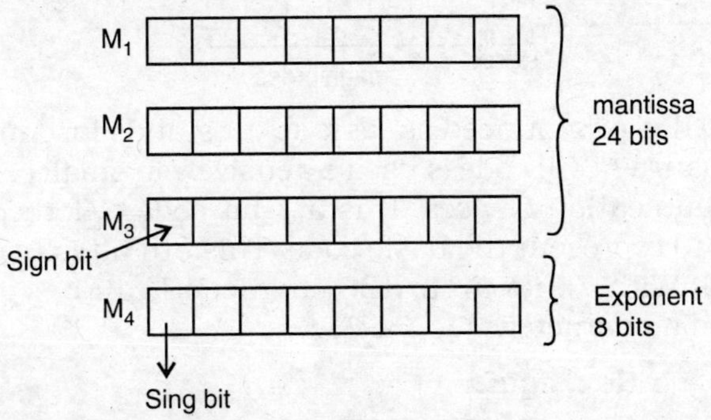

e.g. 5×10^3 may be represented as 32-bit register format.

Solution: The binary conversion of this number is $(5)_{10} = (101)_2$ and binary conversion of exponent (3) is 011.

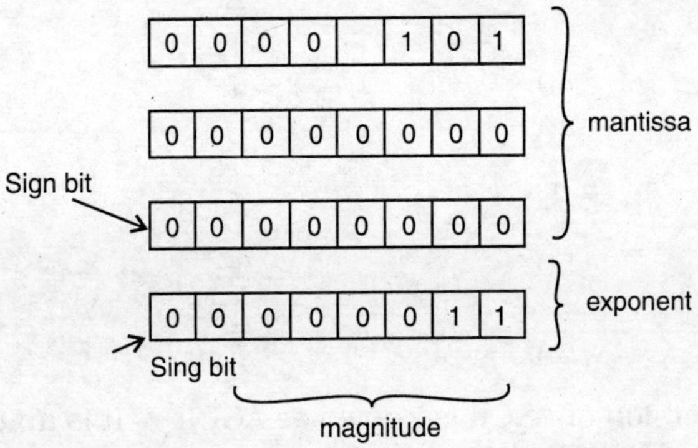

Negative Integer Representation

***(b)* Using is Complement Method:** To represent –5 using 1's complement method first of all find the binary equivalent of +5 *i.e.* 101.

The 8-bits equivalent is

00000101 = +5

Find the 1's complement of this no as

11111010 = –5

Represent this number is register of 8-bits.

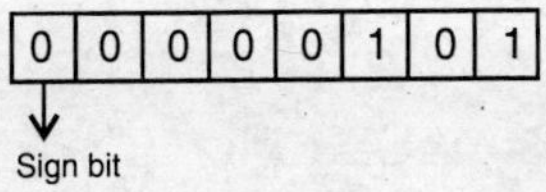

(c) Using 2's Complement Method: To represent –5 using 2's complement method first of all find the binary equivalent of +5 *i.e.* 101.

The 8-bits equivalent is

00000101 = +5

1's complement is 11111010

+1

2's complement is 11111011

The representation of this number is

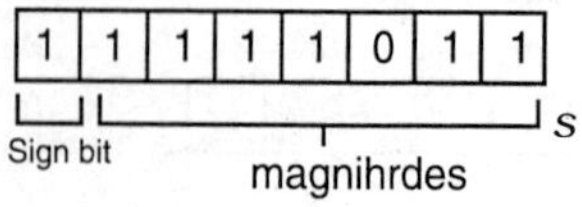

ASCII-7 Code: ASCII is pronounced as "ask-ee". It stands for American standard code for information inter change ASCII code is used extensively in small computers, peripherals, insruments and communication devices. It is a 7-bit code. Microcomputers using 8-bit word length use 7 bits to represent the basic code. This 8th bit is used for parity or it may be permanently 1 or 0. With 7 bits up to 128 characters can be coded. A letter, digit or special symbol is called a character.

Character	***Hex. equi.***		***ASCII-8***
0	30	0	50
|	|	|	|
9	39	9	59
A	41	*A*	A_1
|	|	|	|
I	49	*I*	a_9
J	4*A*	*J*	*AA*
|	|	|	|
O	4*F*	*O*	*AF*
P	50	*P*	B_0
|	|	|	|
Z	5*A*	*Z*	*BA*

ASCII-8: A newer version of ASCII is known as ASCII-8. It is an 8-bit code. With 8 bits code capacity is extended to 256 characters.

EBCDIC Code: It stands for Extended-Binary Code-Decimal Interchange code. It is pronounced as "ebb-see-dick". This code is the standard character code for large computers. It is an 8-bit code.

Actually EBCDIC code can be represented 8-bit for one alphabet letters equivalent to hexadecimal numbers 8 bit be divided into two parts: First is 4 bits are known as zone half and second part is also 4 bits known as digit half. Developed by IBM, EBCDIC code is used in most IBM models and in many other computers.

$$A - I \rightarrow C_1 - C_9$$
$$J - R \rightarrow D_1 - D_9$$
$$S - Z \rightarrow E_2 - E_9$$
$$I - 9 \rightarrow F_0 - F_9$$

BCD Code: This code represents each digit in a decimal number, *i.e.* digits 0 through 9 by its binary equivalent using 4-bits only.

Important point to note here is that BCD digits range from 0000 to 1001 because all combinations above this *i.e.* 1010 to 1111 cannot exist in BCD as the highest digit that can be coded is 9.

Example: $(2363)_{10}$ in BCD form is given as:

2	3	6	3
0010	0011	0110	0011

This type of code is also known as 8-4-2-1 code where 8, 4, 2 and 1 are the weights of the four bits of binary code. Thus it is a weighted code.

Excess-3 Code: It is an important 4-bit code. We can say that it is another form of BCD code. Any decimal number is represented in excess-3 form by adding 3 to each decimal digit and then converting the result into BCD code.

Example: Convert $(29)_{10}$ into excess-3 code

$$\begin{array}{cc} 2 & 9 \\ +3 & +3 \\ \hline 5 & 12 \\ 0101 & 1100 \end{array}$$

$\therefore \quad (29)_{10} = 0101\ 1100$

Gray Code: In this code, a decimal number is represented in binary form in a way such that each code (Gray code) differs from the preceding and succeeding one by a single bit.

Binary to Gray: Steps to perform binary to Gray are as follows:

Step I. Keep MSB bit as it is.

Step II. Starting from MSB, *EX-OR* every bit with next bit and place the result.

Or

The same can be achieved by adding two bits ignore carry.

The *EX-OR* function is shown by symbol ⊕

$$Y = A \oplus B$$

Output Ist bit *X-OR* 2nd bit

***X-OR* Table**

Inputs			*Outputs*
A		*B*	*Y*
0	0	0	
0	1	1	
1	0	1	
1	1	0	

(*a*) **By *EX-O* Ring Method:** Consider decimal 6. The equivalent binary is $(0110)_2$

$$(6)_{10} = (0110)$$

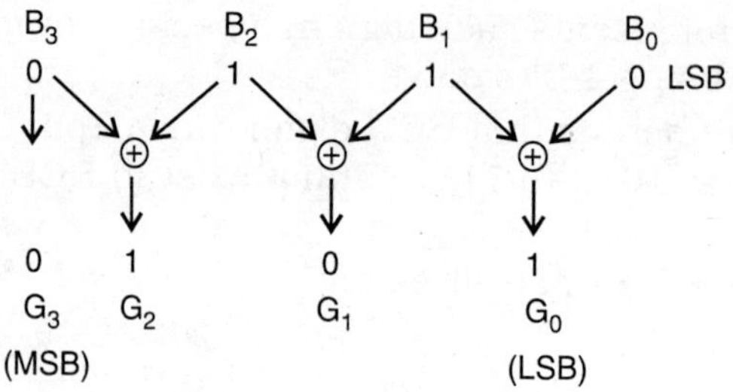

(*b*) **By Addition Method:** Consider the same example as in (*a*)

$$(6)_{10} = (0110)_2$$

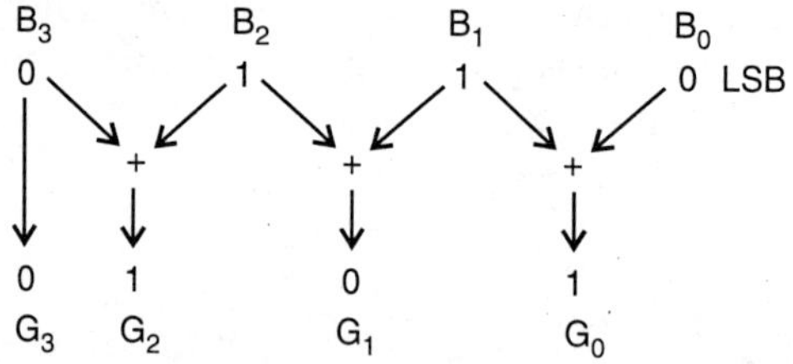

Gray to Binay:

Step I. MSB bit is kept as it is.

Step II. This bit is *X-OR*ed with neat bit from gray code (added).

Step III. The resulting bit in Step-II is *EX-OR*ed with next bit from gray code.

Step IV. Step-III is repeated till you reach LSB.

$$B_3 = G_3$$
$$B_2 = B_3 \oplus G_2$$
$$B_1 = B_2 \oplus G_1$$
$$B_0 = B_1 \oplus G_0$$

Syntax:

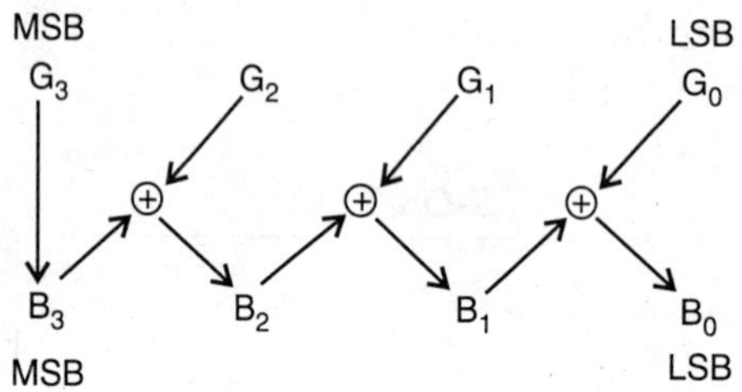

e.g. Convert $(0101)_2$ gray to equivalent binary code.

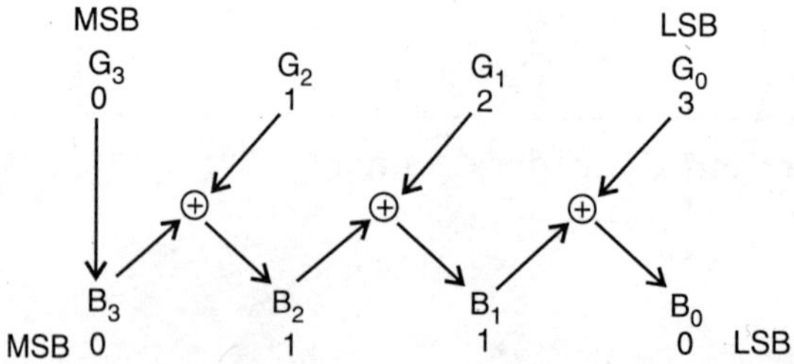

1.2.7. IEEE Standard 754 Floating Point Numbers

IEEE Standard 754 floating-point is the most common representation today for real numbers on computers, including Intel-based PC's, Macintoshes, and most Unix platforms. This article gives a brief overview of IEEE floating-point and its representation. Discussion of arithmetic implementation may be found in the book mentioned at the bottom of this article.

What Are Floating Point Numbers?

There are several ways to represent real numbers on computers. Fixed point places a radix point somewhere in the middle of the digits, and is equivalent to using integers that represent portions of some unit. For example, one might represent 1/100ths of a unit; if you have four decimal digits, you could represent 10.82, or 00.01. Another approach is to use rationals, and represent every number as the ratio of two integers.

Floating-point representation - the most common solution - basically represents reals in scientific notation. Scientific notation represents numbers as a base number and an exponent. For example, 123.456 could be represented as 1.23456×10^2. In hexadecimal, the number 123.abc might be represented as $1.23abc \times 16^2$.

Floating-point solves a number of representation problems. Fixed-point has a fixed window of representation, which limits it from representing very large or very small numbers. Also, fixed-point is prone to a loss of precision when two large numbers are divided.

Floating-point, on the other hand, employs a sort of "sliding window" of precision appropriate to the scale of the number. This allows it to represent numbers from 1,000,000,000,000 to 0.0000000000000001 with ease.

Storage Layout

IEEE floating point numbers have three basic components: the sign, the exponent, and the mantissa. The mantissa is composed of the fraction and an implicit leading digit (explained below). The exponent base (2) is implicit and need not be stored.

The following figure shows the layout for single (32-bit) and double (64-bit) precision floating-point values. The number of bits for each field are shown (bit ranges are in square brackets):

	Sign	***Exponent***	***Fraction***	***Bias***
Single Precision	1 [31]	8 [30-23]	23 [22-00]	127
Double Precision	1 [63]	11 [62-52]	52 [51-00]	1023

The Sign Bit

The sign bit is as simple as it gets. 0 denotes a positive number; 1 denotes a negative number. Flipping the value of this bit flips the sign of the number.

The Exponent

The exponent field needs to represent both positive and negative exponents. To do this, a bias is added to the actual exponent in order to get the stored exponent. For IEEE single-precision floats, this value is 127. Thus, an exponent of zero means that 127 is stored in the exponent field. A stored value of 200 indicates an exponent of (200-127), or 73. For reasons discussed later, exponents of – 127 (all 0s) and + 128 (all 1s) are reserved for special numbers.

For double precision, the exponent field is 11 bits, and has a bias of 1023.

The Mantissa

The mantissa, also known as the significand, represents the precision bits of the number. It is composed of an implicit leading bit and the fraction bits.

To find out the value of the implicit leading bit, consider that any number can be expressed in scientific notation in many different ways. For example, the number five can be represented as any of these:

5.00×10^0
0.05×10^2
5000×10^{-3}

In order to maximize the quantity of representable numbers, floating-point numbers are typically stored in normalized form. This basically puts the radix point after the first non-zero digit. In normalized form, five is represented as $5.0 \times 10^{\circ}$.

A nice little optimization is available to us in base two, since the only possible non-zero digit is 1. Thus, we can just assume a leading digit of 1, and don't need to represent it explicitly. As a result, the mantissa has effectively 24 bits of resolution, by way of 23 fraction bits.

Putting it All Together

So, to sum up:

1. The sign bit is 0 for positive, 1 for negative.
2. The exponent's base is two.
3. The exponent field contains 127 plus the true exponent for single-precision, or 1023 plus the true exponent for double precision.
4. The first bit of the mantissa is typically assumed to be 1.f, where f is the field of fraction bits.

Ranges of Floating-Point Numbers

Let's consider single-precision floats for a second. Note that we're taking essentially a 32-bit number and re-jiggering the fields to cover a much broader range. Something has to give, and it's precision. For example, regular 32-bit integers, with all precision centered around zero, can precisely store integers with 32-bits of resolution. Single-precision floating-point, on the other hand, is unable to match this resolution with its 24 bits. It does, however, approximate this value by effectively truncating from the lower end. For example :

```
11110000 11001100 10101010 00001111     // 32-bit integer
= +1.1110000 11001100 10101010 x 231    // Single-Precision Float
= 11110000 11001100 10101010 00000000   // Corresponding Value
```

This approximates the 32-bit value, but doesn't yield an exact representation. On the other hand, besides the ability to represent fractional components (which integers lack completely), the floating-point value can represent numbers around 2^{127}, compared to 32-bit integers maximum value around 2^{32}.

The range of positive floating point numbers can be split into normalized numbers (which preserve the full precision of the mantissa), and denormalized numbers (discussed later) which use only a portion of the fractions's precision.

	Denormalized	*Normalized*	*Approximate Decimal*
Single Precision	$\pm 2^{-149}$ to $(1 - 2^{-23}) \times 2^{-126}$	$\pm 2^{-126}$ to $(2 - 2^{-23}) \times 2127$	$\pm \sim 10^{-44.85}$ to $\sim 10^{38.53}$
Double Precision	$\pm 2^{-1074}$ to $(1 - 2^{-52}) \times 2^{-1022}$	$\pm 2^{-1022}$ to $(2 - 2^{-52}) \times 2^{1023}$	$\pm \sim 10^{-323.3}$ to $\sim 10^{308.3}$

Since the sign of floating point numbers is given by a special leading bit, the range for negative numbers is given by the negation of the above values.

There are five distinct numerical ranges that single-precision floating-point numbers are not able to represent:

1. Negative numbers less than $-(2 - 2^{-23}) \times 2^{127}$ (*negative overflow*)
2. Negative numbers greater than -2^{-149} (*negative underflow*)
3. Zero
4. Positive numbers less than 2^{-149} (positive underflow)
5. Positive numbers greater than $(2 - 2^{-23}) \times 2^{127}$ (*positive overflow*)

Overflow means that values have grown too large for the representation, much in the same way that you can overflow integers. Underflow is a less serious problem because is just denotes a loss of precision, which is guaranteed to be closely approximated by zero.

Here's a table of the effective range (excluding infinite values) of IEEE floating-point numbers:

	Binary	**Decimal**
Single	$\pm (2 - 2^{-23}) \times 2^{127}$	$\sim \pm 10^{38.53}$
Double	$\pm (2 - 2^{-52}) \times 2^{1023}$	$\sim \pm 10^{308.25}$

Note that the extreme values occur (regardless of sign) when the exponent is at the maximum value for finite numbers (2127 for single-precision, 21023 for double), and the mantissa is filled with 1s (including the normalizing 1 bit).

Special Values

IEEE reserves exponent field values of all 0s and all 1s to denote special values in the floating-point scheme.

Zero

As mentioned above, zero is not directly representable in the straight format, due to the assumption of a leading 1 (we'd need to specify a true zero mantissa to yield a value of zero). Zero is a special value denoted with an exponent field of zero and a fraction field of zero. Note that – 0 and + 0 are distinct values, though they both compare as equal.

Denormalized

If the exponent is all 0s, but the fraction is non-zero (else it would be interpreted as zero), then the value is a denormalized number, which does not have an assumed leading 1 before the binary point. Thus, this represents a number $(-1)^s \times 0.f \times 2^{-126}$, where s is the sign bit and f is the fraction. For double precision, denormalized numbers are of the form $(-1)^s \times 0.f \times 2^{-1022}$. From this you can interpret zero as a special type of denormalized number.

Infinity

The values +infinity and – infinity are denoted with an exponent of all 1s and a fraction of all 0s. The sign bit distinguishes between negative infinity and positive infinity. Being able to denote infinity as a specific value is useful because it allows operations to continue past overflow situations. Operations with infinite values are well defined in IEEE floating point.

Not A Number

The value NaN (*Not a Number*) is used to represent a value that does not represent a real number. NaN's are represented by a bit pattern with an exponent of all 1s and a non-zero fraction. There are two categories of NaN: QNaN (*Quiet NaN*) and SNaN (*Signalling NaN*).

A QNaN is a NaN with the most significant fraction bit set. QNaN's propagate freely through most arithmetic operations. These values pop out of an operation when the result is not mathematically defined.

An SNaN is a NaN with the most significant fraction bit clear. It is used to signal an exception when used in operations. SNaN's can be handy to assign to uninitialized variables to trap premature usage.

Semantically, QNaN's denote indeterminate operations, while SNaN's denote invalid operations.

Special Operations

Operations on special numbers are well-defined by IEEE. In the simplest case, any operation with a NaN yields a NaN result. Other operations are as follows:

Operation	***Result***
n ÷ ±Infinity	0
± Infinity × ± Infinity	±Infinity
± nonzero ÷ 0	±Infinity
Infinity + Infinity	Infinity
± 0 ÷ ± 0	NaN
Infinity − Infinity	NaN
± Infinity ÷ ± Infinity	NaN
± Infinity × 0	NaN

To sum up, the following are the corresponding values for a given representation:

Float Values (b = bias)			
Sign	***Exponent (e)***	***Fraction (f)***	***Value***
0	00..00	00..00	+ 0
0	00..00	00..01 : 11..11	Positive Denormalized Real $0.f \times 2^{(-b+1)}$
0	00..01 : 11..10	XX..XX	Positive Normalized Real $1.f \times 2^{(e-b)}$
0	11..11	00..00	+ Infinity
0	11..11	00..01 : 01..11	SNaN

0	11..11 1	10..00 : 11..11	QNaN
1	00..00	00..00	– 0
1	00..00	00..01 : 11..11	Negative Denormalized Real – $0.f \times 2^{(-b+1)}$
1	00..01 : 11..10	XX..XX	Negative Normalized Real – $1.f \times 2^{(e-b)}$
1	11..11	00..00	– Infinity
1	11..11	00..01 : 01..11	SNaN
1	11..11	10..00 : 11.11	QNaN

1.2.8. Binary Codes

Binary codes are codes which are represented in binary system with modification from the original ones. Below we will be seeing the following:

- Weighted Binary Systems
- Non Weighted Codes

Weighted Binary Systems

Weighted binary codes are those which obey the positional weighting principles, each position of the number represents a specific weight. The binary counting sequence is an example.

Decimal	***8421***	***2421***	***5211***	***Excess-3***
0	0000	0000	0000	0011
1	0001	0001	0001	0100
2	0010	0010	0011	0101
3	0011	0011	0101	0110
4	0100	0100	0111	0111
5	0101	1011	1000	1000
6	0110	1100	1010	1001
7	0111	1101	1100	1010
8	1000	1110	1110	1011
9	1001	1111	1111	1100

8421 Code/BCD Code

The BCD (Binary Coded Decimal) is a straight assignment of the binary equivalent. It is possible to assign weights to the binary bits according to their positions. The weights in the BCD code are 8,4,2,1.

Example: The bit assignment 1001, can be seen by its weights to represent the decimal 9 because:

1 x 8 + 0 x 4 + 0 x 2 + 1 x 1 = 9

2421 Code

This is a weighted code, its weights are 2, 4, 2 and 1. A decimal number is represented in 4-bit form and the total four bits weight is 2 + 4 + 2 + 1 = 9. Hence the 2421 code represents the decimal numbers from 0 to 9.

5211 Code

This is a weighted code, its weights are 5, 2, 1 and 1. A decimal number is represented in 4-bit form and the total four bits weight is 5 + 2 + 1 + 1 = 9. Hence the 5211 code represents the decimal numbers from 0 to 9.

Reflective Code

A code is said to be reflective when code for 9 is complement for the code for 0, and so is for 8 and 1 codes, 7 and 2, 6 and 3, 5 and 4. Codes 2421, 5211, and excess-3 are reflective, whereas the 8421 code is not.

Sequential Codes

A code is said to be sequential when two subsequent codes, seen as numbers in binary representation, differ by one. This greatly aids mathematical manipulation of data. The 8421 and Excess-3 codes are sequential, whereas the 2421 and 5211 codes are not.

Non Weighted Codes

Non weighted codes are codes that are not positionally weighted. That is, each position within the binary number is not assigned a fixed value.

Excess-3 Code

Excess-3 is a non weighted code used to express decimal numbers. The code derives its name from the fact that each binary code is the corresponding 8421 code plus 0011(3).

Example: 1000 of 8421 = 1011 in Excess-3

Gray Code

The gray code belongs to a class of codes called minimum change codes, in which only one bit in the code changes when moving from one code to the next. The gray code is non-weighted code, as the position of bit does not contain any weight. The gray code is a reflective digital code which has the special property that any two subsequent numbers codes differ by only one bit. This is also called a unit-distance code. In digital gray code has got a special place.

Decimal Number	*Binary Code*	*Gray Code*
0	0000	0000
1	0001	0001
2	0010	0011
3	0011	0010

4	0100	0110
5	0101	0111
6	0110	0101
7	0111	0100
8	1000	1100
9	1001	1101
10	1010	1111
11	1011	1110
12	1100	1010
13	1101	1011
14	1110	1001
15	1111	1000

Binary to Gray Conversion

- Gray Code MSB is binary code MSB.
- Gray Code MSB-1 is the XOR of binary code MSB and MSB-1.
- MSB-2 bit of gray code is XOR of MSB-1 and MSB-2 bit of binary code.
- MSB-N bit of gray code is XOR of MSB-N-1 and MSB-N bit of binary code.

1.3 ERROR DETECTION AND CORRECTION

Error detection is most commonly realized using a suitable hash function (or checksum algorithm). A hash function adds a fixed-length tag to a message, which enables receivers to verify the delivered message by recomputing the tag and comparing it with the one provided.

There exists a vast variety of different hash function designs. However, some are of particularly widespread use because of either their simplicity or their suitability for detecting certain kinds of errors (*e.g.*, the cyclic redundancy check's performance in detecting burst errors).

Random-error-correcting codes based on minimum distance coding can provide a suitable alternative to hash functions when a strict guarantee on the minimum number of errors to be detected is desired. Repetition codes, described below, are special cases of error-correcting codes: although rather inefficient, they find applications for both error correction and detection due to their simplicity.

A repetition code is a coding scheme that repeats the bits across a channel to achieve error-free communication. Given a stream of data to be transmitted, the data is divided into blocks of bits. Each block is transmitted some predetermined number of times. For example, to send the bit pattern "1011", the four-bit block can be repeated three times, thus producing "1011 1011 1011". However, if this twelve-bit pattern was received as "1010 1011 1011" - where the first block is unlike the other two - it can be determined that an error has occurred.

Repetition codes are not very efficient, and can be susceptible to problems if the error occurs in exactly the same place for each group (*e.g.*, "1010 1010 1010" in the previous example would be detected as correct). The advantage of repetition codes is that they are extremely simple, and are in fact used in some transmissions of numbers stations.[citation needed].

Parity

The simplest case is by adding a parity bit. Suppose we have a three-bit word (so the bit strings define points in a cube). If we add a fourth bit, we can decree that any time we want to switch a bit in the original three-bit string, we also have to switch the parity bit. If we start with 000 in the left cube, so the full string is 0000, changing any one of the original three bits requires us to change to the other cube: 1001, 1010, and 1100. Now if we change a second bit, we have to move back to the left cube: 0011, 0101, 0110. And if we change the third bit, we move back to the right cube: 0111.

So, there is a Hamming distance of two between any two valid strings. If we get a one-bit error, we know it is an error because it's on one of the invalid vertices.

This can be computed by counting the number of 1's, and making sure it's always even (so this is called even parity). We could have selected exactly the opposite set of vertices as the valid ones, which would have given us odd parity. We picked even parity because we'll be using it in the next step.

Error Correction

The weakness of the parity scheme is that we can tell we had an error, but we can't know which bit is wrong. If we use enough extra bits, we can tell not only that a bit is wrong, but which one it is. Since we need to have enough check bits to spot both an error in the data and in the check bits themselves (after all, they aren't going to be perfect either), we need (log n) + 1 bits (Hamming derives this result much, much more carefully in his paper). The basic idea in what follows is that we'll divide the data bits into log n subsets where each subset contains roughly half of all the bits, and compute the even parity of each subset. If we have an error, we'll be able to tell which bit has the error because it will be uniquely determined by the set of subsets that turn up with bad parity.

(**Note:** in Hamming's paper, the following appears just as unmotivated as it does here. I really have no idea how he derived this technique; he does show that it actually does establish the needed distance between valid bit positions) We'll put the check bits in bit positions which are powers of two, and intersperse the data bits between them. Here's what it looks like if we have eight data bits:

Table 1.1

BitPosition	*PositionNumber*	*CheckBit*	*DataBit*
12	1100		M8
11	1011		M7
10	1010		M6
9	1001		M5
8	1000	C8	
7	0111		M4
6	0110		M3
5	0101		M2
4	0100	C4	
3	0011		M1
2	0010	C2	
1	0001	C1	

Here's how we find the subsets: The data bit positions which contain a 1 in the bit corresponding to a check bit number are used in calculating that check bit. So, looking at the table, data bits M1, M2, M4, M5, and M7 are in rows 3, 5, 7, 9, and 11; those row numbers all contain 20; those data bits are used in calculating check bit C1. We simply set C1 as having the parity of its data bits.

Looking at all the check bits, we get:

```
C1 = M1 ^ M2      ^ M4 ^ M5      ^ M7
C2 = M1      ^ M3 ^ M4      ^ M6 ^ M7
C4 =    ^ M2 ^ M3 ^ M4           ^      ^ M8
C8 =                   ^ M5 ^ M6 ^ M7 ^ M8
```

Now, if we get an error, the parity will be wrong for all of the sets based on that bit. The check bits that turn up wrong will be the bit number of the error!

We can combine ECC with parity. The way we do this, is we take the parity over all the bits in the word (including the check bits). In our bit numbering scheme, we consider Parity as bit 0000.

So, when we look at the parity and check bits, we get the following results:

- If the parity is correct and the check bits are correct, our data is correct.
- If the parity is incorrect, the check bits indicate which bit is wrong. If the check bits indicate that the error is in bit 0000, it's the parity bit itself that is incorrect.
- If the parity is correct but the check bits indicate an error, there is a two-bit error. This can't be corrected.

Checksums and Cyclic Redundancy Checks

This technique has seen a lot more development, by a lot more authors, than error correcting codes have. The basic technique as described here appeared in a paper by Peterson and Brown, which appeared in the January, 1961 issue of the *Proceedings of the Institute of Radio Engineers* (the IRE was, of course, a predecessor organization to the IEEE). Much has been done since on selecting good CRCs.

Once again, we'll start by defining a simple technique, and then define a more complex one that works better.

Checksums

Suppose we have a fairly long message, which can reasonably be divided into shorter words (a 128 byte message, for instance). We can introduce an accumulator with the same width as a word (one byte, for instance), and as each word comes in, add it to the accumulator. When the last word has been added, the contents of the accumulator are appended to the message (as a 129th byte, in this case). The added word is called a checksum.

Now, the receiver performs the same operation, and checks the checksum. If the checksums agree, we assume the message was sent without error.

A related approach would be, instead of performing an actual addition, we can just do a bitwise exclusive-or of the new word with the accumulator. If we do this, we calculate a vertical parity on the data. Notice that in the special case of a one-bit word, this is equivalent to calculating the parity of the buffer!

Performing a vertical parity has two advantages over a real checksum: it can be performed with less hardware if the data is serial, and it will lead us into performing a CRC.

To see how a vertical parity can be performed with less hardware than a checksum, take a look at the next Fig. 1.3.

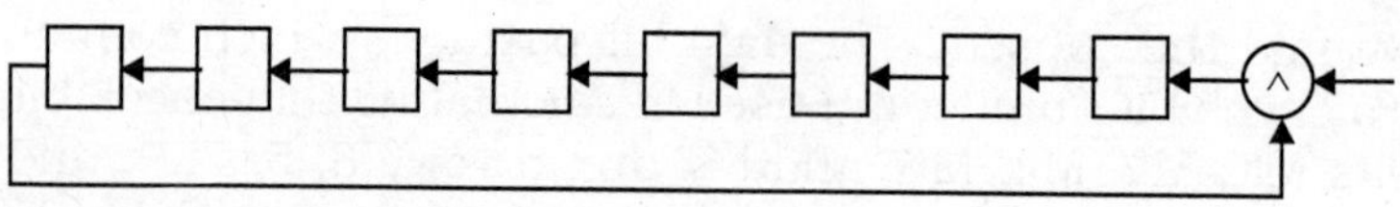

Fig. 1.3.

This figure shows an eight bit shift register and an exclusive-or gate. Initially, the shift register is filled with 0's. As each bit is put into it, the new bit is exclusive-ored with the contents of the eighth cell in the register. When the entire message has been passed through the shift register, it contains the vertical parity.

Here's an example of passing a 32 bit message through the unit:

00000000 11010110101010010100011101101010

(as we start, the shift register is empty

00000001 1010110101010010100011101101010
00000011 010110101010010100011101101010
00000110 10110101010010100011101101010
00001101 0110101010010100011101101010
00011010 110101010010100011101101010
00110101 10101010010100011101101010
01101011 0101010010100011101101010
11010110 101010010100011101101010

(at this point, the shift register contains the first byte of the message

10101100 01010010100011101101010
01011001 1010010100011101101010
10110011 010010100011101101010
01100111 10010100011101101010
11001111 0010100011101101010
10011111 010100011101101010
00111111 10100011101101010
01111111 0100011101101010

(it now contains the vertical parity of the first two bytes of the message)

11111110 100011101101010
11111100 00011101101010
11111001 0011101101010
11110011 011101101010
11100111 11101101010
11001110 1101101010
10011100 101101010
00111000 01101010

(first three bytes)

01110000 1101010
11100001 101010
11000010 01010
10000101 1010
00001010 010
00010100 10
00101001 0
01010010

(and the vertical parity of the whole 32 bit message)

Notice that a checksum or a vertical parity is much more efficient than ECC (in the sense that it doesn't need as many added bits), but it isn't capable of correcting errors.

The problem with checksums is that a 1-bit error turns into a 1-bit code. If you have a burst of noise, the odds are far too good that you'll end up with something that still looks correct, even though it isn't. The next approach, CRC checks, "smears" the results of the parity calculations through the signature, reducing the likelihood of that happening.

Mathematical Digression: Modulo-2 Arithmetic

Taking a bitwise exclusive-or in place of performing an addition is an example of "Modulo-2 Arithmetic," which is one form of "polynomial arithmetic." I've seen one author call it "CRC arithmetic."

Modulo-2 arithmetic is an arithmetic scheme; like most of the oddities that mathematicians like to study it seems completely useless to a non-mathematician at first glance but turns out to have some very practical applications. In this case, the practical application is in developing CRC checks.

The basic idea of modulo-2 arithmetic is just that we are working in binary, but we don't have a carry in addition or a borrow in subtraction. This means:

- Addition and subtraction become the same operation: just a bit-wise exclusive-or. Because of this, the total ordering we expect of integers is replaced by a partial ordering: one number is greater, than another iff its left-most 1 is farther left than the other's. This will have an impact on division, in a moment.
- Multiplication is just like multiplication in ordinary arithmetic, except that the adds are performed using exclusive-ors instead of additions.
- Division is like long division in ordinary arithmetic, except for two differences: the subtractions are replaced by exclusive-ors, and you can subtract any time the leftmost bits line up correctly (since, by the partial ordering described above, they are regarded as equal in this case).

It'll probably help to show examples of modulo-2 multiplication and division:

Multiplication

```
   1101
   0110
  -----
   0000
  11010
 110100
0000000
----------
0101110
```

Division

```
          1101
        --------
    0110)0101110
          0110
         -------
           0111
           0110
          ------
```

```
0011
0000
------
 0110
 0110
 ------
 0000
```

Notice that the first subtraction is possible in modulo-2 arithmetic, while it wouldn't be possible in normal arithmetic.

One last thing to say here is that most of the time, when we perform a modulo-2 addition on two numbers we get an answer of 0 or 1. In this case, we're performing the arithmetic on *each coefficient of the polynomial* modulo-2. Easy to get confused....

Cyclic Redundancy Checks

I'm going to be following some of Peterson & Brown's notation here:

- k is the length of the message we want to send, i.e. the number of information bits.
- n is the total length of the message we will end up sending: the information bits followed by the check bits. Peterson and Brown call this a code polynomial.
- $n - k$ is the number of check bits. It is also the degree of the generating polynomial. The basic (mathematical) idea is that we're going to pick the n-k check digits in such a way that the code polynomial is divisible by the generating polynomial. Then we send the data, and at the other end we look to see whether it's still divisible by the generating polynomial; if it's not then we know we have a error, if it is we hope there was no error.

The way we calculate a CRC is we establish some predefined $n - k + 1$ bit number P (called the Polynomial, for reasons relating to the fact that modulo-2 arithmetic is a special case of polynomial arithmetic). Now we append $n - k$ 0's to our message, and divide the result by P using modulo-2 arithmetic. The remainder is called the Frame Check Sequence. Now we ship off the message with the remainder appended in place of the 0's. The receiver can either recompute the FCS and see if it gets the same answer, or it can just divide the whole message (including the FCS) by P and see if it gets a remainder of 0!

As an example, let's set a 5-bit polynomial of 11001, and compute the CRC of a 16 bit message:

```
      ---------------------
11001)10011101010101100000
      11001
      -----
       1010101010101100000
       11001
        -----
         110001010101100000
        11001
         -----
         00011010101100000
            11001
            -----
            0011101100000
```

```
11001
-------
 100100000
 11001
 --------
  10110000
  11001
  -------
    1111000
    11001
    -------
      11100
      11001
      --------
       0101
```

Notice that when I did the division, I didn't bother to keep track of the quotient; we don't care about the quotient. Our only goal here is to get the remainder (0101), which is the FCS.

CRC's can actually be computed in hardware using a shift register and some number of exclusive-or gates (sounds a bit like the vertical parity calculation, doesn't it?).

The key insight is that we can perform a subtraction any time there is a 1 in the bit that lines up with the most significant bit of the polynomial, and we can perform that subtraction by performing an exclusive-or of the bits corresponding to 1's in all the other places of the polynomial. This lets us implement the CRC calculation by using a shift register similar to the one for vertical parity.

You can see how it's done by comparing the division we performed above to the circuit in the next figure. The figure shows a shift register; the string to be checked is inserted from the right. Whenever a "1" exits the left side of the shift register, it means there is a 1 in the most significant bit of the part of the dividend we're working with; since we're working in modulo-2 arithmetic, this means we can do a subtraction. What this works out to is:

1. The most significant bit will be xored away, so it falls off to the left.
2. For every other bit with a "1" in the divisor, perform an exclusive-or with the corresponding bit in the number being checked.
3. For bits with a "0" in the divisor, do nothing.

The figure below attempts to show this for the example CRC polynomial. Each of the square boxes is a position in the shift register, where a value can be stored. Every round box is a position where we may or may not perform an exclusive-or, depending on the polynomial we're using. You can see the value of the CRC polynomial written above the round boxes.

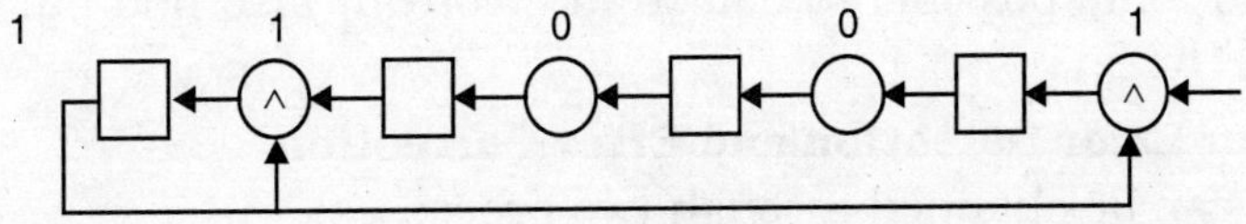

I keep calling this a polynomial, and writing it as a binary number. Frequently, you'll find a CRC polynomial written in polynomial form; the one we've been using would be written as $x^4 + x^3 + x^0$.

So, just a little bit more. First, there is quite a bit of theory behind choosing a "good" CRC polynomial; the choice of polynomial can be tuned to make sure that any burst of some given length can be caught.

Properties of Cyclic Redundancy Checks

The paper lists a few properties of CRCs, which deserve mention:

- If the rightmost place of the generating polynomial were 0, the generating polynomial would be divisible by X. That being the case, any polynomial divisible by P would also be divisible by X, and so the last bit of the check bits would always be 0. That would be useless, so we always have a 1 in the least significant bit of the generating polynomial.

  That's a roundabout way of saying that if you're going to have an $n - k$ bit polynomial, the two outlying bits should be 1's, otherwise you've effectively got a shorter polynomial than that.
- Any error checking code that can always detect a two-bit error can always correct any one-bit error. In the most ridiculous case, we can just check by flipping every bit of the received message; whenever we flip the wrong bit we get a two-bit error, when we flip the right one we get a 0-bit error. Of course, Hamming's scheme is a lot more clever than this!
- Any cyclic code whose generating polynomial is of length $n - k$ will always detect any burst error of length less than $n - k$.

There are a few "classic" CRC polynomials of given lengths which are so sell established that they've been given names.

Name	***Definition***	
	As Polynomial	***As Number***
CRC12	$X^{12} + X| + X^3 + X + 1$	1100000001011
CRC16	$X^{16} + X^{15} + X^2 + 1$	11000000000000101
CRC-CCITT	$X^{16} + X^{12} + X^5 + 1$	10001000000100001
CRC32	$X^{32} + X^{26} + X^{23} + X^{22} + X^{16} + X^{12} + X^{11} + X^{10} + X^8 + X^7 + X^5 + X^4 + X^2 + X + 1$	100000100110000010001110110110111

1.4 HAMMING CODE

As we have already done in parity checker and generator, that by adding a single parity bit to a code we can detect a single error in the code. The error detection is parible as the addition of the parity bit. creater a minimum hamming distance between any two coder.

Richard hamming in 1950, showed that by systematically introducing more parity bits in the code, it is not only possible to detect an error but also find out where the error occured and correct it.

Hamming Coder for Error Detection and Error Correction

Hamming code can be obtained through two processes:

(*i*) Venn diagram

(*ii*) Linear block diagram using hamming algorithm.

Venn diagram

Construction of hamming code can be explained in the following e.g., with the help of venn diagram on a 4 bit data.

Example: The data is 1100.

Data is 4 bits, so parity bit will be 3.

So, 4 bit data + 3 parity bit = 7 bit hamming code.

Hence, the venn diagram contain 3 circles A, B, C which together forms 7 regions. The data is filled in the region AB, ABC, AC, BC.

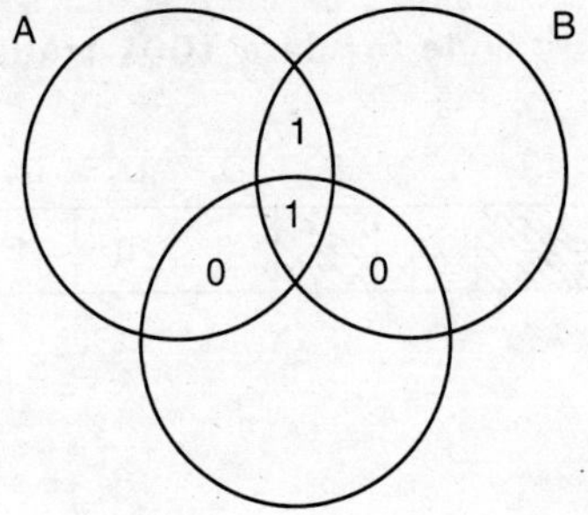

Now, we add parity bit to each of the three empty regions (assuming even parity).

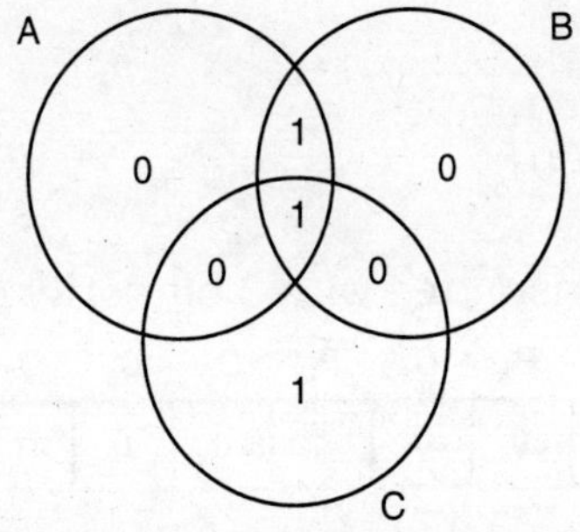

So, the data is 1100 and pority bit is 001 hamming code data is 1100001.

Now, suppose that the bit in the AB region change from 1 to 0. The computer can now see that the circles *A* and *B* have odd parity which is wrong. Now, AB is restore to 0. Thus, correcting the error.

Hamming Algorithm

Hamming algorithm can be used to construct error detecting coder for any size of data. If a data, is stored in the *n* bits *r*-bits parity is added then the new data formed is of length $(n + r)$ bits. This $(n + r)$ bits data is a hamming code.

All the bits where bit number is a power of 2 are parity bits, *i.e.* 2^n where $n = 0, 1, 2, 3,$ ..... and the rest of bits are used for data.

E.g., with a 4-bit data, 3 parity bits are added at 2^0, 2^1 and 2^2 bit position.

Also, 4 bit data + 3 parity bit = 7-bit hamming code.

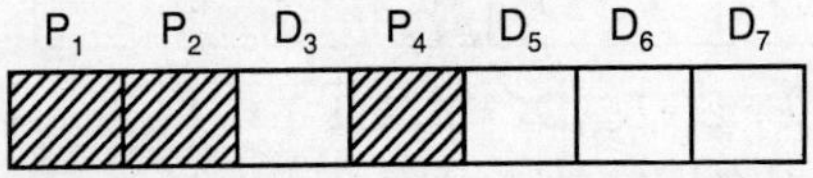

P_1, P_2 and P_4 – parity bits.

D_3, D_5, D_6, D_7 – Data bits.

Rules

Pority Bits	*Bits to be Checked*
P1	1, 3, 5, 7, 9, 11, ...
P2	2, 3, 6, 7, 10, 11, ...
P4	4, 5, 6, 7, 12, 13, 14, 15, ...
P8	8, 9, 10, 11, 12, 13, 14, 15, ...
P16	16, 17, 18, 19, 20, ...

EXAMPLE 1.1. Contruct hamming code for data 1001 transmitted with even pority. Data can be represented as:

P_1	P_2	D_3	P_4	D_5	D_6	D_7
		1		0	0	1

Here, P_1 had bits 1, 3, 5, 7

So, $P_1\ D_3\ D_5\ D_7 = P_1\ 101$

$\Rightarrow \qquad P_1 = 0$

For P_2,

$P_2\ D_3\ D_6\ D_7$

$\Rightarrow 1001 \Rightarrow P_2 = 0$

For P_4,

$$P_4\ P_5\ D_6\ D_7 = P_4\ 001$$

$\Rightarrow \qquad P_4 = 1$

Hence, it bit data 1001 is transmitted with 3 bit pority data 001

P_1	P_2	D_3	P_4	D_5	D_6	D_7
0	0	1	1	0	0	1

When the transmitted code is received, the following procedure is used direct and correct the error (Assuming 7-bit hamming code data with even pority.)

1. Check even pority on positions 1, 3, 5, 7. If it passes then $C_1 = 0$ else $C_1 = 1$.
2. Check even pority on positions 2, 3, 6, 7. If it passes then $C_2 = 0$ else $C_2 = 1$.
3. Check even pority on positions 4, 5, 6, 7. If it passes then $C_4 = 0$ else $C_4 = 1$.

The decimal equivalent of $C_4\ C_2\ C_1$ gives the position of incorrect bit and that bit is to be corrected. If that value is 0 change it to 1 and vice versa.

If $C_4\ C_2\ C_1 = 0$ then there is no error in the code.

E.g., A seven bit hamming code is received as 1010111 with even potity. What is the correct code?

Received data

P_1	P_2	D_3	P_4	D_5	D_6	D_7
1	0	1	0	1	1	1

$$C_1 = P_1\ D_3\ D_5\ D_7 = 1\ 1\ 1\ 1 = 1$$
$$C_2 = P_2\ D_3\ D_6\ D_7 = 0\ 1\ 1\ 1 = 1$$
$$C_4 = P_4\ D_5\ D_6\ D_7 = 0\ 1\ 1\ 1 = 1$$

The position of error is $C_4\ C_2\ C_1 = 1\ 1\ 0 = 6$

hence, the error is in 6th bit.

Correct code is obtained by changing 1 to 0 in 6th porition.

P_1	P_2	D_3	P_4	D_5	D_6	D_7
1	0	1	0	1	0	1

Limitation of Hamming Code

It can detect and correct only single bit error.

1.5 DECIMAL NUMBER SYSTEM

There are ten digits 0 to 9. The base of the decimal number system is 10.

Value of each digit depends on its position in the number as described below:

Value of the 1st digit of the number from the right side = 1st digit × 10.

Value of the 2nd digit of the number from the right side = 2nd digit × 10^1.

Value of the 3rd digit of the number from the right side = 3rd digit × 10^2.

Value of the *n*th digit of the number from the right side = *n*th digit × 10^{n-1}.

= *n*th digit × $(\text{base})^{n-1}$.

Example. 4598 = 4000 + 500 + 90 + 8 = $4 \times 10^2 + 5 \times 10^2 + 9 \times 10^1 + 8 \times 10^0$

1.6 BINARY NUMBER SYSTEM

In the binary number system there are only two digits 0 and 1. The binary digits are called *bits*. The base of the binary number system is 2. In binary system zero is represented by 0 and one by 1. After this there is no digit for two. Therefore, two is written as 10 and three as 11. Again for four we have to write 100. In this way we proceed further.

The weight of each bit of a number depends on its position as described below:

Value of the 1st bit of the number from the right side = 1st bit × 2

Value of the 2nd bit of the number from the right side = 2nd bit × 2^1

Value of the 3rd bit of the number from the right side = 3rd bit × 2^2

Value of the 3rd bit of the number from the right side = 4th bit × 2^3

Value of the *n*th bit of the number from the right side = *n*th bit × 2^{n-1}

= *n*th bit × $(\text{base})^{n-1}$.

Example. 1011 (binary number) = $1 \times 2^3 + 0 \times 2^2 + 1 \times 2^0$

= 8 + 0 + 2 + 1 = 11 (decimal number)

1.7 OCTAL NUMBER SYSTEM

The base of the octal number system is 8. The digits are 0 to 7 same as those for the decimal number system. The number 8 of the decimal number system is represented by 10, 9 by 11 and so on.

EXAMPLE 1.2. Convert the octal number 645 to equivalent decimal number.

Solution: 645 (octal) = $6 \times 8^2 + 4 \times 8^1 + 5 \times 8^0 = 6 \times 64 + 4 \times 8^1 + 5 \times 1$

= 384 + 32 + 5 = **421 (decimal)**

EXAMPLE 1.3. Convert the octal number 5273 to its equivalent decimal number.

Solution: 5273 (octal) = $5 \times 8^3 + 2 \times 8^2 + 7 \times 8^1 + 3 \times 8^0$

= 5 × 512 + 2 × 64 + 7 × 8 + 3 × 1 = **2747 (decimal)**

1.8 HEXADECIMAL NUMBER SYSTEM

The base of the hexadecimal number system is 16. The digits are 0 to 9 same as those for that decimal number system and digits from 10 to 15 are represented by alphabets A to P respectively.

Table 1.3. Numbers with different bases

Decimal (base 10)	*Binary (base 2)*	*Octal (base 8)*	*Hexadecimal (base 16)*
00	0000	00	0
01	0001	01	1
02	0010	02	2
03	0011	03	3
04	0100	04	4
05	0101	05	5
05	0110	06	6
07	0111	07	7
08	1000	10	8
09	1001	11	9
10	1010	12	A
11	1011	13	B
12	1100	14	C
13	1110	15	D
14	1110	16	E
15	1111	17	F

EXAMPLE 1.4. Convert the hexadecimal number 549 to its decimal equivalent.

Solution: 9 is the 1st digit from the right; its weight is 9×16^0

A is the 2nd digit from the right; its weight is $A \times 16^1$

5 is the 3rd digit from right, its weight is 5×16^2

$$549(\text{hex}) = 5 \times 16^2 + A \times 16^1 + 9 \times 16^1 = 5 \times 256 + 10 \times 16 + 9$$
$$= 1280 + 160 + 9 = \mathbf{1449\ (decimal)}$$

EXAMPLE 1.5. Convert the hexadecimal number 1A5D to its decimal equivalent.

Solution:
$$1A5D(\text{hex}) = 1 \times 16^3 + A \times 16^2 + 5 \times 16^1 + D + 16^0$$
$$= 4096 + 10 \times 256 + 80 + 13 \times 1 = 4096 + 2560 + 80 + 13$$
$$= \mathbf{6749\ (decimal)}$$

1.9 CONVERSION OF DECIMAL NUMBER TO BINARY NUMBER

In a binary number the weights of bits from the right side are 1, 2, 4, 8, 16 and so on. This concept can be utilized for the conversion of a decimal number to a binary number. A procedure is to be developed to examine which multiples of 2 are present in the decimal number. For example, 12 (decimal number) has one 8 and one 4. It has no 2 and 1. In other words it has zero 2 and zero 1. Its binary equivalent is 1100. Based on this concept for decimal to binary conversion, the decimal number is divided by 2 successively. At each stage the quotient and remainder are noted down. The quotient of one stage is divided by 2 at the next stage. The procedure is repeated till the quotient becomes zero. If the remainder of the 1st stage is R_1, remainder of the 2nd stage R_2, and so on, the binary number is given by

$$\text{Binary number} = R_i\, R_{i-1},\, R_{i-2} \,....\, R_3\, R_2\, R_1$$

where R_f = remainder of *i*th stage.

The following examples will illustrate the conversion of decimal numbers to binary numbers:

EXAMPLE 1.6. Find the binary equivalent of the decimal number 39.

Solution:

	Quotient	Remainder	Remark
Divide 39 by 2	19	1 (LSB)	There is one 1 and nineteen 2s.
Divide 19 by 2 (*This is equivalent to division by 4*)	9	1	There is one 2 and nine 4s.
Divide 9 by 2 (*Is is equivalent to division by 8*)	4	1	There is one 4 and four 8s.
Divide 4 by 2 (*Is is equivalent to division by 16*)	2	0	There is zero 8 and two 16s.
Divide 2 by 2 (*Is is equivalent to division by 32*)	1	0	There is zero 16 and one 32.
Divide 1 by 2 (*Is is equivalent to division by 64*)	0	1 (MSB)	There is one 32 and zero 64.

The binary number is 100111.

The last remainder is MSB and the 1st remainder LSB.

1.9.1 Conversion of a Decimal Number to Octal Number

For the conversion of a decimal number to an octal number, the technique of division by 8 can be used.

EXAMPLE 1.7. Convert 61 (decimal) to its equivalent octal number

Solution:

8	61	Remainder
8	7	5(least significant digit)
	0	(most significant digit)

The octal number = 75.

EXAMPLE 1.8. Convert 461 (decimal) to its equivalent octal number

Solution:

8	461	Remainder
8	57	5 (least significant digit)
8	7	1
	0	7 (most significant digit)

The octal number = 715.

EXAMPLE 1.9. Convert the decimal 2747 to its equivalent octal number

Solution:

8	2747	Remainder
8	343	3 (LSD)
8	42	7
8	5	2
	0	5 (MSD)

The octal number = 5273.

1.9.2 Conversion of a Decimal Number to Hexadecimal Number

For the conversion of a decimal number to a hexadecimal number, the technique of division by 16 can be used.

EXAMPLE 1.10. Convert 35 (decimal) to its equivalent octal number

Solution:

16	35	Remainder
16	2	3
	0	2

The hexadecimal number is **23.**

EXAMPLE 1.11. Convert the decimal number 10767 to a hexadecimal number.

Solution:

16	10767	Remainder
16	672	15 = F
16	42	0
16	2	10 = A
	0	2

The hexadecimal number is **2A0F.**

1.9.3 Conversion of Binary Number to Octal Number

The conversion of a binary number to octal number is very easy. The base of an octal number is 8.

As $8 = 2^3$, for binary to octal conversion groups of 3 bits are formed in the binary number from the right. After forming the groups each 3-bit binary group is replaced by its octal equivalent.

EXAMPLE 1.12. Convert the binary number 101111100 to its octal equivalent.

Solution: $(101111100)_2 = (101)\ (111)\ (100) = (5)\ (7)\ (4) = (574)_5 =$ **574 (octal).**

EXAMPLE 1.13. Convert the binary number 1101110 to its octal equivalent.

Solution: $(1101110)_2 = (1)\ (101)\ (110) = (001)\ (101)\ (110)$

$= (1)\ (5)\ (6) = (156)_6 =$ **156 (octal).**

1.9.4 Conversion of an Octal Number to Binary Number

For the conversion of an octal number to a binary number each digit of the given octal number is converted to its 3-bit binary equivalent.

EXAMPLE 1.14. Convert the octal number 436 to binary

Solution: $(436)_8 =$ (100) (011) (110)

4 3 6

= **100011110 (binary).**

EXAMPLE 1.15. Convert 527 (octal) to binary.

Solution: 527(octal) = (101) (010) (111)

5 2 7

= **101010111 (binary).**

1.9.5 Conversion of a Binary Number to Hexadecimal Number

The conversion of a binary number to hexadecimal number is easy. The base of a hexadecimal number is 16. As $16 = 2^4$, for binary to hexadecimal conversion groups of four bits are made in the binary number from the least significant bit (from the right). After forming the group each 4-bit binary group is replaced by its hexadecimal equivalent.

EXAMPLE 1.16. Convert the binary number 010111011001 to its equivalent hexadecimal number.

Solution: $(010111011001)_2$ = (0101) (1101) (1001) = (5) (D) (9) = $(5\text{D}9)_{16}$ = **5D9 (hex).**

Suffix 2 and 16 indicate the base of the binary and hexadecimal number respectively.

EXAMPLE 1.17. Convert the binary number 1101011010 to its equivalent hexadecimal number.

Solution: Making the groups of 4 bits from the right in the given binary number we have the groups as follows:

$(1101011010)_2$ = (11) (1010) (1010).

Here the group of the most significant binary bits contains only 2 binary bits. This can be extended to 4 binary bits by putting zeros in MSB positions. If MSBs are extended by zeros the number is not affected. Thus the given number is grouped as follows:

$(1101011010)_2$ = (0011) (0101) (1010) = (3) (5) (A) = $(35\text{A})_{16}$ = **35A (hex).**

1.9.6 Conversion of a Hexadecimal Number to Binary Number

For the conversion of a hexadecimal number to binary number each digit of the given hexadecimal number is converted to its 4-bit binary equivalent.

EXAMPLE 1.18. Convert the hexadecimal number 5CB8 to its equivalent

Solution: $(5\text{CB}8)_{16}$ = (0101) (1100) (1011) (1000)

5 C B 8

= $\mathbf{(0101110010111000)_2}$.

EXAMPLE 1.19. Convert the hexadecimal number 4F2D to its binary equivalent

Solution: $(4\text{F}2\text{D})_{16}$ = (0100) (1111) (0010) (1101)

4 F 2 D

= $\mathbf{(0100111100101101)_2}$.

Hexadecimal number are used with microprocessors and microcomputers for the convenience of the programmer. But the microprocessor accepts the binary equivalent of a hexadecimal number. There is a provision for such machine language conversion in the microprocessor-based system.

1.9.7 Conversion of Octal Number to Hexadecimal Number and Vice Versa

The conversion of a hexadecimal number to octal number or *vice-versa* can easily be done through the binary method as illustrated by the following examples:

EXAMPLE 1.20. Convert the hexadecimal number 4DF to its octal equivalent.

Solution: 4DF(hex) = (0100) (1101) (1111)

4 D F

= **010011011111 (binary).**

Now the equivalent binary number is divided into groups of 3 bits from the right to obtain its octal equivalent.

010011011111 (binary) = (010) (011) (011) (111)

= (2) (3) (3) (7)

= **2337 (octal).**

EXAMPLE 1.21. Convert 5A (hexadecimal) to octal.

Solution: 5A (hex) = (0101) (1010)

5 A

= 01011010 (binary)

01011010 (binary) = (01) (011) (010)

= (001) (011) (010)

= (1) (3) 2)

= **132 (octal).**

EXAMPLE 1.22. Convert the octal number 456 to hexadecimal.

Solution: 456 (octal) = (100) (101) (110)

4 5 6

= 100101110 (binary)

Now the equivalent binary number is divided into groups of 4 bits to get the equivalent hexadecimal number.

100101110 (binary) = (1) (0010) (1110)

= (0001) (0010) (1110)

(1) (2) (E)

= **12 E (hex).**

EXAMPLE 1.23. Convert 27 (octal) to hexadecimal.

Solution: 27 (octal) = (010) (111)

2 7

= 010111 (binary)

010111 (binary) = (01) (0111)

= (0001) (0111)

(1) (7)

= **17 (hex).**

Advantages of the hexadecimal system over octal system:

(*i*) For long computer word length the binary representation of hexadecimal number is shorter.

(*ii*) For the word length divisible by 4 but not by 3 the hexadecimal system is more convenient. The octal number system will require the extension of zero bits ahead of the most significant bit for such a case.

Advantages of the octal number system over the hexadecimal number system:

(*i*) In the hexadecimal number system letters are used to represent decimal numbers 10 to 15. The numbers using letters resemble with words or names. Thus, the use of letters in hexadecimal numbers creates confusion. There is no such confusion in the octal number system as the octal number system uses only digits 1 to 7.

(*ii*) The conversion of octal numbers to decimal numbers is simpler as an octal number contains only digits. The letters used in hexadecimal numbers have to be converted into equivalent decimal numbers during the conversion process.

1.10 CONVERSION OF REAL NUMBER

In the decimal number system the digits after the decimal point have weights explained below:

$$0.538 = 0.5 + 0.03 + 0.008 = 5 \times \frac{1}{10} + 3 \times \frac{1}{100} + 8 \times \frac{1}{1000}$$
$$= 5 \times 10^{-1} + 3 \times 10^{-4} + 8 \times 10^{-3}.$$

Similarly, in the binary number system the weights of binary bits after the binary point are as follows:

$$0.1011 = 1 \times 2^{-1} + 0 \times 2^{-2} + 1 \times 2^{-3} + 1 \times 2^{-4}$$
$$= 1\ \frac{1}{2} + 0 \times \frac{1}{4} + 1 \times \frac{1}{8} + 1 \times \frac{1}{16} = 0.5 + 0 + 0.125 + 0.0625$$
$$= \mathbf{0.6875\ (decimal).}$$

EXAMPLE 1.24. Convert the binary real number 1011.011 to a decimal real number.

Solution: The above binary real number consists of two parts-an integer and a fraction. The decimal equivalent are written for both the integer and the fraction and they are added to give the decimal real number.

$$1011.011 = (1 \times 2^3 + 0 \times 2^2 + 1 \times 2^1 + 1 \times 2^0) + (0 \times 2^{-1} + 1 \times 2^{-2} + 1 \times 2^{-3})$$
$$= (8 + 0 + 2 + 1) + \left(0 \times \frac{1}{2} + \frac{1}{4} + \frac{1}{3}\right) = 11 + (0 + 0.25 + 0.125)$$
$$= \mathbf{11.375\ (decimal\ real\ number).}$$

1.10.1 Conversion of Decimal Fraction to Binary Fraction

The conversion of decimal fraction to a binary fraction is obtained using the technique of successive multiplication by 2. The integer part is noted down after each multiplication and the new fraction is used for further multiplication. The following examples will illustrate the procedure:

EXAMPLE 1.25. Convert the decimal fraction 0.375 to its binary equivalent.

Solution:

Fraction	*Fraction × 2*	*Remainder Fraction*	*Integer*
0.375	0.750	0.75	0 (MSB)
0.75	1.50	0.50	1
0.50	1.00	0.00	1 (LSB)
0.375 (decimal = 011 (binary)			

EXAMPLE 1.26. Convert 10.7 (decimal) to its binary equivalent.

Solution: For the above decimal real number the binary equivalent are obtained for the First 10 (decimal) is converted to its equivalent binary number.

		Remainder
2	10	
2	5	0 (LSB)
2	2	1
2	1	0
	0	1 (MSB)

10 (decimal) = 1010 (binary)

Now 0.7 (decimal) is converted to its equivalent binary fraction.

EXAMPLE 1.27. Convert the decimal number 25 to its binary equivalent.

Solution: The remainder can be written in a simpler form as shown below:

		Remainder
2	25	
2	12	1 (LSB)
2	6	0
2	3	0
2	1	1
	0	1 (MSB)

∴ Binary number = 1101.

1.11 BINARY ADDITION

Table given below shows the rules for binary addition. When 1 is added to 1, the sum is 10 (binary) = 2 (decimal).

Table 1.4. Binary Addition

A	*B*	*A + B*
0	0	1
0	1	1
1	0	1
1	1	10

EXAMPLE 1.28.

```
               decmal
   1 1 0 1       13
 + 0 0 0 1      + 1
 ---------     ----
   1 1 1 0       14
```

EXAMPLE 1.29.

```
               decmal
   0 0 1 1        3
 + 0 1 1 1        7
 ---------     ----
   1 0 1 0       10
```

1.12 BINARY SUBTRACTION

Table given below shows the rules for binary addition.

Table 1.5. Binary Subtraction

A	*B*	*A – B*
0	0	0
0	1	1
1	0	1
1	1	0

EXAMPLE 1.30.

	decmal
1 1 0 1	13
– 0 0 1 1	– 3
1 0 1 0	10

EXAMPLE 1.31.

	decmal
1 1 0 1	13
+ 0 1 1 0	– 6
0 1 1 1	7

EXAMPLE 1.32. Convert the decimal number $(52)_{10}$ to an equivalent binary number.

Division	*Integer quotient*	*Remainder*
52/2	**26**	**0**
26/2	**13**	**0**
13/2	**6**	**1**
6/2	**3**	**0**
3/2	**1**	**1**
1/2	**0**	**1**

Solution: The binary number is formed from the remainder terms as $(110100)_2$.

EXAMPLE 1.33. Convert the decimal fraction $(0.4375)_{10}$ to a binary fraction.

Solution:

Multiplication	*Integer*	*Fraction Portion*
0.4375 × 2	0	0.8750
0.8750 × 2	1	0.7500
0.7500 × 2	1	0.5000
0.5000 × 2	1	0.0000

The converted binary fraction is formed from the integer portion as $(0.111)_2$.

EXAMPLE 1.34. Convert the decimal number $(10.75)_{10}$ to its binary equivalent.

Solution: The integer portion and fraction portion are converted separately and combined to form the complete binary number.

Division	*Integer quotient*	*Remainder*
10/2	5	0
5/2	2	1
2/2	1	0
1/2	0	1

$\therefore \quad (10)_{10} = (1010)_2$

Converting the fraction portion.

Multiplication	*Integer portion*	*Fraction portion*
0.75 × 2	1	0.500
0.500 × 2	1	0.000

Hence $(0.75)_{10} = (0.11)_2$

The complete conversion is

$$(10.75)_{10} = (1010)_2 + (0.11)_2 = \mathbf{(1010.11)_2}$$

EXAMPLE 1.35. Convert decimal 41 to binary.

Solution:

	Integer	Remainder
2	41	
2	20	1
2	10	0
2	5	0
2	2	1
2	1	0
	0	1

↑ 101001

EXAMPLE 1.36. Convert $(0.975)_{10}$ to binary.

Solution:

integer	fraction	Coefficient
0.6875 × 2 = 1	+ 0.3750	$a_{-1} = 1$
0.3750 × 2 = 0	+ 0.7500	$a_{-2} = 0$
0.7500 × 2 = 1	+ 0.5000	$a_{-3} = 1$
0.5000 × 2 = 1	+ 0.0000	$a_{-4} = 1$

$(0.6875)_{10} = (0.1011)_2$

EXAMPLE 1.37. From the octal equivalent of $(100)_{10}$

Solution:

Division	*Integer quotient*	*Remainder*
100/38	12	4
12/8	1	4
1/8	0	1

The octal number is formed from the remainder terms as $(144)_8$.

EXAMPLE 1.38. Find the octal equivalent of $(100.250)_{10}$

Solution:

Multiplication	*Integer portion*	*Fraction portion*
0.250 × 8	2	0.000

Hence $(0.250)_{10} = (0.2)_8$

The complete conversion is

$$(0.250)_{10} = (144)_8 + (0.2)_8 = \mathbf{(44.2)_8}$$

1.13 COMPLIMENTS

Complements are used in digital computers for simplifying the subtraction operation and for logical manipulations. There are types of complements for each base-r system:

(*i*) the r's complements;

(*ii*) the $(r - 1)$'s complement.

When the value of the base is substituted, the two types receive the names 2's and 1's complement for binary numbers, or 10's and 9's complement for decimal numbers.

(*i*) *r*'s Complement

Given a positive number N in base r with an integer part of n digits, the r's complement of N is defined as:

$$r_n - \text{N for} \neq 0$$

and $$0 \text{ for N} = 0$$

The following numerical examples will clarify the definition.

The 10's complement of $(52520)_{10}$ is

$$10^5 - 52520 = 47480$$

The 2's complement of $(101100)_2$ is

$$(2^6)_{10} - (101100)_2 = (1000000 - 101100)_2 = 010100$$

The 2's complement of $(0.0110)_2$ is

$$(1 - 0.0110)_2 = 0.1010.$$

(*ii*) The (*r* – 1)'s Complement

Given a positive number N in base r with an integer of n digits and a fraction part of m digits, the $(r - 1)$'s complement of N is defined as $r^n - r^m - \text{N}$. The following numerical examples will clarify the definition:

The 9's complement of $(52520)_{10}$ is

$$(10^5 - 1 - 52520) = 99999 - 52520 = 47479.$$

No fraction part, so

$$10^{-m} = 10^0 = 1$$

The 9's complement of $(0.3267)_{10}$ is

$$(1 - 10^{-4} - 0.3267) = 0.9999 - 0.3267 = 0.6732.$$

The 1's complement of $(101100)_2$ is

$$(2^6 - 1) - (101100) = (111111 - 101100)_2 = 010011.$$

The 1's complement of $(0.0110)_2$ is

$$(1 - 2^{-4})_{10} - (0.0110)_2 = (0.1111 - 0.0110)_2 = 0.1001.$$

1.13.1 9's Complement

The obtain the 9's complement of a decimal number each digit of the number is subtracted from 9. Example,

9's complement of 45 is

$$(99 - 45) = 54.$$

9's complement of 523 is

$$(999 - 523) = 476.$$

1.13.2 10's Complement

The 10's complement of a decimal number = its 9's complement + 1.

The 10's complement of 45 is

$$54 + 1 = 55.$$

The 10's complement of 534 is

$$476 + 1 = 477.$$

Now add the number and units 10's complement and examine the sum.

Example:

45	(decimal number)
+ 55	(its 10's complement)
00	

Carry = 1

In this case the decimal number is of two digits. If we consider the sum of the number and its 10's complement only upto two digits, it is zero. In other words we are neglecting the carry of the last stage.

Example:

5 2 3	(decimal number)
+ 4 7 7	(its 10's complement)
0 0 0	

Carry = 1

The decimal number 523 is of three digits. If we consider the sum of the number and its 10's complement only upto three digits, it is zero. Thus the 10's complement gives the negative value of the number.

The 10's complement of a decimal number = – decimal number.

1.13.3 1's Complement

The 1's complement in the binary number system is similar to the 9's complement in the decimal system. The 1's complement of a binary number is obtained by subtracting each bit of the number from 1. The 1's complement of 01 is 10. The 1's complement of 111 is 000. Thus the 1's complement of a binary number can by changing the bit 1 to 0 and 0 to 1.

EXAMPLE 1.39. Find 1's complement of 100110.

Solution: The 1's complement of 100110 = 011001

EXAMPLE 1.40. Find 1's complement of 0000.

Solution: The 1's complement of 0000 is 1111.

EXAMPLE 1.41. Find 1's complement of 11111.

Solution: The 1's complement of 11111 is 00000.

1.13.4 2's Complement

The 2's complement in the binary number system is similar to the 10's complement in the decimal number system.

The 2's complement of binary number = its 1's complement + 1.

EXAMPLE 1.42. Find 2's complement of 10011.

Solution: The 2's complement of 10011 = 01100 + 1 = 01101

EXAMPLE 1.43. ***Find 2's complement of 111.***

Solution: The 2's complement of 111 = 000 + 1 = 001

EXAMPLE 1.44. Find 2's complement of 0000.

Solution: The 2's complement of 0000 = 1111 + 1 = 0000

Now examine the sum when a binary number is added to its 2's complement.

EXAMPLE 1.45. Binary number = 1 0 0 1

Solution: Its 1's complement = 0 1 1 0

It 2's complement = 0 1 1 0 + 1 = 0 1 1 1

Number + its 2's complement = 1 0 0 1

+ 0 1 1 1

0 0 0 0

Carry

The last carry is lost if the pressor is of 4 bits or sum is considered only upto 4 bits. For 8-bit processor the number and 2's complement will be written in 8 bits as given below:

Number = 00001001

1's complement = 11110110

2's complement = 11110110 + 1 = 11110111

Number = 00001001

+ 2's complement = 11110111

00000000

Carry

The last carry will be lost, if the 8 bit sum is considered.

EXAMPLE 1.46. Add + 4 and – 4.

Solution: + 4(decimal) = 00000100

– 4 = Its 2's complement + 1 = 11111011 + 1 = 11111100

+ 4 = 00000100

+ – 4 = 11111100

0 00000000

EXAMPLE 1.47. Add + 5 and – 7.

Solution: 7 = 00000111

– 7 = 11111000 + 1 = 11111001

+ 5 = 00000101

+ – 7 = 11111001

– 2 = 11111110

1.13.5 Subtraction with *r*'s Complements

The subtraction of two positive numbers (M – N), both of base *r*, may be done as follows:

1. Add the minuend M to the r's complement of the subtrahend N.
2. Inspect the result obtained in step 1 for an end carry:
   (*a*) If an end carry occurs, discard it.
   (*b*) If an end carry does not occur, take the *r*'s complement of the number obtained in step 1 and place a negative sign in front.

EXAMPLE 1.48. Using 10's complement, subtract 72532 – 3250.

Solution:

M = 72532   72532
N = 03250
10's complement of N = 96750   + 96750
∴ end carry → 1   69282

69282

EXAMPLE 1.49. Subtract $(3250 - 72532)_{10}$

Solution:

M = 03250   03250
N = 72532
10's complement of N = 27468   + 27468
No carry   30718

Hence 69282 = – (10's complement of **30718**)

1.13.6 Subtraction with (*r* – 1)'s Complement

The subtraction of M – N, both positive numbers in base *r*, may be calculated in the following manner:

1. Add the minuend M to the (r – 1)'s complement of the subtrahend N.
2. Inspect the result obtained in step 1 for an end carry.
   (*a*) If an end carry occurs, add to 1 to the least significant digit (end-around carry).
   (*b*) If an end carry does not occur, add 1 to the (*r* – 1)'s complement of the number obtained in step 1 and place a negative sign in front.

EXAMPLE 1.50. Use 1's complement to perform M – N with given binary number.

Solution: (*a*)

M = 1010100   1010100
N = 1000100
1's complement of N = 0111011   + 0111011
end around carry   1_ 0001111
\+ 1
0010000

(*b*)

M = 1000100   1000100
N = 1010100
1's complement of N = 0101011   0101011
no carry   _1101111

Hence 10000 = – (1's complement of 1101111)

Binary codes for digit require a minimum of four bits. Numerous different codes can be obtained by arranging four or more bits in ten distinct possible combinations. A few possibilities are shown in Table 1.6 below.

Table 1.6. Binary codes for the decimal digits

Decimal digit	***(BCD) 8421***	***Excess-3***	***8-4-2-1***	***2421***	***(Binary) 5043210***
0	0000	0011	0000	0000	0100001
1	0001	0100	0111	0001	0100010
2	0010	0101	0110	0010	0100100
3	0011	0110	0101	0011	0101000
4	0100	0111	0100	0100	0110000
5	0101	1000	1011	1011	1000001
6	0110	1001	1010	1100	1000010
7	0111	1010	1001	1101	1000100
8	1000	1011	1000	1110	1001000
9	1001	1100	1111	1111	1010000

EXAMPLE 1.51. Use (*r* – 1)'s complement subtraction to find the difference $(111)_2 - (010)_2$

Solution: Form the 1's complement of the subtrahend:

$$\begin{array}{r} 111 \\ -\,010 \\ \hline (101)_2 \end{array}$$

The resulting addition gives an end-carry, thus, the end-around carry operation is performed to form the answer:

$$\begin{array}{r} 111 \\ +\,101 \\ \hline (1100)_2 \end{array}$$

and

$$\begin{array}{r} 100 \\ +\,1 \\ \hline (101)_2 \end{array}$$

EXAMPLE 1.52. Apply (*r* – 1)'s complement subtraction to find the difference $(010)_2 - (111)_2$

Solution: The 1's complement of the subtrahend is

$$\begin{array}{r} 111 \\ -\,111 \\ \hline (000)_2 \end{array}$$

Performing the addition,

$$\begin{array}{r} 010 \\ +\,000 \\ \hline (010)_2 \end{array}$$

Since no end-carry resulted in the addition, the answer is formed as the negative at the 1's complement of $(010)_2$, or

$$\begin{array}{r} 111 \\ -010 \\ \hline -(101)_2 \end{array}$$

EXAMPLE 1.53. Use r's complement subtraction to find the difference $(111)_2 - (010)_2$

Solution: Forming the 2's complement of the subtrahend:

$$\begin{array}{r} 111 \\ -010 \\ \hline 101 \\ +1 \\ \hline (110)_2 \end{array}$$

The resulting addition gives an end-carry which is disregarded to give the answer, or

$$\begin{array}{r} 111 \\ +110 \\ \hline (1101)_2 \end{array}$$

Hence $(111)_2 - (010)_2 = (101)_2$

Example 1.54. By r's complement subtraction, determine $(010)_2 - (111)_r$

Solution: The 2's complement of the subtrahend is

$$\begin{array}{r} 111 \\ -111 \\ \hline 000 \\ +1 \\ \hline (001)_2 \end{array}$$

Performing the addition,

$$\begin{array}{r} 010 \\ +001 \\ \hline (011)_2 \end{array}$$

Since the addition produced no end-carry, the 2's complement of the sum and a leading negative sign are formed:

$$\begin{array}{r} 111 \\ -011 \\ \hline 100 \\ +1 \\ \hline -(101)_2 \end{array}$$

SOLVED PROBLEMS

Q.1. Represent the binary number 1010111101110111 in octa form or hexadecimal form ?

Solution: Binary number = 1010 1111 0111 0111

Hexadecimal = A F 7 7 = $(AF77)_{16}$

Octal form = 1010111101110111

$\Rightarrow 1\ 2\ 7\ 5\ 6\ 7$

$= (127567)_8.$

Q.2. Describe the alpha numeric representation.

Solution: An alpha numeric character set is a set of elements that includes the 10 decimal digits, the 26 letters of the alphabets and a number of special characters, such as, \$, + and = etc. such a set contains 32 and 64 elements (if only upper case letters are included) or between 64 to 128 (if both upper case and lower case letters are included). In the first case, the binary code will require six bits and in second case seven bits the standard alphanumeric binary code is the ASCII (American Standard Code for Information Interchange) which uses seven bits to code 128 characters.

Some examples in alphanumeric

a = 97 (ASCII)

A = 65 (ASCII)

O = 48 (ASCII).

Q.3. Find the value of number 12389 in

(*a*) 10's complement form

(*b*) 9's complement form

(*c*) 1's complement form

(*d*) 2's complement form.

Solution: Given number = 12389

9's complement = 99999 – 12389

= 87610

10's complement = 100000 – 12389

= 87611

1's complement binary of 1238 is

= 11000001100101

1's complement = 00111110011010

2's complement = 00111110011011

Q.4. Describe the fixed point representation.

Solution: Positive integers including zero can be represented as unsigned numbers. However to represent negative integers left most position (bit) is 1.

The fixed point method assumes that the binary point is always fixed in one position. Two position most widely used are:

(*i*) A binary point in the extreme left of the register to make the stored number a fraction and

(*ii*) A binary point in the extreme right to make the stored number and integer.

Q.5. What do you mean by overflow ? Describe the overflow detection.

Solution: When the two numbers of *n* digits each are added and the sum occupies *n* digits, we say over flow occurred.

The detection of an overflow after addition of two binary numbers depends on whether the numbers are considered to be signed or unsigned when two unsigned numbers are added, an overflow is detected from the end carry out of the most significant position. In the case of sign numbers, the left most bit always represent the sign, and negative numbers are in 2's complement form.

When two signed numbers are added the sign bit is treated as part of the number and the end carry does not indicate an overflow.

An overflow cannot occur after an addition if one number is positive and other is negative, since adding a positive number to a negative number produces a result that is smaller than larger of the two original numbers.

An overflow occur if the two numbers added are both positive or both negative.

+ 70	01000110
+ 80	01010000
+ 150	10010110

Since sum of two number is +150, it exceeds the capacity of the 8-bit register.

–70	10111010
–80	10110000
–150	01001010

Overflow Detection: An overflow condition can be detected by observing the carry into the sign bit position and carry out of the sign bit position. If these two carry are not equal, an overflow condition is produced. If two carries are applied to an exclusive –*OR* gate, an overflow will be detected when the output of gate is equal to 1.

Q.6. Represent the decimal number

(*a*) 27, (*b*) 396, (*c*) 4096

Into binary form in

(*i*) Binary code

(*ii*) BCD code

(*iii*) Excess-3 code

(*iv*) Gray code

(*v*) Octal code

(*vi*) Hexadecimal code.

Solution: (*a*) (*i*)27 = 11011

(*ii*) 27 = 00100111

(*iii*) 27 = 01011010

(*iv*) 00110100

(*v*) 27 = $(33)_8$ = 011011

(*vi*) 27 = $(1B)_{16}$ = 00011011

(*b*) (*i*) 396 = 110001100

(*ii*) 396 = 001110010110

(*iii*) 396 = 011011001001

(*iv*) 396 = 001011010101

(*v*) 396 = $(614)_8$ = 110001100

(*vi*) 396 = $(186)_{16}$ = 000110001100

(*c*) (*i*) 4096 = 1000000000000

(*ii*) 4096 = 0100000010010110

(*iii*) 4096 = 0111001111001001

(*iv*) 4096 = 0110000011010101

(*v*) 4096 = $(10000)_8$ = 001000000000000

(*vi*) 4096 = $(1000)_{16}$ = 000100000000000

Q.7. What is floating point notation ? What is benefit of these ?

Solution: The usual method used by computers to represent real number is floating point notation. The key concept is that a real number is represented by a numbers called a mantissa, times a base raised to an integer power called an exponent. The base is usually fixed and mantissa and exponent vary to represent different real numbers.

e.g. If the base is fixed at 10, the number 378.53 could be represented as 37853×10^{-2}. The mantissa is 37853 and the exponent is –2. Other possible representations are $.37853 \times 10^3$. A real number is represented by a 32-bit string consisting of 24-bit mantissa followed by an 8-bit exponent.

The base is fixed at 10. Both the mantissa and exponent are two's complement binary integers.

e.g. 24 bit binary representation of 38753 is 000000001001011101100001 and 8 it two's complement binary representation of –2 is 11111110. So, the representation of 387.53 is 0000 0000 1001 0111 0110 0001.

The advantage of floating point notation is that it can be used to represent numbers with extremely small absolute values. For example, in the notation presented priviously, the largest number can be represented is $(2^{23} - 1) \times 10^{127}$, which is very large number indeed.

The small positive number that can be represented is 10^{-128} which is quite small.

Q.8. Convert the following binary number to decimal:

(*a*) 101110, (*b*) 1110101, (*c*) 110110100.

Solution: (*a*) $(101110)_2 = 32 + 8 + 4 + 2 = 46$

(*b*) $(1110101)_2 = 64 + 32 + 16 + 4 + 1 = 117$

(*c*) $(110110100)_2 = 256 + 128 + 32 + 16 + 4 = 436$.

Q.9. Convert the following numbers with the indicated bases to decimal:

(*a*) $(12121)_3$ (*b*) $(4310)_5$

(*c*) $(50)_7$ (*d*) $(198)_{12}$.

Solution: (*a*) $(12121)_3 = 3^4 + 2 \times 3^3 + 3^2 + 2 \times 3 + 1 = 81 + 54 + 9 + 6 + 1 = 151$.

(*b*) $(4310)_5 = 4 \times 5^3 + 3 \times 5^2 + 5 = 500 + 75 + 5 = 580$.

(*c*) $(50)_7 = 5 \times 7 = 35$.

(*d*) $(198)_{12} = 12^2 + 9 + 12 + 8 = 144 + 108 + 8 = 260$.

Q.10. Convert the following decimal numbers to binary:

(*a*) 1231, (*b*) 673, (*c*) 1998.

Solution: (*a*) $(1231) = 1024 + 128 + 64 + 15 = 2^{10} + 2^7 + 2^6 + 2^3 + 2^2 + 1$

$= (10011001111)_2$

(*b*) $673 = 512 + 128 + 32 + 1 = 2^9 + 2^7 + 2^5 + 1 = (1010100001)_2$

(*c*) $1998 = 1024 + 512 + 256 + 128 + 64 + 8 + 4 + 2$

$= 2^{10} + 2^9 + 2^8 + 2^7 + 2^6 + 2^3 + 2^2 + 1 = (11111001110)_2$.

Q.11. Convert the following decimal numbers to the bases indicated.

(*a*) 7562 to octal

(*b*) 1998 to hexadecimal

(*c*) 175 to binary.

Solution: (*a*) $(7562)_{10} = (16612)_8$

(*b*) $(1938)_{10} = (792)_{16}$

(*c*) $(175)_{10} = (10101111)_2$.

Q.12. Convert the hexadecimal number F3A7C2 to binary and octal.

Solution: $(F3A7C2)_{16} = (111100111010011111000010)_2 = (74723702)_8$.

Q.13. What is the radix of the numbers if the solution to the quadratic equation $x^2 - 10x + 31 = 0$ is $x = 5$ and $x = 8$?

Solution:
$$(x^2 - 10x + 31) = [(x-5)(x-8)]_{10} = x^2 - (5+8)_{10}x + (40)_{10}x$$
$$(10)_r = (13)_r$$
$$\text{Radix} = 13$$
Also,
$$(31)_r = 3 \times 13 + 1 = (40)_{10}$$
$$\text{Radix} = 13.$$

Q.14. Show the value of all bits of a 12-bit register that hold the number equivalent to the decimal 215 in

(a) binary
(b) binary coded octal
(c) binary coded hexadecimal
(d) binary coded decimal.

Solution: $(215)_{10} = 128 + 64 + 16 + 7 = (11010111)_2$

(a) 000 011 010 111
(b) 000 011 010 111 binary coded octal
0 3 2 7
(c) 0000 1101 0111 binary coded hexadecimal
0 0 7
(d) 0010 0001 0101 binary coded decimal
2 1 5

Q.15. Show the bit configuration of a 24-bit register when its content represents the decimal equivalent of 295:

(a) binary
(b) in BCD
(c) in ASCII using eight bits with even parity.

Solution: $(295)_{10} = 256 + 32 + 7 = (100100111)_2$

(a) 0000 0000 0000 0001 0010 0111
(b) 0000 0000 0000 0010 1001 0101
(c) 10110010 00111001 00110101.

Q.16. Decode the following ASCII code:

1001010 1001111 1001000 1001110 0100000 1000100 1001111 100010

Solution: JOHN DOE.

Q.17. Obtain the 9's complement of the following eight digit decimal numbers:

(a) 12349876 **(b) 00980100**
(c) 90009951 **(d) 00000000.**

Solution: (a) 87650123 (b) 99019899
(c) 09990048 (d) 99999999.

Q.18. Obtain the 10's complement of the following six digit decimal numbers:

(a) 876100 **(b) 909343**
(c) 900000 **(d) 000000.**

Solution: (a) 876100 (b) 909343
(c) 900000 (d) 000000.

Q.19. Obtain the 1's and 2's complements of the following six digit decimal numbers:

(*a*) 10101110 **(*b*) 10000001**

(*c*) 10000000 **(*d*) 00000001**

(*e*) 00000000.

Solution: 1's Complement:

(*a*) 01010001 (*b*) 01111110

(*c*) 01111111 (*d*) 11111110

(*e*) 11111111.

2's Complement:

(*a*) 01010010 (*b*) 01111111

(*c*) 10000000 (*d*) 11111111

(*e*) 00000000.

Q.20. Perform the subtraction with the following unsigned decimal numbers by taking the 10's complement of the subtrahend.

(*a*) 5250 – 1321 **(*b*) 1753 – 8640**

(*c*) 20 – 100 **(*d*) 1200 – 250.**

Solution: (*a*)

```
     5250
     8629
   ------
  1) 3929
```

(*b*)

```
     1753
     1360
   ------
  0) 3113
    –6887  10's complement
```

(*c*)

```
      020
    + 900
   ------
   0) 920
     –080  10's complement
```

(*d*)

```
     1200
   + 9750
   ------
  0) 0950
```

Q.21. Perform the subtraction with the following unsigned binary numbers by taking the 2's complement of the subtrahend.

(*a*) 11010 – 10000 **(*b*) 11010 – 1181**

(*c*) 11010 – 110000 **(*d*) 1010100 – 1010100.**

Solution: (*a*)

```
       11010
       10000
     -------
    1) 01010
(26 – 16 = 10)
```

(*b*)

```
       11010
       10011
     -------
    1) 01101
(26 – 13 = 13)
```

(c)
```
        000100
        010000
      ---------
      1) 010100
  (4 – 48 = –44)
```

(d)
```
       1010110
       0101100
      ---------
     1) 0000000
  (84 – 84 = 0)
```

Q.22. Perform the arithmetic operations (+42) + (–13) and (–42) – (–13) in binary using signed 2's complement representation for negative numbers.

Solution:

+42 = 0101010		+13 = 0001101	
–42 = 1010110		–13 = 1110011	
(+42) = 0101010		(–42) = 1010110	
(–13) = 1110011		(+13) = 0001101	
(+29) = 0011101		(–29) = 1100011	

Q.23. Represent the binary number 1010 1111 0111 0111 in Octal form or Hexadecimal form ?

Solution: Binary number = 1010 1111 0111 0111

Hexadecimal = A F 77

Octal form = 1010 111 101 110 111

$\Rightarrow$ 1 2 7 5 6 7 = $(127567)_8$.

Q.24. Describe the alpha numeric representation ?

Solution: An alpha numeric character set is a set of elements that includes the 10 decimal digits, the 26 letters of the alphabet and a number of special characters, such as $, +, and =. Such a set contains 32 and 64 elements (if only upper case letters are included) or between 64 to 128 (if both upper case and lower case letters are included). In the first case the binary code will require six bits and in second case seven bits the standard alpha numeric binary code is the ASCII (American Standard Code for Information Interchange) which uses seven bits to code 128 characters.

Some example in alpha numeric

a = 97 (in ASCII)

A = 65 (in ASCII)

O = 48 (in ASCII)

Q.25. Find the value of number 12389 in

(*a*) 10's complement form

(*b*) 9's complement form

(*c*) 1's complement form

(*d*) 2's complement form

Solution: Given number = 12389

9's complement = 99999 – 12389 = 87610

10's complement = 100000 – 12389 = 87611

1's complement binary number of 12389 is = 11000001100101

1's complement = 00111110011010

2's complement = 00111110011010

1

00111110011011

Q.26. Describe the fixed-point representation ?

Solution: Positive integer including zero can be represented as unsigned. However we represent negative integers leftmost position (bit) is 1.

The fixed point method assumes that the binary pint is always fixed in one position. The two position most widely used are:

(*i*) a binary point in the extreme left of the register to make the stored number of fraction, and

(*ii*) a binary point in the extreme right of the register to make the stored number and integer.

Q.27. What do you mean by overflow ? Describe the over flow detection.

Solution: When two numbers of a digits each are added and the sum occupies *n* digits, we say over flow occurred.

The detection of an overflow after the addition of two binary numbers depends on whether the numbers are considered to be signed or unsigned when two unsigned number is added, an overflow is detected from the end carry out of the most significant position. In the case of sign numbers, the leftmost bit always represent the sign, and negative numbers are in 2's complement form.

When two signed numbers are added, the sign bit is treated as part of the number and the end carry does not indicate an overflow.

An overflow cannot occur after an addition if one number is positive and other is negative, since adding a positive number to a negative number produces a result that is smaller than larger of the two original numbers.

An overflow occur, if the two numbers added are both positive or both negative.

+ 70	0 1000110
+ 80	0 1010000
+ 150	1 0010110

Since sum of two number is + 150, it exceeds the capacity of the 8-bit registers.

Carries: 10

– 70	1 0111010
– 80	1 0110000
– 150	0 1001010

Over Flow Detection: An over flow condition can be detected by observing the carry into the sign bit position and carry out of the sign bit position. If these two carry are not equal, an overflow condition is produced. If two carries are applied to an exclusive — OR gate, an overflow will be detected when the output of the gate is equal to 1.

Q.28. Represent the decimal number

(*a*) 27

(*b*) 396

(*c*) 4096

in binary form in

(*i*) Binary code

(*ii*) d

(iii) d

(iv) Gray code

(v) Octal code

(vi) Hexadecimal code

Solution:

(a) (i) 27 = 11011

(ii) 27 = 00100111

(iii) 27 = 01011010

(iv) 0011 0100

(v) 27 = $(33)_8$ = 011 011

(vi) 27 = $(1B)_{16}$ = 0001 1011

(b) (i) 396 = 11000 1100

(ii) 396 = 0011 1001 0110

(iii) 396 = 0110 1100 1001

(iv) 396 = 0010 1101 0101

(v) 396 = $(614)_8$ = 110 001 100

(vi) 396 = $(18C)_{16}$ = 0001 1000 1100

(c) (i) 4096 = 10000 0000 0000

(ii) 4096 = 0100 0000 1001 0110

(iii) 4096 = 0111 0011 1100 1001

(iv) 4096 = 0111 0011 1100 1001

(v) 4096 = $(10000)_8$ = 001 000 000 000000

(vi) 4096 = $(1000)_{16}$ = 0001 0000 0000 0000

Q.29. What is floating point notation ? What is benefit of these ?

Solution: The usual method used by computers to represent real number is floating point notation. The key concept is that a real number is represented by a number, called a mantissa, times a base raised to an integer power called an *exponent*. The base is usually fixed and the mantissa and exponent vary to represent different real numbers.

E.g., if the base is fixed at 10, the number 378.53 could be represented as 38753 × 10^{-2}, The mantissa is 38753 and the exponent is – 2. Other possible representations are .38753 × 10^3 and 387.53 × 10^0.

A real number is represented by a 32 bit string consisting of a 24-bit mantissa followed by an 8-bit exponent. The base is fixed at 10. Both the mantissa and exponent are two's complement binary integers.

e.g. 24 bit binary representation of 38753 is 000000001001011111100001 and 8 bit two's complement binary representation of – 2 is 11111110. So, the representation of 387.53 is 0000 0000 1001 0111 0110 0001.

The advantage of floating point notation is that it can be used to represent numbers with extremely large or extremely small absolute values. For example, in the notation presented previously, the largest number can be represented is $(2^{23} - 1) \times 10^{127}$. Which is very large number indeed.

Q.30. Convert the following binary numbers to decimal:

101110; 1110101; and 110110100

Solution:

$(101110)_2$ = 32 + 8 + 4 + 2 = 46

$(1110101)_2$ = 64 + 32 + 16 + 4 + 1 = 117

$(110110100)_2$ = 256 + 128 + 32 + 16 + 4 = 436.

Q.31. Convert the following numbers with the indicated bases to decimal:

$(12121)_3$; $(4310)_5$; $(50)_7$; and $(138)_{12}$

Solution:
$$(12121)_3 = 3^4 + 2 \times 3^3 + 3^2 + 2 \times 3 + 1 = 81 + 54 + 9 + 6 + 1 = 151$$
$$(4310)_5 = 4 \times 5^3 + 3 \times 5^2 + 5 = 500 + 75 + 5 = 580$$
$$(50)_7 = 5 \times 7$$
$$(198)_{12} = 12^2 + 9 \times 12 + 8 = 144 + 108 + 8 = 260.$$

Q.32. Convert the following decimal numbers to binary:

1231; 673 and 1998

Solution:
$$(1231)_{10} = 1024 + 128 + 64 + 15 = 2^{10} + 2^7 + 2^6 + 2^3 + 2^2 + 1$$
$$= (10011001111)_2$$
$$(673)_{10} = 512 + 128 + 32 + 1 = 2^9 + 2^7 + 2^5 + 1 = (1010100001)_2$$
$$(1998)_{10} = 1024 + 512 + 256 + 128 + 64 + 8 + 4 + 2$$
$$= 2^{10} + 2^9 + 2^8 + 2^6 + 2^3 + 2^2 + 1 = (11111001110)_2$$

Q.33. Convert the following decimal numbers to the bases indicated.

(*a*) 7562 to octal

(*b*) 1938 to hexadecimal

(*c*) *175 to binary*

Solution:

(*a*) $(7562)_{10} = (16612)_8$

(*b*) $(1938)_{10} = (792)_{19}$

(*c*) $(175)_{10} = (10101111)_2$

Q.34. Convert the hexadecimal number F3A7C2 to binary and octal.

Solution: $(F3A7C2)_{16} = (11110011\ 10100111\ 1100\ 0010)_2 = (74723702)_8$

Q.35. What is the radix of the numbers if the solution to the quadratic equation

$$x^2 - 10x + 31 = 0$$

is $x = 5$ and $x = 8$?

Solution: $(x^2 - 10x + 31)_r = [(x - 5)(x - 8)]_{10} = x^2 - (5 + 8)_{10}\, x + (40)_{10}\, x$

$\therefore$ $(10)_r = (13)_{10}$

Radix = 13

Also $(31)_r = 3 \times 13 + 1 = (40)_{10}$

Radix = 13.

Q.36. Show the value of all bits of a 12-bit register that hold the number equivalent to decimal 215 in

(*a*) binary

(*b*) binary-coded octal;

(*c*) binary-coded hexadecimal

(*d*) binary coded decimal (BCD).

Solution: (215) 1D $= 128 + 64 + 16 + 7 = (110\ 10\ 111)_2$

(*a*) 000011010111	Binary
(*b*) 000 011 010 111 0 3 2 7	Binary coded octal
(*c*) 0000 1101 0111 0 2 7	Binary coded hexadecimal
(*d*) 0010 0001 0101 2 1 5	Binary coded decimal

Q.37. Show the bit configuration of a 24-bit register when its content represent the decimal equivalent of 295:

(*a*) in binary

(*b*) in BCD

(*c*) in ASCII using eight bits with even parity.

Solution: $(295)_{10} = 256 + 32 + 7 + (100100111)_2$

(*a*) 0000 0000 0000 0001 0010 0111

(*b*) 0000 0000 0000 0010 1001 0101

(*c*) 10110010 00111001 00110101

Q.38. Decode the following ASCII Code:

1001010 1001111 1001000 1001110 0100000 1000100 1001111 1000101

Solution: JOHN DOE

Q.39. Obtain the 9's complement of the following eight-digit decimal numbers:

12349876; 00980100; 90009951 and 00000000

Solution: 87650123; 99019899; 09990048; 99999999

Q.40. Obtain the 10's complement of the following six-digit decimal numbers:

123900, 090657; 100000; and 000000

Solution: 876100; 909343; 900000; 000000.

Q.41. Obtain the 1's and 2's complements of the following eight-digit binary numbers :

10101110, 10000001, 10000000, 00000001; and 00000000.

Solution: 01010001; 01111110; 01111111; 11111110; 11111111

01010010; 01111111; 10000000; 11111111; 00000000

Q.42. Perform the subtraction with the following unsigned decimal numbers by taking the 10's complement of the subtrahend.

(*a*) 5250 – 1321

(*b*) 1753 – 8640

(*c*) 20 – 100

(*d*) 1200 – 100

Solution:

(*a*)
```
      5250
    + 8629
    ------
   1) 3929
```

(*b*)
```
      1753
    + 1350
    ------
   0) 3113
    – 6887     10's complement
```

(*c*)
```
       020
     + 900
    ------
    0) 920
     – 080     10's complement
```

(*d*)
```
      1200
    + 9750
    ------
   1) 0950
```

Q.43. Perform the subtraction with the following unsigned binary numbers by taking the 2's complement of the subtrahend.

(*a*) 11010 – 10000

(*b*) 11010 – 1101

(*c*) 100 – 110000

(*d*) 1010100 – 1010100

Solution:

(*a*)
```
     11010
     10000
  --------
  1) 01010
(26 – 16 = 10)
```
(*b*)
```
     11010
     10011
  --------
  0) 01101
(26 – 13 = 13)
```
(*c*)
```
     000100
     010000
  ---------
  1) 010100
   – 101100
(4 – 48 = – 44)
```
(*d*)
```
     1010110
     0101100
  ----------
  1) 0000000
(84 – 84 = 0)
```

Q.44. Perform the arithmetic operations (+ 42) + (– 13) and (– 42) – (– 13) in binary using signed – 2's complement representation for negative numbers.

Solution:

+ 42 =	0101010	+ 13 =	0001101
– 42 =	1010110	– 13 =	1110011
(+ 42)	0101010	(– 42)	1010110
(–13)	1110011	(+13)	0001101
(+ 29)	0011101	(– 29)	1100011

Q.45. Perform the arithmetic operations (+70) + (+80) and (–70) + (+80) with binary numbers in signed 2's complement representation. Use eight bits to accommodate each number together with its sign. Show that overflow occurs in both cases, that the last two carries are unequal and that there is a sign reversal.

Solution: 01 ← last two carries → 10

+ 70	01000110	– 70	10111010
+ 80	01010000	– 80	10110000
+ 150	10010110	– 150	01101010
greater ↑ than + 1271	↑ negative	less than ↑ – 128	↑ Positive

Q.46. Perform the following arithmetic operations with the decimal numbers using signed – 10's complement representation for negative numbers.

(*a*) (– 638) + (+ 758)

(*b*) (– 638) – (+ 185)

Solution:

(a)	(– 638)	9362
	(+ 785)	+ 0785
	(+ 147)	0147
(b)	(– 638)	9362
	(+ 185)	+ 9815
	(+ 823)	9177

Q.47. A 36-bit floating point binary number has eight bits plus sign for the exponent and 26 bits plus sign for the mantissa. The mantissa is a normalized fraction. Numbers in the mantissa and exponent are in signed magnitude representation. What are the largest and smallest positive quantities that can be represented, excluding zero ?

Solution:

	Mantissa	Exponent	
	+ | 26 bits	S | 8 bits	36 bits
Largest:	+ 0.1111...1	+ 11111111	
	$1 - 2^{-26}$	+ 255	$(1 - 2^{-26}) \times 2^{+255}$
Smallest:	+ 0.1000...0	– 11111111	
(normalized)	2^{-1}	– 255	2^{-256}

Q.48. Represent the number (+ 46.5) 10 as a floating point number with 24 bits.

Solution: The normalized tractions mantissa has 16 bits and the exponents has 8 bits.

$$46.5 = 32 + 8 + 4 + 2 + 0.5 = (101110.12)_2$$

Sign

0 101110100000000	00000110	
24 – bit mantissa	8 – bit exponent	(+ 6)

Q.49. The Gray code is sometimes called a reflected code because the bit values are elected on both sides of any 2^n value. For example, as shown in Table given below the values of the three low-order bits are reflected over a line drawn between 7 and 8. Using this property of the Gray code, obtain:

(a) The Gray code numbers for 16 through 31 as a continuation of Table given below.

(b) The excess-3 Gray code for decimals 10 to 19.

Table. 1.6 Bit Gray Code

Binary code	*Decimal equivalent*	*Binary Code*	*Decimal equivalent*
0000	0	1100	8
0001	1	1101	9
0011	2	1111	10
0010	3	1110	11
0110	4	1010	12
0111	5	1011	13
0101	6	1001	14
0100	7	1000	15

Solution:

(*a*)

Decimal	*Gray code*
16	11000
17	11001
18	11011
19	11010
20	11110
21	11111
22	11101
23	11100
24	10100
25	10101
26	10111
27	10110
28	10010
29	10011
30	10001
31	10000

(*b*)

Decimal	*Excess-3*	*Gray*
9	0010	1010
10	0110	1010
11	0110	1110
12	0110	1101
13	0110	1101
14	0110	1100
15	0110	0100
16	0110	0101
17	0110	0111
18	0110	0110
19	0110	0010
20	0111	0010

Q.50. Represent decimal number 8620 in

(*a*) BC D;

(*b*) excess-3 code

(*c*) 2421 code;

(*d*) as a binary number.

Solution: Decimal number = 8620

(*a*) *BCD* 1000 0110 0010 0000

(*b*) *xs-3* 1011 1001 0101 0011

(*c*) 2421 1110 1100 0010 0000

(*d*) Binary 10000110101100 (8192 + 256 + 28 + 32 + 8 + 4)

Q.51. List the 10 BCD digits with an even parity in the leftmost position (total of five bits per digit). Repeat an odd-parity bit.

Solution:

Decimal	**BCD with even parity**	**BCD with odd parity**
0	0000	10000
1	10001	00001
2	10010	00011
3	00011	10100
4	10100	00101
5	00101	10110
6	00110	10110
7	10111	00111
8	11000	01000
9	0100	11001

Q.52. Represent decimal 3984 in the 2421 code. Complement all bits of the coded number and show that the result is the 9's complement of 3984 in the 2421 code.

Solution: 3984 = 0011 1111 1110 0100

1100 0000 0001 1011 = 6015

Q.53. Show that the exclusive -OR function $x = A \oplus B \oplus C \oplus D$ is an odd function. One way to show this is to obtain the truth table for $y = A \oplus B$ and for $z = C \oplus D$ and then formulate the truth table for x = y x z. Show that x = 1 only when the total number of 1's in *A*, *B*, *C* and *D* is odd.

Solution:

A	B	Y = A ⊕ B
0	0	0
0	1	1
1	0	1
1	1	0

C	D	z = C ⊕ D
0	0	0
0	1	1
1	0	1
1	1	0

y	z	x = y ⊕ z
0	0	0
0	1	1
1	0	0
1	1	0

AB = 00 or 11
CD = 01 or 10
AB = 01 or 10
CD = 00 or 11

A B C D
0001, 0010, 1101, 1110
0100, 0111, 1000, 1011
(Always odd numbers of 1's)

Q.54. Derive the circuits for a 3-bit parity generator and 4-bit parity checker using an even parity bit. (The circuits is given below, use odd parity).

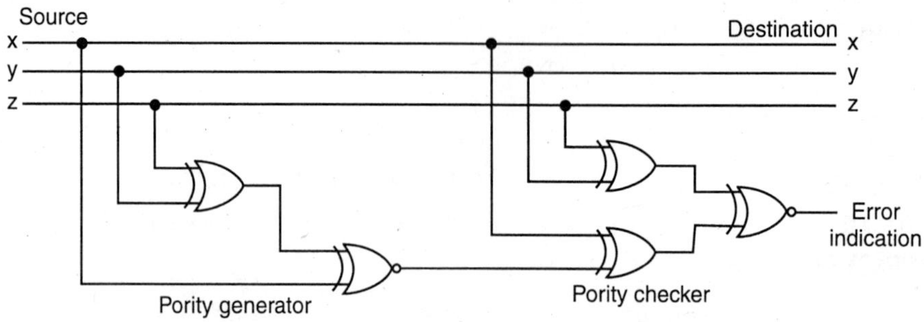

Fig. 1.4. Error detection with odd parity bit

Solution: Same as in figure given above without the complemented circles in the outputs of the gates.

$$P = x \oplus y \oplus z$$

$$\text{Error} = x \oplus y \oplus z \oplus P.$$

SUMMARY

1. Numbers are represented by a string of digit symbols
2. The binary number system uses the base 2, the octal 8 and hexadecimal 16.
3. Remainder method is the most commonly used method for carrying out conversions from decimal system to other systems. It involves the following steps:
4. Divide the given number of decimal system by the radix R of the proposed system. From this, get a quotient Q_1 and a remainder R_1.

5. Dividing the quotient Q_1 by the radix R again to get quotient Q_2 and a remainder R_2.
6. Divide the quotient Q_2 by the radix R again to get quotient Q_3 and a remainder R_3.
7. American Standard Code for Information Interchange uses seven bits to code 128 characters
8. Floating-point representation has inherent demerits in the sense that the same number can be represented in many ways. For example, and all represents the same number. This necessitates a representation that is unique or normal for m and can be adopted by each system.
9. The mantissa is said to be normalized if the digits to the right of radix point is not zero, that is, no lending zero appear in the magnitude part of the number. Hence, for above number 0.3´1015 is the normal form for m or we can say that 0.3´1015 is a normalized floating point number.
10. Complements are the operations employed for the simplification of the subtraction and logical manipulation as well as for negative numbers representation.
11. In a base-r system, two types of complements are possible: (*i*) the r's complement and (*ii*) (*r* – 1)'s complement. For example: for binary numbers there are 2's and 1's complements and decimal numbers have 10's and 9's complements.
12. A set of elements that comprises of the 10 decimal digits, the 26 letters of alphabets, a number of special characters, such as $, @, and = etc. form a alphanumeric character set. The standard alphanumeric binary code is the ASCII (American Standard Code for Information Interchange), which uses seven bits to code 128 characters.
13. The decimal digits in ASCII can be converted to BCD (Binary Coded Decimal) by removing the three high-orders bits, 011. Since registers can only hold binary information, the code must be in binary form.

EXERCISE

(I) Fill in the Blanks (Solved)

1. $(11011.1001)_2$ = (..)$_8$
2. ASCII stands for ....

(II) True or False (Solved)

1. Memory read micro-operation takes 10 clock cycles.
2. A microprogram is a set of microoperations.

(III) Fill in the Blanks (Unsolved)

1. The channel through which CPU and memory accomplish using ....
2. Mirco-operation for adding content of memory location 5 to the content of register R1 is ....
3. The 10's complement of (5432)10 is ....
4. $(9876)_{10} - (5432)_{10}$ using 9's complement is ....
5. Acronym IEEE stands for ....

(IV) True or False (Unsolved)

1. Accumulator is a decoder.
2. Control unit is a part of ALU.

(V) Multiple Choice Questions

1. Find the decimal equivalent of the following binary numbers:
   (*i*) 101011 (*ii*) 111101101 (*iii*) 100001 (*iv*) 111111
2. Find the decimal equivalent of the following binary numbers:
   (*i*) 0.01110 (*ii*) 111.00110 (*iii*) 1101.0011 (*iv*) 1111.0001

3. Convert the following binary numbers into decimal and hexadecimal equivalents:
   (*i*) 45.525 (*ii*) 87.1250
4. Find the 1's and 2's complement of the following binary numbers:
   (*i*) 1010110 (*ii*) 0110110 (*iii*) 1101.0011 (*iv*) 1101.1100
5. Perform the subtraction of following decimal numbers using (a) 10's complement (b) 9's complement. Check the answer by direct subtraction.
   (*i*) 5432-1234 (*ii*) 876.45-283.4 (*iii*) 5438-9432 (*iv*) 124.43-876.85

(V) Detailed Questions

1. What is the advantage of using hexadecimal numbers?
2. In a number system, the symbols used are 0, 1, 2, 3, 4. What is the base of the system?
3. Find the 10's complement of (9456)10.
4. Write first 20 decimal numbers in base 5.
5. Perform the arithmetic operations with binary numbers given below and with negatives numbers unsigned 2's complement representation. Use seven bits to accommodate each number together with its sign
   (*a*) (+ 15) + (+ 14), (*b*) (– 15) + (– 14), (*c*) (– 15) + (+ 14)
6. Describe the hardware algorithm for addition with an example
7. Multiply the following numbers using booth multiplication algorithm
   (*a*) + 14 * – 13, (*b*) – 14 * + 3; use 5 bit register
8. Design 2–bit multiplier by 3 bit array multiplier
9. Divide the number 0110000010 by 10001
10. Describe the flow chart of division algorithm
11. Why floating point number operations are move complicated than fixed point numbers operations.
12. Describe addition algorithm for addition of floating point numbers with example.
13. Perform operations ash r A and ash l A where the content of register A is 0111100001100000
14. Show the block diagram of BCD adder. Hence show the block diagram of adding decimal numbers using BCD address.
15. Describe strobe control in asynchronous data transfer.
16. What is hand shaking ? What advantage it has over strobe control ?
17. What are different modes of data transfer to and from peripherals ?
18. Discuss the need for direct memory access.
19. What is need of input/output processor ? Give block diagram of CPU with input/output processor.

Answers

(I) Fill in the Blanks (Solved)

1. 33.41
2. American Standard Code for Information Interchange.

(II) True or False (Solved)

1. False
2. True

Chapter

2

Digital Computer Generation and Types

Chapter Outline

- Introduction
- History of Computers
- Computer Generations
- Classification of Computer
- Types of Computers
- The Functional Unit of a Computer System

2.0 INTRODUCTION

The history of computers starts out about 2000 years ago, at the birth of the abacus, a wooden rack holding two horizontal wires with beads strung on them. When these beads are moved around, according to *programming* rules memorized by the user, all regular arithmetic problems can be done. Another important invention around the same time was the *Astrolabe, used for navigation. Blaise Pascal* is usually credited for building the first digital computer in 1642. It added numbers entered with dials and was made to help his father, a tax collector. In 1671, Gottfried Wilhelm von *Leibniz invented a computer that was built in 1694. It could add, and, after changing some things around, multiply. Leibniz invented a special stepped gear mechanism for introducing the addend digits, and this is still being used. The prototypes made by Pascal and Leibniz* were not used in many places, and considered weird until a little more, than a century later, when Thomas of Colmar (A.K.A. Charles Xavier Thomas) created the first successful *mechanical calculator* that could add, subtract, multiply, and divide. A lot of improved desktop calculators by many inventors followed, so that by about 1890, the range of improvements included :

- Accumulation of partial results
- Storage and automatic reentry of past results (A memory function)

- Printing of the results

Each of these required manual installation. These improvements were mainly made for commercial users, and not for the needs of science.

Babbage

While Thomas of Colmar was developing the *desktop calculator*, a series of very interesting developments in computers was started in Cambridge, England, by Charles *Babbage* (left, of which the computer store "*Babbages*, now GameStop, is named), a mathematics professor. In 1812, *Babbage* realized that many long calculations, especially those needed to make mathematical tables, were really a series of predictable actions that were constantly repeated. From this he suspected that it should be possible to do these automatically. He began to design an automatic mechanical calculating machine, which he called a *difference engine*. By 1822, he had a working model to demonstrate with. With financial help from the British government, *Babbage* started fabrication of a difference engine in 1823. It was intended to be steam powered and fully automatic, including the printing of the resulting tables, and commanded by a fixed instruction program. The difference engine, although having limited adaptability and applicability, was really a great advance. *Babbage* continued to work on it for the next 10 years, but in 1833 he lost interest because he thought he had a **better idea** – the construction of what would now be called a general purpose, fully program-controlled, automatic mechanical digital computer. *Babbage* called this idea an **Analytical Engine**. The ideas of this design showed a lot of foresight, although this couldn't be appreciated until a full century later. The plans for this engine required an identical decimal computer operating on numbers of 50 decimal digits (or words) and having a storage capacity (memory) of 1,000 such digits. The built-in operations were supposed to include everything that a modern general – purpose computer would need, even the all important **Conditional Control Transfer Capability** that would allow commands to be executed in any order, not just the order in which they were programmed. The analytical engine was soon to use **punched cards** (similar to those used in a Jacquard loom), which would be read into the machine from several different **Reading Stations**. The machine was supposed to operate automatically, by steam power, and require only one person there. *Babbage's* computers were never finished. Various reasons are used for his failure. Most used is the lack of precision machining techniques at the time. Another speculation is that *Babbage* was working on a solution of a problem that few people in 1840 really needed to solve. After *Babbage*, there was a temporary loss of interest in automatic digital computers. Between 1850 and 1900 great advances were made in mathematical physics, and it came to be known that most observable dynamic phenomena can be identified by *differential equations* (which meant that most events occurring in nature can be measured or described in one equation or another), so that easy means for their calculation would be helpful. Moreover, from a practical view, the availability of steam power caused manufacturing (boilers), transportation (steam engines and boats), and commerce to prosper and led to a period of a lot of engineering achievements. The designing of railroads, and the making of steamships, textile mills, and bridges required *differential calculus* to determine such things as:

- center of gravity
- center of buoyancy
- moment of inertia

• stress distributions

Even the assessment of the power output of a steam engine needed mathematical integration. A strong need thus developed for a machine that could rapidly perform many repetitive calculations.

Use of Punched Cards by Hollerith

A step towards automated computing was the development of punched cards, which were first successfully used with computers in 1890 by *Herman Hollerith* (left) and James Powers, who worked for the *US. Census Bureau.* They developed devices that could read the information that had been punched into the cards automatically, without human help. Because of this, reading errors were reduced dramatically, work flow increased, and, most importantly, stacks of punched cards could be used as easily accessible memory of almost unlimited size. Furthermore, different problems could be stored on different stacks of cards and accessed when needed. These advantages were seen by commercial companies and soon led to the development of improved punch-card using computers created by *International Business Machines* (*IBM*), Remington (yes, the same people that make shavers), Burroughs, and other corporations. These computers used electromechanical devices in which electrical power provided mechanical motion - like turning the wheels of an adding machine. Such systems included features to :

• feed in a specified number of cards automatically
• add, multiply, and sort
• feed out cards with punched results

As compared to today's machines, these computers were slow, usually processing 50 - 220 cards per minute, each card holding about 80 decimal numbers (characters). At the time, however, punched cards were a huge step forward. They provided a means of I/O, and memory storage on a huge scale. For more than 50 years after their first use, punched card machines did most of the world's first business computing, and a considerable amount of the computing work in science.

Electronic Digital Computers

The start of World War II produced a large need for computer capacity, especially for the military. New weapons were made for which **trajectory tables** and other essential data were needed. In 1942, John P. Eckert, *John W. Mauchly* (*left*), and their associates at the *Moore school of Electrical Engineering of University of Pennsylvania* decided to build a high - speed electronic computer to do the job. This machine became known as *ENIAC* (Electrical Numerical Integrator And Calculator) The size of *ENIAC*'s numerical "word" was 10 decimal digits, and it could multiply two of these numbers at a rate of 300 per second, by finding the value of each product from a multiplication table stored in its memory. *ENIAC* was therefore about 1,000 times faster then the previous generation of relay computers. *ENIAC* used 18,000 vacuum tubes, about 1,800 square feet of floor space, and consumed about 180,000 watts of electrical power. It had punched card input/output, 1 multiplier, 1 divider/square rooter, and 20 adders using decimal ring counters, which served as adders and also as quick-access (.0002 seconds) read-write register storage. The executable instructions making up a program were embodied in the separate "units" of ENIAC, which were plugged together to form a "route" for the flow of information. These connections had to be redone after each computation, together with presetting function tables and switches. This "wire your own" technique was inconvenient (for obvious reasons), and with only some latitude could *ENIAC* be considered programmable. It was, however, efficient in handling the particular programs for which it had been designed. *ENIAC* is commonly accepted as the first successful high - speed Electronic Digital Computer (EDC) and was used from 1946 to 1955. A controversy developed in 1971, however, over the patentability of *ENIAC*'s basic digital concepts, the claim being made that another physicist, *John V. Atanasoff* (*left*) had already used basically the same ideas in a simpler vacuum - tube device he had built in the 1930's while at *Iowa State College*. In 1973 the courts found in favor of the company using the Atanasoff claim.

Fig. 2.1.

The Modern Stored Program EDC

Fascinated by the success of *ENIAC*, the mathematician *John Von Neumann* (*left*) undertook, in 1945, an abstract study of computation that showed that a computer should have a **very simple, fixed physical structure**, and yet be able to execute any kind of computation by means of a **proper programmed control** without the need for any change in the unit itself. *Von Neumann* contributed a new awareness of how practical, yet fast computers should be organized and built. These ideas, usually referred to as the stored - program technique, became essential for future generations of high - speed digital computers and were universally adopted.

The stored - program technique involves many features of computer design and function besides the one that it is named after. In combination, these features make very - high - speed operation attainable. A glimpse may be provided by considering what 1,000 operations per second means. If each instruction in a job program were used once in consecutive order, no human programmer could generate enough instruction to keep the computer busy. Arrangements must be made, therefore, for parts of the job program (called subroutines) to be used repeatedly in a manner that depends on the way the computation goes. Also, it would clearly be helpful if instructions could be changed if needed during a computation to make them behave differently.

Von Neumann met these two needs by making a special type of machine instruction, called a **Conditional control transfer** - which allowed the program sequence to be stopped and started again at any point - and by storing all instruction programs together with data in the same memory unit, so that, when needed, instructions could be arithmetically changed in the same way as data. As a result of these techniques, computing and programming became much faster, more flexible, and more efficient with work. Regularly used subroutines did not have to be reprogrammed for each new program, but could be kept in "libraries" and read into memory only when needed. Thus, much of a given program could be assembled from the subroutine library.

The all - purpose computer memory became the assembly place in which all parts of a long computation were kept, worked on piece by piece, and put together to form the final results. The computer control survived only as an "errand runner" for the overall process. As soon as the advantage of these techniques became clear, they became a standard practice.

Fig. 2.2.

The first generation of modern programmed electronic computers to take advantage of these improvements were built in 1947. This group included computers using Random - Access - Memory (RAM), which is a memory designed to give almost constant access to any particular piece of information. . These machines had punched - card or punched tape I/O devices and RAM's of 1,000 - word capacity and access times of .5 Greek MU seconds (.5*10-6 seconds). Some of them could perform multiplications in 2 to 4 MU seconds.

Physically, they were much smaller than *ENIAC*. Some were about the size of a grand piano and used only **2,500** electron tubes, a lot less then required by the earlier *ENIAC*. The first - generation stored - program computers needed a lot of maintenance, reached probably about 70 to 80% Reliability Of Operation (ROO) and were used for 8 to 12 years. They were usually programmed in ML, although by the mid 1950's progress had been made in several aspects of advanced programming. This group of computers included *EDVAC* (above) and *UNIVAC* (right) the first commercially available computers.

2.1 HISTORY OF COMPUTERS

Significant Developments which led to Modern Electronic Computer Year

1845 Boolean Algebra was developed by George Boole.

1890 Herman Hollerith used Punched Cards for storing data.

1900 Valdemar Poulson developed recording devices like tapes and drums coated with thin films of magnetic oxide.

1906 Lee do Forest invented Thermoionic valve. Values were the first devices to be used in the modern digital computers.

1937 Alan Turing, A British Mathematician, showed how a problem can be solved with logical steps. These logical steps could be broken into a set of small and simple instructions.

1938 Claude Shannon, an electrical engineer established the fact that electronic switching circuits can be employed to perform logic and arithmetic operations.

1944 Professor Howard Aiken developed a fully automatic electrically driven machine. This was called Automatic Sequence Controlled Calculator (ASCC) or Harvard Mark I. During the Second World War electronic valve based calculating machines were used in 1941. These were called Z3 and Z4. These machines were produced in Germany by Konrad Zuse. Later on they were destroyed by Allied Bombing. In 1943, another valve based machine called COLOSSUS was used at the British Intelligence Establishment at Bletchley Park.

In 1946, J. Presper Eckert and John Mauchly at Pennsylvania University completed a machine called Electronic Numerical Integrator and Calculator (ENIAC). This machine could perform 5000 additions per second.

In 1946 itself, John Von Neumann set out a summary of the design requirements for the modern electronic computer. The main points were:

- Binary codes should be used for representation of data.
- Computer should be capable of processing both data and instructions.

This includes modification of programs.

2.2 COMPUTER GENERATIONS

The development of computers has passed through various stages. This development is divided into five categories:

- First Generation (1942 to 1960)
- Second Generation (1960 to 1965)
- Third Generation (1965 to 1975)
- Fourth Generation (1975 to date)
- Fifth Generation (Coming Generation of Computers)

Each generation of computers is characterized by major technological development that fundamentally changed the way computers operate, resulting in increasingly smaller, cheaper, more powerful and more efficient and reliable devices. Read about each generation and the developments that led to the current devices that we use today.

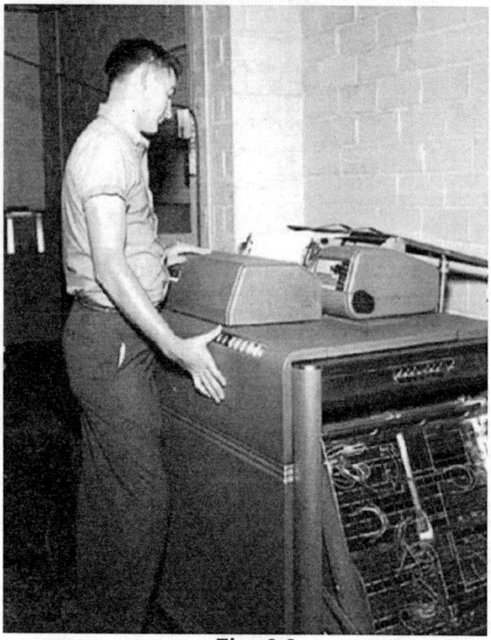

Fig. 2.3.

First Generation - 1940-1956: Vacuum Tubes

The first computers used vacuum tubes for circuitry and magnetic drums for memory, and were often enormous, taking up entire rooms. A magnetic drum,also referred to as drum, is a metal cylinder coated with magnetic iron-oxide material on which data and programs can be stored. Magnetic drums were once use das a primary storage device but have since been implemented as auxiliary storage devices.

Fig. 2.4.

The tracks on a magnetic drum are assigned to channels located around the circumference of the drum, forming adjacent circular bands that wind around the drum. A single drum can have up to 200 tracks. As the drum rotates at a speed of up to 3,000 rpm, the device's read/write heads deposit magnetized spots on the drum during the write operation and sense these spots during a read operation. This action is similar to that of a magnetic tape or disk drive.

They were very expensive to operate and in addition to using a great deal of electricity, generated a lot of heat, which was often the cause of malfunctions. First generation computers relied on machine language to perform operations, and they could only solve one problem at a time. Machine languages are the only languages understood by computers. While easily understood by computers, machine languages are almost impossible for humans to use because they consist entirely of numbers. *Computer Programmers*, therefore, use either ***high level programming languages*** or an *assembly language programming.* An assembly language contains the same instructions as a machine language, but the instructions and variables have names instead of being just numbers.

Programs written in ***high level programming languages*** retranslated into assembly language or machine language by a compiler. Assembly language program retranslated into machine language by a program called an assembler (*assembly language compiler*).

Every CPU has its own unique machine language. Programs must be rewritten or recompiled, therefore, to run on different types of computers. Input was based on punch card and paper tapes, and output was displayed on printouts.

The UNIVAC and ENIAC computers are examples of first-generation computing devices. The UNIVAC was the first commercial computer delivered to a business client, the U.S. Census Bureau in 1951.

Acronym for Electronic Numerical Integrator And Computer, the world's first operational electronic digital computer, developed by Army Ordnance to compute World War II ballistic firing tables. The ENIAC, weighing 30 tons, using 200 kilowatts of electric power and consisting of 18,000 vacuum tubes,1,500 relays, and hundreds of thousands of resistors,capacitors, and inductors, was completed in 1945. In addition to ballistics, the ENIAC's field of application included weather prediction, atomic-energy calculations, cosmic-ray studies, thermal ignition,random-number studies, wind-tunnel design, and other scientific uses. The ENIAC soon became obsolete as the need arose for faster computing speeds.

Second Generation - 1956-1963: Transistors

Transistors replaced vacuum tubes and ushered in the *second generation computer.* Transistor is a device composed of semiconductor material that amplifies a signal or opens or closes a circuit. Invented in 1947 at Bell Labs, transistors have become the key ingredient of all digital circuits, including computers. Today's ***latest microprocessor*** contains tens of millions of microscopic transistors.

Prior to the invention of transistors, digital circuits were composed of vacuum tubes, which had many disadvantages. They were much larger, required more energy, dissipated more heat, and were more prone to failures. It's safe to say that without the invention of transistors, computing as we know it today would not be possible.

The transistor was invented in 1947 but did not see widespread use in computers until the late 50s. The transistor was far superior to the vacuum tube,allowing computers to become smaller, faster, cheaper,more energy-efficient and more reliable than their first-generation predecessors. Though the transistor still generated a great deal of heat that subjected the computer to damage, it was a vast improvement over the vacuum tube. Second-generation computers still relied on punched cards for input and printouts for output.

Fig. 2.5.

Second-generation computers moved from cryptic binary machine language to symbolic, or assembly, languages,which allowed programmers to specify instructions in words. High-level programming languages were also being developed at this time, such as early versions of COBOL and FORTRAN. These were also the first computers that stored their instructions in their memory, which moved from a magnetic drum to magnetic core technology.

The first computers of this generation were developed for the atomic energy industry.

Third Generation - 1964-1971: Integrated Circuits

The development of the integrated circuit was the hallmark of the third generation of computers. Transistors were miniaturized and placed on silicon chips, called semiconductors, which drastically increased the speed and efficiency of computers.

A nonmetallic chemical element in the carbon family of elements. Silicon - atomic symbol "Si" - is the second most abundant element in the earth's crust, surpassed only by oxygen. Silicon does not occur uncombined in nature. Sand and almost all rocks contain silicon combined with oxygen, forming silica. When silicon combines with other elements, such as iron, aluminum or potassium, a silicate is formed. Compounds of silicon also occur in the atmosphere, natural waters, many plants and in the bodies of some animals.

Fig. 2.6.

Silicon is the basic material used to make computer chips, transistors, silicon diodes and other electronic circuits and switching devices because its atomic structure makes the element an ideal semiconductor. Silicon is commonly doped, or mixed,with other elements, such as boron, phosphorous and arsenic, to alter its conductive properties.

A chip is a small piece of semiconducting material (usually silicon) on which an integrated circuit is embedded. A typical chip is less than ¼-square inches and can contain millions of electronic components (transistors). Computers consist of many chips placed on electronic boards called printed circuit boards. There are different types of chips. For example, CPU chips (also called microprocessors) contain an entire processing unit, whereas memory chips contain blank memory.

Semiconductor is a material that is neither a good conductor of electricity (like copper) nor a good insulator (like rubber). The most common semiconductor materials are silicon and germanium. These materials are then doped to create an excess or lack of electrons.

Computer chips, both for CPU and memory, are composed of semiconductor materials. Semiconductors make it possible to miniaturize electronic components, such as transistors. Not only does miniaturization mean that the components take up less space, it also means that they are faster and require less energy.

History Behind It All

Instead of punched cards and printouts, users interacted with third generation computers through keyboards and monitors and interfaced with an operating system, which allowed the device to run many different applications at one time with a central program that monitored the memory. Computers for the first time became accessible to a mass audience because they were smaller and cheaper than their predecessors.

Fourth Generation - 1971-Present: Microprocessors

The microprocessor brought the fourth generation of computers, as thousands of integrated circuits we rebuilt onto a single silicon chip. A silicon chip that contains a CPU. In the world of personal computers, the terms microprocessor and CPU are used interchangeably. At the heart of all personal computers and most workstations sits a microprocessor. Microprocessors also control the logic of almost all digital devices, from clock radios to fuel-injection systems for automobiles.

Fig. 2.7.

Three basic characteristics differentiate microprocessors:

- **Instruction Set:** The set of instructions that the microprocessor can execute.
- **Bandwidth:** The number of bits processed in a single instruction.
- **Clock Speed:** Given in megahertz (MHz), the clock speed determines how many instructions per second the processor can execute.

In both cases, the higher the value, the more powerful the CPU. For example, a 32-bit microprocessor that runs at 50MHz is more powerful than a 16-bitmicroprocessor that runs at 25MHz.

What in the first generation filled an entire room could now fit in the palm of the hand. The Intel 4004 chip, developed in 1971, located all the components of the computer - from the central processing unit and memory to input/output controls - on a single chip.

Abbreviation of central processing unit, and pronounced as separate letters. The CPU is the brains of the computer. Sometimes referred to simply as the processor or central processor, the CPU is where most calculations take place. In terms of computing power,the CPU is the most important element of a computer system.

On large machines, CPUs require one or more printed circuit boards. On personal computers and small workstations, the CPU is housed in a single chip called a microprocessor.

Two typical components of a CPU are:

- The Arithmetic Logic Unit (ALU), which performs arithmetic and logical operations.
- The control unit, which extracts instructions from memory and decodes and executes them, calling on the ALU when necessary.

In 1981 IBM introduced its first computer for the home user, and in 1984 Apple introduced the Macintosh. Microprocessors also moved out of the realm of desktop computers and into many areas of life as more and more everyday products began to use microprocessors.

As these small computers became more powerful, they could be linked together to form networks, which eventually led to the development of the Internet. Fourth generation computers also saw the development of GUI's, the mouse and handheld devices.

Fifth Generation – Present and Beyond: Artificial Intelligence

Fifth generation computing devices, based on artificial intelligence, are still in development,though there are some applications, such as voice recognition, that are being used today.

Fig. 2.8.

Artificial Intelligence is the branch of computer science concerned with making computers behave like humans. The term was coined in 1956 by John McCarthy at the Massachusetts Institute of Technology. Artificial intelligence includes:

- **Games Playing:** Programming computers to play games such as chess and checkers.
- **Expert Systems:** Programming computers to make decisions in real-life situations (for example, some expert systems help doctors diagnose diseases based on symptoms).
- **Natural Language:** Programming computers to understand natural human languages.
- **Neural Networks:** Systems that simulate intelligence by attempting to reproduce the types of physical connections that occur in animal brains.
- **Robotics:** Programming computers to see and hear and react to other sensory stimuli.

Currently, no computers exhibit full artificial intelligence (that is, are able to simulate human behavior). The greatest advances have occurred in the field of games playing. The best computer chess programs are now capable of beating humans. In May,1997, an IBM super-computer called Deep Blue defeated world chess champion Gary Kasparov in a chess match.

In the area of robotics, computers are now widely used in assembly plants, but they are capable only of very limited tasks. Robots have great difficulty identifying objects based on appearance or feel, and they still move and handle objects clumsily.

Natural-language processing offers the greatest potential rewards because it would allow people to interact with computers without needing any specialized knowledge. You could simply walk up to a computer and talk to it. Unfortunately, programming computers to understand natural languages has proved to be more difficult than originally thought. Some rudimentary translation systems that translate from one human language to another are in existence, but they are not nearly as good as human translators.

There are also voice recognition systems that can convert spoken sounds into written words, but they do not understand what they are writing; they simply take dictation. Even these systems are quite limited -- you must speak slowly and distinctly.

In the early 1980s, expert systems were believed to represent the future of artificial intelligence and of computers in general. To date, however, they have not lived up to expectations. Many expert systems help human experts in such fields as medicine and engineering, but they are very expensive to produce and are helpful only in special situations.

Today, the hottest area of artificial intelligence is neural networks, which are proving successful in an umber of disciplines such as voice recognition and natural-language processing.

There are several programming languages that are known as AI languages because they are used almost exclusively for AI applications. The two most common are LISP and Prolog.

2.3 CLASSIFICATION OF COMPUTER

Until recently computers were classifieds as microcomputers, super minicomputers, mainframes, and supercomputers. Technology, however, has changed and this classification is no more relevant. Today all computers used microprocessors as their CPU. Thus classification is possible only through their mode of use. Based on mode of use we can classify computers as Palms, Laptop PCs, Desktop PCs and Workstations. Based on interconnected computers we can classify computers we can classify them as distributed computers and parallel computers.

Fig. 2.9.

Palm PCs or Simputer

With miniaturization and high-density packing of transistor on a chip, computers with capabilities nearly that of PCs which can be held in a palm have emerged. Palm accept handwritten inputs using an electronic pen which can be used to write on a Palm's screen (besides a tiny keyboard), have small disk storage and can be connected to a wireless network. One has to train the system on the user's handwriting before it can be used as a mobile phone, Fax, and e-mail machine. A version of Microsoft operating system called Windows-CE is available for Palm.

An Indian initiative to meet the needs of rural population of developing countries is called Simputer. Simputer is a mobile handheld computer with input through icons on a touch sensitive overly on the LCD display panel. A unique feature of Simputer is the use of free open source OS called GNU/Linux. The cost of ownership is thus low as there is no software cost for OS. Another unique feature of Simputer not found in Palm, is a smart card reader/writer, which increases the functionality of the Simputer including possibility of personalization of a single Simputer for several users.

Laptop PCs

Laptop PCs (also known as notebook computers) are portable computers weighing around 2 kgs. They have a keyboard, flat screen liquid crystal display, and a Pentium or Power PC processor. Color displays are available. They normally run using WINDOWS OS. Laptops come with hard disk (around 40 GB), CDROM and floppy disk. They should run with batteries and are thus designed to conserve energy by using power efficient chips. Many Laptops can be connected to a network. There is a trend towards providing wireless connectivity to Laptops so that they can read files from large stationary computers. The most common use of Laptop computers is used for word processing, and spreadsheet computing. As Laptops use miniature components which have to consume low power and have to be packaged in small volumes.

Personal Computers (PCs)

The most popular PCs are desktop machines. Early PCs had Intel 8088 microprocessors as their CPU. Currently (2004), Intel Dual Core is the most popular processor. The machines made by IBM are called IBM PCs. Other manufacturers use IBM's specifications and design their own PCs. They are known as IBM compatible PCs. IBM PCs mostly use MS-Windows, WINDOWS -XP or GNU/Linux as Operating System. IBM PCs, nowadays (2004) have 64 to 256 MB main memory, 40 to 80 GB of Hard Disk and a floppy disk or flash ROM. Besides these a 650 MB CDROM is also provided in PCs intended for multimedia use. Another company called Apple also makes PCs. Apple PCs are known as Apple Macintosh. They use Apple's proprietary OS, which is designed for simplicity of use. Apple Macintosh machines used Motorola 68030 microprocessors but now use Power PC 603 processor. IBM PCs are today the most popular computers with millions of them in use throughout the world.

Workstations

Workstations are also desktop machines. They are, however, more powerful providing processor speeds about 10 times that of PCs. Most workstations have a large colour video display unit (19 inch monitors). Normally they have main memory of around 256 MB to 4 GB and Hard Disk of 80 to 320 GB. Workstations normally use RISC processors such as MIPS (SIG), RIOS (IBM), SPARC (SUN), or PA-RISC (HP). Some manufacturers of Workstations are Silicon Graphics (SIG), IBM, SUN Microsystems and Hewlett Packard (HP). The standard Operating System of Workstations is UNIX and its derivatives such as AIX (IBM), Solaris (SUN), and HP-UX (HP). Very good graphics facilities and large video screens are provided by most Workstations. A system called X WINDOWS is provided by Workstations to display the status of multiple processes during their execution. Most Workstations have built-in hardware to connect to a Local Area Network (LAN). Workstations are used for executing numeric and graphic intensive applications such as those, which arise in Computer Aided Design, simulation of complex systems and visualizing the results of simulation.

Servers

While manufacturers such as IBM, SUN and Silicon Graphics have been manufacturing high performance workstations the speed of Intel Pentium Processors has been going up. In 2004, Pentium with clock speed 3 GHz are available. They can support several GB main memories. Thus the difference between high end PCs and Workstations is vanishing. Today companies such as SUN make Intel based workstations.

While Workstations are characterized by high performance processors with large screens for interactive programming, servers are used for specific purpose such as high performance numerical computing (called compute server), web page hosting, database store, printing, etc. interactive large screens are not necessary. Compute servers have performance processors with large main memory, database servers have big on-line disk storage (100s of GB) and print servers support several high speed printers.

Mainframes Computers

There are organizations such as banks and insurance companies process large number of transactions on-line. They require computers with very large disks to store several Terabytes of data and transfer data from disk to main memory at several hundred Megabytes/sec. The processing power needed from such computers is hundred million transactions per second. These computers are much bigger and faster than workstations and several hundred times more expensive. They normally use proprietary operating systems, which usually provide high expensive services such as user accounting, file security and control. They are normally much more reliable when compared to Operating System on PCs. These types of computers are called mainframes. These are a few manufacturers of mainframes (*e.g.*, IBM and Hitachi). The number of mainframe users has reduced as many organizations are rewriting their systems to use networks of powerful workstations.

Supercomputers

Supercomputers are the fastest computers available at any given time and are normally used to solve problems, which require intensive numerical computations. Examples of such problems are numerical weather prediction, designing supersonic aircrafts, design of drugs and modeling complex molecules. All of these problems require around 10^16 calculations to be performed. These problems will be solved in about 3 hours by a computer, which can carry out a trillion floating point calculations per second. Such a computer is classifieds as supercomputer today (2004). By about the year 2006 computers which can carry out 10^15 floating point operations per second on 64 bit floating point numbers would be available and would be the ones which be called supercomputers. Interconnecting several high speed computers and programming them to work cooperatively to solve problems build supercomputers. Recently applications of supercomputers have expanded beyond scientific computing, they are now used to analyze large commercial database, produced animated movies and play games such as chess.

Besides arithmetic speed, a computer to be classified as a supercomputer should have a large main memory of around 16 GB and a secondary memory of 1000 GB. The speed of transfer of data from secondary memory to the main memory should be at least a tenth of the memory to CPU data transfer speed. All supercomputers use parallelism to achieve their speed.

2.4 TYPES OF COMPUTERS

Definition of a Computer

A computer can be defined as an electronic device that receives input from the user through a mouse or keyboard, processes it in some fashion and displays the result on a screen. It can also be assumed as any device that has a microprocessor in it.

Types of Computers

Fundamentally, there are two types of computers:

- analog
- digital

Analog computers solve problems by using continuously changing data such as:

- temperature
- pressure
- voltage

Digital computers solves problems by manipulating discrete binary digits (1s and 0s).

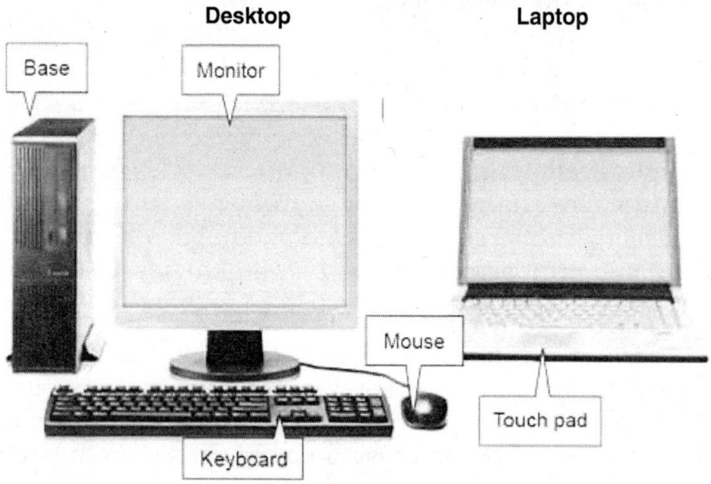

Fig. 2.10.

3 Main Types of Computers

- mainframes
- minicomputers
- microcomputers

They differ in size, speed of operation, amount of data that can be stored, and the number of simultaneous users.

1. microcomputers
2. minicomputers
3. mainframe computers

personal or desktop computer designed for general use by a single person mid-range servers supporting smaller number of simultaneous users, typically 50 to 100 allow many simultaneous users home and all types of businesses larger businesses industry, research, and university computing centers entertainment, communication, personal databases, and spreadsheets, word processing, accounting, inventory control, research accounting, billing, and inventory records typically huge databases, complex mathematical operations, processing millions of transactions everyday.

Types of Computers Based on Size

Desktop: A personal computer that is not designed for portability. The computer is set up in a permanent location. Most desktops offer more power, storage and versatility for less cost.

Laptop: These are also called notebooks. Laptops are portable computers that integrate the display, keyboard, a pointing device or trackball, processor, memory and hard drive all in a battery-operated package slightly larger than an average hardcover book.

Palmtop: More commonly known as Personal Digital Assistants (PDAs):

- Palmtops are tightly integrated computers that often use flash memory instead of a hard drive for storage.
- These computers usually do not have keyboards but rely on touch screen technology for user input.
- Palmtops are typically smaller than a paperback novel, very lightweight with a reasonable battery life. A slightly larger and heavier version of the palmtop is the handheld computer.

Workstation: A desktop computer that has a more powerful processor, additional memory and enhanced capabilities for performing aspecial group of task, such as 3D Graphics or game development.

Server: A computer that has been optimized to provide services to other computers over a network. Servers usually have powerful processors, lots of memory and large hard drives.

Supercomputer: This type of computer usually costs hundreds of thousands or even millions of dollars. Although some supercomputers are single computer systems, most are comprized of multiple high performance computers working in parallel as a single system. The best known supercomputers are built by Cray Supercomputers.

Wearable: The latest trend in computing is wearable computers. Common computer applications (e-mail, database, multimedia, calendar/scheduler) are integrated into watches, cell phones, even clothing!

Basic Units of a Computer

Modular units usually consist of the following separate components:

- The Computer
- The Monitor
- The Keyboard
- The Mouse
- Peripherals

Central Processing Unit (CPU)

Central Processing Unit (CPU): The heart of the computer, this is the component that actually executes instructions organized in programs ("software") which tell the computer what to do.

Storage Devices

Memory(fast, expensive, short-term memory): Enables a computer to store, at least temporarily, data, programs, and intermediate results.

Mass storage device (slower, cheaper, long-term memory): Allows a computer to permanently retain large amounts of data and programs between jobs. Common mass storage devices include disk drives and tape drives.

Input Devices

1. Keyboard
2. Mouse
3. Tracker ball
4. Joystick
5. Light pen
6. Digial camera
7. Microphone

8. Touch screen
9. Video Digitiser
10. Graphics tablet
11. Scanner
12. OCR (Optical Character Recognition)

Output Devices

1. Monitors or Visual Display Units (VDUs)
2. Printers
3. Plotters

Other devices

1. Speakers
2. LCD projectors

2.5 THE FUNCTIONAL UNIT OF A COMPUTER SYSTEM

Processing Unit: This is the responsible for processing all the various operations that goes on in the system unit. It is referred to as the brain of the computer system, without it the computer will be valueless.

Example of processing unit is the Central Processing Unit (CPU).

The processing unit can be divided into three sections:

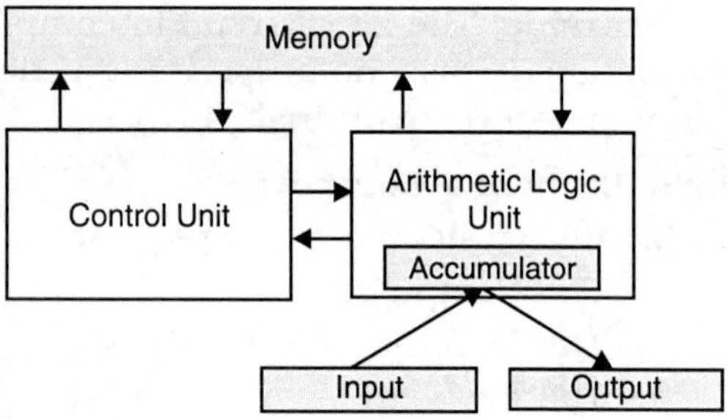

Fig. 2.11.

(i) Control Unit
(ii) Arithmetic and logical unit (ALU)
(iii) Memory Unit.

Schematic of the von Neumann architecture. The *Control Unit* and *Arithmetic Logic Unit* form the main components of the *Central Processing Unit (CPU)*.

The **von Neumann architecture** is a design model for a stored-program *digital computer* that uses a

The Control Unit: This unit coordinates all the various operations of the input and output units in the system unit precisely the Central Processing Unit (CPU).

Functions of the control unit:

(i) It directs the sequence of operations,
(ii) It interprets the instructions, of a program, in storage unit and produces signals the command circuits to execute the instructions.
(iii) It directs the flow of all activities in the computer system.

The arithmetic and logical unit: This unit is responsible for performing all the various Arithmetic operation of addition, subtraction, multiplication, division and relational operations Such as not equal to (^{1}) greater than (>) less than (<) greater than or equal to (=) and logical operation. Etc.

The memory unit: This unit otherwise known to be the main or primary memory (storage). It stores information (*i.e.*, instruction data intermediate and final result of processing) that arrives via the input unit so that this information is made available to the appropriate Quarters for further processing after which they can be presented to the user via the Output unit.

The memory unit divided into two:

(*i*) Random Access Memory (RAM)

(*ii*) Read Only Memory (ROM)

Other versions of ROM are:

- PROM(Programmable Read Only Memory)
- EPROM.(Erasable Programmable Read Only Memory)

SOLVED EXAMPLES

Q.1 Define the history of Computers.

Ans. History of Computers: Significant Developments which led to Modern Electronic Computer Year.

1845 Boolean Algebra was developed by George Boole.

1890 Herman Hollerith used Punched Cards for storing data.

1900 Valdemar Poulson developed recording devices like tapes and drums coated with thin films of magnetic oxide.

1906 Lee do Forest invented Thermoionic valve. Values were the first devices to be used in the modern digital computers.

1937 Alan Turing, A British Mathematician, showed how a problem can be solved with logical steps. These logical steps could be broken into a set of small and simple instructions.

1938 Claude Shannon, an electrical engineer established the fact that electronic switching circuits can be employed to perform logic and arithmetic operations.

1944 Professor Howard Aiken developed a fully automatic electrically driven machine. This was called Automatic Sequence Controlled Calculator (ASCC) or Harvard Mark I. During the Second World War electronic valve based calculating machines were used in 1941. These were called Z3 and Z4. These machines were produced in Germany by Konrad Zuse. Later on they were destroyed by Allied Bombing. In 1943, another valve based machine called COLOSSUS was used at the British Intelligence Establishment at Bletchley Park.

1946 In 1946, J. Presper Eckert and John Mauchly at Pennsylvania University completed a machine called Electronic Numerical Integrator and Calculator (ENIAC). This machine could perform 5000 additions per second.

In 1946 itself, John Von Neumann set out a summary of the design requirements for the modern electronic computer. The main points were:

- Binary codes should be used for representation of data.
- Computer should be capable of processing both data and instructions.

This includes modification of programs.

Q.2. Briefly Explain Computer Generations

Ans. Computer Generations: The development of computers has passed through various stages. This development is divided into five categories:

- First Generation (1942 to 1960)
- Second Generation (1960 to 1965)
- Third Generation (1965 to 1975)
- Fourth Generation (1975 to date)
- Fifth Generation (Coming Generation of Computers)

Each generation of computers is characterized by major technological development that fundamentally changed the way computers operate, resulting in increasingly smaller, cheaper, more powerful and more efficient and reliable devices. Read about each generation and the developments that led to the current devices that we use today.

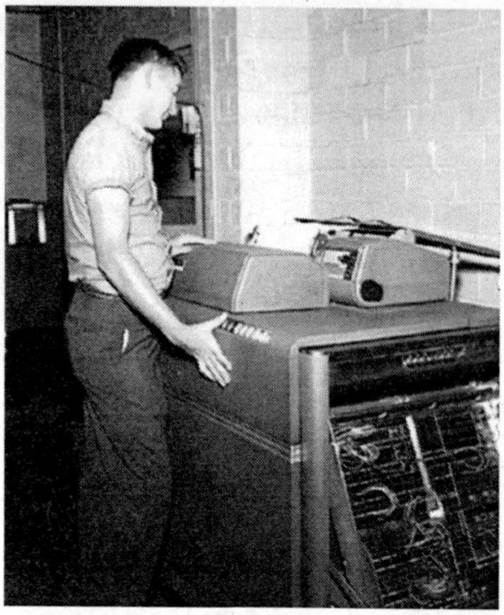

Fig. 2.12.

Q.3 Explain First Generation.

Ans. First Generation — 1940-1956: Vacuum Tubes

The first computers used vacuum tubes for circuitry and magnetic drums for memory, and were often enormous, taking up entire rooms. A magnetic drum,also referred to as drum, is a metal cylinder coated with magnetic iron-oxide material on which data and programs can be stored. Magnetic drums were once use das a primary storage device but have since been implemented as auxiliary storage devices.

Fig. 2.13.

The tracks on a magnetic drum are assigned to channels located around the circumference of the drum, forming adjacent circular bands that wind around the drum. A single drum can have up to 200 tracks. As the drum rotates at a speed of up to 3,000 rpm, the device's read/write heads deposit magnetized spots on the drum during the write operation and sense these spots during a read operation. This action is similar to that of a magnetic tape or disk drive.

They were very expensive to operate and in addition to using a great deal of electricity, generated a lot of heat, which was often the cause of malfunctions. First generation computers relied on machine language to perform operations, and they could only solve one problem at a time. Machine languages are the only languages understood by computers. While easily understood by computers, machine languages are almost impossible for humans to use because they consist entirely of numbers. Computer Programmers, therefore, use either **high level programming languages** or an assembly language programming. An assembly language contains the same instructions as a machine language, but the instructions and variables have names instead of being just numbers.

Programs written in **high level programming languages** retranslated into assembly language or machine language by a compiler. Assembly language program retranslated into machine language by a program called an assembler (assembly language compiler).

Every CPU has its own unique machine language. Programs must be rewritten or recompiled, therefore, to run on different types of computers. Input was based on punch card and paper tapes, and output was displayed on printouts.

The UNIVAC and ENIAC computers are examples of first-generation computing devices. The UNIVAC was the first commercial computer delivered to a business client, the U.S. Census Bureau in 1951.

Acronym for Electronic Numerical Integrator And Computer, the world's first operational electronic digital computer, developed by Army Ordnance to compute World War II ballistic firing tables. The ENIAC, weighing 30 tons, using 200 kilowatts of electric power and consisting of 18,000 vacuum tubes,1,500 relays, and hundreds of thousands of resistors,capacitors, and inductors, was completed in 1945. In addition to ballistics, the ENIAC's field of application included weather prediction, atomic-energy calculations, cosmic-ray studies, thermal ignition,random-number studies, wind-tunnel design, and other scientific uses. The ENIAC soon became obsolete as the need arose for faster computing speeds.

Q 4. Explain Second Generation.

Ans. Second Generation — 1956-1963: Transistors: Transistors replaced vacuum tubes and ushered in the second generation computer. Transistor is a device composed of semiconductor material that amplifies a signal or opens or closes a circuit. Invented in 1947 at Bell Labs, transistors have become the key ingredient of all digital circuits, including computers. Today's **latest microprocessor** contains tens of millions of microscopic transistors.

Fig. 2.14.

Prior to the invention of transistors, digital circuits were composed of vacuum tubes, which had many disadvantages. They were much larger, required more energy, dissipated more heat, and were more prone to failures. It's safe to say that without the invention of transistors, computing as we know it today would not be possible.

The transistor was invented in 1947 but did not see widespread use in computers until the late 50s. The transistor was far superior to the vacuum tube,allowing computers to become smaller, faster, cheaper,more energy-efficient and more reliable than their first-generation predecessors. Though the transistor still generated a great deal of heat that subjected the computer to damage, it was a vast improvement over the vacuum tube. Second-generation computers still relied on punched cards for input and printouts for output.

Second-generation computers moved from cryptic binary machine language to symbolic, or assembly, languages,which allowed programmers to specify instructions in words. High-level programming languages were also being developed at this time, such as early versions of COBOL and FORTRAN. These were also the first computers that stored their instructions in their memory, which moved from a magnetic drum to magnetic core technology.

The first computers of this generation were developed for the atomic energy industry.

Q.5 Explain Third Generation.

Ans. Third Generation — 1964-1971: Integrated Circuits: The development of the integrated circuit was the hallmark of the third generation of computers. Transistors were miniaturized and placed on silicon chips, called semiconductors, which drastically increased the speed and efficiency of computers.

A nonmetallic chemical element in the carbon family of elements. Silicon - atomic symbol "Si" - is the second most abundant element in the earth's crust, surpassed only by oxygen. Silicon does not occur uncombined in nature. Sand and almost all rocks contain silicon combined with oxygen, forming silica. When silicon combines with other elements, such as iron, aluminum or potassium, a silicate is formed. Compounds of silicon also occur in the atmosphere, natural waters,many plants and in the bodies of some animals.

Silicon is the basic material used to make computer chips, transistors, silicon diodes and other electronic circuits and switching devices because its atomic structure makes the element an ideal semiconductor. Silicon is commonly doped, or mixed,with other elements, such as boron, phosphorous and arsenic, to alter its conductive properties.

A chip is a small piece of semi conducting material(usually silicon) on which an integrated circuit is embedded. A typical chip is less than ¼-square inches and can contain millions of electronic components(transistors). Computers consist of many chips placed on electronic boards called printed circuit boards. There are different types of chips. For example, CPU chips (also called microprocessors) contain an entire processing unit, whereas memory chips contain blank memory.

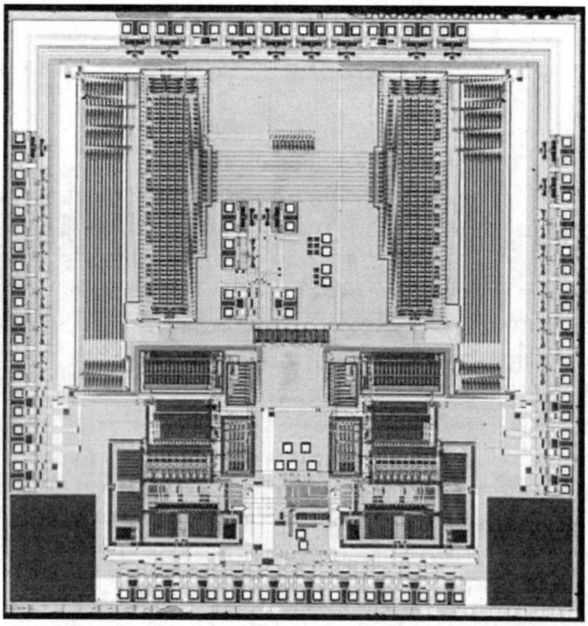

Fig. 2.15.

Semiconductor is a material that is neither a good conductor of electricity (like copper) nor a good insulator (like rubber). The most common semiconductor materials are silicon and germanium. These materials are then doped to create an excess or lack of electrons.

Computer chips, both for CPU and memory, are composed of semiconductor materials. Semiconductors make it possible to miniaturize electronic components, such as transistors. Not only does miniaturization mean that the components take up less space, it also means that they are faster and require less energy.

Instead of punched cards and printouts, users interacted with third generation computers through keyboards and monitors and interfaced with an operating system, which allowed the device to run many different applications at one time with a central program that monitored the memory. Computers for the first time became accessible to a mass audience because they were smaller and cheaper than their predecessors.

Q.6 Explain Fourth Generation.

Ans. Fourth Generation — 1971-Present: Microprocessors: The microprocessor brought the fourth generation of computers, as thousands of integrated circuits we rebuilt onto a single silicon chip. A silicon chip that contains a CPU. In the world of personal computers,the terms microprocessor and CPU are used interchangeably. At the heart of all personal computers and most workstations sits a microprocessor. Microprocessors also control the logic of almost all digital devices, from clock radios to fuel-injection systems for automobiles.

Fig. 2.16.

Three basic characteristics differentiate microprocessors:

- **Instruction Set**: The set of instructions that the microprocessor can execute.
- **Bandwidth**: The number of bits processed in a single instruction.
- **Clock Speed**: Given in megahertz (MHz), the clock speed determines how many instructions per second the processor can execute.

In both cases, the higher the value, the more powerful the CPU. For example, a 32-bit microprocessor that runs at 50 MHz is more powerful than a 16-bitmicroprocessor that runs at 25 MHz.

What in the first generation filled an entire room could now fit in the palm of the hand. The Intel 4004 chip, developed in 1971, located all the components of the computer – from the central processing unit and memory to input/output controls – on a single chip.

Abbreviation of central processing unit, and pronounced as separate letters. The CPU is the brains of the computer. Sometimes referred to simply as the processor or central processor, the CPU is where most calculations take place. In terms of computing power,the CPU is the most important element of a computer system.

On large machines, CPUs require one or more printed circuit boards. On personal computers and small workstations, the CPU is housed in a single chip called a microprocessor.

Two typical components of a CPU are:

- The Arithmetic Logic Unit (ALU), which performs arithmetic and logical operations.
- The control unit, which extracts instructions from memory and decodes and executes them, calling on the ALU when necessary.

In 1981 IBM introduced its first computer for the home user, and in 1984 Apple introduced the Macintosh. Microprocessors also moved out of the realm of desktop computers and into many areas of life as more and more everyday products began to use microprocessors.

As these small computers became more powerful, they could be linked together to form networks, which eventually led to the development of the Internet. Fourth generation computers also saw the development of GUI's, the mouse and handheld devices.

Q.7 Explain Fifth Generation.

Ans. Fifth Generation – Present and Beyond: Artificial Intelligence: Fifth generation computing devices, based on artificial intelligence, are still in development,though there are some applications, such as voice recognition, that are being used today.

Artificial Intelligence is the branch of computer science concerned with making computers behave like humans. The term was coined in 1956 by John McCarthy at the Massachusetts Institute of Technology. Artificial intelligence includes:

- **Games Playing**: Programming computers to play games such as chess and checkers.

- **Expert Systems**: Programming computers to make decisions in real-life situations (for example, some expert systems help doctors diagnose diseases based on symptoms).
- **Natural Language**: Programming computers to understand natural human languages.
- **Neural Networks**: Systems that simulate intelligence by attempting to reproduce the types of physical connections that occur in animal brains.
- **Robotics**: Programming computers to see and hear and react to other sensory stimuli.

Currently, no computers exhibit full artificial intelligence (that is, are able to simulate human behavior). The greatest advances have occurred in the field of games playing. The best computer chess programs are now capable of beating humans. In May,1997, an IBM super-computer called Deep Blue defeated world chess champion Gary Kasparov in a chess match.

In the area of robotics, computers are now widely used in assembly plants, but they are capable only of very limited tasks. Robots have great difficulty identifying objects based on appearance or feel, and they still move and handle objects clumsily.

Natural-language processing offers the greatest potential rewards because it would allow people to interact with computers without needing any specialized knowledge. You could simply walk up to a computer and talk to it. Unfortunately, programming computers to understand natural languages has proved to be more difficult than originally thought. Some rudimentary translation systems that translate from one human language to another are in existence, but they are not nearly as good as human translators.

There are also voice recognition systems that can convert spoken sounds into written words, but they do not understand what they are writing; they simply take dictation. Even these systems are quite limited — you must speak slowly and distinctly.

In the early 1980s, expert systems were believed to represent the future of artificial intelligence and of computers in general. To date, however, they have not lived up to expectations. Many expert systems help human experts in such fields as medicine and engineering, but they are very expensive to produce and are helpful only in special situations.

Today, the hottest area of artificial intelligence is neural networks, which are proving successful in an number of disciplines such as voice recognition and natural-language processing.

There are several programming languages that are known as AI languages because they are used almost exclusively for AI applications. The two most common are LISP and Prolog.

Q.8 Briefly Explain the Classification of Computer

Ans. Classification Of Computer: Until recently computers were classifieds as microcomputers, super minicomputers, mainframes, and supercomputers. Technology, however, has changed and this classification is no more relevant. Today all computers used microprocessors as their CPU. Thus classification is possible only through their mode of use. Based on mode of use we can classify computers as Palms, Laptop PCs, Desktop PCs and Workstations. Based on interconnected computers we can classify computers we can classify them as distributed computers and parallel computers.

Palm PCs or Simputer: With miniaturization and high-density packing of transistor on a chip, computers with capabilities nearly that of PCs which can be held in a palm have emerged. Palm accept handwritten inputs using an electronic pen which can be used to write on a Palm's screen (besides a tiny keyboard), have small disk storage and can be connected to a wireless network. One has to train the system on the user's handwriting before it can be used as a mobile phone, Fax, and e-mail machine. A version of Microsoft operating system called Windows-CE is available for Palm.

An Indian initiative to meet the needs of rural population of developing countries is called Simputer. Simputer is a mobile handheld computer with input through icons on a touch sensitive overly on the LCD display panel. A unique feature of Simputer is the use of free open source OS called GNU/Linux. The cost of ownership is thus low as there is no software cost for OS. Another unique feature of Simputer not found in Palm, is a smart card reader/writer, which increases the functionality of the Simputer including possibility of personalization of a single Simputer for several users.

Laptop PCs: Laptop PCs (also known as notebook computers) are portable computers weighing around 2 kgs. They have a keyboard, flat screen liquid crystal display, and a Pentium or Power PC processor. Color displays are available. They normally run using WINDOWS OS. Laptops come with hard disk (around 40 GB), CDROM and floppy disk. They should run with batteries and are thus designed to conserve energy by using power efficient chips. Many Laptops can be connected to a network. There is a trend towards providing wireless connectivity to Laptops so that they can read files from large stationary computers. The most common use of Laptop computers is used for word processing, and spreadsheet computing. As Laptops use miniature components which have to consume low power and have to be packaged in small volumes.

Personal Computers (PCs): The most popular PCs are desktop machines. Early PCs had Intel 8088 microprocessors as their CPU. Currently (2004), Intel Dual Core is the most popular processor. The machines made by IBM are called IBM PCs. Other manufacturers use IBM's specifications and design their own PCs. They are known as IBM compatible PCs. IBM PCs mostly use MS-Windows, WINDOWS –XP or GNU/Linux as Operating System. IBM PCs, nowadays (2004) have 64 to 256 MB main memory, 40 to 80 GB of Hard Disk and a floppy disk or flash ROM. Besides these a 650 MB CDROM is also provided in PCs intended for multimedia use. Another company called Apple also makes pCs. Apple PCs are known as Apple Macintosh. They use Apple's proprietary OS, which is designed for simplicity of use. Apple Macintosh machines used Motorola 68030 microprocessors but now use Power PC 603 processor. IBM PCs are today the most popular computers with millions of them in use throughout the world.

Workstations: Workstations are also desktop machines. They are, however, more powerful providing processor speeds about 10 times that of PCs. Most workstations have a large color video display unit (19 inch monitors). Normally they have main memory of around 256 MB to 4 GB and Hard Disk of 80 to 320 GB. Workstations normally use RISC processors such as MIPS (SIG), RIOS (IBM), SPARC (SUN), or PA-RISC (HP). Some manufacturers of Workstations are Silicon Graphics (SIG), IBM, SUN Microsystems and Hewlett Packard (HP). The standard Operating System of Workstations is UNIX and its derivatives such as AIX (IBM), Solaris (SUN), and HP-UX (HP). Very good graphics facilities and large video screens are provided by most Workstations. A system called X WINDOWS is provided by Workstations to display the status of multiple processes during their execution. Most Workstations have built-in hardware to connect to a Local Area Network (LAN). Workstations are used for executing numeric and graphic intensive applications such as those, which arise in Computer Aided Design, simulation of complex systems and visualizing the results of simulation.

Servers: While manufacturers such as IBM, SUN and Silicon Graphics have been manufacturing high performance workstations the speed of Intel Pentium Processors has been going up. In 2004, Pentium with clock speed 3 GHz are available. They can support several GB main memories. Thus the difference between high end PCs and Workstations is vanishing. Today companies such as SUN make Intel based workstations.

While Workstations are characterized by high performance processors with large screens for interactive programming, servers are used for specific purpose such as high performance numerical computing (called compute server), web page hosting, database store, printing, etc. interactive large screens are not necessary. Compute servers have performance processors with large main memory, database servers have big on-line disk storage (100s of GB) and print servers support several high speed printers.

Mainframes Computers: There are organizations such as banks and insurance companies process large number of transactions on-line. They require computers with very large disks to store several Terabytes of data and transfer data from disk to main memory at several hundred Megabytes/sec. The processing power needed from such computers is hundred million transactions per second. These computers are much bigger and faster than workstations and several hundred times more expensive. They normally use proprietary operating systems, which usually provide high expensive services such as user accounting, file security and control. They are normally much more reliable when compared to Operating System on PCs. These types of computers are called mainframes. These are a few manufacturers of mainframes (*e.g.*, IBM and Hitachi). The number of mainframe users has reduced as many organizations are rewriting their systems to use networks of powerful workstations.

Supercomputers: Supercomputers are the fastest computers available at any given time and are normally used to solve problems, which require intensive numerical computations. Examples of such problems are numerical weather prediction, designing supersonic aircrafts, design of drugs and modeling complex molecules. All of these problems require around 10^16 calculations to be performed. These problems will be solved in about 3 hours by a computer, which can carry out a trillion floating point calculations per second. Such a computer is classifieds as supercomputer today (2004). By about the year 2006 computers which can carry out 10^15 floating point operations per second on 64 bit floating point numbers would be available and would be the ones which be called supercomputers. Interconnecting several high speed computers and programming them to work cooperatively to solve problems build supercomputers. Recently applications of supercomputers have expanded beyond scientific computing, they are now used to analyze large commercial database, produced animated movies and play games such as chess.

Besides arithmetic speed, a computer to be classified as a supercomputer should have a large main memory of around 16 GB and a secondary memory of 1000 GB. The speed of transfer of data from secondary memory to the main memory should be at least a tenth of the memory to CPU data transfer speed. All supercomputers use parallelism to achieve their speed. In Sec. 12.9 we discuss the organization of parallel computers.

Q.9 Briefly Explain the Types of Computer.

Ans. Types Of Computers

Definition of a Computer: A computer can be defined as an electronic device that receives input from the user through a mouse or keyboard, processes it in some fashion and displays the result on a screen. It can also be assumed as any device that has a microprocessor in it.

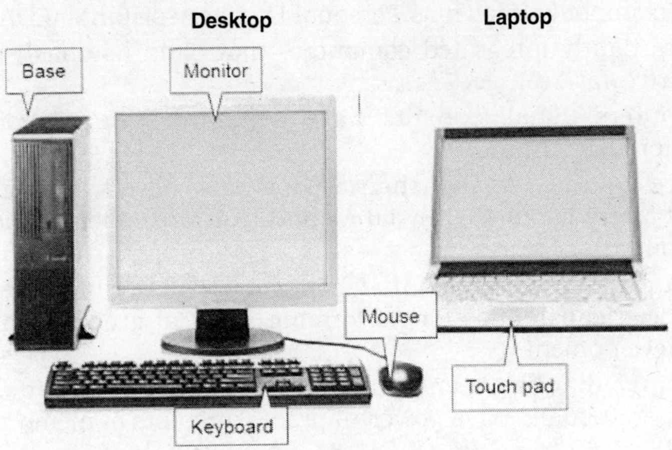

Fig. 2.17.

Types of Computers: Fundamentally, there are two types of computers

- analog
- digital

Analog computers solve problems by using continuously changing data such as

- temperature
- pressure
- voltage

Digital computers solves problems by manipulating discrete binary digits (1s and 0s).

3 Main Types of Computers

- mainframes
- minicomputers
- microcomputers

They differ in size, speed of operation, amount of data that can be stored, and the number of simultaneous users.

1. microcomputers
2. minicomputers
3. mainframe computers

Personal or desktop computer designed for general use by a single person mid-range servers supporting smaller number of simultaneous users, typically 50 to 100 allow many simultaneous users home and all types of businesses larger businesses industry, research, and university computing centers entertainment, communication, personal databases, and spreadsheets, word processing, accounting, inventory control, research accounting, billing, and inventory records typically huge databases, complex mathematical operations, processing millions of transactions everyday.

Types of Computers Based on Size

Desktop: A personal computer that is not designed for portability. The computer is set up in a permanent location. Most desktops offer more power, storage and versatility for less cost.

Laptop: These are also called notebooks. Laptops are portable computers that integrate the display, keyboard, a pointing device or trackball, processor, memory and hard drive all in a battery-operated package slightly larger than an average hardcover book.

Palmtop: More commonly known as Personal Digital Assistants (PDAs):

- Palmtops are tightly integrated computers that often use flash memory instead of a hard drive for storage.
- These computers usually do not have keyboards but rely on touch screen technology for user input.
- Palmtops are typically smaller than a paperback novel, very lightweight with a reasonable battery life. A slightly larger and heavior version of the palmtop is the handheld computer.

Workstation: A desktop computer that has a more powerful processor, additional memory and enhanced capabilities for performing aspecial group of task, such as 3D Graphics or game development.

Server: A computer that has been optimized to provide services to other computers over a network. Servers usually have powerful processors, lots of memory and large hard drives.

Supercomputer: This type of computer usually costs hundreds of thousands or even millions of dollars. Although some supercomputers are single computer systems, most are comprised of multiple high performance computers working in parallel as a single system. The best known supercomputers are built by cray supercomputers.

Wearable: The latest trend in computing is wearable computers. Common computer applications (e-mail, database, multimedia, calendar/scheduler) are integrated into watches, cell phones, even clothing!

Q.10. Briefly Explain the Basic Unit of Computer.

Ans. Basic Units of a Computer

Modular units usually consist of the following separate components:

- The Computer
- The Monitor
- The Keyboard
- The Mouse
- Peripherals

Central Processing Unit (CPU):

Central Processing Unit (CPU): The heart of the computer, this is the component that actually executes instructions organized in programs ("software") which tell the computer what to do.

Storage Devices

Memory(fast, expensive, short-term memory): Enables a computer to store, at least temporarily, data, programs, and intermediate results.

Mass storage device (slower, cheaper, long-term memory): Allows a computer to permanently retain large amounts of data and programs between jobs. Common mass storage devices include disk drives and tape drives.

Input Devices

1. Keyboard
2. Mouse
3. Tracker ball
4. Joystick

5. Light pen
6. Digial Camera
7. Microphone
8. Touch screen
9. Video Digitiser
10. Graphics tablet
11. Scanner
12. OCR (Optical Character Recognition)

Output Devices

1. Monitors or Visual Display Units (VDUs)
2. Printers
3. Plotters
4. Speakers
5. LCD projectors

Q.11. Explain Functional Unit of Computer.

Ans. The Functional Unit of a Computer System: Processing Unit: This is the responsible for processing all the various operations that goes on in the system unit. It is referred to as the brain of the computer system, without it the computer will be valueless.

Example of processing unit is the Central Processing Unit (CPU).

The processing unit can be divided into three sections:

(i) Control Unit
(ii) Arithmetic and logical unit (ALU)
(iii) Memory Unit.

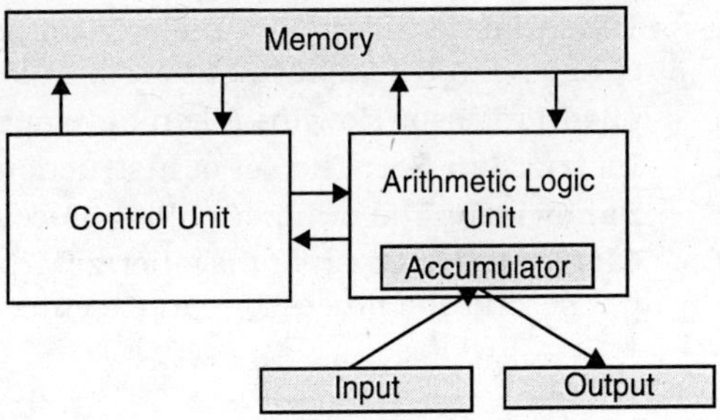

Fig. 2.18.

Schematic of the von Neumann architecture.
The control unit and arithmetic logic unit form the main components of the Central Processing Unit (CPU)

The **von Neumann architecture** is a design model for a stored-program digital computer that uses a

The Control Unit: This unit coordinates all the various operations of the input and output units in the system unit precisely the Central Processing Unit (CPU).

Functions of the control unit

(i) It directs the sequence of operations,
(ii) It interprets the instructions, of a program, in storage unit and produces signals the command circuits to execute the instructions.
(iii) It directs the flow of all activities in the computer system.

The arithmetic and logical unit: This unit is responsible for performing all the various Arithmetic operation of addition, subtraction, multiplication, division and relational operations. Such as not equal to (¹) greater than (>) less than (<) greater than or equal to (e") and logical operation. Etc.The memory unit: This unit otherwise known to be the main or primary memory (storage). It stores information (*i.e.* instruction data intermediate and final result of processing) that arrives via the input unit so that this information is made available to the appropriate Quarters for further processing after which they can be presented to the user via the Output unit.

The memory unit divided into two:

(*i*) Random Access Memory (RAM)

(*ii*) Read Only Memory (ROM)

Other versions of ROM are:

- PROM(Programmable Read Only Memory)
- EPROM.(Erasable Programmable Read Only Memory).

SUMMARY

1. 1845 Boolean Algebra was developed by George Boole.
2. 1890 Herman Hollerith used Punched Cards for storing data.
3. 1900 Valdemar Poulson developed recording devices like tapes and drums coated with thin films of magnetic oxide.
4. 1906 Lee do Forest invented Thermoionic valve. Values were the first devices to be used in the modern digital computers.
5. 1937 Alan Turing, A British Mathematician, showed how a problem can be solved with logical steps. These logical steps could be broken into a set of small and simple instructions.
6. 1938 Claude Shannon, an electrical engineer established the fact that electronic switching circuits can be employed to perform logic and arithmetic operations.
7. 1944 Professor Howard Aiken developed a fully automatic electrically driven machine.
8. **Instruction Set:** The set of instructions that the microprocessor can execute.
9. **Bandwidth:** The number of bits processed in a single instruction.
10. **Clock Speed:** Given in megahertz (MHz), the clock speed determines how many instructions per second the processor can execute.

EXERCISE

(I) Fill in the Blanks (Solved)

1. Time Period of Third Generation Computer was ....
2. The processors in mainframe are slower than those in ....
3. Speed of mainframes is measured in ....
4. A sequence of instructions is called as ....
5. Technology used in fifth generation computer is .... and .... technologies.

(II) True or False (Solved)

1. Boolean Logic (Algebra) was invented by Lady Augusta Ada Byran
2. The technology of digital computers is based upon bi-stable electronic components.
3. Many of the today's minicomputers can't accommodate even 150 users working from individual terminals.
4. Memory unit can also hold the data for permanent storage too.
5. The technology of Analog computer is based upon bi-stable electronic components.

(III) Fill in the Blanks (Unsolved)

1. The computer organisation often compared with the ....
2. A computer can store .... amounts of information.
3. Computer technology has evolved rapidly since ....

4. Memory capacity of second generation computer is .... characters.
5. Most minicomputers are .... systems of larger size and processing power as compared to microcomputer.

(IV) True or False (Unsolved)

1. Computer is an electronic device or combination of electronic devices.
2. The computer's physical processing circuits rarely make errors.
3. Human intervention is required once the instructions and data are given.
4. Technology used in IIIrd generation computer is Microprocessor using Very Large Scale Integration (VLSI).
5. Digital Computers can Operates on both digits and alphabets.

(V) Detailed Questions

1. Everybody knows that in Today's world computer is a need. Can Computer be dangerous to human being? How?
1. Define and describe the model of a digital computer.
2. Explain salient features of Analog, Digital and Hybrid Computers.
3. What is a supercomputer? List various uses of supercomputers.
4. Write the differences between mainframe computer and minicomputer.

Answers

(I) Fill in the Blanks (Solved)

1. 1963 to 1975.
2. Supercomputer
3. Megaflops.
4. Program.
5. (*a*) VLSI, (*b*) Parallel processing

(II) True or False (Solved)

1. False
2. True
3. False
4. False
5. False.

Chapter

3

Register Transfer and Micro-operations

3.0 INTRODUCTION

This chapter will introduce you about computer registers representation notation and transfer of data from one register to another. In this chapter we also tried to explain the concept of bus and how this bus is useful. For transferring data from one location to another. Here are different types of micro-operations, these micro-operations perform some arithmetic operations on the numeric data stored in the registers.

Chapter Outline

3.1 REGISTER TRANSFER LANGUAGE

The symbolic notation used for data transfer from one register to another register is called register transfer language. The sequence of micro-operations in a digitial computer can be explained by describing every operation in the form of words. A word is a 16-bit register. This method is generally not used because of time consuming process. So we use certain symbols to describe the sequence of transfers between registers and micro-operations associated with the register transfers, using the RTL (Register transfer language), using the register transfer language we can list the micro-operations sequences among the registers of digital computer system.

3.2 REGISTER TRANSFER

We have discussed the RTL in previous section now, we will consider how the register transfer is performed using RTL. We Represent Computer Registers by capital letters to denote the function of the register. For example, the register that holds an address for the memory unit is usually called a memory address register represented by *M*[*AR*]. Similar example are PC (Performance Counter), IR (Instruction Register) and R_1 (Processor register). We show the individual flip-flop in *n*-bit register by giving numbers in sequence from 0 through $n - 1$, starting from 0 in the right most position and increasing the numbers toward the left.

The computer needs processor register for manipulating data and a register for holding a memory address. These requirements are listed in Table 3.1, with a brief description of their functions and number of bits that they contain.

TABLE 3.1. (List of Register for the Basic Computer)

Register Symbol	*Number of Bits*	*Register Name*	*Function*
DR	16	Data Register	Holds memory operand
AR	12	Address Register	Holds address for memory
AC	16	Accumulator	Processor register
IR	16	Instruction Register	Holds instruction code
PC	12	Program Counter	Holds address of instruction
TR	16	Temporary Register	Holds temporary data
INPR	8	Input Register	Holds input character
OUTR	8	Output Register	Holds output character

A 16-bit register is divided into two parts. Low byte (bits 0 to 7) is assigned the symbol and high byte (8 to 15) is assigned the symbol *H*. The name of 16-bit register is DR, AC, IR and TR. The symbol AC (L) represents the low order byte and AC(H) designates the high order byte. The statement $R_2 \leftarrow R_1$ refers the transfer of the content of register R_1 into register R_2. It should be noted that the content. If source register R_1 does not change after the transfer. In real applications, the transfer occurs only under a predetermined control condition. This can be shown by means of an if then statement:

If $P = 1$ then $R_2 \leftarrow R_1$

It means that the transfer of contents of R_1 into R_2 depends on *P*. If $P = 1$ then the transfer can be performed. The control signal is written as follows:

$$P: R_2 \leftarrow R_1$$

3.3 BUS AND BUS ARCHITECTURE

The concept of Bus in digital computer is ame as in our daily life. For describing the concept of bus we will take the example of simple bus in which Mr. Pankaj wants to go from Sikandrabad to Ghaziabad. By bus which has one driver Mr. Mukesh and a conductor Mr. Vijay be seeing this example, here Mr. Pankaj is the dates and Mr. Mukesh and Mr. Vijay are the control signal and Ghaziabad is the destination address and Sikandrabad is the source address.

Similarly, bus has three signals in digital computer basically it is not the physical entity but it is the physical circuit based on the concept of logic gates and logic circuits. Since a computer has many registers, paths must be provided to transfer information from one register to another. If separate lines are used between each register and all other registers, number of wires will be excessive in the system. A more efficient scheme for transferring information between registers in a multiple. Register configuration is a, common bus system.

A bus structure consists of set of common lines, one for the each bit of register, through which binary information is transferred one at a time, one for control signal which is usual for performing the data transfer, and one for address of data where it will be transferred.

A common bus system can be constructed using multiplexers (a combinational circuit which is discussed in previous chapter). These multiplexers select the source register whose binary information is then placed on the bus. A bus system will multiplex registers of bit each to produce an n-line common bus. The number of multiplexers required to construct the bus is equal to n, where n is the number of bits in each register. The size of each multiplexer must be KXI since it multiplexers K data lines. A bus system can be constructed with three-state-gates instead of multiplexers. But here will try to construct a bus system using multiplexers as shown in Fig. 3.1.

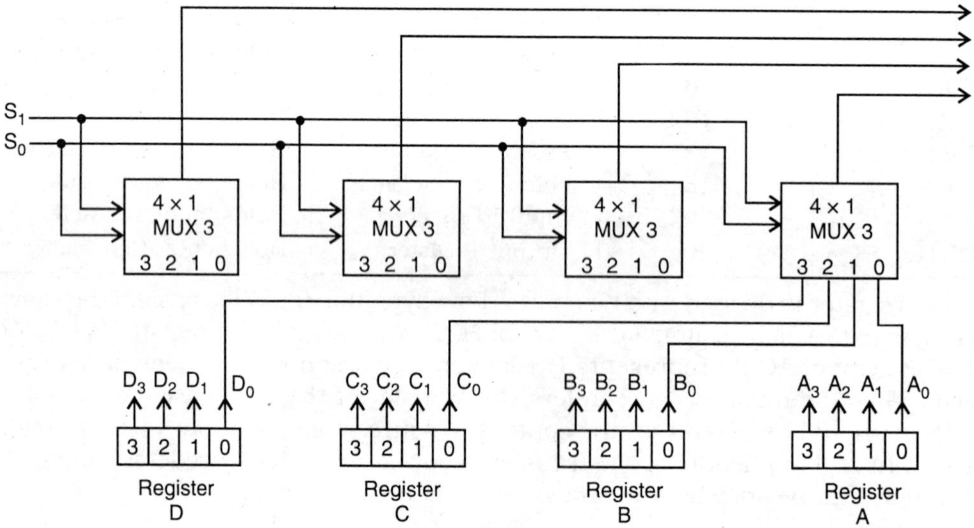

Fig. 3.1.

Here 4 multiplexers MUX 0, 1, 2, 3 are responsible for transferring individual bits of registers, as MUX 0 will carry A_0, B_0, C_0 and D_0 and MUX F will carry A_1, B_1, C_1 and D_1 and MUX 2 will carry A_2, B_2, C_2 and D_2 and MUX 3 is responsible for A_3, B_3, C_3 and D_3. This depends of selection lines as shown in the following Table 3.2.

TABLE 3.2.

S_0	S_1	*Register Collected*
0	0	*A*
0	1	*B*
1	0	*C*
1	1	*D*

3.3.1 Introduction to the concept of a bus

A bus, in computing, is a set of physical connections (cables, printed circuits, etc.) which can be shared by multiple hardware components in order to communicate with one another.

The purpose of buses is to reduce the number of "pathways" needed for communication between the components, by carrying out all communications over a single data channel. This is why the metaphor of a "data highway" is sometimes used.

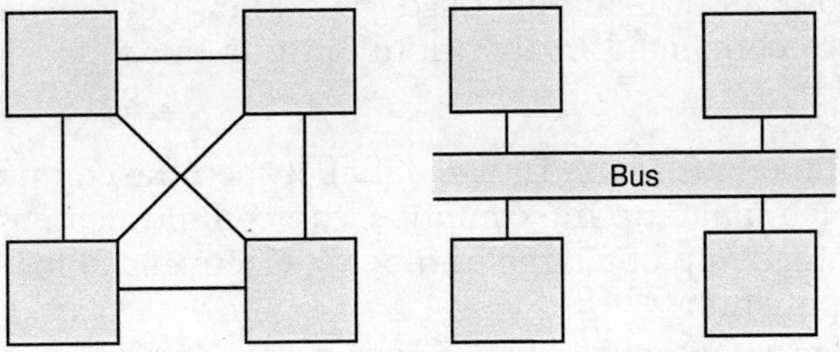

Fig. 3.2.

If only two hardware components communicate over the line, it is called a hardware port (such as a serial port or parallel port).

3.3.2 Characteristics of a bus

A bus is characterised by the amount of information that can be transmitted at once. This amount, expressed in bits, corresponds to the number of physical lines over which data is sent simultaneously. A 32-wire ribbon cable can transmit 32 bits in parallel. The term "width" is used to refer to the number of bits that a bus can transmit at once.

Additionally, the bus speed is also defined by its frequency (expressed in Hertz), the number of data packets sent or received per second. Each time that data is sent or received is called a cycle.

This way, it is possible to find the maximum transfer speed of the bus, the amount of data which it can transport per unit of time, by multiplying its width by its frequency. A bus with a width of 16 bits and a frequency of 133 MHz, therefore, has a transfer speed equal to:

16 * 133.106 = 2128*106 bit/s,
or 2128*106/8 = 266*106 bytes/s
or 266*106 /1000 = 266*103 KB/s
or 259.7*103 /1000 = 266 MB/s

3.3.3 Bus subassembly

In reality, each bus is generally constituted of 50 to 100 distinct physical lines, divided into three subassemblies:

- The **address bus** (sometimes called the *memory bus*) transports memory addresses which the processor wants to access in order to read or write data. It is a unidirectional bus.
- The **data bus** transfers instructions coming from or going to the processor. It is a bidirectional bus.
- The **control bus** (or *command bus*) transports orders and synchonisation signals coming from the control unit and travelling to all other hardware components. It is a bidirectional bus, as it also transmits response signals from the hardware.

3.3.4 The primary buses

There are generally two buses within a computer:

- The **internal bus** (sometimes called the *front-side bus*, or *FSB* for short). The internal bus allows the processor to communicate with the system's central memory (the *RAM*).

- The **expansion bus** (sometimes called the *input/output bus*) allows various motherboard components (USB, serial, and parallel ports, cards inserted in PCI connectors, hard drives, CD-ROM and CD-RW drives, etc.) to communicate with one another. However, it is mainly used to add new devices using what are called **expansion slots** connected to the input/outpur bus.

3.3.5 The chipset

A **chipset** is the component which routes data between the computer's buses, so that all the components which make up the computer can communicate with each other. The **chipset** originally was made up of a large number of electronic chips, hence the name. It generally has two components:

- The **NorthBridge** (also called the *memory controller*) is in charge of controlling transfers between the processor and the RAM, which is way it is located physically near the processor. It is sometimes called the **GMCH**, forr *Graphic and Memory Controller Hub.*
- The **SouthBridge** (also called the *input/output controller* or *expansion controller*) handles communications between peripheral devices. It is also called the **ICH** (*input/output Controller Hub*). The tem **bridge** is generally used to designate a component which connects two buses.

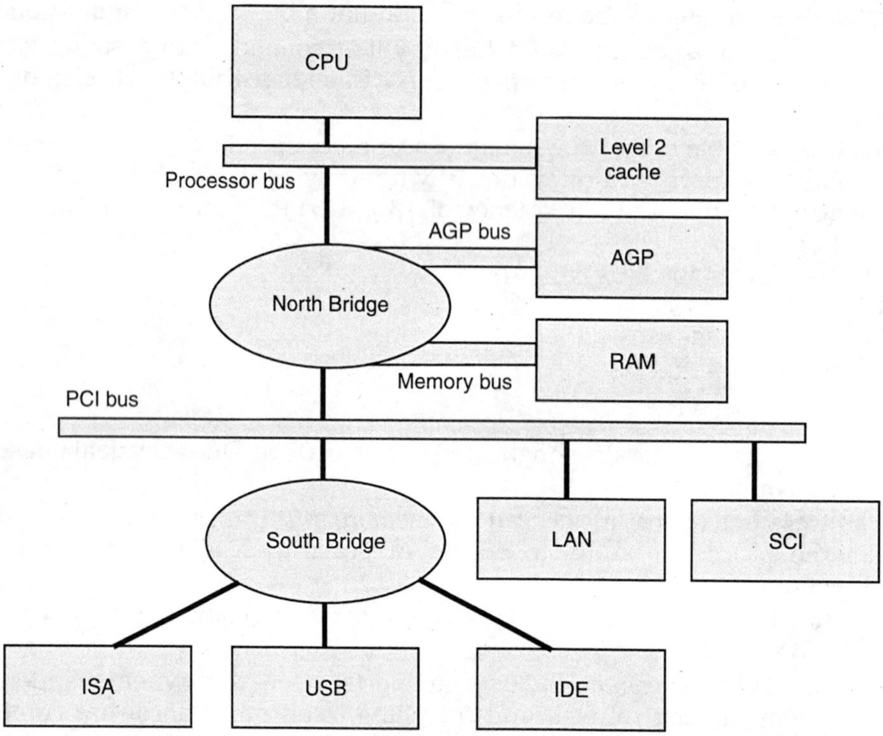

Fig. 3.3.

It is interesting to note that, in order to communicate, two buses must have the same width. The explains why RAM modules sometimes have to be installed in pairs (for example, early Pentium chips, whose processor buses were 64-bit, required two memory modules each 32 bits wide).

Here is a Table which gives the specifications for the most commonly used buses:

TABLE 3.3.

Standard	*Bus width (bits)*	*Bus speed (MHz)*	*Bandwidth (MB/sec)*
ISA 8-bit	8	8.3	7.9
ISA 16-bit	16	8.3	15.9
EISA	32	8.3	31.8
VLB	32	33	127.2
PCI 32-bit	32	33	127.2
PCI 64-bit 2.1	64	66	508.6
AGP	32	66	254.3
AGP (x2 Mode)	32	66x2	528
AGP (x4 Mode)	32	66x4	1056
AGP (x8 Mode)	32	66x8	2112
ATA33	16	33	33
ATA100	16	50	100
ATA133	16	66	133
Serial ATA (S-ATA)	1		180
Serial ATA II (S-ATA2)	2		380
USB	1		1.5
USB 2.0	1		60
FireWire	1		100
FireWire 2	1		200
SCSI-1	8	4.77	5
SCSI-2 - Fast	8	10	10
SCSI-2 - Wide	16	10	20
SCSI-2 - Fast Wide 32 bits	32	10	40
SCSI-3 - Ultra	8	20	20
SCSI-3 - Ultra Wide	16	20	40
SCSI-3 - Ultra 2	8	40	40
SCSI-3 - Ultra 2 Wide	16	40	80
SCSI-3 - Ultra 160 (Ultra 3)	16	80	160
SCSI-3 - Ultra 320 (Ultra 4)	16	80 DDR	320
SCSI-3 - Ultra 640 (Ultra 5)	16	80 QDR	640

3.3.6 Expansion Bus

Expansion buses (sometimes called *peripheral buses*) are buses that have connectors that allow you to add expansion cards (peripherals) to a computer. There are different types of standard internal buses that are characterised by:

- their shape
- the number of connector pins
- the type of signals (frequency, data, etc.)

3.3.7 ISA Bus

The original version of the **ISA bus** (*Industry Standard Architecture*) that appeared in 1981 with PC XT was an 8-bit bus with a clock speed of 4.77 MHz.

In 1984, with the appearance of PC AT (the *Intel 286* processor), the bit was expanded into a 16-bit bus and the clock speed went from 6 to 8 MHz and finally to 8.33 MHz, offering a maximum transfer rate of 16 Mb/s (in practice only 8 Mb/s because one cycle out of every two was used for addressing).

The ISA bus permitted **bus mastering**, *i.e.* it enabled controllers connected directly to the bus to communicate directly with the other peripherals without going through the processor. One of the consequences of bus mastering is **direct memory access (DMA)**. However, the ISA bus only allows hardware to address the first 16 megabytes of *RAM*.

Up until the end of the 1990s, almost all PC computers were equipped with the ISA bus, but it was progressively replaced by the *PCI bus*, which offered a better performance.

- **8-bit ISA Connector:**

- **16-bit ISA Connector:**

3.3.8 MCA Bus

The **MCA bus** (*Micro Channel Architecture*) is an improved proprietary bus designed by IBM in 1987 to be used in their PS/2 line of computer. This 16 to 32-bit bus was incompatible with the ISA bus and could reach a throughput of 20 Mb/s.

3.3.9 EISA Bus

The **EISA bus** (*Extended Industry Standard Architecture*) was developed in 1988 by a consortium of companies (AST, Compaq, Epson, Hewlett-Packard, NEC, Olivetti, Tandy, Wyse and Zenith) in order to compete with the MCA proprietary bus that was launched by IBM the previous year. The EISA bus used connectors that were the same size as the ISA connector but with 4 rows of contacts instead of 2, for 32-bit addressing.

The EISA connectors were deeper and the additional rows of contacts were placed below the rows of ISA contacts. Thus, it was possible to plug an ISA expansion board into an EISA connector. However, they did not plug as deep into the connector (because of the bezels) and thus only used the top rows (ISA) of contacts.

3.3.10 Local Bus

Traditional input/output buses, such as ISA, MCA our EISA buses, are directly connected to the main bus and there are forced to work at the same frequency. However, some I/O peripherals need a very low bandwidth while other need higher bandwidths. Therefore there are **bottlenecks** on the bus. In order to solve this problem, the "**local bus**" architecture offers to take advantage of the system bus, or Front Side Bus (FSB), by interfacing directly with it.

3.3.11 VLB Bus

In 1992, the **VESA Local Bus (VLB)** was developed by the *VESA* (*Video Electronics Standard Association* under the aegis of the company *NEC*) in order to offer a local bus dedicated to graphics systems. The VLB is a 16-bit ISA connector with an added 16-bit connector:

The VLB bus is a 32-bit bus initially intended to work a bandwidth of 33 MHz (the bandwidth of the first PC 486s at that time). The VESA local bus was used on the following 486 models (40 and 50 MHz, respectively) as well as on the very first Pentium processors, but it was quickly replaced by the *PCI bus*.

3.3.12 The PCI Bus

The **PCI bus** (*Peripheral Component Interconnect*) was developed by Intel on 22 June 1992. Contrary to the *VLB* bus, it is not so much a traditional local bus but rather an intermediate bus located between the *processor bus* (*NorthBridge*) and the *input/output bus* (*SouthBridge*).

3.4 MEMORY TRANSFER

Memory transfer means reading data from memory or writing data into memory. A read operation imples transfer of information to the outside environment from a memory word, whereas storage of information into the memory is defined as write operation. Symbolizing a memory word by the letter *M*, it is selected by the memory address during the transfer which is a specification for transfer operations. The address is specified by enclosing it in square brackets following the letter *M*.

3.4.1 Memory Read

1. Put memory address in the memory address register (MAR).
2. Read the data of the location. Generally this is achieved by putting the data in MAR on address bus along with a memory read control signal on the control bus. The resultant of memory read is put into the data bus which in turn stores the read data in the Data Register (DR). This whole operation can be shown as

$$\text{DR} \leftarrow \text{M [MAR]}$$

3.4.2 Memory Write

1. Put the desired memory address in memory address register (MAR) and the data to be written in the data register (DR).
2. Write the data into the location MAR puts the address bus and DR puts the data on data bus to be written into memory location addressed by MAR.

$$\text{M [MAR]} \leftarrow \text{DR}$$

3.5 MICRO- OPERATIONS

A micro-operation is an elementary operation which is performed on the data stored in registers. We can classify the micro-operations into four categories.

1. **Register transfer:** Transfer binary information from one register to another.
2. **Arithmetic:** Perform arithmetic operations on numeric data stored in registers.
3. **Logic:** Perform bit manipulation operation on numeric data stored in registers.
4. **Shift:** Perform shift operations on data stored in registers.

3.5.1 Register Transfer

When the bits of one register are transferred to another register. It should be noted that the information does not change during this micro-operation. A register transferred micro-operation may be designed as

$$R_1 \leftarrow R_2$$

This implies that the transfer the content of register R_2 to register R_1. A common internal data bus of processor is used for connecting various registers. In general, the size of this data bus should be equal to the number of bits in a general regster. Memory transfer can be achieved by a system bus.

3.5.2 Arithmetic Micro-operation

These micro-operations perform some basic arithmetic operations on the numerical data stored in the registers. These basic operations may be addition, subtraction, incrementing a number, decrementing a number.

Addition:

$$R_3 \leftarrow R_1 + R_2$$

This means the addition of R_1 and R_2 registers is transferred to R_3.

Subtraction:

$R_3 \leftarrow R_1 - R_2$ is implemented as

$R_3 \leftarrow R_1$ + (1's complement of R_2)

$R_3 \leftarrow R_1$ + (2's complement of R_2)

Increment:

$$R_1 \leftarrow R_1 + 1$$

Decrement:

$$R_1 \leftarrow R_1 - 1$$

3.5.3 Logic Micro-operations

These operations are performed on the binary data stored in the register. For a logic micro-operation each bit of a register is treated as separate variable.

For example if R_1 and R_2 are 8 bits registers.

R_1 contains 10010011

R_2 contains 01010101

R_1 and R_2 00010101

Some of the common logic operations are AND, OR and NOT, EX-OR, NOR and NAND.

3.5.4 Shift Micro-operations

Shift micro-operation can be used for serial transfer of data. They are used generally with the arithmetic, logic and other data-processing operations. The contents of a register can be shfited to the left or the right. During a shift right operation the serial input transfers a bit into the left most position. the serial input transfers a bit into the right most position during a shift-left operation. There are three types of shifts, logical, circular and arithmetic.

3.5.4.1 Logical Shift

A logical shift operation transfers 0, through the serial input, we use the symbols shl and shr for logical shift left and shift right micro-operations.

e.g. If we have a register R_1 =		1	0	1	0	1	0	1	1
After left shift	1	0	1	0	1	0	1	1	0
After right shift	0	1	0	1	0	1	0	1	1

3.5.4.2 Circular Shift

The circular shift is also known as rotate operation. It circulates the bits of the register around the two ends and there is no loss of information. This is accomplished by connecting the serial output of the shift to its signal input. We use the symbols cil and cir for the circular shift left and circular shift right.

e.g. If we have register R_1 =

1	0	1	0	1	0	1	1

1 0 1 0 1 0 1 1

After left shift

0	1	0	1	0	1	1	1

R_1 =

0	1	0	1	0	1	1	1

After right shift R_1 =

1	0	1	0	1	0	1	1

=

	1	0	1	0	1	0	1

=

1	1	0	1	0	1	0	1

3.5.4.3 Arithmetic Shift

An arithmetic shift micro-operation shifts a signed binary number to the left or right. The effect of an arithmetic shift left operation is to multiply the binary number by 2. Similarly is to multiply the binary number by 2. Similarly an arithmetic shift right divides the number by 2.

e.g. If we have a register that contains the value of 6 as R_1 = 0110

After Left shift 6 × 22 = 12 = 1100

After Right shift $\frac{6}{2}$ = 3 = 0011

3.6 BUS ARBITRATION

Bus arbitration is defined as the process by which the requesting processor (By any device) may be granted the access to the bus.

A computer system has several buses to perform the transfer of information between the components of computer. The processors access the memory or other common resources through the system bus. If the system buses are free that means no processors is currently utilizing the bus, the requesting processor may be granted the system buses immediately. But if some processor is currently utilizing the bus, the requesting processor has to wait. Also, at the same time the system buses are requested by some other processor. At this situation, the arbitration is the only way to resolve this multiple contention for the shared resources.

There are two approaches to bus arbitration:

3.6.1 Serial Arbitration

3.6.2 Parallel Arbitration

3.6.1 Serial Arbitration

Serial arbitration procedure is known as daisy chain arbitration. According to the position of the processors connected to the system bus a priority is assigned to them and when multiple devices request the use of bus at the same time the shared resources are granted to the device that is having highest-priority. A daisy chain connection of arbiters is shown in Fig. 3.4.

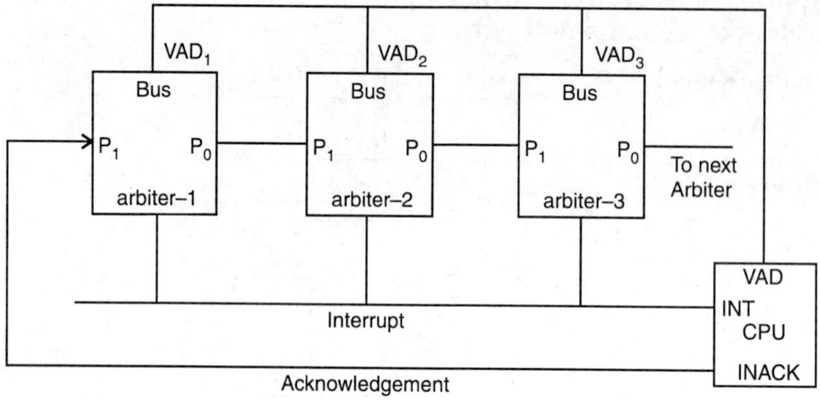

Fig. 3.4. Daisy chaining arbiteration

In above Fig. 3.4. there are three BUS arbiters arbiter 1, arbiter 2 and arbiter 3. Each processor has its own bus arbiter logic with Priority-in (*PI*) and Priority-out (*PO*) lines. The processor that is closest to the priority line is assigned the highest priority followed by lowest priority processor upto the lowest priority that is placed last in the chain. Each processor bus has its own VAD (Vector Address) to inform its location to CPU and there is an interrupt line, through which all the processor bus sends its interrupt request and there is another line that is INACK (Interrupt Acknowledge) to inform the highest priority devic that, it can interrupt or it can process. First of all the turn is given to first processor bus to interrupt CPU. When any device has its turn then its PI value must be '1' and PO value '0'.

If the device do not want to interrupt then it sets its $PI = 0$ and $PO = 1$, when its $PO = 1$, hence the *PI* of next device is equal to 1 and its $PO = 0$. If means that the first device transfers its priority to the next device. If the next device also do no want to interrupt then it will also transfer the control to next device by performing the same operation as done by first device, this process continues until all the devices in the chain do not get turn to interrupt to CPU.

3.6.2 Parallel Arbitration

The Fig. 3.5 shows the parallel arbiteration. This technique uses an external priority encoder and decoder. Each of the bus arbiter has a bus request output line and bus acknowledge input line.

The input to the priority encoder is the bus request output lines from each arbiter and output of the encoder generates 2 bit codes. These 2-bit codes represents the highest priority processor among these requesting the bus. These 2-bit acknowledge input line of the highest priority processor to take over the control of the system buses when the processor takes the control of the system bus, it can then conduct the required bus transfer.

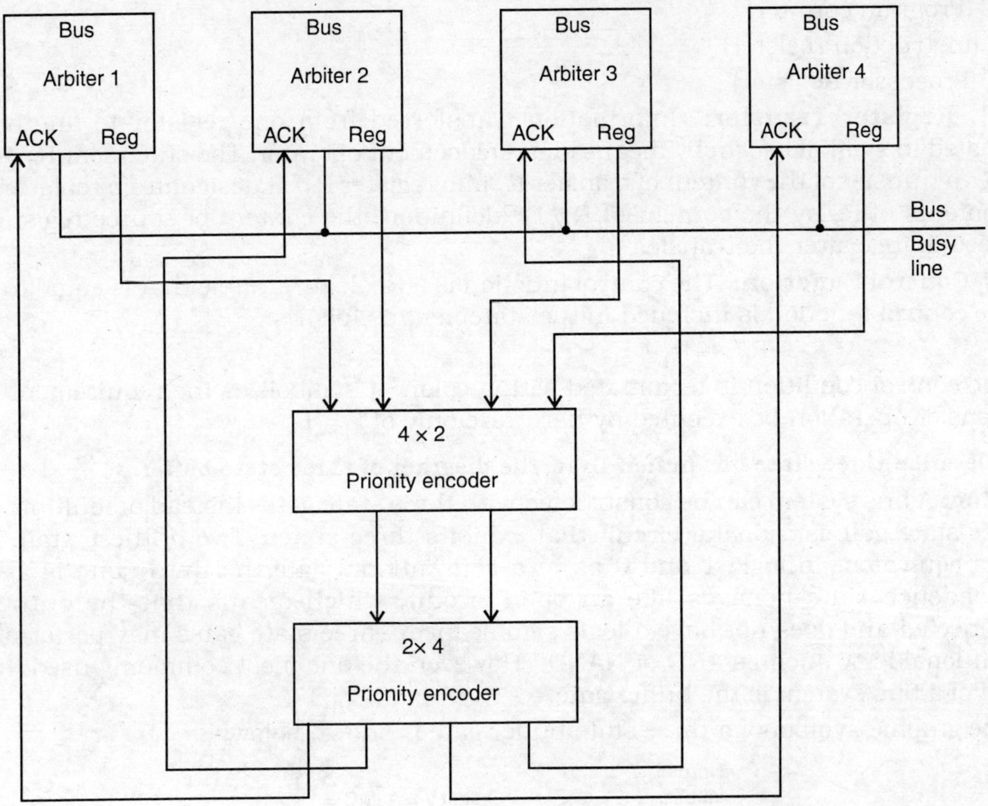

Fig. 3.5. Parallel Arbiteration

SOLVED PROBLEMS

Q.1. Define the following terms in brief:

(a) Register transfer language

(b) Micro-operation

(c) Registers

(d) Register Transfer

(e) Control function

Solution: (a) Register transfer language: The symbolic notation used to describe the micro-operation transfers among registers is called a register transfer language. The term register transfer implies the availability of hardware logic circuits that can perform a stated micro-operation and transfer the result of the operation to the same or another registers the word "language" is borrowed from programmes who apply this term to programming language.

(b) Micro-operation: The operations executed on data stored in registers are called micro-operations. A micro-operation is an elementary operation performed on the information stored in one or more registers. The result of operation may replace the previous binary information of a register or may be transferred to another register, *e.g.* shift, count, clear and load.

(c) Registers: The register that gold the address of the memory units is called a memory address register and is designated by the name MAR other designations for registers are

PC (Program counter)

IR (Instruction register)

R_1 (Processor Register)

(d) Register Transfer: Information transferred from one register to another is designated in symbolic form by means of a replacement operator. The statement $R_2 \leftarrow R_1$, denotes a transfer of the content of transfer R_1 into register R_2. It designates a replacement the contents of R_2 by the content of R_1. By definition, the content of source register R_1 does not change after the transfer.

(e) Control Function: The control function is a Boolean variable that is equal to 1 or 0. The control function is included in the statement as follows:

P, $R_2 \leftarrow R_1$

The control condition is terminated with a colon, it symbolizes the requirement that the transfer operation be executed by hardware only of P = 1.

Q.2 Describe three state bus buffer. Draw the diagram of three state buffer.

Solution: A bus system can be constructed with three state gates instead of multiplexers. A three state gate is a digital circuit that exhibits three states. Two of these states are signals equivalent to logic 1 and 0 as in a conventional gate the third state is a high independence state behaves like an open circuit, which means that the output is disconnected and does not have a logic significance. Three state gates may perform any conventional logic such as AND or NAND. However, the one most commonly used in the design of a bus system is the buffer gate.

The graphic symbol of a three state buffer gate is shown below:

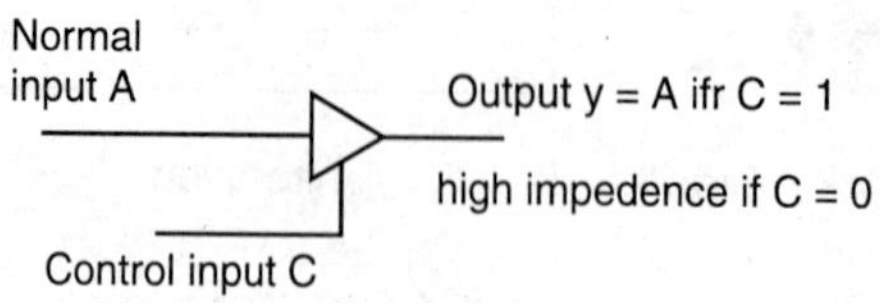

Fig. 3.6.

To construct a common bus for four registers of *n*-bit using three state buffers, we need n circuit with four buffers. Each group of four buffers receives one significant bit from the four registers. Each common output produces one of the lines for the common bus for a total of *n* lines only one decoder is necessary to select between the four registers.

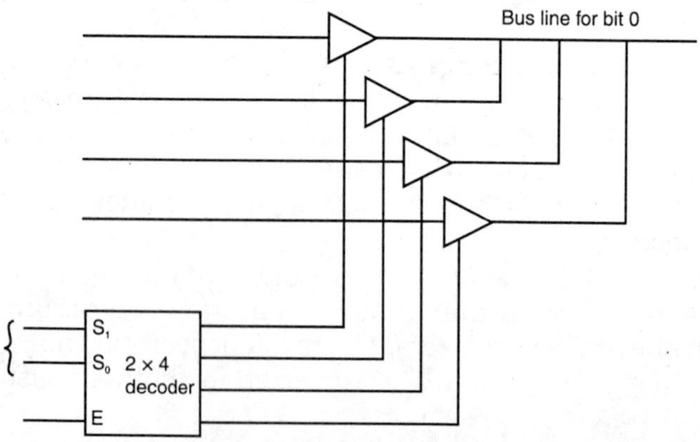

Fig. 3.7. Bus line with three state buffers

Q.3. Consider the following circuit A = $a_2a_1a_0$ and B = $b_2b_1b_0$ are three bit binary numbers input to the circuit. The output is Z is z_3z_2,z_0. R_0 R_1 and are registers with loading clock shown. The registers are loaded with their input data with the falling edge of the clock pulse. The bits of the input numbers A, B and full address are shown in the circuit.

(*a*) Find the output 2 for the inputs

A = 110

B = 101

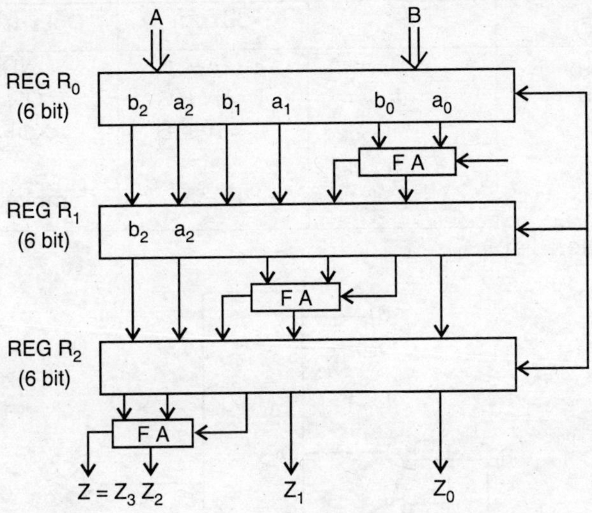

Fig. 3.8.

Solution: Given value

$$A = 110 = a_2a_1a_0$$
$$B = 101 = b_2b_1b_0$$
$$a_0b_0 = 01$$
$$a_1b_1 = 10$$
$$a_2b_2 = 11$$

For bit $a_0b_0 = 01$

and third bit = 01

∴ First bit is 010

$$S = 0 \oplus 1 \oplus 0 = 1$$
$$C = 0.1 + (0 \oplus 1)\,0$$
$$= 0 + 0$$
$$= 0$$

Second bit is 100

$$S = 1 \oplus 0 \oplus 0 = 1$$
$$C = 1.0 + (1 \oplus 0).0 = 0 \quad z_1 = 1$$

Third bit is 110

$$S = 1 \oplus 1 \oplus = 0 \quad z_2 = 0$$
$$C = 1.1 + (1 \oplus 1)\,0 = 1 \quad z_3 = 1$$
$$z = z_3z_2z_1z_0 = 1011$$

Q.4. Describe the hardware implementation of logic-micro-operation. Draw the diagram of the stage of logic circuit used with AND, Or, XOR and NOT GATE with the help of function Table.

Solution: The hardware implementation of logic micro-operations requires that logic gates be inserted for each bit or pair of bits in the register to perform the required logic function. Although there are 16 logic micro-operations, most computers use only four AND, OR XOR and complement.

S_1	S_0	*Output*	*Operation*
0	0	$E = A \wedge B$	AND
0	1	$E = A \vee B$	OR
1	0	$E = A \oplus B$	XOR
1	1		
1	1	$E = A$	Complement

Function Table

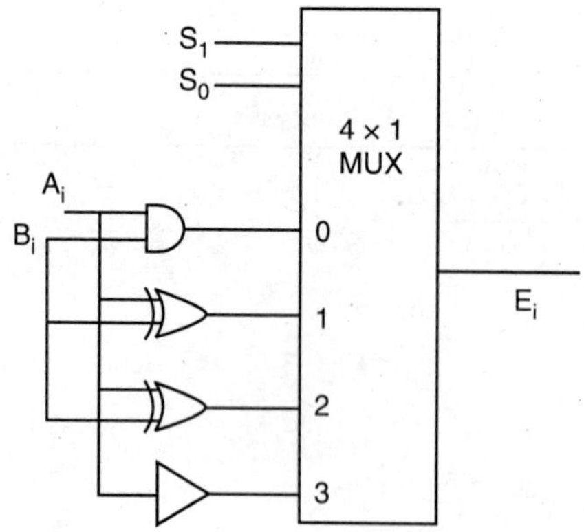

Logic diagram of one store of logic circuit

Fig. 3.9.

One stage of circuit that generates the four basic logic micro-operation. It consists of four gates and multiplexer. Each of the four logic operations is generated through a gate that perform the required logic the outputs of gates are applied to the data inputs of the multiplexer. The two selection inputs S_1 and S_0. Choose one of the data inputs of the multiplexer and direct its value to the output. For a logic circuit with *n* bits the diagram mist be repeated *n* times for $i = 0, 1, 2, 3, n$ the selection variables are applied to all stages.

Q.5. Define the term and describe its operation:

(*a*) **Selective set**
(*b*) **Selective complement**
(*c*) **Selective clear**
(*d*) **Mask operation**
(*e*) **Insert operation**
(*f*) **Clear operation**

Solution: (*a*) Selective set: The selective set operation sets to 1. The bits in the register A where there are corresponding is in register B. It does not affect that have O's in the B.

e.g.,	1010	A before
	1100	B (logical operand)
	1110	A (after)

The two left most bits of B are 1's B are 1's, so the corresponding bits of A are set to 1. The two bits of A with corresponding O's in B remain unchanged. Therefore the OR micro-operation can be used to selectively get bits of a register.

***(b)* Selective complement:** The selective complement operation complements bits in A where there are corresponding 1's in B. It does not affect bit positions that have O's in B.

e.g.,	1010	A before
	1100	B (logic operand)
	0110	

Therefore exclusive–Or operation can be used to selectively complement bits of register.

***(c)* Selective clear:** Selective clear operation clears to O. The bits in A only where there are corresponding 1's in B.

e.g.,	1010	A before
	1100	B (logic operand)
	0010	A after

The boolean operation performed on the individual bits in AB^1. The corresponding logic operation

$$A \leftarrow A \cap B^1$$

***(d)* Mask operation:** The mask operation is similar to the selective clear operation except that the bits of A are cleared only above. There are corresponding O's in B. The mask operation is an AND micro operation as seen from the following numerical example

1010	A before
1100	B (mask)
1000	A after masking

***(e)* Insert operation:** The insert operation inserts a new value into a group of bits this is done by first masking the bits and then OR in them the required value.

e.g.,	0110	1010	A before
	0000	1111	B (mask)
	0000	1010	After masking
	0000	1010	A before
	1001	0000	B (insert)
	1001	1010	A after insertion

***(f)* Clear operation:** Clear operation compares the words in A and B and produces an all O's result if the two numbers are equal. This operation is achieved by an exclusive–OR micro-operation

e.g.,	1010	A
	1010	B
	0000	

$$A \leftarrow A \oplus B$$

Q.6. Show the bock diagram of the hardware that implements the following register transfer statements

$$yT_1 : R_2 \leftarrow R_1, R_1 \leftarrow R_2$$

Solution:

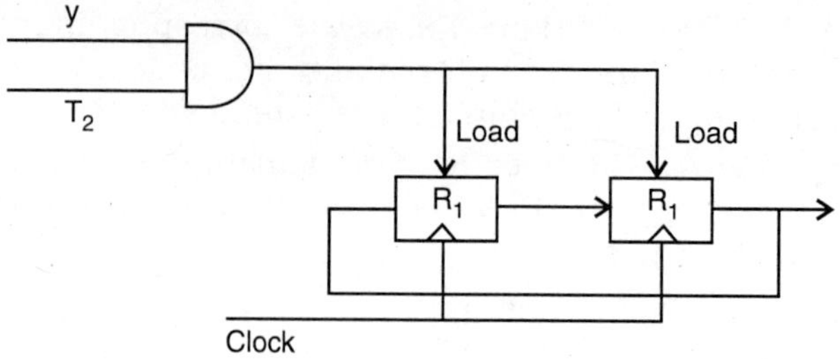

Fig. 3.10.

Q.7. The outputs of four registers R_0, R_1, R_2 and R_3 are connected through 4–to–1 line multiplexers to the inputs of a fifth register, R_5. Each register is eight bits long. The required transfers are dictated by four timing variables to through T_3 as follows:

$$T_0 : R_5 \leftarrow R_0$$
$$T_1 : R_5 \leftarrow R_1$$
$$T_2 : R_5 \leftarrow R_2$$
$$T_3 : R_5 \leftarrow R_3$$

The timing variables are mutually exclusive, which means that only one variable is equal to 1 at any given time, while the other three are equal to 0. Draw a block diagram showing the hardware implementation of register transfers, include the connection S necessary from the four timing variables to the selection inputs of the multiplexers and to load input of register R_5.

Solution:

T_0	T_1	T_2	T_3	S_1	S_0	R_3 load
0	0	0	0	×	×	0
1	0	0	0	0	0	1
0	1	0	0	0	1	1
0	0	1	0	1	0	1
0	0	0	1	1	1	1

$$S_1 = T_2 + T_3$$
$$S_0 = T_1 + T_3$$
$$\text{load} = T_0 + T_1 + T_2 + T_3$$

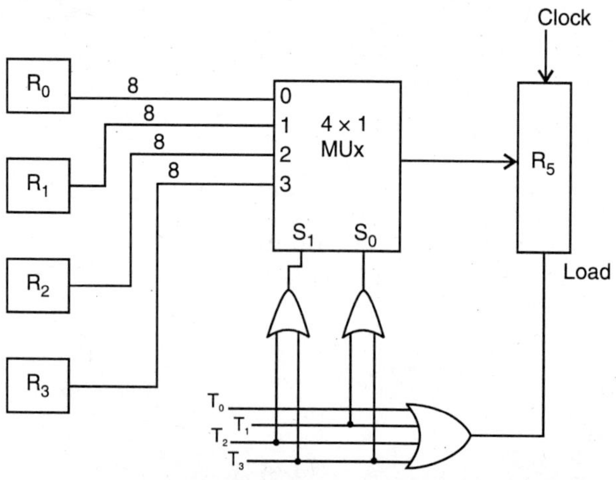

Fig. 3.11.

Q.8. Represent the following conditional, control statement by two register transfer statements with control functions.

If (P = 1) then
$(R_1 \leftarrow R_2)$
else if (Q = 1) then
$(R_1 \leftarrow R_3)$

Solution:

$$P : R_1 \leftarrow R_2$$
$$P^1Q : R_1 \leftarrow R_3$$

Q.9. What has to the bus system of figure given below to be able to transfer information from any register to any other register ? Specifically. Show the connections that must be included to provide a path from the outputs of register C to the inputs of register A.

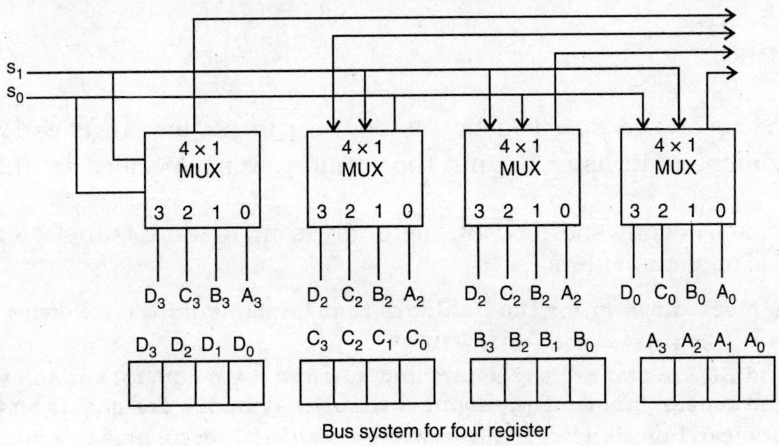

Fig. 3.12. Bus system for Four register

Solution: Connect the 4–line common bus to the fair inputs of each register. Provide a "load" control input in each register provide a clock input for each register to transfer from register C to register A.

Apply S_1S_0 = 10 (to select C for the bus)

Enable the load input of A and apply a clock pulse.

Q.10. Draw a diagram of a bus system similar to the one shown in the figure below but use three-state buffers and a decoder instead of multiplexers.

Solution:

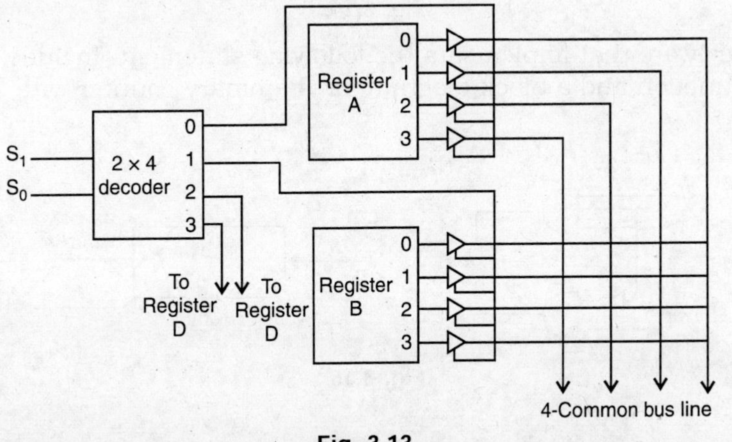

Fig. 3.13.

Q.11. A digital computer has a common bus system for 16 registers of 32 bits each. The bus is constructed with multiplexers.

(*a*) How many selection inputs are there in each multiplexers?

(*b*) What size of multiplexers are needed?

(*c*) How many multiplexers are there in the bus?

Solution: (*a*) 4 selection lines to select one of 16 register

(*b*) 16 × 1 multiplexers are needed.

(*c*) 32 multiplexers, one for each bit of the registers.

Q.12. The following transfer statements specify a memory. Explain the memory operation in each case.

(*a*) $R_2 \leftarrow M[AR]$

(*b*) $M[AR] \leftarrow R_3$

(*c*) $R_5 \leftarrow M[R_5]$

Solution:

(*a*) Read memory word specified by the address in AR into register R_2

(*b*) Write content of Register R_3 into the memory word specified by the address in AR

(*c*) Read memory word specified by the address in R_5 and Transfer content to R_5 (destroys previous value)

Q.13. Draw the block diagram for the hardware that implements the following statements:

$$x + yz : AR \leftarrow AR + BR$$

Where AR and BR are two n-bit registers and *x*, *y* and *z* are control variables. Include the logic gates for the control function. (Remember that the symbol + designates an OR operation in a control or Boolean function but that it represents an arithmetic plus in a micro-operation)

Solution:

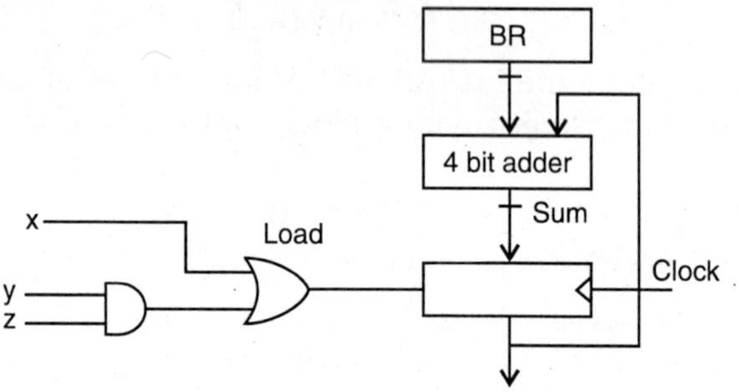

Fig. 3.14.

Show the hardware that implements the following statement. Include the logic gates for the control function and a block diagram for the binary counter with a count enable input.

$$xyT_0 + T_1 + y^1T_2 : AR \leftarrow AR + 1$$

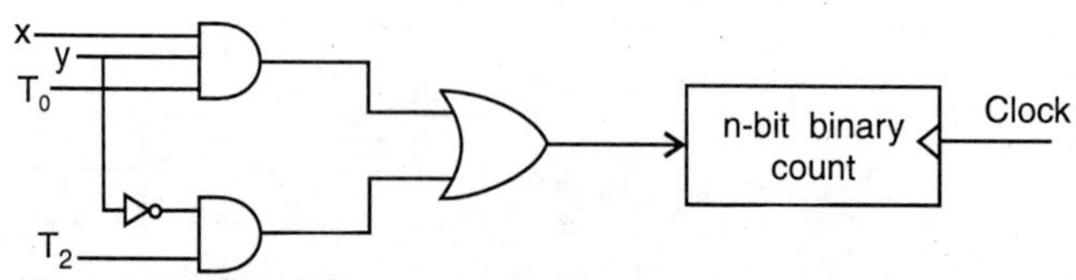

Fig. 3.15.

Q.14. Consider the following register transfer statements for two 4–bit register R_1 and R_2

$$xT : R_1 \leftarrow R_2$$
$$x^1T : R_1 \leftarrow R_2$$

Every time that variable $T = 1$, either the content of R_2 is added to the content of R_1 if $x = 1$, or the content of R_2 is transferred to R_1 if $x = 0$. Draw a diagram showing the hardware implementation of the two statements. Use block diagrams for the two. 4-bit registers, 4-bit adder and a quotable 2-to-1 line multiplexer that selects the inputs to R_1. In the diagram, show how the control variables x and T select the inputs of the multiplexers and the load input of register R_1.

Solution:

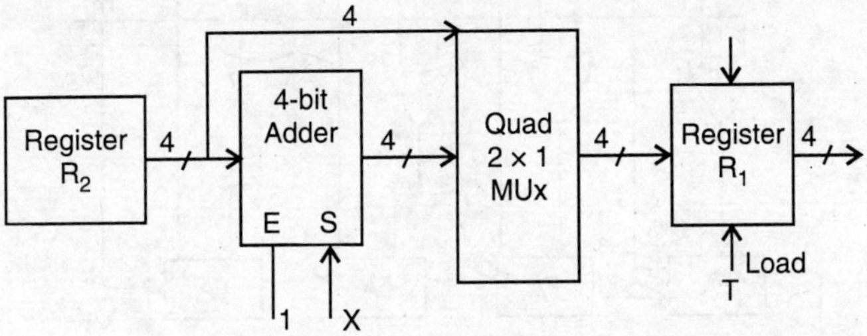

Fig. 3.16.

Q.15. Using a 4-bit counter with parallel load and a 4-bit adder as in figure given below, draw a block diagram that shows how to implement the following statements

$$x : R_1 \leftarrow R_1 + R_2 \qquad \text{Add } R_2 \text{ to } R_1$$
$$x^1y : R_1 \leftarrow R_1 + 1 \qquad \text{Increment } R_1$$

Where R_1 is a counter with parallel load and R_2 is a 4-bit register.

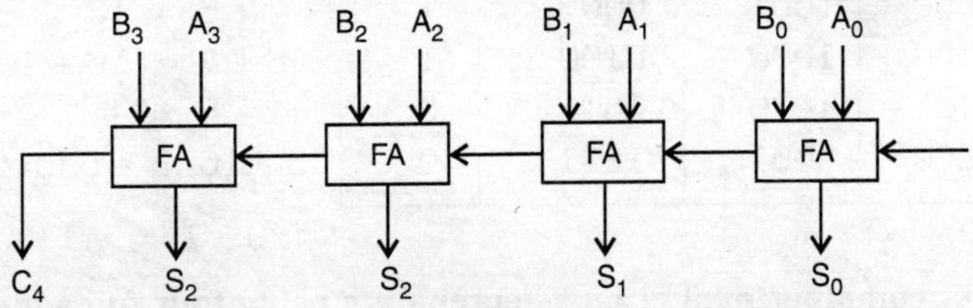

Fig. 3.17.

Solution:

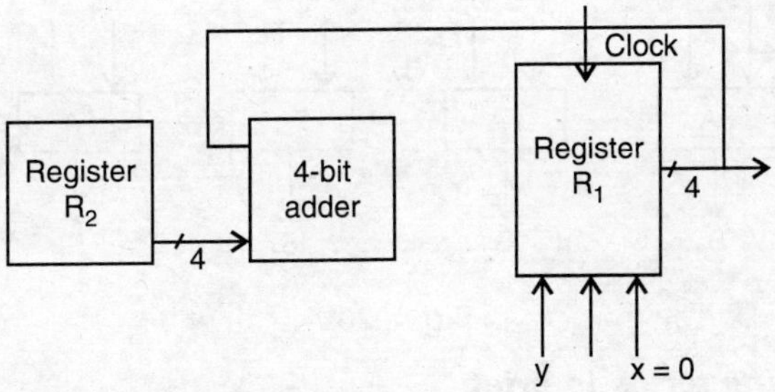

Fig. 3.18.

Q.16. The adder-subtracter circuit of figure given below has the following values for input mode M and data inputs A and B. In each case, determine the values of the outputs: S_3, S_2, S_2, S_1, S_0 and C4

	M	A	B
(a)	0	0111	0110
(b)	0	1000	1001
(c)	1	1100	1000
(d)	1	0101	1010
(e)	1	0000	0001

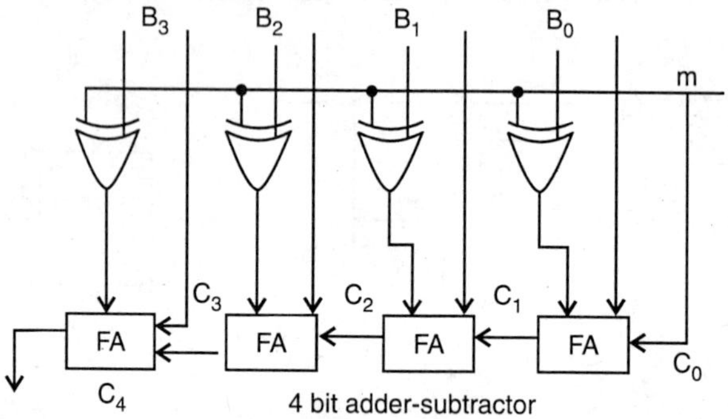

Fig. 3.19.

Solution:

M	A	B	Sum	C_4	
0	0111	0110	1101	0	
0	1000	1001	0001	1	7 + 6 = 13
1	1100	1000	0100	1	8 + 8 = 16 + 1
1	0101	1010	1011	0	12 – 8 = 4
1	0000	0001	1111	0	5 – 10 = – 5 (in 2's comp. form)

0 – 1 = – 1 (in was comp. form)

Q.17. Design a 4-bit combinational circuit decrements using four full adder circuits.

Solution: A 's complement of 1 = P + 1111

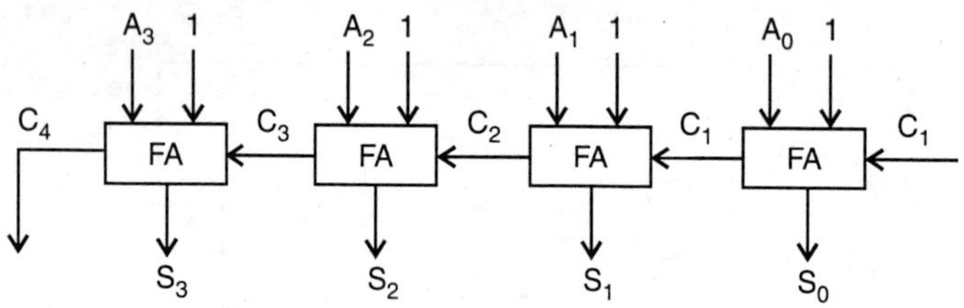

Fig. 3.20.

Q.18. ***Define the following terms in brief:***

(a) **Register transfer language.**

(*b*) **Micro-operation.**

(*c*) **Registers.**

(*d*) **Register transfer.**

(*e*) **Control function.**

Solution: (*a*) Register transfer language: The symbolic notation used to describe the micro-operation transfers among registers is called *register transfer language.* The term register transfer unpiles the availability of hardware logic circuits that can perform a stated micro-operation and transfer the result of the operation to the same or another registers. The word "language" is borrowed from programmes who apply this term to programming language.

(*b*) Micro-operation: The operations executed on data stored in registers are called *micro-operations.* A micro-operation is an elementary operation performed on the information stored in one or more registers. The result of the operation may replace the previous binary information of a register or may be transferred to another register.

e.g. shift, count clear and load.

(*c*) Registers: The register that hold the address of the memory units in usually called a memory address register and is designated by the name MAR. Other designations for registers are:

PC (for program counter),
IR (for instruction register) and
R_1 (for processor register).

(*d*) Register transfer: Information transferred from one register to another is designated in symbolic form by means of a replacement operator. The statement

$$R_2 \leftarrow R_1$$

denotes a transfer of the content of register R_1 into register R_2. It designates a replacement of the content of R_2 by the content of R_1. By definition, the content of the source register R_1 does not change after the transfer.

(*e*) Control function: A control function is a Boolean variable that is equal to 1 or 0. The control function is included in the statement as follows:

$$P\text{: } R_2 \leftarrow R_1$$

The control condition is terminated with a colon. It symbolizes the requirement that the transfer operation be executed by the hardware only if $P = 1$.

Q.19. Describe three state bus buffer. Draw the diagram of three state buffer.

Solution: A bus system can be constructed with three state gates instead of multiplexers. A three state gate is a digital circuit that exhibits three states. Two of these states are signals equivalent to logic 1 and 0 as in a conventional gate the third state is a high impedance state. The high impedance state behaves like an open circuit, which means that the output is disconnected and does not have a logic significance. Three state gates may perform any conventional logic, such as AND or NAND. However, the one most commonly used in the design of a bus system is the buffer gate.

The graphic symbol of a three state buffer gate is shown below:

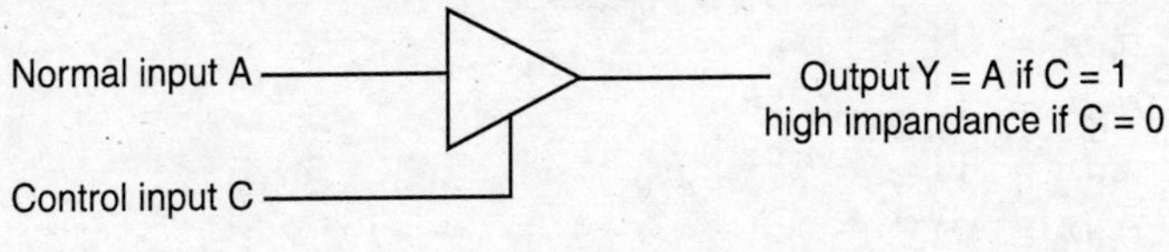

Fig. 3.21.

To construct a common bus for four registers of *n* bit each using three state buffers. We need *n* circuit with four buffers. Each group of four buffers receives one significance bit from the four registers. Each common output produces one of the lines for the common bus for a total of *n* lines. Only one decoder is necessary to select between the four registers.

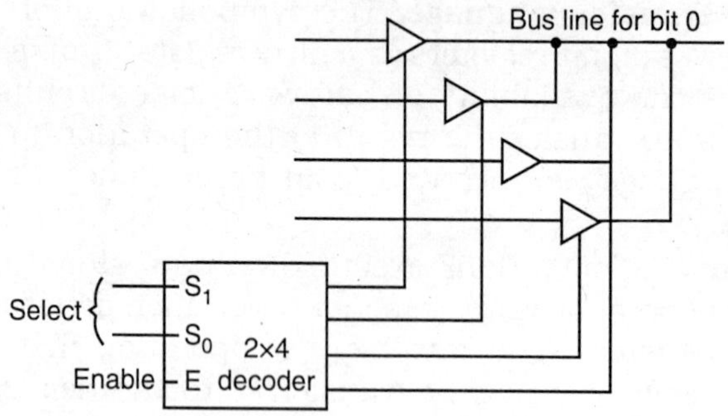

Fig. 3.22. Bus line with three state buffers

Q.20. Consider the following circuit $A = a_2a_1a_0$ and $B = b_2b_1b_0$ are three bit binary numbers input to the circuit. The output is z is $z_2z_2z_1z_0$ R_0 R_1 and R_2 are registers with loading clock shown. The registers are loaded with their input data with the falling edge of a clock pulse. The bits of input numbers A, B and full address are shown in the circuit.

(a) Find the output z for the inputs

$$A = 110$$
$$B = 101$$

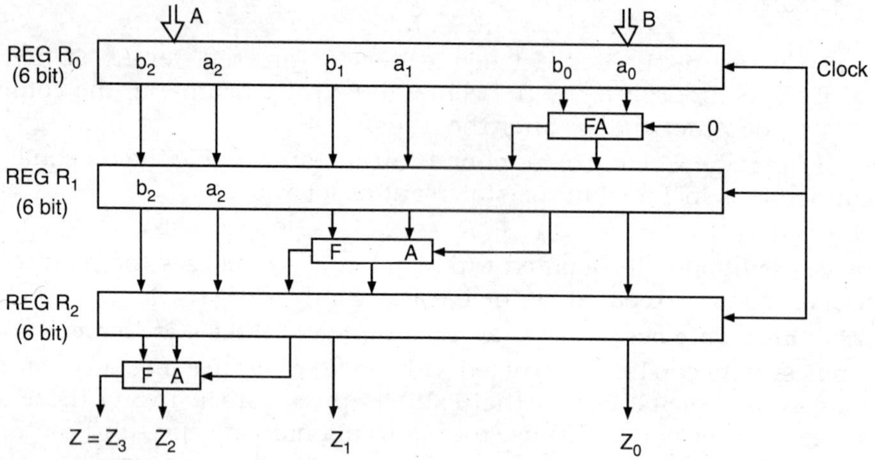

Fig. 3.23.

Solution: Given value

$$A = 110 = a_2a_1a_0$$
$$B = 101 = b_2b_1b_0$$
$$a_0b_0 = 01$$
$$a_1b_1 = 10$$
$$a_2b_2 = 11$$

For bit $a_0b_0 = 01$

and third bit = 0

∴ **First bit is 010**

$$S = 0 \oplus 1 \oplus 0 = 1 \quad z_0 = 1$$
$$C = 0.1 + (1 \oplus 0)\, 0 = 0 + 0 = 0$$

Second bit is 100

$S = 1 \oplus 0 \oplus 0 = 1$ $\quad z_1 = 1$

$C = 1.0 + (1 \oplus 0) . 0 = 0$

Third bit is 110

$S = 1 \oplus 1 \oplus 0 = 0$ $\quad z_2 = 0$

$C = 1.1 + (1 \oplus 1) 0 = 1$ $\quad z_3 = 0$

$z = z_3 z_2 z_1 z_0 = \mathbf{1011}$

Q.21. Describe the hardware implementation of logic micro-operation. Draw the diagram of one stage of logic circuit used with AND, OR, XOR and NOT GATE with the help of function table.

Solution: The hardware implementation of logic micro-operations requires that logic gates be inserted for each bit or pair of bits in the register to perform the required logic function. Although there are 16 logic micro-operations, most computers use only four AND, OR.

S_1	S_2	*Output*	*Operation*
0	0	$E = AAB$	AND
0	1	$E = AVB$	OR
1	0	$E = A \oplus B$	XOR
1	1		
1	1	$E = \overline{A}$	Complement

Fig. 3.24. Function Table

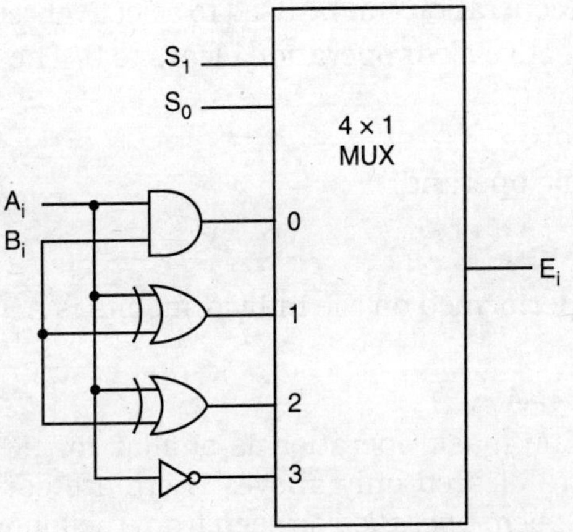

Fig. 3.25. Logic diagram of one stage of logic circuit

One stage of circuit that generates the four basic logic micro-operations. It consists of four gates and multiplexers. Each of the four logic operations is generated through a gate that perform the required logic. The outputs of the gates are applied to the data inputs of the multiplexers. The two selection inputs S_1 and S_0. Choose one of the data inputs of the multiplexer and direct its value to the output. For a logic circuit with *n* bits, the diagram must be repeated *n* times for $i = 0, 1, 2, 3, nl$. The selection variables are applied to all stages.

Q.22. Define the term and describe its operations:

(a) Selective set
(b) Selective implement
(c) Selective clear
(d) Mask operation
(e) Insert operation
(f) Clear operation.

Solution: (a) Selective Set: The selective set operation sets to 1. The bits in register 1 where there are corresponding 1's in register *B*. It does not affect the positions that have 0's in the *B*. *e.g.*

1010	*A* before
1100	*B* (logic operand)
1110	*A* after

The two leftmost bits of *B* are 1's, so the corresponding bits of *A* are set to 1. The two bits of *A* with corresponding 0's in *B* remain unchanged. Therefore the OR micro-operation can be used to selectively set bits of a register.

(b) Selective Complement: The selective complement operation complements bits in *A* where there are corresponding 1's in *B*. It does not affect bit positions that have 0's in *B*. *e.g.*

1010	*A* before
1100	*B* (logic operand)
0110	*A* after

therefore exclusive — OR operation can be used to selectively complement bits of register.

(c) Selective Clear: Selective clear operation clears to 0. The bits in *A* only where there are corresponding 1's in *B*. *e.g.*

1010	*A* before
1100	*B* (logic operand)
0010	*A* after

The Boolean operation performed on the individual bits is *AB'*. The corresponding logic operation is

$$A \leftarrow A \cap \overline{B}$$

(d) Mask Operation: The mask operation is similar to the selective clear operation except that the bits of *A* are cleared only above. There are corresponding 0's in *B*. The mask operations is an *AND micro-operation* as seen from the following numerical example :

1010	*A* before
1100	*B* (mask)
0110	*A* (after masking)

(e) Insert Operation: The insert operation inserts a new value into a group of bits this is done by first masking the bits and *then ORing* them the required value. *e.g.*

0110	1010	*A* before
0000	1111	*B* (mask)
0000	1010	*A* after masking

0000	1010	*A* before
1001	0000	*B* (insert)
1001	1010	*A* after insertion

(f) Clear operation: Clear operation compares the words in *A* and *B* and produces an all 0's result if the two numbers are equal. This operation is achieved by an *exclusive-OR* micro-operation. *e.g.*,

1010	A
1010	B
0000	$A \leftarrow A \oplus B$

Q.23. Show the block diagram of the hardware that implements the following register transfer statement.

$$yT_2\text{: } R_2 \leftarrow R_1,\ R_1 \leftarrow R_2$$

Solution:

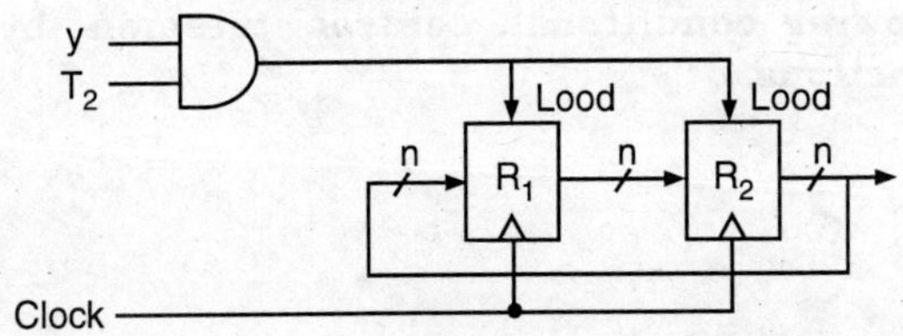

Fig. 3.26.

Q.24. The output of four registers R_0, R_1, R_2 and R_3 are connected through 4-to-1-line multiplexers to the inputs of a fifth register, R_5. Each register is eight bits long. The required transfers are dictated by four timing variables T_0 through T_3 as follows :

$$T_0\text{: } R_5 \leftarrow R_0$$
$$T_1\text{: } R_5 \leftarrow R_1$$
$$T_2\text{: } R_5 \leftarrow R_2$$
$$T_3\text{: } R_5 \leftarrow R_3$$

The timing variables are mutually exclusive, which means that only one variable is equal to 1 at any given time, while the other three are equal to 0. Draw a block diagram showing the hardware implementation of the register transfers. Include the connections necessary from the four timing variables to the selection inputs of the multiplexers and to load input of register R_5.

Solution:

T_0	T_1	T_2	T_3	S_1	S_0	R_3 load
0	0	0	0	×	×	0
1	0	0	0	0	0	1
0	1	0	0	0	1	1
0	0	1	0	1	0	1
0	0	0	1	1	1	1

$$S_1 = T_2 + T_3$$
$$S_0 = T_1 + T_3$$
$$\text{load} = T_0 + T_1 + T_2 + T_3$$

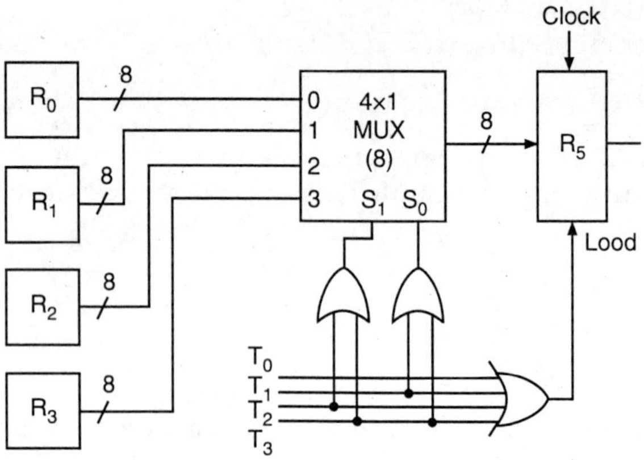

Fig. 3.27.

Q.25. Represent the following conditional, control statement by two register transfer statement with control functions.

If (P = 1) then

($R_1 \leftarrow R_2$)

else if (Q = 1) then

($R_1 \leftarrow R_3$)

Solution:

$$P\text{: } R_1 \leftarrow R_2$$
$$P'Q\text{: } R_1 \leftarrow R_3$$

Q.26. What has to be done to the bus system of figure given blow to be able to transfer information from any register to any other register? Specifically, show that connections that must be included to provide a path from the outputs of register C to the inputs of register A.

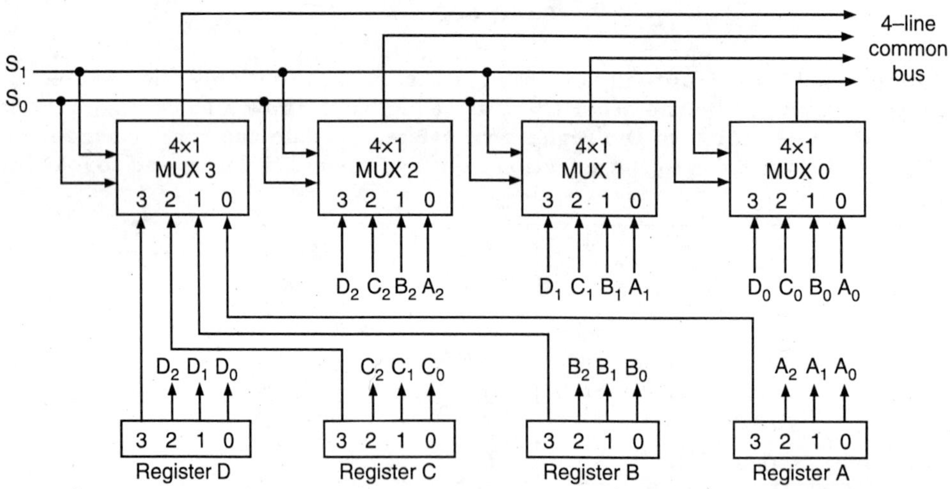

Fig. 3.28. Bus system for four registers

Solution: Connect the 4-line common bus to the four inputs of each register. Provide a "load" control input in each register. Provide a clock input for each register.

To transfer from register C to register A:

Apply $S_1 S_0 = 10$ (to select C for the bus)
Enable the load input of A and apply a clock pulse.

Q.27. Draw a diagram of a bus system similar to the one shown in the figure below but use three-state buffers and a decoder instead of the multiplexers.

Solution:

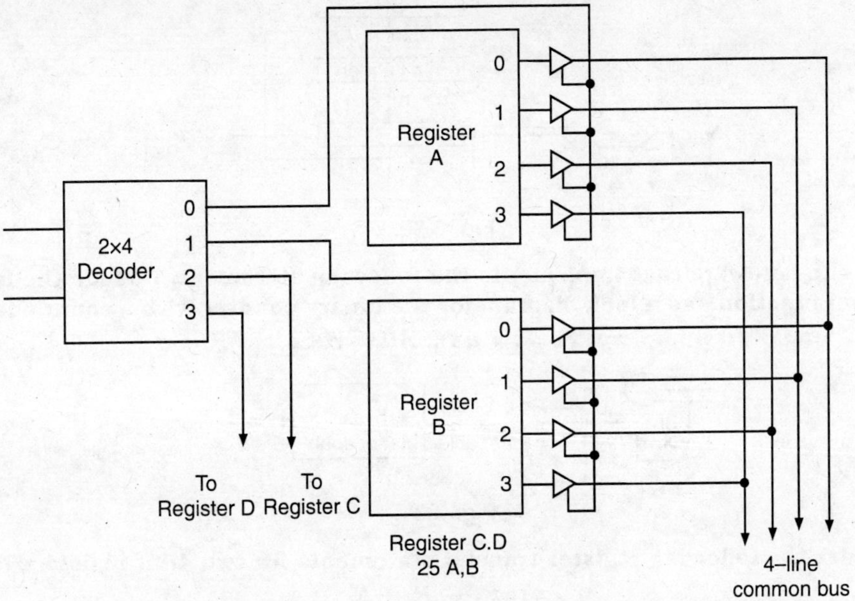

Fig. 3.29.

Q.28. A digital computer has a common bus system for 16 registers of 32 bits each. The bus is constructed with multiplexers.

(a) How many selection inputs are there in each multiplexer ?

(b) What size of multiplexers are needed ?

(c) How many multiplexers are there in the bus ?

Solution:

(a) 4 selection lines to select one of 16 registers.

(b) 16 × 1 multiplexers are needed.

(c) 32 multiplexers, one for each bit of the registers.

Q.29. The following transfer statements specify a memory. Explain the memory operation in each case.

(a) $R_2 \leftarrow$ M[AR]

(b) M[AR] $\leftarrow R_3$

(c) $R_5 \leftarrow M[R_5]$

Solution:

(a) Read memory word specified by the address in AR into register R_2.

(b) Write content of register R_3 into the memory word specified by the address in AR.

(c) Read memory word specified by the address in R_3 and Transfer to R_3 (destroys previous value).

Q.30. Draw the block diagram for the hardware that implements the following statements:

$$x + yz: AR \leftarrow AR + BR$$

where *Ar* and *BR* are two *n*-bit registers and *x*, *y* and *z* are control variables. Include the logic gates for the control function. (Remainder that the symbol + designates an OR operation in a control or Boolean function but that it represents an arithmetic plus in a micro-operation.)

Solution:

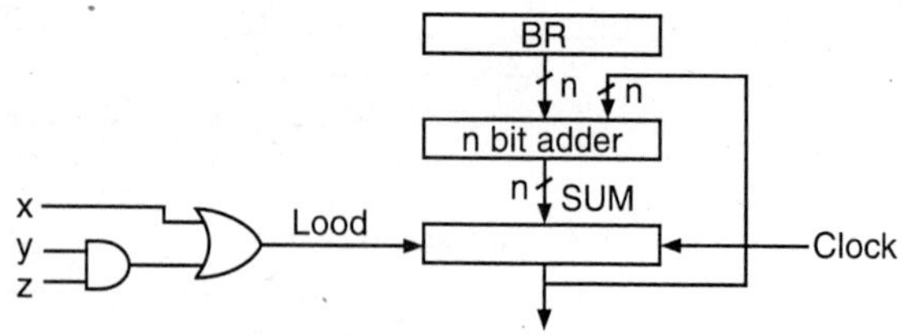

Fig. 3.30.

Q.31. Show the hardware that implements the following statement. Include the logic gates for the control function and a lock diagram for the binary counter with a count enable input

$$xyT_0 + T_1 + y'T_2: AR \leftarrow AR + 1$$

Solution:

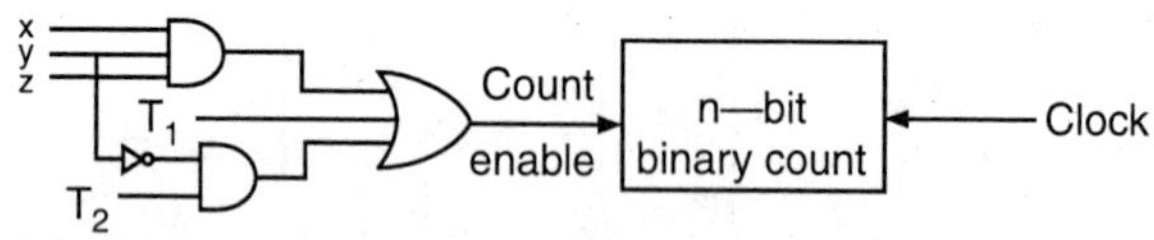

Fig. 3.31.

Q.32. Consider the following register transfer statements for two 4-bit registers R_1 and R_2.

$$xT: R_1 \leftarrow R_1 + R_2$$
$$xT: R_1 \leftarrow R_2$$

Every time that variable $T = 1$, either the content of R_2 is added to the content of R_1 if $x = 1$, or the content of R_2 is transferred to R_1 if $x = 0$. Draw a diagram showing the hardware implementation of the two statements. Use block diagrams for the two 4-bit registers, 4-bit adder, and a quadruple 2-to-1-line multiplexer that selects the inputs to R_1. In the diagram, show how the control variables and T select the inputs of the multiplexers and the load input of register R_1.

Solution:

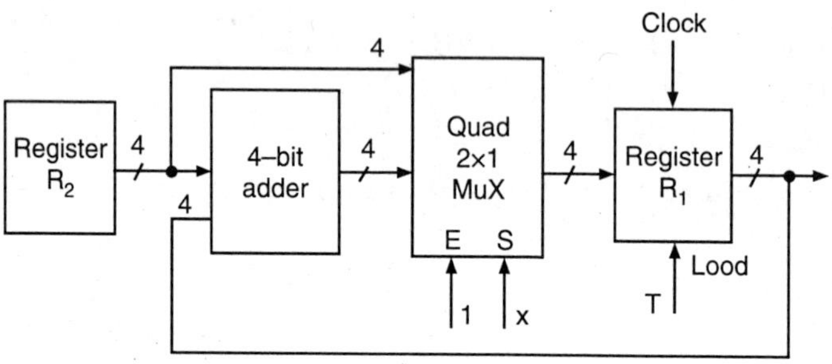

Fig. 3.32.

Q.33. Using a 4-bit counter with parallel load and a 4-bit adder as in figure given below, draw a block diagram that shows how to implement the following statements :

$$x: R_1 \leftarrow R_1 + R_2 \quad \text{Add } R_2 \text{ to } R_1$$
$$x'y: R_1 \leftarrow R_1 + R_1 \quad \text{Increment } R_1$$

where R_1 is a counter with parallel load and R_2 is a 4-bit register.

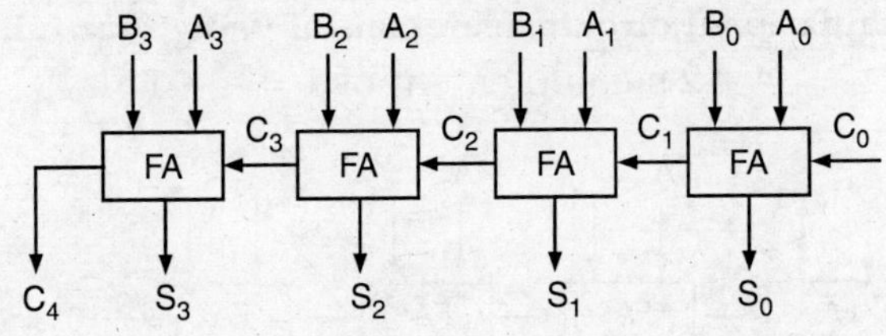

Fig. 3.33. 4-bit binary adder

Solution:

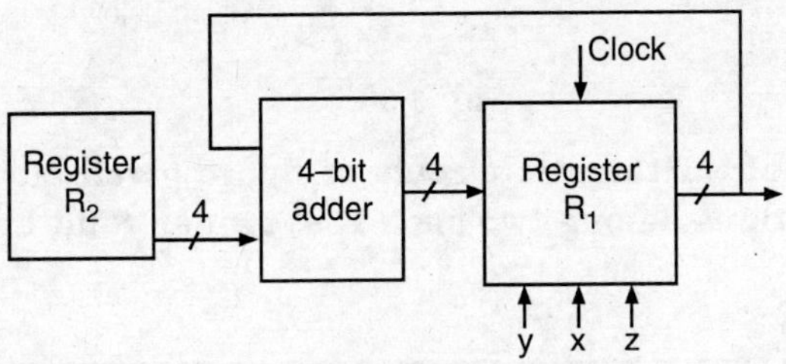

Fig. 3.34.

Q.34. The adder-subtracter circuit of figure given below has the following values for input mode *M* and data inputs *A* and *B*. In each case, determine the values of the outputs: S_3, S_2, S_1, S_0 and C_4.

	M	*A*	*B*
(*a*)	0	0111	0110
(*b*)	0	1000	1001
(*c*)	1	1100	1000
(*d*)	1	0101	1010
(*e*)	1	0000	0001

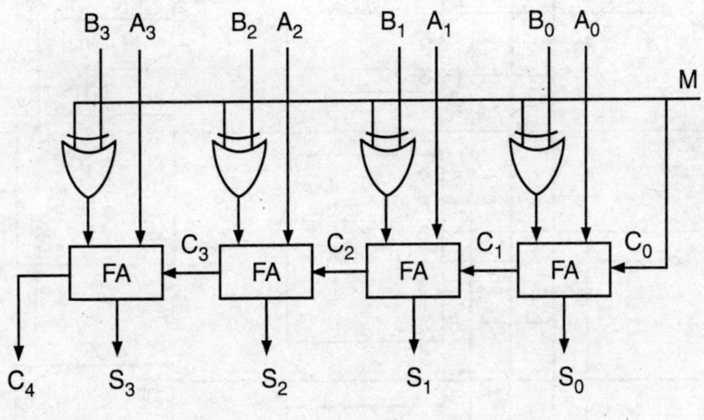

Fig. 3.35. 4-bit adder-subtracter

Solution:

M	*A* *B*	*Sum*	C_4	
0	0111 + 0110	1101	0	7 + 6 = 13
0	1000 + 1001	0001	1	8 + 9 = 16 + 1
1	1100 – 1000	0100	1	12 – 8 = 4
1	0101 – 1010	1011	0	5 – 10 = 5 (in 2's complement)
1	0000 – 0001	1111	0	0 – 1 = – 1 (in 2's complement)

Q.35. Design a 4-bit combinational circuit decremental using four full adder circuits.

Solution: $A - 1 = A + 2\text{'s complement of } 1 = A + 1111$

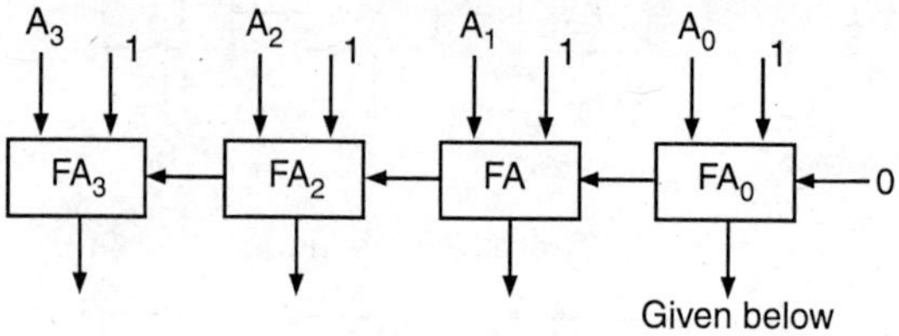

Fig. 3.36.

Q.36. Assume that the 4-bit arithmetic circuit of figure given blow is enclosed in on IC package. Show the connections among two such ICs form an 8-bit arithmetic circuit.

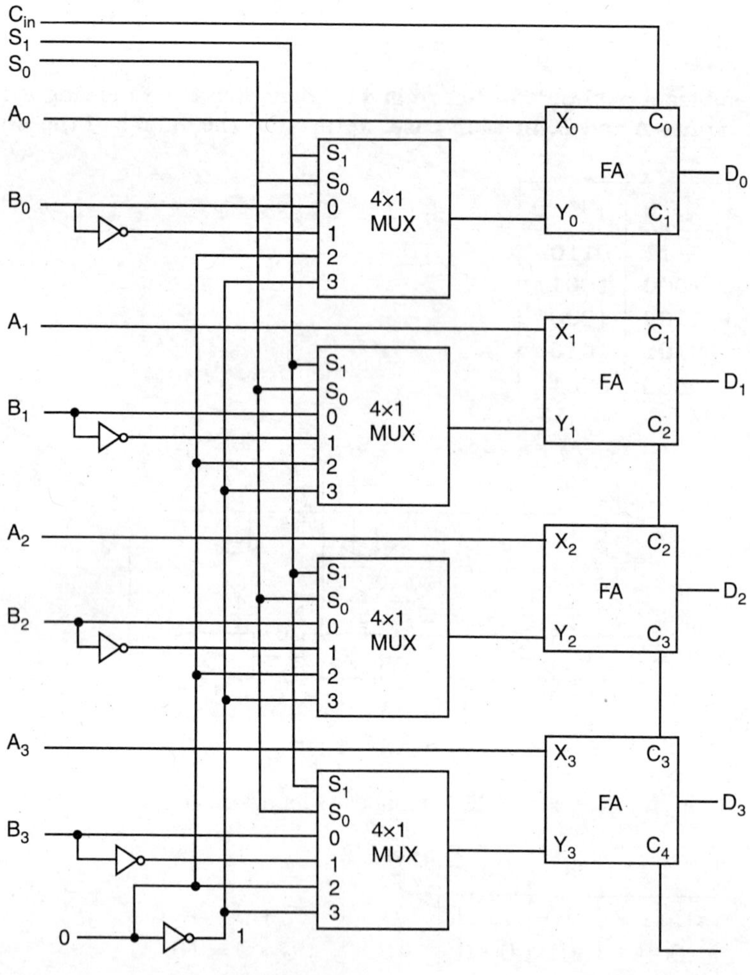

Fig. 3.37. 4-bit arithmetic circuit

Solution:

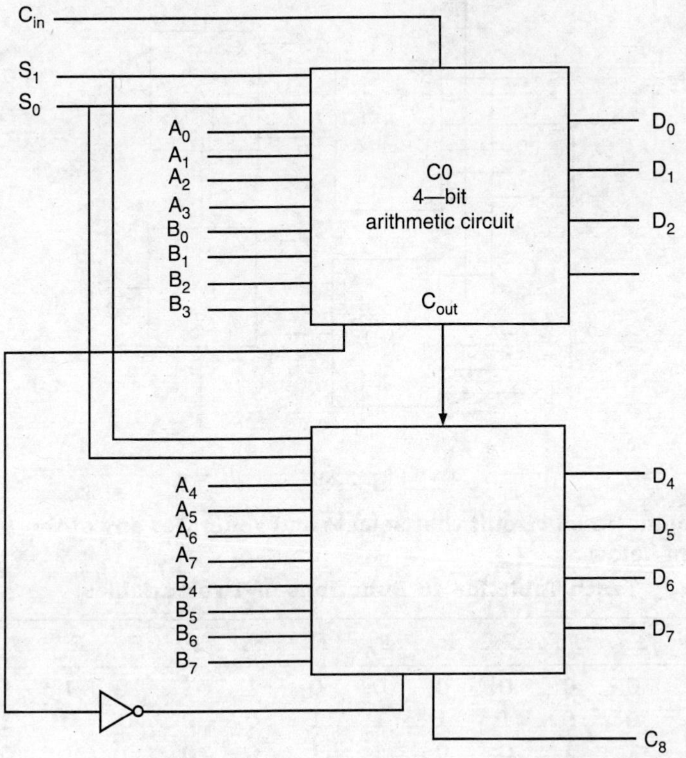

Fig. 3.38.

Q.37. Design an arithmetic circuit with one selection variable *S* and two n-bit data inputs *A* and *B*. The circuit generates the following four arithmetic operations in conjunction with the input carry C_{in}. Draw the logic diagram for the first two stages.

S	$C_{in} = 0$	$C_{in} = 1$
0	$D = A + B$ (add)	$D = A + 1$ (increment)
1	$D = A - 1$ (decrement)	$D = A + \overline{B} + 1$ (subtract)

Solution:

S	C_{in}	X	Y	
0	0	A	B	$(A + B)$
0	1	A	0	$(A + 1)$
1	0	A	1	$(A - 1)$
1	1	A	B	$(A - B)$

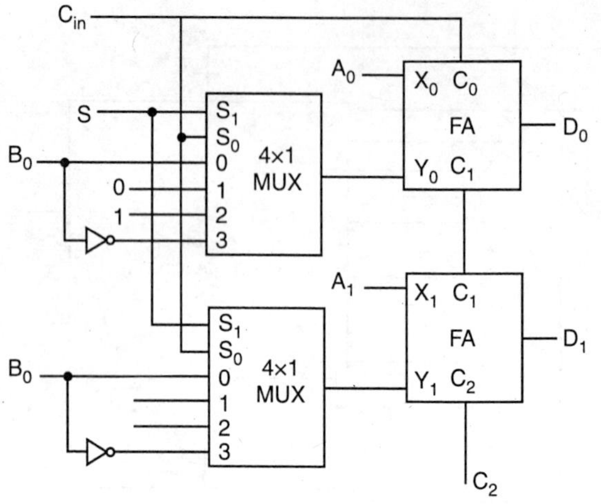

Fig. 3.39.

Q.38. Derive a combinational circuit that selects and generates any of the 16 logic functions listed in Table given below.

Truth Table for 16 Functions of Two Variables.

x	y	F_0	F_1	F_2	F_3	F_4	F_5	F_6	F_7	F_8	F_9	F_{10}	F_{11}	F_{12}	F_{13}	F_{14}	F_{15}
0	0	0	0	0	0	0	0	0	0	1	1	1	1	1	1	1	1
0	1	0	0	0	0	1	1	1	1	0	0	0	0	1	1	1	1
1	0	0	0	1	1	0	0	1	1	0	0	1	1	0	0	1	1
1	1	0	1	0	1	0	1	0	1	0	1	0	1	0	1	0	1

Solution:

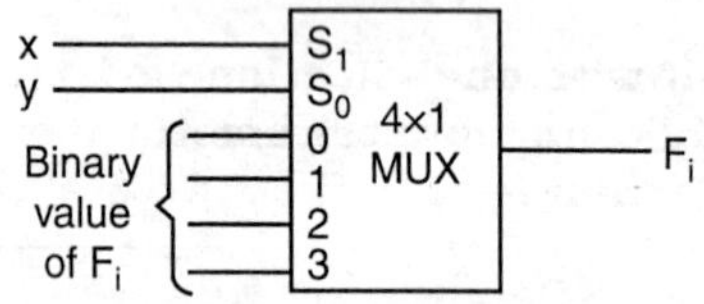

Fig. 3.40.

Q.39. Design a digital circuit that performs the four logic operations of exclusive OR, exclusive NOR, NOR and NAND. Use two selection variables. Show the logic diagram of one typical stage.

Solution:

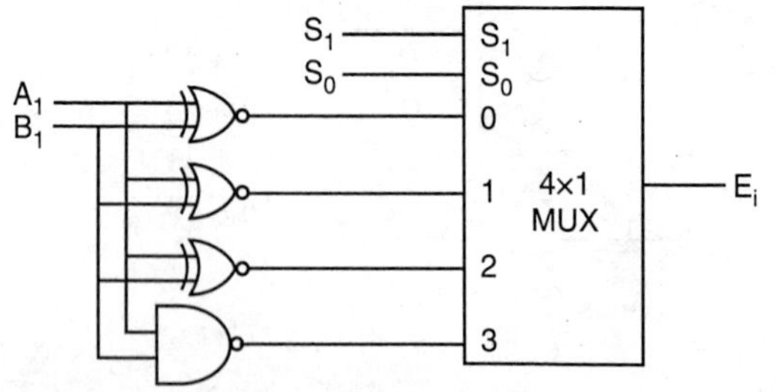

Fig. 3.41.

Q.40. Register *A* holds the 8-bits binary 11011001. Determine the *B* operand and the logic micro-operation to be performed in order to change the value in *A* to:

(a) 01101101

(b) 11111101

Solution:

(a) $A = 1\ 1\ 0\ 1\ 1\ 0\ 0\ 1$

$B = 1\ 0\ 1\ 1\ 0\ 1\ 0\ 0 \quad \oplus$

$A \leftarrow A \oplus B \quad 0\ 1\ 1\ 0\ 1\ 1\ 0\ 1$

(b) $A = 1\ 1\ 0\ 1\ 1\ 0\ 0\ 1$ (OR)

$B = 1\ 1\ 1\ 1\ 1\ 1\ 0\ 1$

$1\ 1\ 1\ 1\ 1\ 1\ 0\ 1 \quad A \leftarrow A \vee B$

Q.41. The 8-bit registers AR, BR, CR and DR initially have the following values:

***AR* = 11110010**

***BR* = 11111111**

***CR* = 10111001**

***DR* = 11101010**

Determine the 8-bit values in each register after the execution of the following sequence of microoperations.

$AR \leftarrow AR + BR$	**Add *BR* to *AR***
$CR \leftarrow CR \wedge DR,\ BR \leftarrow + 1$	**AND *DR* to *CR*, increment *BR***
$AR \leftarrow AR - CR$	**Subtract *CR* from *AR***

Solution:

(a) $AR = 1\ 1\ 1\ 1\ 0\ 0\ 1\ 0$

$BR = 1\ 1\ 1\ 1\ 1\ 1\ 1\ 1$ (+)

$AR = 1\ 1\ 1\ 1\ 0\ 0\ 0\ 1$ $\quad BR = 11111111,\ CR = 10111001,\ DR = 11101010$

(b) $CR = 1\ 0\ 1\ 1\ 1\ 0\ 0\ 1$ $\quad BR = 11111111$

$DR = 1\ 1\ 1\ 0\ 1\ 0\ 1\ 0$ (AND)

$CR = 1\ 0\ 1\ 0\ 1\ 0\ 0\ 0$ $\quad BR = 00000000,\ AR = 11110001,\ DR = 11101010$

(c) $AR = 1\ 1\ 1\ 1\ 0\ 0\ 0\ 1$ (–)

$CR = 1\ 0\ 1\ 0\ 1\ 0\ 0\ 0$

$AR = 0\ 1\ 0\ 0\ 1\ 0\ 0\ 1$ $\quad BR = 00000000,\ CR = 10101000,\ DR = 11101010$

Q.42. An 8-bit register contains the binary value 10011100. What is the register value after an arithmetic shift right ? Starting from the initial number 10011100, determine the register value after an arithmetic shift left, and state whether there is an overflow.

Solution: $R = 10011100$

Arithmetic shift right: 11001110

Arithmetic shift left: 00111000 overflow because a negative number changed to positive.

Q.43. Starting from an initial value of R = 1101101, determine the sequence of binary values in R after a logical shift-left, followed by a circular shift-right, followed by a logical shift-right and a circular shift-left.

Solution: $R = 11011101$

Logical shift left: 10111010

Circular shift right: 01011101

Logical shift right: 00101110

Circular shift left: 01110100

Q.44. What is the value of output *H* in figure given below if input is ..... $I_R = 1$ and $I_L = 0$?

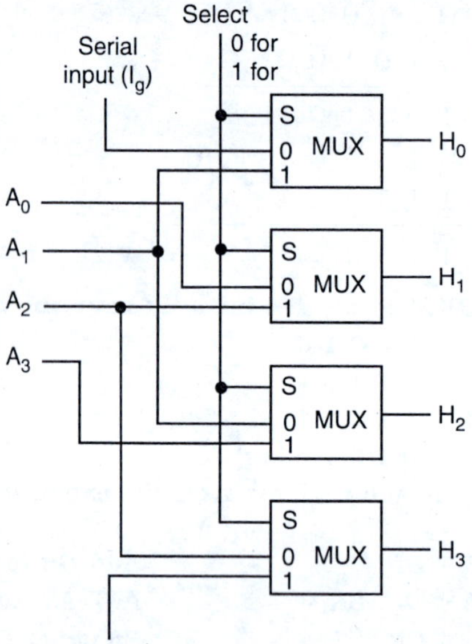

Fig. 3.42. 4-bit combinational circuit shifter

Function Table

Select	***Output***			
S	H_0	H_1	H_2	H_3
0	I_R	A_0	A_1	A_2
1	A_1	A_2.	A_3	I_L

Solution: S = 1 Shift left

	A_0	A_1	A_2	A_3	I_L	
	1	0	0	1	0	
H = 0		0	1	0		Shif left

SUMMARY

1. Computer registers are designated by capital letters.
2. Symbols are a handy tool to represent the micro-operation sequences in registers and the control functions that initiate them, in a lucid & concise form. Such a symbolic notation is referred to as register transfer language.
3. A bus structure consists of a set of common lines, one for each bit of a register, through which binary information is transferred one at a time. Control signals determine which register is selected by the bus during each particular register transfer.
4. A micro-operation is an elementary operation performed with the data stored in registers.
5. These micro-operations perform some basic arithmetic operations on the numeric data stored in the registers.
6. A read operation implies transfer of information to the outside environment from a memory

word, whereas storage of information into the memory is defined as write operation. Symbolizing a memory word by the letter M, it is selected by the memory address during the transfer which is a specification for transfer operations. The address is specified by enclosing it in square brackets following the letter M. For example, the read operation for the transfer of a memory unit M from an address register AR to another data register DR can be illustrated as:

Read: DR M[AR]

7. A write operation implies transfer of information from the outside world to a memory word. For example, Write: M[AR]<-DR.
8. A micro-operation is an elementary operation which is performed on the data stored in registers. We can classify the micro-operations into four categories:
9. Register transfer: transfer binary information from one register to another.
10. Arithmetic: perform arithmetic operations on numeric data stored in registers.
11. Logic: perform bit manipulation operation on non-numeric data stored in registers.
12. Shift: perform shift operations on data stored in registers.

EXERCISE

(I) Fill in the Blanks (Solved)

1. .... is shift left micro-operation.
2. To address 35 memory words at least .... number of address bits are required.

(II) True or False (Solved)

1. Memory read micro-operation takes 10 clock cycles.
2. A microprogram is a set of micro-operations.

(III) Fill in the Blanks (Unsolved)

1. The channel through which CPU and memory accomplish using ....
2. Mirco-operation for adding content of memory location 5 to the content of register R1 is ....

(IV) True or False (Unsolved)

1. Accumulator is a decoder.
2. Control unit is a part of ALU.

(V) Detailed Questions

1. Starting from an initial value of R=11011101, determine the sequence of binary values in R after a logical shift-left, followed by a circular shift-right followed by a logical shift right and a circular shift left.
2. What is wrong with the following register transfer statements?

   (*i*) $xT: AR \leftarrow AR, AR \leftarrow 0$ (*ii*) $yT: R1 \leftarrow R2, R1 \leftarrow R3$

   (*iii*) $zT: PC \leftarrow AR, PC \leftarrow PC + 1$.

Answers

(I) Fill in the Blanks (Solved)

1. Shl 2. 6

(II) True or False (Solved)

1. False 2. True

Chapter

4

Central Processing Unit (Computer Arithmetic)

Chapter Outline

- Introduction
- Arithmetic and Logic Unit operations
- Arithmetic Algorithms
- IEEE (Institute Electronics and Electrical Engineering) Standard for Binary
- Some Applications of Logic Micro-Operations
- Faster Algorithm and their Implementation
- 4-bit Carry Look Ahead Adder
- Addition and Subtraction Algorithm
- Multiplication Algorithm
- Floating Point Arithmetic
- Addtion and Subtraction of Floating Point Numbers

4.0. INTRODUCTION

The most common and efficient use of computers is in the case of computation work. Inspite of it data manipulation is required in almost every type of computer work. The data manipulation is mainly done with the help of arithmetic operations. The four basic arithmetic operations are addition, subtraction, division and multiplication. If required we can derive other results from these basic operations.

Arithmetic processing unit in central processing unit is used to execute all these operations. The arithmetic instructions are performed generally on binary or decimal data. Fixed-point numbers are used to represent integers or fractions. We can have signed or unsigned negative numbers. Fixed-point addition is the simplest arithmetic operation.

A processor has an arithmetic processor (as a sub part of it) that executes arithmetic operations. The data type, assumed to reside in processor, registers during the execution of an arithmetic instruction. Negative numbers may be in a signed magnitude or signed complement representation. There are three ways of representing negative fixed point – binary numbers signed magnitude, signed 1's complement or signed 2's complement. Most computers use the signed magnitude representation for the mantissa.

4.1 ARITHMETIC AND LOGIC UNIT OPERATIONS

An arithmetic logic unit (ALU) is a multi-operation combinational logic digital function. It can perform a set of basic arithmetic operations and set of logic operations. The ALU has a number of selection lines to select a particular operation in the unit. The selection lines are decoded within the ALU so that *K* selection variables can specify up to 2*K* distinct operations.

Figure 4.1 shows the block diagram of 4-bit ALU. The four data inputs from *A* combined with the four inputs from *B* to generate an operation at the *F* outputs. The mode select input (S_2) distinguishes between arithmetic and logic operations. The two function select inputs S_1 and S_0 specify the particular arithmetic or logic operation to be generated with three selection variables, it is possible to specify four arithmetic operations ($S_2 = 0$) and four logic operations ($S_2 = 1$). The input and output carries have meaning only during an arithmetic operation.

When $S_1 = 0$ the circuit perform arithmetic ($S = 0$, ADD)
and ($S = 1$, sub) operation.
When $S_1 = 1$ the circuit perform logic OR ($S = 0$) and logic *AND* ($S = 1$) operation.

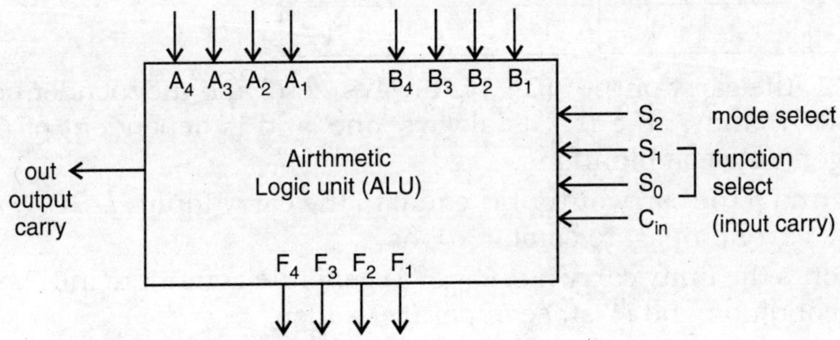

Fig. 4.1.

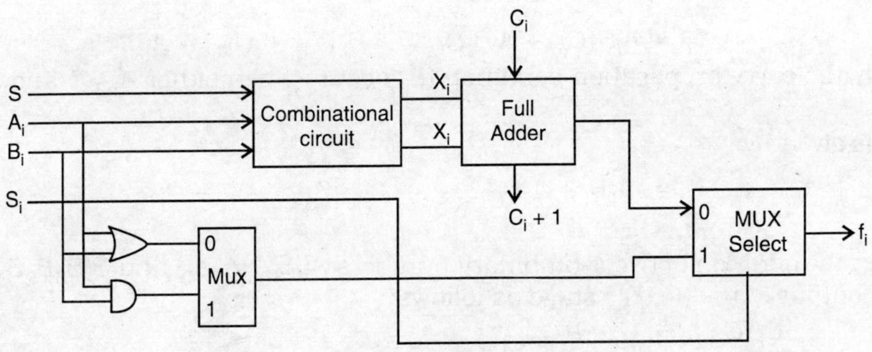

Fig. 4.2.

4.1.1 Faster Algorithms and their Implementation

Fast Adder: In the parallel binary adder, the carry generated by the ith adder is fed as carry input to the $(i + 1)^{th}$ adder. In this adder, the output (C_{out} $S_3S_2S_1S_0$) is available only after the carry is propagated through each of the adders, *i.e.* from *LSB* to *MSB* adder through intermediate adders. Hence the addition process can be considered to be complete only after the above carry propagation delay through adders, which is proportional to the number of stages in it. One of the methods of making this faster is look ahead carry addition, which eliminates the ripple carry delay. This method is based on the carry generating and carry propagating functions of the full adder.

4-bit carry Look Ahead Adder: The carry look ahead adder is based on the principle of looking at the lower order bits of the augend and addend, if high order carry is generated. This adder reduces the carry delay by reducing the number of gates through which a carry signal must propagate. The explain its operation, consider the truth table of full adder.

TABLE 4.1.

Row	A	B	C_{in}	S	C_{out}	
0	0	0	0	0	0	No carry generation
1	0	0	1	1	0	i.e. $C_{out} = 0$
2	0	1	0	1	0	Carry propagation
3	0	1	1	0	1	i.e. $C_{out} = C_{in}$
4	1	0	0	1	0	
5	1	0	1	0	1	
6	1	1	0	0	1	Carry generation
7	1	1	1	1	1	i.e. $C_{out} = 1$

In row 0 and 1, the carry output (C_{out}) is always 'zero' and independent of carry input (C_{in}), while in rows 6 and 7, the C_{out} is always 'one' and is dependent of C_{in}. These are known as carry generate combinations.

In row 2, 3, 4 and 5 the carry output is equal to the carry input, *i.e.* $C_{out} = 1$ only when $C_{in} = 1$. These are carry propagate combinations.

Let G_x represents the unity carry (*i.e.* $C_{out} = 1$) generate condition and P_i represents the carry propagate condition and i^{th} stage of parallel adder.

From truth table, G_i is obtained by summing up the combinations corresponding to 6th and 7th rows as follows:

$$G_i = A_iB_i C_{in} + A_iB_i\overline{C}_{in} = A_iB_i (C_{in} + \overline{C}_{in}) = A_iB_i$$

Similarly the carry propagation condition (P_i) occurs when either $A_i = 1$ and $B_i = 0$ or vice-versa.

Now P_i is given by

$$P_i = A_i\overline{B}_i + \overline{A}_iB_i$$

$$P_i = P_i \oplus B_i$$

Consider the addition of two 4-bit binary numbers $A(A_3A_2A_1A_0)$ and $B(B_3B_2B_1B_0)$. The unity carry output of the $(i-1)^{th}$ stage as follows:

$$C_i\,(C_{out}) = G_i + P_iC_{i-1}$$

where C_{i-1} for LSB stage is C_{in} which is assumed to be zero. In a 4-bit binary adder. Four stages of addition are required to add A_0B_0, A_1B_1, A_2B_2 and A_3B_3. Therefore, for $i = 0, 1, 2$ and 3 the C_i's are given by

$$C_0 = G_0 + P_0\, C_{in}$$

where $G_0 = A_0B_0$, $P_0 = A_0 \oplus B_0$ and $C_{in} = 0$

$$C_1 = G_1 + P_1C_0 = G_1 + P_1(G_0 + P_0C_0)$$

where $G_1 = A_1B_1$ and $P_1 = A_1 \oplus B_1 = G_1 + P_1G_0 + P_0P_1C_{in}$

$$C_2 = G_2 + P_2C_1 = G_2 + P_2(G_1 + P_1G_0 + P_0P_1\, C_{in})$$

$$= G_2 + P_2G_1 + P_1P_2G_0 + P_0P_1P_2\, C_{in}$$

where $G_2 = A_2B_2$ and $P_2 = A_2 \oplus B_2$

$$C_3 = G_3 + P_3C_2 = G_3 + P_3(G_2 + P_2G_1 + P_2P_1G_0 + P_0P_1P_2C_{in})$$
$$= G_3 + P_3G_2 + P_2P_3G_1 + P_1P_2P_3G_0 + P_0P_1P_2P_3\ C_{in}$$

where $G_3 = A_3B_3$ and $P_3 = A_3 \oplus B_3$

Sum of A and B is given by

$$S = C_3S_3S_2S_1S_0$$

where $S_i = A_i \oplus B_i \oplus C_{i-1}$ for $i = 0, 1, 2, 3$

i.e.
$$S_0 = A_0 \oplus B_0 \oplus C_{in}$$
$$S_1 = A_1 \oplus B_1 \oplus C_0$$
$$S_2 = A_2 \oplus B_2 \oplus C_1$$
$$S_3 = A_3 \oplus B_3 \oplus C_2$$

Using above equations a 4-bit carry look ahead adder can be realized as following figure. From the diagra, one can easily understand that the addition of two 4-bit numbers can be done by carry look ahead adder in a four gate propagation time. It is important to nose here that the addition of n-bit binary numbers takes the same four gate propagation delay.

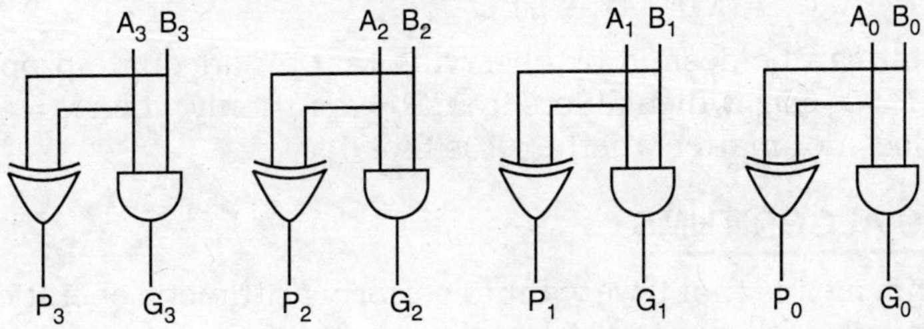

Fig. 4.3.

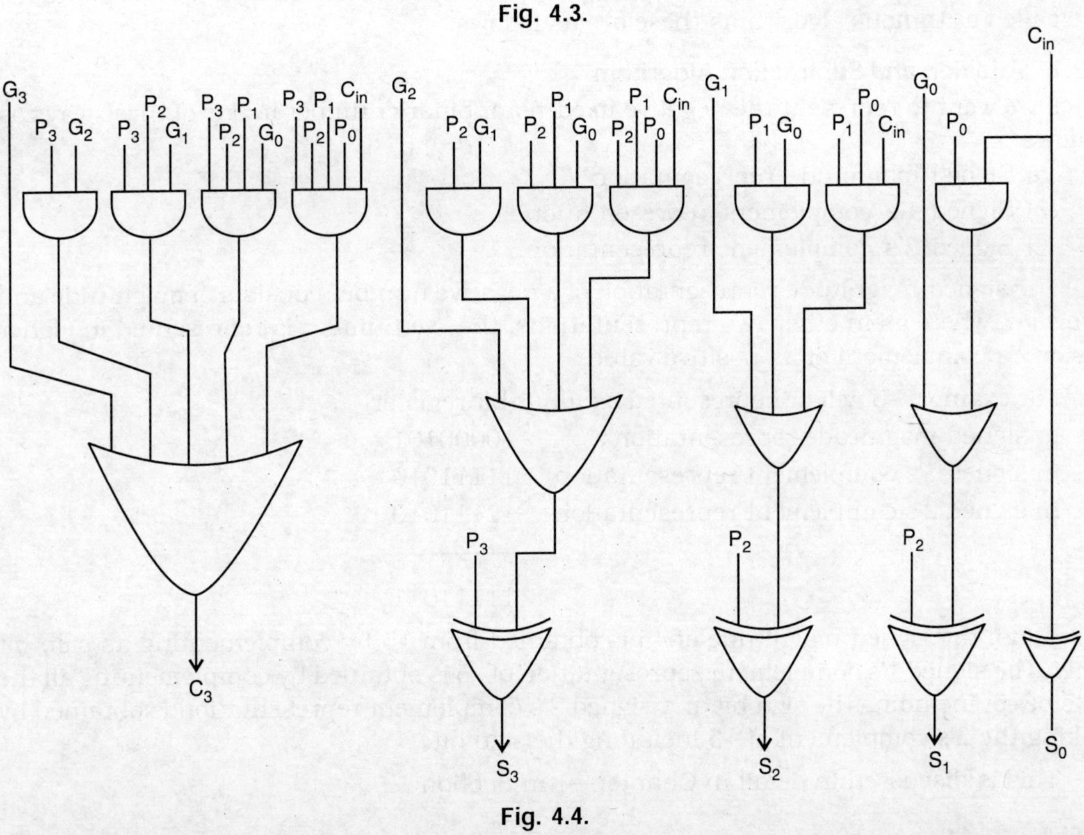

Fig. 4.4.

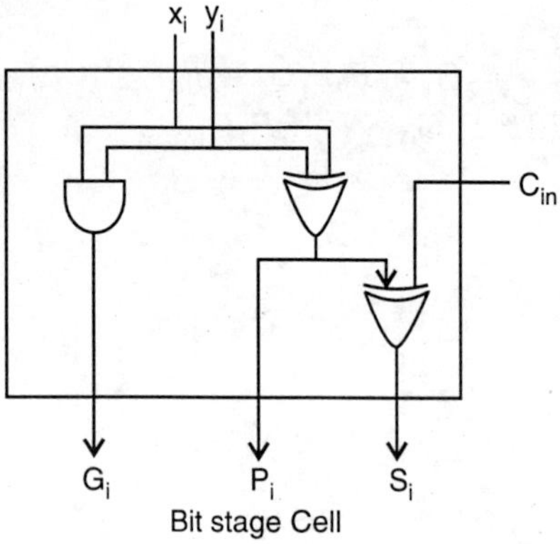

Fig. 4.5.

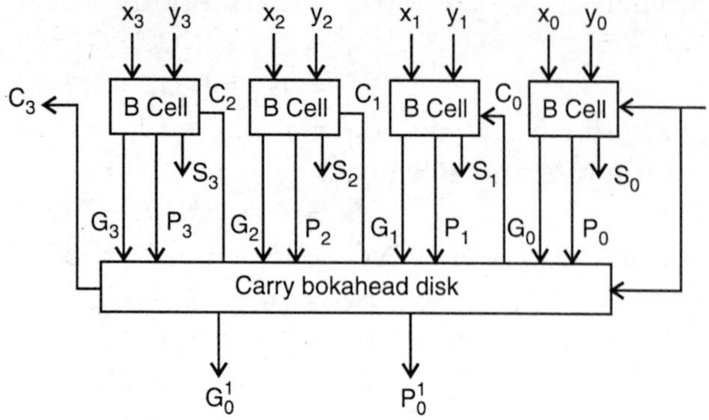

Fig. 4.6.

$$S_i = x_i \oplus y_i \oplus C_{i-1}$$
$$C_i = x_i y_i + y_i C_{i-1} + C_{i-1} x_i$$
$$C_i = x_i y_i + (x_i + y_i)\, C_{i-1}$$

A simpler circuit can be desired by observing that an adequate propagate. Function can be realized as $P_i = x_i \oplus y_i$, which differs from $P_i = x_i + y_i$ only when $x_i = y_i = 1$. But in this case $G_i = 1$, so it does not matter whether P_i is 0 or 1.

4.2 ARITHMETIC ALGORITHMS

Arithmetic algorithm means that if we want to perform arithmetic operation on any binary operation, then we must follow some rules, these rules to perform arithmetic operations are called arithmetic algorithms these are as follows:

4.2.1 Addition and Subtraction Algorithm

If we want to represent the negative fixed-point binary number in one of three ways as follows:

(*a*) Signed magnitude representation
(*b*) Signed 1's complement representation
(*c*) Signed 2's complement representation.

The signed-magnitude representation of a negative number consists of magnitude and –ve sign, where as in other two representations, the –ve number is represented in eigher 1's or 2's complement of its positive value.

For example –5 will be represented as: (in 8-bit register)

In signed-magnitude representation	10000101
In signed 1's complement representation	11111010
In signed 2's complement representation	11111010
	+ 1
	11111011

Thus, the signed magnitude of –5 is obtained from +5 by complementing its sign bit only. The signed 1's complement representation of –5 is obtained by complementing all the bits of +5, including the sign bit and signed 2's complement representation is obtained by taking the 2's complement of +5 including the sign bit.

This is discussed in detail in Chapter—1 in Section 1.5.

4.2.1.1 Addition and Subtraction with Signed-magnitude Data

It we assume that two numbers have their magnitude A and B. If the signed numbers are to be added or subtracted, in there are eight different conditions depending on the operation to be performed and the sign of magnitude of numbers, according to following table 4.2.:

TABLE 4.2. Addition and Subtraction of Signed Magnitude Numbers

Operation	Add magnitude	Subtract Magnitude		
		When A > B	When A = B	When A < B
$(+A) + (+B)$	$+(A + B)$			
$(+A) + (-B)$		$+(A - B)$	$+(A - B)$	$-(B - A)$
$(-A) + (+B)$		$-(A - B)$	$+(A - B)$	$+(B - A)$
$(-A) + (-B)$	$-(A + B)$			
$(+A) - (+B)$		$+(A - B)$	$+(A - B)$	$-(B - A)$
$(+A) - (-B)$	$+(A + B)$			
$(-A) - (+B)$	$-(A + B)$			
$(-A) - (-B)$		$-(A - B)$	$+(A - B)$	$+(B - A)$

In above table there are two rules or algorithms for addition and subtraction algorithms as follows:

4.2.1.2 Addition Algorithm

While the sign of A and B are identical, add the two magnitudes and attach the sign of A to the result. When the signs of A and B are different; compare the magnitudes and subtract the smaller magnitude from the larger. Choose the sign of the result to be same as of A if $A > B$ or the complement of sign A if $A < B$. If two magnitudes are same, subtract B from A and assign the sign of result +ve.

4.2.1.3 Subtraction Algorithm

When the sign of A and B are different, add the two magnitudes and attach the sign of A to the result. When the sign of A and B are identical, compare the magnitudes and subtract the smaller from the larger. Choose the sign of the result to be same as that of A if $A > B$ or complement of the sign of A if $A < B$. If both magnitudes are equal, subtract B from A and make the sign of the result +ve.

4.2.1.4 Hardware Implementation of Addition and Subtraction Algorithm

For implementation of H/w of addition and subtraction, there must be two registers, for storing two numbers. Let X and Y be two registers that holds the magnitude of two numbers and X_s and Y_s are two flip-flps. For storing the sign of two numbers respectively. The result of the operation may be transferred to the third register or an also be transferred into X and X_s. Thus X and X_s together from an accumulator register.

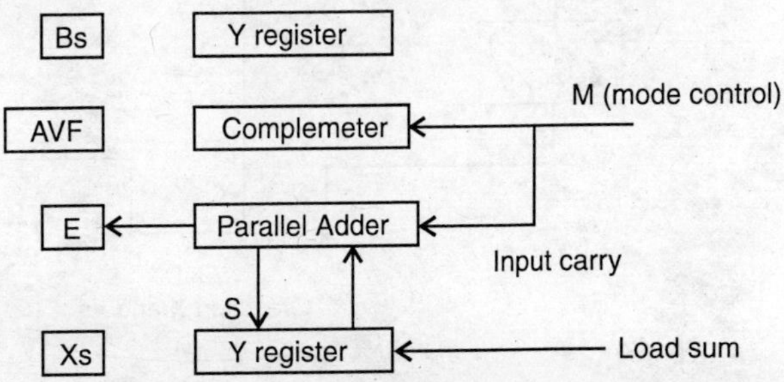

Fig. 4.7. Block diagram for addition and subtraction

The block diagram consists of two registers *X* and *X* alongwith the associated sign flip-flops X_s and Y_s as shown above figure subtraction is to be accomplished by adding *X* with the 2's complement of *Y* and the output carry is transferred into flip-flop *E*. Also there is an add-overflow flip-flop *AVF* that holds the overflow bit *A* and *B* are added.

Parallel adder performs the addition of *X* and *Y* and the sum output *S* of the adder is applied to the input of *X* register. Also there is digital circuit complementer that provides an output of *Y* or the complement of *Y* depending on the state of the mode control *M*. Also *M* signal is applied to the input carry of the adder and will work as:

If $M = 0$, the output of *Y* is the 1's complement of *Y* which is applied to the adder, the input carry is 1 and hence the output of the adder is $X + \overline{Y} + 1$ which is equal to *X* plus 2's complement of *Y*, which is equivalent to $X - X$.

4.2.1.5 Flow Chart for Addition and Subtraction Algorithms

Flow chart is the diagramatic representation of algorithm. This is same as we discussed the working of hardware of addition and subtraction algorithms, which is as follows:

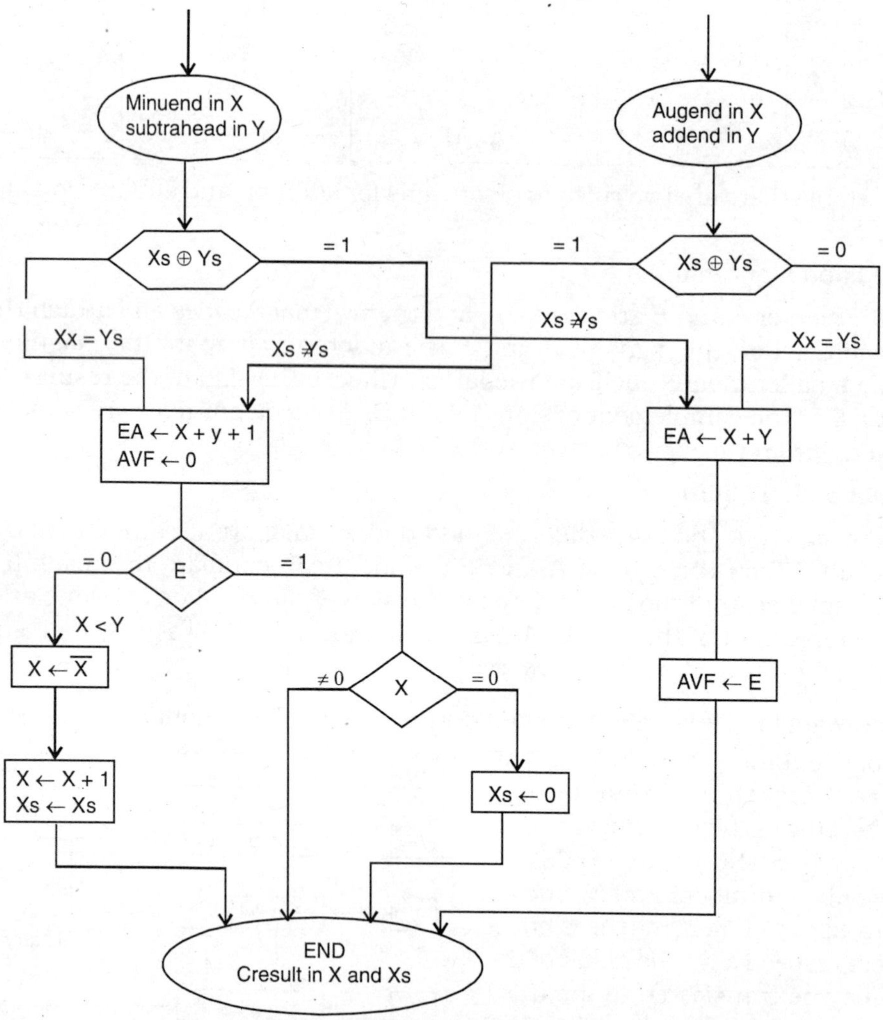

Fig. 4.8. Flow chart for addition and subtraction

4.2.2 Multiplication Algorithm

In general, if we want to multiply decimal numbers 23 by 12. It can be done as:

$$
\begin{array}{rl}
23 & \rightarrow \text{Multiplication} \\
\times\ 12 & \rightarrow \text{(Multiplier)} \\
\hline
46 & \rightarrow \text{(First partial product)} \\
+\ 230 & \rightarrow \text{(Second partial product)} \\
\hline
276 & \rightarrow \text{(Final result)}
\end{array}
$$

To multiply 23 by 12, the result will be 276, that is in decimal.

If we want to multiply the binary numbers then, it can be done as follows:

$$
\begin{array}{rcl}
23 & = & 10111 \\
12 & = & \times\, 01100 \\
\end{array}
$$

$$
\begin{array}{rl}
00000 & \rightarrow \text{First partial product} \\
00000 & \rightarrow \text{2nd partial product} \\
10111 & \rightarrow \text{3rd partial product} \\
10111 & \rightarrow \text{4th partial product} \\
00000 & \rightarrow \text{5th partial product} \\
\hline
100010100 & \rightarrow \text{Final product}
\end{array}
$$

After addition the result is 100010100 or 276 in decimal number system.

Thus in above procedure, if multiplier bit is 1 then the multiplicand is as it is, copied and if the multiplier bit is 0, then there is a series of zeros (equal to number of bits of a multiplicand). Thus every partial product should be shifted left by one position every time and finally, after addition all partial product, we get final result. The sign of result is determined by the sign of multiplicand and multiplier. If they are same, the sign of result is +ve, if not then the sign of result is –ve.

4.2.2.1 Flow Chart for Binary Multiplication

Booth algorithm for +ve numbers.

The above figure gives the flow chart of multiply algorithm. Initially the multiplicand is in register Y and the multiplier is in Q with their respective signs in Y_s and Q_s. The signs are compared and both X and Q corresponds to the sign of the product.

Also registers X and Q are cleared, *i.e.* X and Q are set to be zero, and the sequence counter (C) is set to the number of bits in the multiplier an after initialization, the low order bit of the multiplier in Q_n is tested. If $Q_n = 1$ then the multiplicand in Y is added to the partial product in X and if $Q_n = 0$, nothing is done. Register $E \times Q$ is then shifted once to the right to form new partial product. Sequence counter C is decremented by 1 and its new value is checked. If $C \neq 0$ the final product is in both X and Q, with X holding most significant bits and Q holding the least significant bits.

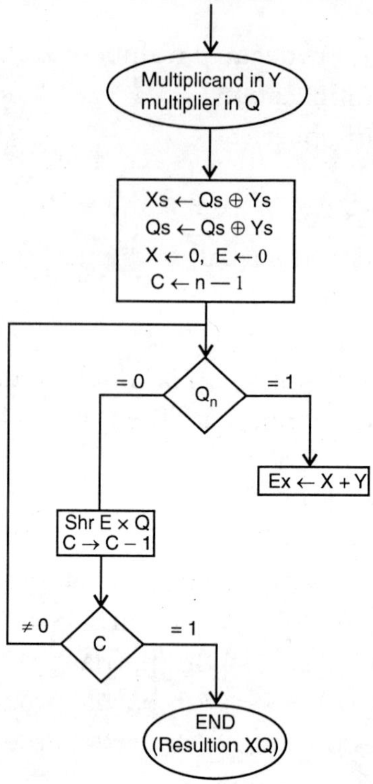

Fig. 4.9. Flow chart for binary multiplication

To clarify the hardware multiplication process, we have given a Table 4.3. as follows:

TABLE 4.3. Numerical Example for Multiplication

Multiplicand Y = 10111	*E*	*X*	*Q*	*C*
Multiplier in *Q*	0	00000	01100	101
Q_n = 0; shift right $E \times Q$	0	00000	0010	100
Q_n = 0; shift right $E \times Q$	0	00000	00011	011
Q_n = 1; add *Y*		10111		
	q			
First partial product:	0	10111		
Shift right $E \times Q$	0	01011	10001	010
Q_n = 1; Add *Y*		10111		
Second partial product:	1	00010		
Shift right $E \times Q$	0	10001	01000	001
Q_n = 0; shift right $E \times Q$	0	01000	10100	000
Final product is in *XQ* = 0100010100				

4.2.2.2 Booth Multiplication for –ve Numbers

In previous section, we have seen that the multiplication of positive multiplier and positive multiplicand. But, in that situation, probem arises when we perform with negative multiplier or multiplicand or both are negative.

If there is a situation, where both multiplier and multiplicand are negative, then we must follow the following flow chart and procedure to perform multiplication:

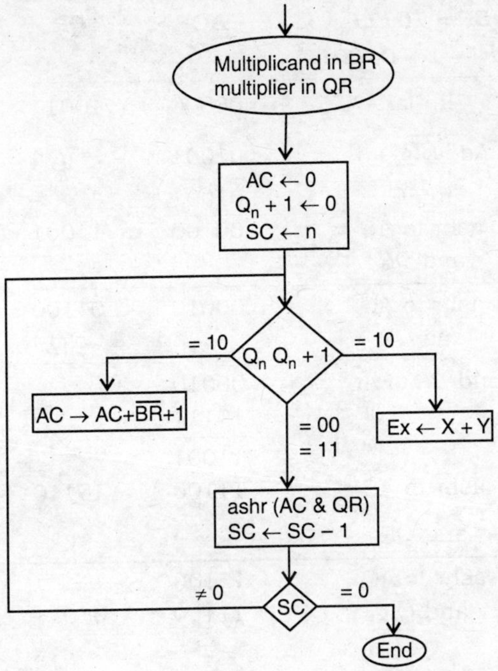

Fig. 4.10. Both multiplication for signed 2's complement numbers

4.2.2.3 Working of Flow Chart

The flow chart of Booth multiplication of signed 2's complement numbers is shown in above figure. In figure, the multiplicand is stored in *BR*, and multiplier is stored in *QR*. Initially, we should store the valve of *AC* as '0', $Q_n + 1$ as '0' and the value of *SC* is equivalent to the number of bits of multiplier (*i.e.* in *QR*) as '*n*'.

Now, in this procedure we would compare the value of Q_n and Q_{n+1}:

1. If $Q_n Q_{n+1}$ = 10, then we add the 2's complement of *BR* (*i.e.* $\overline{BR}$ + 1) and apply arithmetic shift right operation to (*AC* and *QR*) and decrement the sequence counter *SC* as $SC \leftarrow SC - 1$.
2. If $Q_n Q_{n+1}$ = 01, then we add *BR* to the *AC* and store it into *AC* and apply arithmetic shift right operation to *AC* and *QR* and decrement the counter as $SC \leftarrow SC - 1$.
3. If $Q_n Q_{n+1}$ = 00 or 11, we would not add any thing to the *AC* and we should only apply arithmetic right shift operation on *AC* and *QR*, for every value of $Q_n Q_{n+1}$ (*i.e.* 00 or 11) and would decrement the counter $SC \leftarrow SC - 1$.

This procedure will continue until the *SC* becomes zero and final answer is stored in *AC* and *QR AC* is the most significant bits and *QR* is the least significant bits.

Example:

TABLE 4.4.

Step	Q_n Q_{n+1}	$BR = 10111$ $\overline{BR} + 1 = 01001$	AC	QR	Q_{n+1}	SC
1.	1 0	Initial	00000	10011 ↓	0	101
		Add $\overline{BR}$ + 1	01001	(Q_n)		
		to *AC*	01001			
		ashr to *AC*	00100	11001	1	100
		and *QR*				
2.	1 1	ashr to *AC*	00010	01100	1	011
		and *QR*		(Q_n)		
3.	0 1	add *BR* to *AC*	00010			
			10111			
			11001			
		ashr to *AC*	11100	10110 ↓	0	010
		and *QR*		(Q_n)		
4.	0 0	ashr to *AC*	11100			
		and *QR*	11110	01011 ↓	0	001
				(Q_n)		
5.	1 0	add $\overline{BR}$ + 1	11110			
		to *AC*	01001			
			00111			
		ashr to *AC*	00011	10101	1	000
		AC and *QR*				

Note: In arithmetic shift if we want to right shift then zero will insert at *MSB* position and shift one position to right every digit. And if we want to shift left then zero will insert at *LSB* position and shift every digit one position to left. In both situations, the sign bit remains unchanged. As in step-3 and step-4 of above example.

4.3 IEEE (INSTITUTE ELECTRONICS AND ELECTRICAL ENGINEERING) STANDARD FOR BINARY

The most important floating-point representation is defined in IEEE standard 754 [IEEE 85]. This standard was developed to facilitate the portability of programs from one processor to another and to encourage the development of sophisticated numerically oriented programs.

The standard has been widely adopted and is used on virtually all processors and arithmetic coprocessors.

The IEEE standard defines both a 32-bit single and a 64-bit double format in following figure:

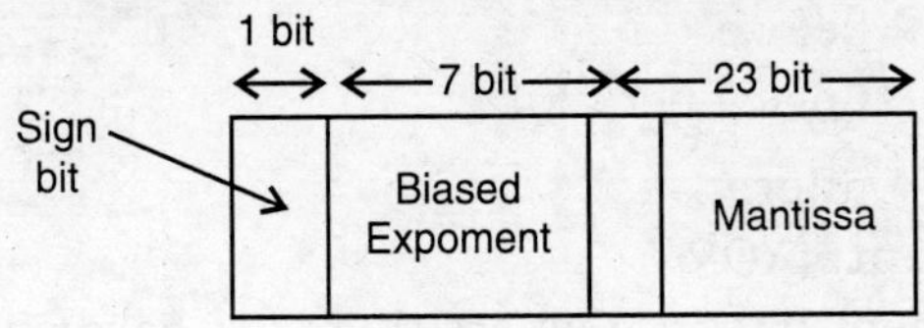

Fig 4.11. (a)

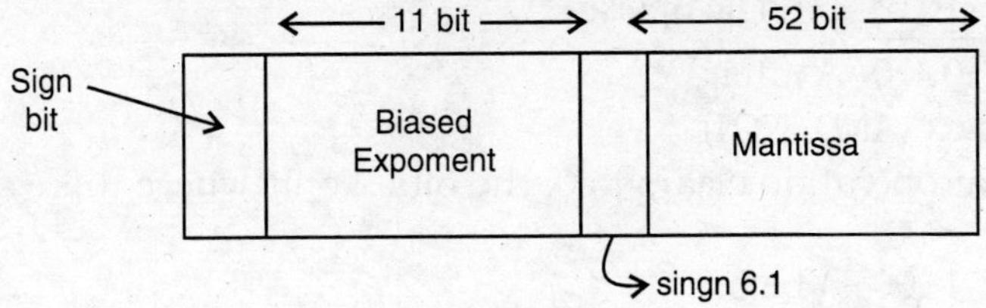

Fig. 4.11. (b)

With 8-bit and 11-bit exponents, respectively. The base is 2. In addition, the standard defines two extended formats single and double, whose exact format is implementation dependent. The extended formats include additional bits in the exponent (extended range) and in the significant (extended precision). The extended formats are to be used for middle calculations with their greater precision.

Not all bit patterns in te *IEEE* formats are interpreted in the usual way instead some bit patterns are used to represent special bits.

The detailed format of *IEEE* floating point is discussed in Chapter—1, in Section 1.10.

4.3.1 Floating Point Arithmetic

As we have studied in Chapter—1, the format of floating point numbers. In this section, we will study, how arithmetic operations are performed on floating point numbers.

4.3.1.1 Addition and Subtraction of Floating Point Numbers

During addition or subtraction, the two floating-point operands are kept in *AC* and *BR*. The sum or difference is formed in the *AC*. The algorithm can be divided into four consecutive parts:

1. Check for zeros.
2. Align the mantissas.
3. Add or subtract mantissa
4. Normalize the result.

A floating point number cannot be normalized, if is zero. Normalization means that decimal point is placed at most significant number before non-zero value *i.e.* $.015 \times 10^5$ is not normalized, if we normalize it in it will be $.15 \times 10^3$ in normalized form.

The alignment of mantissa can be carried out before to the operation. Alignment means, if we want to add two numbers as 1.5×10^5 and 16.5×10^3 then we will have to align mantissas as shifting the decimal point to decrease the power of first number or shift the decimal.

4.4 SOME APPLICATIONS OF LOGIC MICRO-OPERATIONS

1. Selective Set: (*OR*)

This operation sets to 1, the its in Register *A*, where there are corresponding 1's in Register *B*.

e.g.

1010	*A* before
1101	*B* (logic operand)
1111	*A* after

2. Selective Complement: (*XOR*)

This operation complement bits in *A* where there are corresponding 1's in *B*.

e.g.

1100	*A* before
1010	*B* (logic operand)
0110	*A* after

3. Selective Clear: (*AND, NOT*)

The selective clear operation clears to 0, the bits *A* only where there are corresponding 1's is *B*

e.g.

1100	*A* before
1010	*B* (logic operand)
0100	*A* after

4. Mask Operation: (*AND*)

The mask operation is similar to the selective clear operation except that the bits of *A* are cleared only where there are corresponding 0's in *B*. It can be seen the following example.

e.g.

1010	*A* before
1100	*B* (logic operand)
1000	*A* after

5. Insert Operation: This operation is used to insert a new value into a group of bits. This can be done by first masking the bits and then *OR*ing them with the required value. For example suppose that an *A* register contains 8 bits, 01101010. To replace the four unwanted bits.

e.g.

01101010	*A* before
00001111	*B* (mask)
00001010	*A* after masking

and then inser the new value

00001010	*A* before
10010000	*B* (insert)
10011010	*A* (after insertion)

SOLVED PROBLEMS

Q.1. Define the following:

(*a*) Control memory

(*b*) Control word

(*c*) Sequencer

(*d*) Pipeline register.

Solution: (*a*) Control Memory: Control units that use dynamic micro-programming employ a writable control memory. This type of memory can be used for writing, but is used mostly for reading. A memory that is a part of a control unit is received to as control memory.

***(b)* Control Word:** The control variables at any time can be represented by a string of 1's and 2's called control word.

***(c)* Sequencer:** The next address generator is sometimes called a micro-program sequencer, as it determines the address sequence that is read from control memory the address of the next micro-instruction can be specified is several ways, depending on the sequencer inputs.

***(d)* Pipeline Register:** The control data register holds the present micro-instruction while the next address is computed and read from memory. The data register is sometimes called pipelined register.

Q.2. How many types of control organizations are there? Which are is most beneficial for computer system and why?

Solution: There are two major types of control organization

(*i*) Hard wired control

(*ii*) Micro-programmed control.

Hard wired control is more advantage as compared to micro-program control. In the hardware organization, the control logic is implemented with gates flip-flops, decoders and other digital circuits. It has the advantage that it can be optimized to produce a fast mode of operation.

Mostly computer based on the reduced instruction set computer (RISC) architecture concept use hardwired control rather than a control memory with a micro-program. A hardwired control as the name implies; requires changes in the wiring among various components, if the design has to be modified or charged.

In the micro-programmed control organization, the control information is stored in a control memory. The control memory is programmed to initiate the required sequence of micro-operation. The main advantage of micro-programmed control is the fact that once the hard wired configuration is established, there should be need for further hardware or wiring changes.

Q.3. Define the computer instruction format. What is possible memory-reference instructions, describe it.

Solution: In computer instruction format there are thre fields

15	14 12	11 0
I	opcode	Address

It consists of 2 byte length [(0 – 15) bit]

(*i*) Indirect addressing (*I*): Its length is 1 bit field.

(*ii*) Opcode: It is 4 bit operation code.

(*iii*) Address: It is 11 bit address field.

There are 16 possible memory-reference instructions we have described four of them.

Symbol	***Opcode***	***Description***
ADD	0000	$AC \leftarrow AC + M[EA]$
BRANCH	0001	If $(AC > 0)$ then $(PC \leftarrow EA)$
STORE	0010	$M[EA] \leftarrow AC$
EXCHANGE	0011	$AC \leftarrow M[EA]$, $M[EA] \leftarrow AC$

* The ADD instruction adds the content of the operand sound in the effective address to the content of *AC*.

* The BRANCH instruction causes a branch to the effective address if the operand in AC is negative. The program proceeds with the next consecutive instruction if *AC* is not negative. The *AC* is negative if its sign bit is 1.
* STORE instruction transfers the content of *AC* into memory word specified by the effective address.
* EXCHANGE the instruction swaps the data between *AC* and the memory word specified by the effective address.

Each computer instruction must be programmed.

Q.4. Describe the micro-instruction format for the control memory.

Solution: Micro-instruction format is 20 bits in length. It divided into four functional parts. The three fields F_1, F_2 and F_3 specify micro-operations for the computer. The *CD* field selects status bits condition:

* The *BR* field specifies the type of branch to be used.
* The *AD* field contains are branch address. The address fiels is seven bits wide, since the control memory has 128 = 27 words.

3	3	3	2	2	7
F_1	F_2	F_3			AD

F_1, F_2, F_3: Micro-operation fields

CD: Condition for branching

BR: Branch field

AD: Address field

* Micro-operation are sub-divided into three fields of three bits each. These bits in each field are encoded to specify seven distinct micro-operations. This gives 21 micro-operations. If fewer than three micro-opration are used, one or more of the fields will use the binary code 000 for no operation.

Symbols and Binary code for micro-instruction fields

F_1	*Micro-operation*	*Symbol*
000	None	NOP
001	AC ← AC + DR	ADD
010	AC ← 0	CLRAC
011	AC ← AC + 1	INCAC
100	AC ← DR	DRTAC
101	AR ← DR (0 – 10)	DRTAR
110	AR ← PC	PCTAR
111	M[AR] ← DR	WRITE

F_2	*Micro-operation*	*Symbol*
000	None	NOP
001	AC ← AC – DR	SUB
010	AC ← AC + DR	OR
011	AC ← AC ∧ DR	AND
100	DR ← M[AR⁻]	READ
101	DR ← AC	ACTOR
110	DR ← DR + 1	INCOR
111	DR (0 – 10) ← PC	PCTOR

F_3	*Micro-operation*	*Symbol*
000	None	NOP
001	AC ← AC ⊕ DR	XOR
010	AC ← $\overline{\text{AC}}$	COM
011	AC ← SHLAC	SHL
100	AC ← SHRAC	SHR
101	PC ← PC + 1	INCPC
110	PC ← AR	ARTPC
111	Reserved	

CD	Condition	Symbol	Comments
00	Always = 1	U	Unconditional branch
01	DR (15)	I	Indirect Addres bit
10	AC (15)	S	Sign bit of AC
11	AC = 0	Z	Zero value in AC

BR	Symbol	Function
00	JMP	CAR ← ADD, if condition = 1
01	CALL	CAR ← CAR + 1, if condition = 0
10	RET	CAR ← AD, SBR ← CAR + 1, if condition = 1
11	MAP	CAR ← CAR + 1, if condition = 0
		CAR ← SBR (Return of subround
		CAR (2 – 5) ← DR (11 – 14)
		CAR (0, 1, 6) ← 0

Q.5. A logic network has two data inputs *A* and *B* and two control inputs C_0 and C_1. It implements the function *F* according to the following table:

C_1	C_2	F
0	**0**	$\overline{A + B}$
0	**1**	$A + B$
1	**0**	$A \oplus B$

Implement the circuit using one 4 to one multiplier one 2 input exclusive *OR* gate, one 2 input *AND* gate and 2 input *OR* gate and one inverter.

Solution:

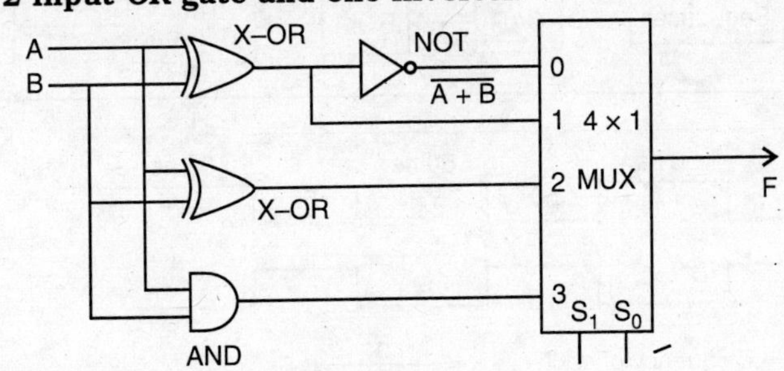

Fig. 4.12.

Q.6. Write the difference between a micro-processor and a micro-program? Is it possible to design a micro-processor without a micro-program? Are all micro-programmed computers also micro-processors?

Solution: A micro-processor is a small size CPU [Computer on a chip]. Micro-program is a program for a sequence of micro-operations. The control unit of a micro-processor can be hard wired or micro-programmed, depending on the specific design. A micro-programmed computer does not have to be a micro-processor.

Q.7. Explain the difference between hard wired control and micro-programmed control. Is it possible to have a hard wired control associated with control memory.

Solution: Hard wired control by definition does not contain a control memory.

Q.8. Define the following:

(*a*) Micro-operation

(*b*) Micro-instruction

(*c*) Micro-program

(*d*) Micro-code.

Solution: (*a*) **Micro-operation:** It is an elementary digital computer operation.

(*b*) **Micro-instruction:** It is an instruction stored in control memory.

(*c*) **Micro-program:** It is a sequence of micro-instructions.

(*d*) **Micro-code:** It is same as micro-program.

Q.9. The micro-programmed control organization shown in figure given below has the following propagation delay times. 40 *ns* to generate the next address, 10 *ns* to transfer the address into the control address register, 40 *ns* to a access the control memory ROM, 10 *ns* to transfer the micro-instruction into the control data register and 40 *ns* to perform the required micro-operations specified by the control word. What is the maximum clock frequency that the control can use? What would the clock frequency be if the control data register is not used?

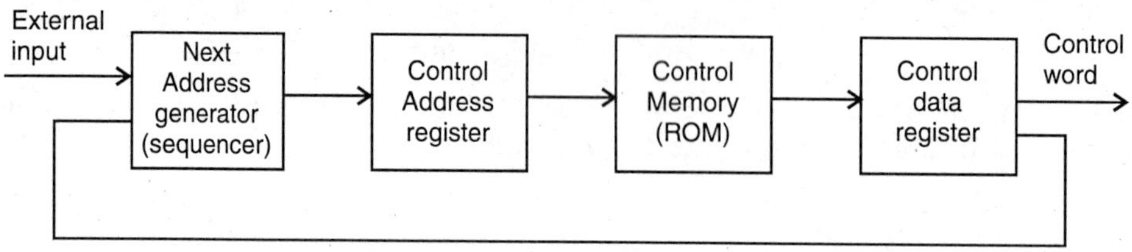

Fig. 4.13.

Solution:

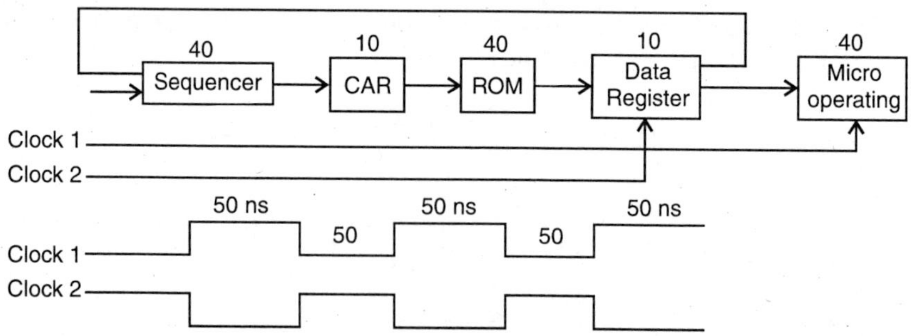

$$\text{Frequency of each clock} = \frac{1}{100 \times 10^{-9}}$$

Fig. 4.14.

$$\frac{1000}{100} \times 10^6 = 10 \text{ MHz}$$

If the data register is moved, we can use a single phase clock with a frequency

$$= \frac{1}{90 \times 10^{-9}} = 11.1 \text{ MHz.}$$

Q.10. The system shown in figure given below uses a control memory of 1024 words of 32 bits each. The micro-instruction has three fields as shown in the diagram. The micro-operations fields has 16 bits.

(a) How many bits are there in the branch address field and the select field?

(b) If there are 16 status bits in the system, how many bits of the branch logic are used to select a status bit?

(c) How many bits are left to select an input for the multiplexers?

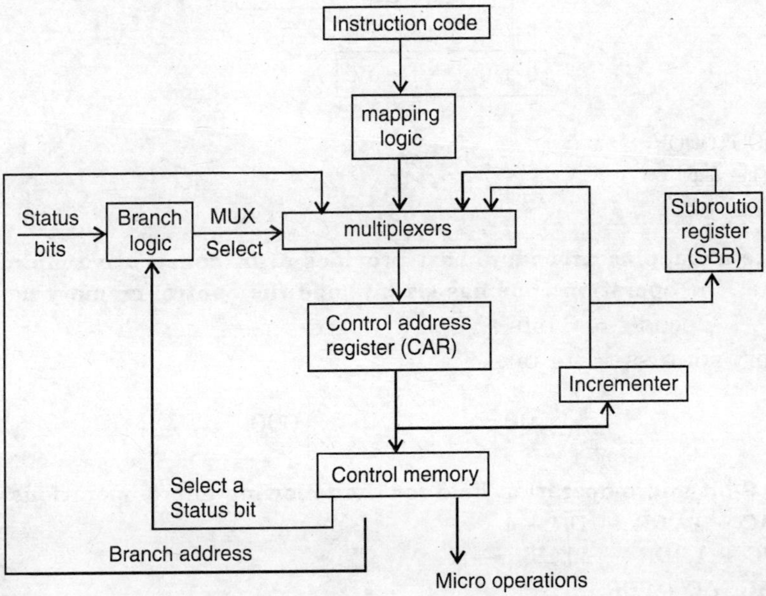

Fig. 4.15.

Solution: Control memory = $2^{10} \times 32 = 32$ bits

(a)

6	10	16
Select	Address	Micro-operations

(b) 4 bits

(c) 2 bits.

Q.11. The control memory in figure has 4096 words of 24 bits each.

(a) How many bits are there in each of the inputs shown going into the multiplexers?

(b) How many bits are there in the control address register?

(c) What are the number of inputs in each multiplexer and how many multiplexers are needed?

Solution: Control memor= $2^{12} \times 24$

(a) 12 bits

(*b*) 12 bits

(*c*) 12 multiplexers, each of size 4 = to –1 line.

Q.12. Using the mapping procedure described in figure given below give the first micro-instruction address for the following operation code

(*a*) 0010

(*b*) 1011

(*c*) 1111

Computer instruction mapping bits

Micro-instruction address

Mapping from instruction code to micro-instruction address.

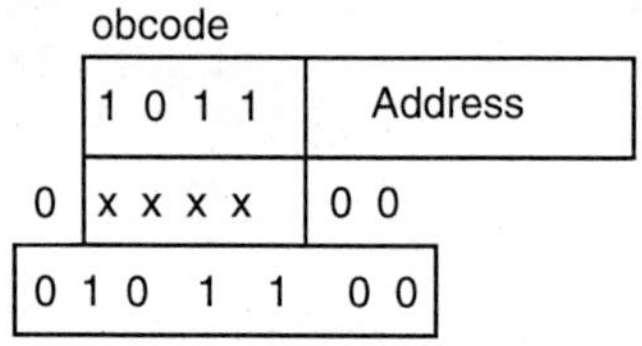

Solution: (*a*) 0001000 = 8

(*b*) 0101100 = 44

(*c*) 0111100 = 60.

Q.13. Formulate a mapping procedure that provides eight consecutive micro-instructions. For each routine. The operation code has six bits and the control memory has 2048 words.

Solution: Opcode = 6 bits

Control memory address = 11 bits

00 | × × × × × × | 000
   | × × × × × × |

Q.14. Give the 9-bit micro-operation field for the following micro-operations

(*a*) $AC \leftarrow AC + 1$, $DR \leftarrow DR + 1$

(*b*) $PC \leftarrow PC + 1$, $DR \leftarrow M[AR]$

(*c*) $DR \leftarrow AC$, $AC \leftarrow DR$.

Solution:

	F_1	F_2	F_3			
(*a*)	011	110	000	INCAL	INCDR	NOP
(*b*)	000	100	101	NOP	READ	INCPC
(*c*)	100	101	000	DRTAC	ACTDR	NOP

Q.15. Convert the following symbolic micro-operations to register transfer statements and to binary.

(*a*) READ, INCPC

(*b*) ACTDR, DRTAC

(*c*) ARTPC, DRTAC, WRITE.

Solution:

				Binary		
(*a*)	READ	$DR \leftarrow M[AR]$	$F_2 = 100$	000	100	101
	DRTAC	$AC \leftarrow DR$	$F_3 = 101$			

(*b*) ACTDR	DR ← AC	F_2 = 101	100 101	
DRTAC	AC ← DR	F_1 = 100		
(*c*) ARTPC	PC ← AR	F_3 = 100		
DRTAC	AC ← DR	F_1 = 100	Impossible	
WRITE	M[AR] ← DR	F_1 = 111	Both use F_1	

Q.16. What is the use of booth multiplication algorithm? Describe hardware for booth diagram.

Solution: Booth algorithm gives a procedure for multiplying binary integers in signed 2's complement representation. It operates, when strings of 0's in the multiplier require no addition but just shifting and a string of 1's in the multiplier from bit weight 2^k to weight 2^m can be treated as $2^{k+1} - 2^m$.

e.g. Binary numer 001110 (+14) has a string of 1's from 2^3 to 2^1. $K = 3$, $m = 1$.

The number can be represented as $2^{KH} - 2^m = 2^4 - 2^1 = 14$.

Therefore multiplication $M \times 14$, where M is multiplicand and 14, the multiplier can be done as $M \times 2^4 - M \times 2^1$.

Thus the product can be obtained by shifting the binary multiplicand M four times to the left and subtracting M shifted left once.

Booth algorithm requires examination of the multiplier bits and shifting of the partial product. Prior to the shifting, the multiplicand may be added to the partial product, subtract from the partial product or left unchanged according to the following rules.

Hardware Ror Both algorithrn

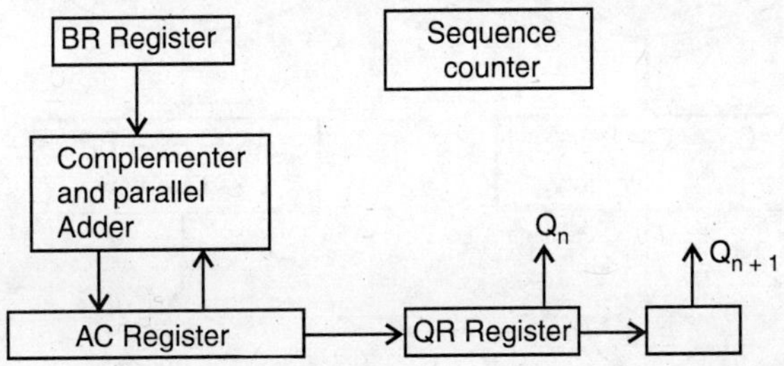

Fig. 4.16.

Q.17. Show the addition and subtraction with signed 2's complement.

Solution: The left most bit of a binary number represents the sign bit: 0 for positive and 1 for negative. The sign bit is 1, the entire number is represented in 2's complement form.

e.g.

+33 represented as: 00100001

–33 represented as: 1101111

The addition of two numbers in signed 2's complement form consists of adding the numbers with the sign bit treated same as the other bits of the number. The carry out of the sign bit position is discarded.

*The subtraction consists of first taking the 2's complement of the subtrahend and then adding it to the minuend.

Hardware for signed 2's complement addition and subtraction:

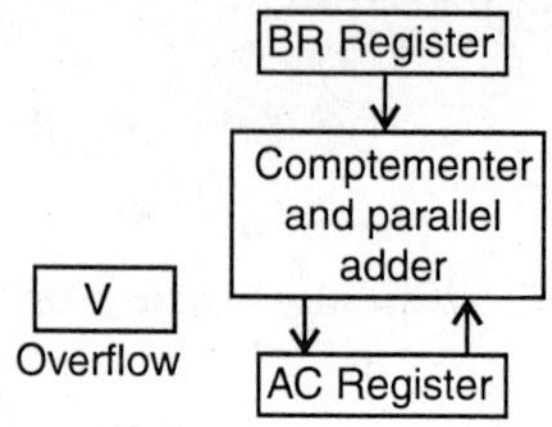

Fig. 4.17.

Q.18. Define over flow. How we detected over flow?

Solution: When two numbers of n digits each are added and the sum occupies $n + 1$ digits, we say that an over flow occured. The effect of an over flow on the sum of two signed 2's complement number.

An over flow can be detected by inspecting the last two carries out of the addition. When two carries are applied to an exclusive *OR* gate, the over flow is detected when the output of the gate is equal to 1.

Q.19. Show adding and subtracting numbers is signed 2's complement representation.

Solution:

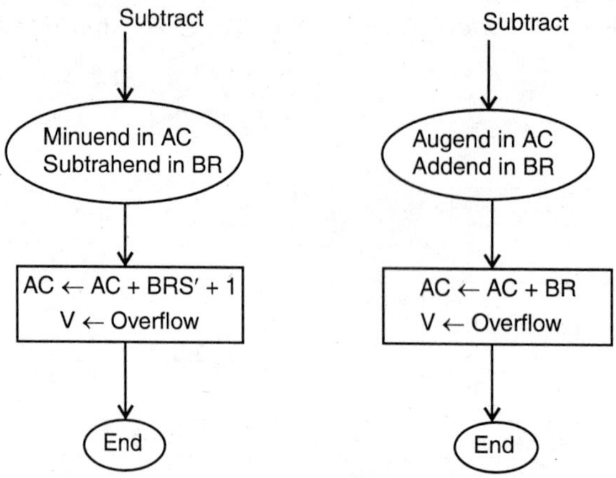

Fig. 4.18.

Q.20. Describe the register configuration for floating point operation.

Solution: The register configuration for floating point operations is quite similar to the layout for fixed point operations. The register for floating point operation is shown below.

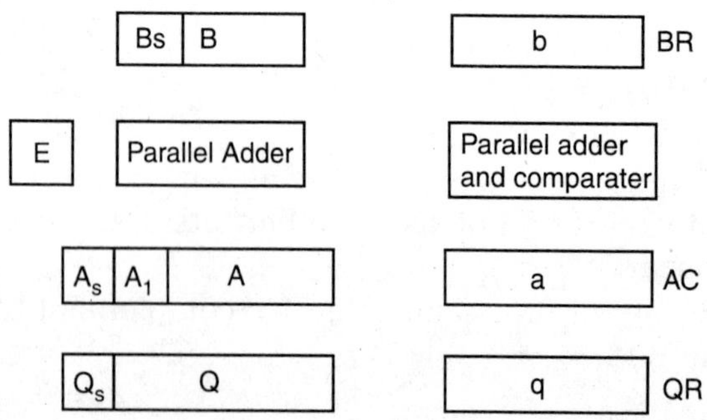

Fig. 4.19.

There are three registers *BR*, *AC* and *QR*. Each register is sub-divided into two parts the mantissa part has the same upper case letter symbols as in fixed point representation. The exponent part uses the corresponding lower case letter symbol. It is assumed that each floating point number has a mantissa in a signed magnitude representation and a biased exponent. Thus the *AC* has a mantissa whose sign is in A_s and magnitude that is in *A*. The exponent is in the part of the register denoted in the lower case letter symbol *a*. The diagram shows explicitly the most significant of *A*. lebeled by A_1. The bit in this position must be a_1 for the number to be normalized. Symbol *AC* represent entire register that is contentation of A_S, *A* and *a*.

Similarly, register *BR* is sub-divided into B_S, *B* and *b* and Q_R into Q_S, *a* and *q*. A parallel adder adds the two mantissa and transfers the sum into *A* and carry into *E*. A separate parallel adder is used for exponents.

Q.21. The complement shown in figure below is not needed if instead of performing $A + B^1 + 1$ we perform $B + A^1$. Derive an algorithm in flow chart form for addition and subtraction of fixed point binary numbers in signed magnitude representation with the magnitudes subtracted by the two micro-operations $A \leftarrow \overline{A}$ and $EA \leftarrow A + B$.

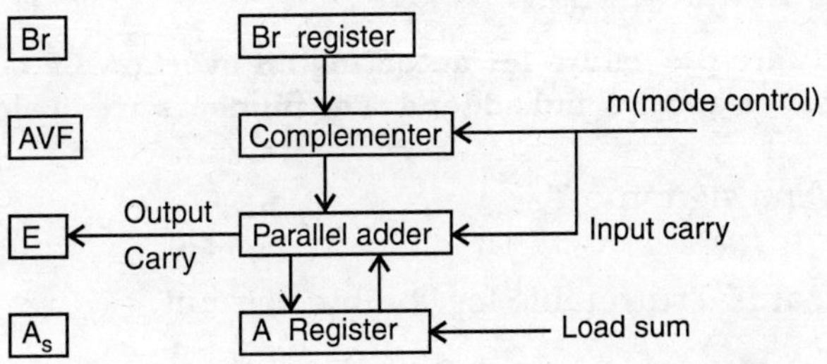

Fig. 4.20.

Solution:

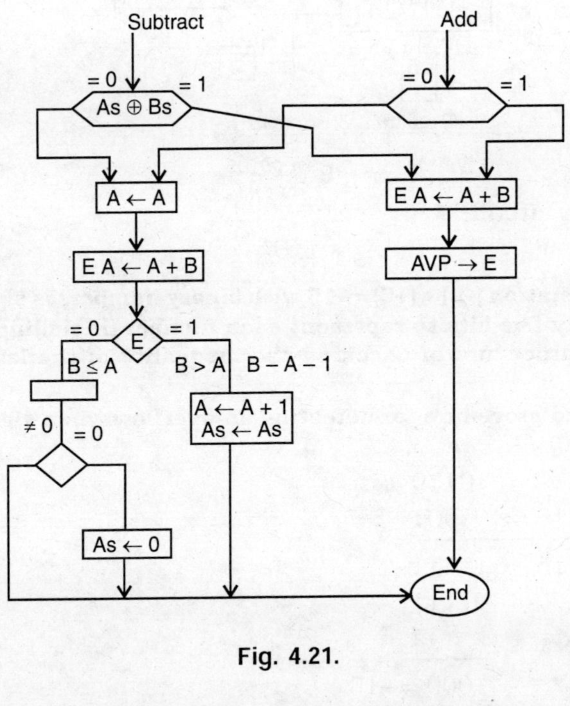

Fig. 4.21.

Q.22. Perform the arithmetic operations below with binary numbers and with negative numbers in signed 2's complement representation. Use seven bits to accommodate each number together with its sign. In each case, determine if there is an overflow by checking the carries into and out of the sign bit position.

(*a*) (+35) + (+40)

(*b*) (–35) + (–40)

Solution: (*a*) +35 = 0100011

→ 75 = 1001011

+40 = 0101000

Here, $F = 0$ and $E = 1$

$F \oplus E = 1$ overflow.

(*b*) +351011101

–401011000

–700110101

Here, $F = 1$ and $E = 0$

$F \oplus E = 1$ overflow.

Q.23. Formulate a hardware procedure for detecting an overflow by comparing the sign of the sum with the signs of the augend and addend. The numbers are in signed 2's complement representation.

Solution: Transfer augend sign into T_s

Then add: $AC \leftarrow AC + BR$

As will hare sign of sum. Truth table for combine circuit.

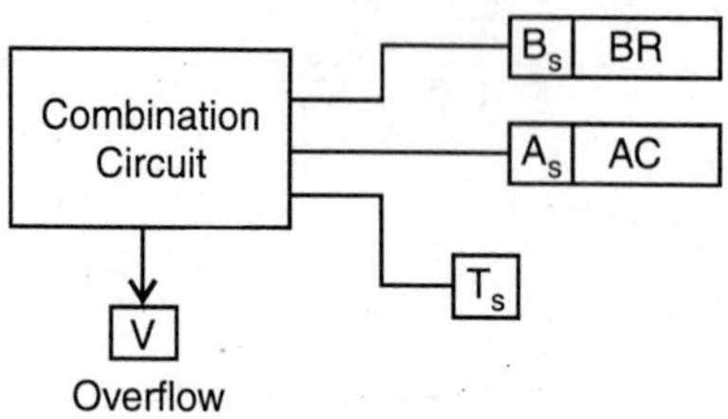

Fig. 4.22.

Boolean function for circuit:

$$V = T_S{}^1 \,.\, B_S{}^1 A_S + T_S B_S A_S{}^1.$$

Q.24. (*a*) Perform the operation (–9) + (+6) = –15 with binary numbers is signed –1's complement representation using only five bits to represent each number (including the sign). Show that the overflow detection procedure of checking the inequality of the last two carries fails in this case.

(*b*) Suggest a modified procedure for detecting an overflow when signed –1's complement numbers are used.

Solution: (*a*) –9 = 10110

+6 = 01001

–15 = 11111

01111

+1

1000 = –15

$F = 1, E = 0 \leftarrow$ Carries

$E \oplus F = 1$ but there should be no overflow since result is –15.

(*b*) The procedure $V \leftarrow E \oplus F$ is valid for 1's complement numbers provided we check the result.

$$\frac{0}{A_s} \quad \frac{1111...11}{A} \quad \text{when } V = 1$$

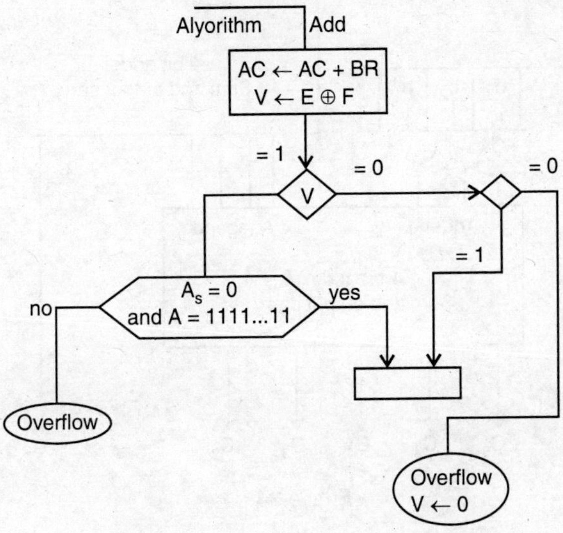

Fig. 4.23.

Q.25. Prove that the multiplication of two *n*-digits numbers in base *r* gives a product no more than 2*n* digits in length. Show that this statement implies that no overflow can occur in the multiplication operation.

Solution: Maximum value of numbers = $r^n - 1$

It is necessary to show that maximum product is less than or equal r^{2n-1}

$$\begin{aligned} \text{Maximum product} &= (r^n - 1)(r^n - 1) \\ &= r^{2n} - 2r^n + 1 \\ &\leq z^{2n} - 1 \end{aligned}$$

which gives $\quad 2 \leq 2\, r^n$

or $\quad 1 \leq r^n$

This is always true since $r \geq z$ and $n \geq 1$.

Q.26. Show that adding *B* after the operation *A* + *B* + 1 restores the original value of *A*. What should be done with the end carry?

Solution: $A + B + 1$ performs $A + 2^n - B = 2^n + A - B$

Adding B: $(2^K + A - B) + B = 2^n + A$

Remove end-carry 2^n to obtain A.

Q.27. Design an array multiplier that multiplies two 4-bit numbers. Use *AND* gates and binary address.

Solution:

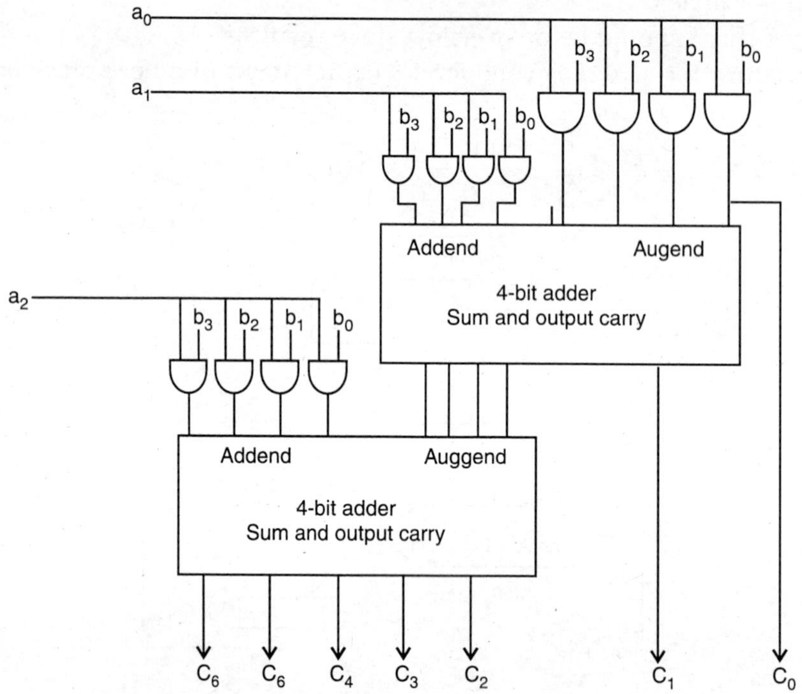

Fig. 4.24.

Q.28. What do you mean by isolated? Define memory mapped input/output?

Solution: Many computers use common bus to transfer information between memory and input/output and the CPU. The distinction between a memory transfer and input/output transfer is made through seprate read and write lines. The CPU specifies whether the address on the address line is for memory word or for an interface register by enabling one of two possible read or write lines. The input/output read input/output write control lines are enabled during an input/output transfer.

The memory read and memory write control lines are enabled during a memory transfer. This configuration isolates all input/output interface addresses from the addresses assigned to memory and is referred to as the isolated input/outputmethod for assigning addresses in a common bus.

In the isolated input/output configuration, the CPU has distinct input and output instructions and each of these instructions is associated with the address of an interface register. The isolated input/output method isolates memory and input/output address, so that memory address values are not affected by interface address assignment since each has its own address space.

In memory mapped input/output, it use the same address space for both memory and input/output. This is the case in computers that employ only one set of read and write signals and do not distinguish between memory and input/output address.

Computers with memory mapped input/output can use memory type instructions to access input/output data with memory mapped input/output. All instructions that refer to memory are also available for input/output.

Q.29. What is the difference between serial and parallel data transmission.

Solution: The transfer of data between two units may be done in parallel or serial. In parallel data transmission each bit of the message has its own path and the total messages is transmitted through *n* seprate conductor paths.

In serial data transmission, each bit in the message is sent in sequence one at a time. This method requires the use of one pair of conductors or one conductor and a common ground. Parallel transmission is faster but requires many wires. It is used for short distances and where speed is important. Serial transmission is slower but is less expensive. Since it requires only one pair of conductors. Serial transmission can be synchronous or asynchronous.

Q.30. What is the difference between synchronous and asynchronous data transfer in case of serial transfer?

Solution: In synchronous transmission, the two units share a common clock frequency and bits are transmitted continuously at the rate dictated by the clock pulses. In long distance serial transmission, each unit is driven by a seprate clock of the same frequency. Synchronization signals are transmitted periodically between two units to keep their clock in step with each other.

In asynchronous transmission binary information is sent only when it is available and the line remains idle when there is no information to be transmitted. This is contrast to synchronous transmission, where bits must be transmitted continuously to keep clock frequency in both units synchronized with each other.

A serial asynchronous data transmission technique used in many interactive terminals employs special bits that are inserted at both ends of the character code with this technique. Each character consists of three parts:

(*a*) Start bit
(*b*) Character bit
(*c*) Stop bits

The convention is that the transmitted resets at the 1 state when no characters are transmitted. The first bit called the start bit, is always 0 and is used to indicate the beginning of a character. The last bit called the stop bit is always a 1.

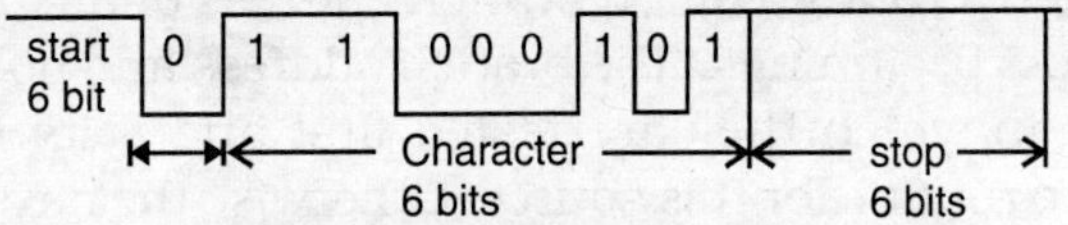

Asynchronous serial transmission

Fig. 4.25.

Q.31. Describe different types of modes of transfer?

Solution: Data transfer between the central computer and input/output devices may be handled in a variety of modes. Some modes use the CPU as an intermediate path, others transfer the data directly to and from the memory unit. Data transfer to and from peripherals may be handled in one of the possible modes.

(*i*) Programmed input/output
(*ii*) Interrupt initiated input/output
(*iii*) Direct memory access (DMA).

(*i*) Programmed input/output: Programmed input/output operations are the result of *I/O* instructions written in the computer program. Each data item transfer is initiated by an instruction in the program usually, the transfer is to and from a CPU register and peripheral. Other instructions are needed to transfer the data to and from CPU and memory. In the programmed input/output method the CPU stage in program loop until the input/output method the CPU stage in program loop until the input/output unit indicates that it is ready for data transfer. Transfer of data under programmed input/output between CPU and peripheral.

***(ii)* Interrupt Initiated input/output**: When an interface determine that the device is ready for the data transfer, it generates an interrupt request to the computer upon detecting the external interrupt signal. The CPU, momentarily stops the task, it is processing, branches to a servic program to process the input/output transfer and then returns to the task it was originally performing.

***(iii)* Direct Memory Access:** In Direct Memory Access (DMA), the interface transfers data into and out of the memory unit through the memory bus. The CPU initiates the transfer by supplying the interface with the starting address and number of words needed to be transferred and then proceeds to execute other tasks when the transfer is made. The DNA requests memory cycles through the memory bus when the request is granted by the memory controller, the DMA transfers the data directly into memory.

Q.32. Define the following:

(*a*) Priority interrupt

(*b*) Polling

(*c*) Vector interrupt.

Solution: (*a*) Priority interrupt: A priority interrupt is a system that establishes a priority over the various sources to determine which condition is to be serviced first when two or more requests arrive simultaneously. The system may also determine which conditions are permitted to interrupt the computer while another interrupt is being serviced. Higher priority interrupt levels are assigned to requests which, if delayed or interrupted could have serious consequences.

(*b*) Polling: Establishing the priority of symultaneous interrupt can be done by software or hardware. A polling procedure is used to identify the highest priority sources by software means. In this method there is one common branch address or all interrupts. The program that takes care of interrupts beginning at the branch address and poll the interrupt sources in sequences. The order in which they are tested first and if its interrupt signal is on, control branches to service routine for this source. Otherwise the next lower-priority sources is tested and so on.

(*c*) Vector Interrupt: The way that the processor chooses the branch address of the service routine varies from one unit to another. In principle there are two methods for accomplishing this.

(*i*) **Non-vector interrupt:** In a non-vector interrupt, the branch address is assigned to fixed location in memory.

(*ii*) **Vector interrupt:** In a vecor interrupt, the source that interrupts supplies the branch information to the computer. This information is called the interrupt vector. In some computers the interrupt vector is the first address of the *I/O* service routine.

Q.33. Give peripheral devices that produce an acceptable output for a person to understand.

Solution:

(*i*) Character printer
(*ii*) Line printer
(*iii*) Laser printer
(*iv*) Digital printer
(*v*) Graphics display
(*vi*) Voice output
(*vii*) Digital to analog converter
(*viii*) Instrument indicator.

Q.34. Indicate whether the following constitute a control status or data transfer commands

(*a*) Skip next instruction if flag is set
(*b*) Seek a given record on a magnetic disk
(*c*) Check if input/output device is ready
(*d*) Move printer paper to beginning of next page
(*e*) Read interface status register.

Solution: (*a*) Status command — Checks status of flag bit
(*b*) Control command — Moves magnetic head in disk
(*c*) Status command — Check if device power is on
(*d*) Control command — Moves paper position
(*e*) Data input command — Reads value of a register.

Q.35. A CPU with a 20 MHz clock is connected to a memory unit whose access time is 40 *ns*. Formulate a read and write timing diagrams using a READ strobe and a WRITE strobe. Include the address in the timing diagram.

Solution:

$$20 \text{ MHz} = 20 \times 10^6 \text{ Hz}$$

$$T = \frac{10^{-6}}{20} = 50\ nx$$

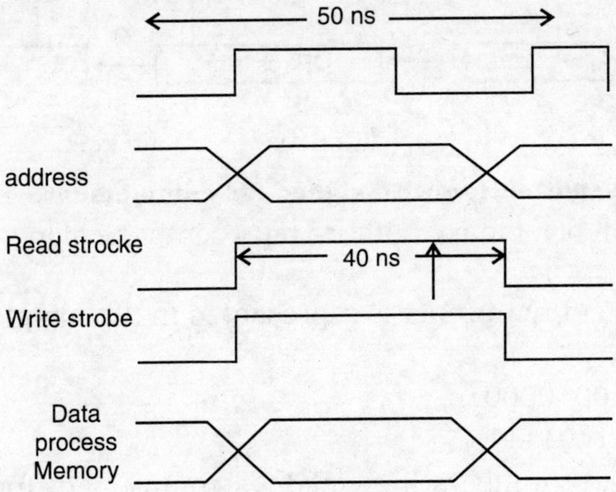

Fig. 4.26.

Q. 36. Give six status conditions for the setting of individual bits in the status register of an asynchronous communication interface.

Solution: Status Conditions:

(*i*) Output flag to indicate when transmitter register is empty.

(*ii*) Input flag to indicate when receiver register is full.

(*iii*) Enable interrupt if any flag is set.

(*iv*) Parity error.

(*v*) Framing error.

(*vi*) Overrun error.

Q.37. What is use of booth multiplication algorithm? Describe Hardware for booth algorithm.

Solution: Booth algorithm gives a procedure for multiplying binary integers in signed 2's complement representation. It operates, when strings of 0's in the multiplier require no addition but just shifting and a string of 1's in the multiplier from bit weight 2^k weight 2^m can be treated as $2^{k+1} - 2^m$.

e.g. Binary number **001110 (+ 14)** has a string of 1's from 2^3 to 2^1, $k = 3$, $m = 1$.

The number can be represented as $2^{k+1} - 2^m - 2^m = 2^4 - 2^1 = 14$

Therefore multiplication $M \times 14$, where M is multiplicand and 14 the multiplier,can be done as $M \times 2^4 - M \times 2$.

Thus the product can be obtained by shifting the binary multiplicand M four times to the left and subtracting M shifted left one.

Booth algorithm requires examination of the multiplier bits and shifting of the partial product. Prior to the shifting, the multiplicand may be added to the partial product, subtracted from the partial product or left unchanged according to the following rules.

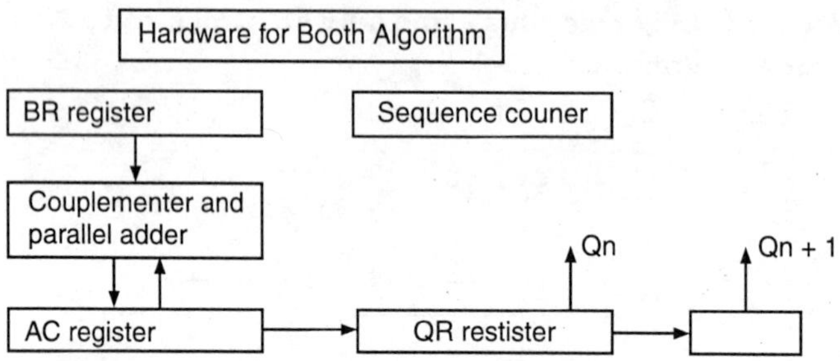

Fig. 4.27.

Q.38. Show addition and subtraction with signed 2's complement.

Solution: The leftmost bit of a binary number represents the sign bit: 0 for positive and 1 for negative.

The sign bit is 1, the entire number is represented in 2's complement form.

e.g.

+ 33 represented as: 00100001

– 33 represented as: 1101111

* The addition of two numbers in signed 2's complement from consists of adding the numbers with the sign bit treated same as the other bits of the number. The carry out of sign bit position is discarded.
* The subtraction consists of first taking the 2's complement of the subtrahend and then adding it to the minuend.

Hardware for signed 2's complement addition and subtraction

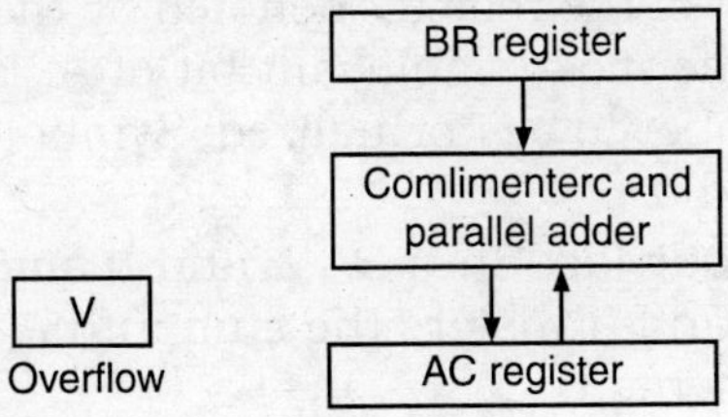

Fig. 4.28.

Q.39. Define over flow. How we detected over flow?

Solution: When two numbers of n digits each are added and the sum occupies $n + 1$ digits, we say that an overflow occurred. The effect of an overflow on the sum of two signed 2's complement number.

An overflow can be detected by inspecting the last two carries out of the addition. When two carries are applied to an exclusive — OR gate, the over flow is detected when the output of the gate is equal to 1.

Q.40. Show adding and subtracting numbers in signed 2's complement representation.

Solution:

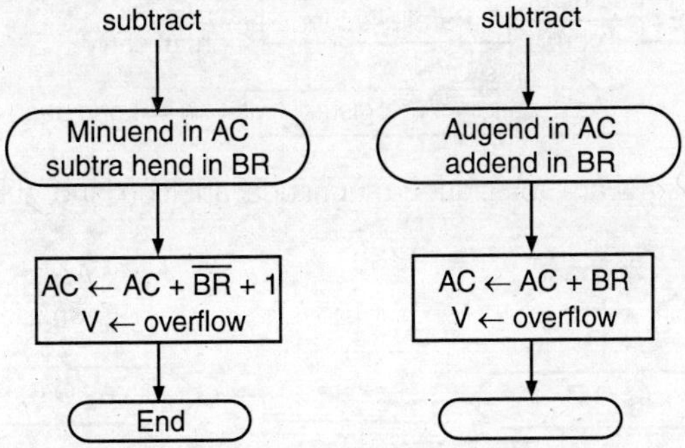

Fig. 4.29.

Q.41. Describe the register configuration for floating print operation.

Solution: The register configuration for floating point operations is quite similar to the layout for fixed point operations. The register organization for floating point operation is shown below:

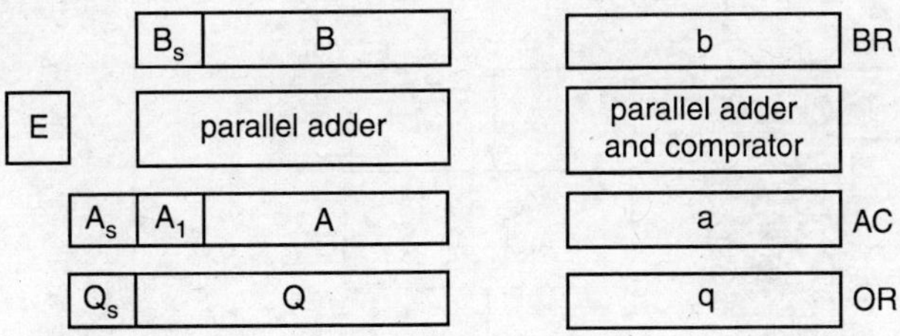

Fig. 4.30.

There are three registers *BR*, *AC* and *QR*. Each register is subdended into two parts. The mantissa part has the same upper case letter symbols as in fixed point representation. The exponent part uses the corresponding lower case letter symbols. It is assumed that each floating point number has a mantissa in signed magnitude representation and a biased exponent. Thus the *AC* has a mantissa whose sign is in A_S and a magnitude that is in *A*. The exponent is in the part of the register denoted by the lower case letter symbols *a*. The diagram shows explicitly the most significant bit of *A*, labeled by A_1. The bit in this position must be a_1 for the number to be normalized. Symbol *AC* represent entire register, that is contestation of A_S, *A* and *a*.

Similarly, register *BR* is subdivided into B_S, *B* and *b* and *QR* into *Q* and *q*. A parallel adder adds the two mantissas and transfers the sum into *A* and carry into *E*. A separate parallel adder is used for exponents.

Q.42. The complement shown in figure below is not needed if instead of performing $A + \overline{B} + 1$ we perform $B + \overline{A}$ (B plus the 1's complement of A). Derive an algorithm in flow chart form for addition and subtraction of fixed-point binary numbers in signed-magnitude representation with the magnitudes subtracted by the two micro-operations $A \leftarrow \overline{A}$ and $EA \leftarrow A + B$.

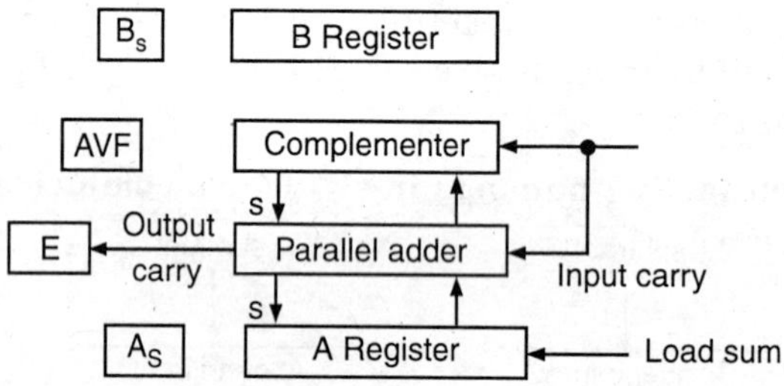

Fig. 4.31. Hardware for signed-magnitude addition and subtraction

Solution:

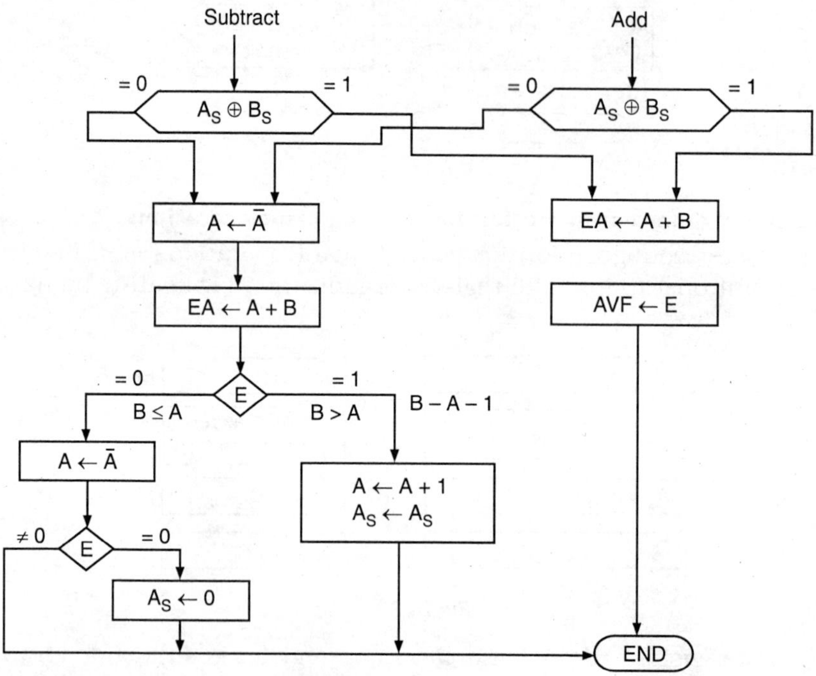

Fig. 4.32.

Q.43. Mark each individual path in the flow chart of figure given below by a number and then indicate the overall path the algorithm takes when the following signed magnitude numbers are computed. In each case give the value of AVF. The left most bit in the following numbers represents the sign bit.

(*a*) **0 101101 + 0 011111**
(*b*) **1 011111 + 1 101101**
(*c*) **0 101101 + 0 011111**
(*d*) **0 101101 + 0 101101**
(*e*) **1 011111 + 0 101101**

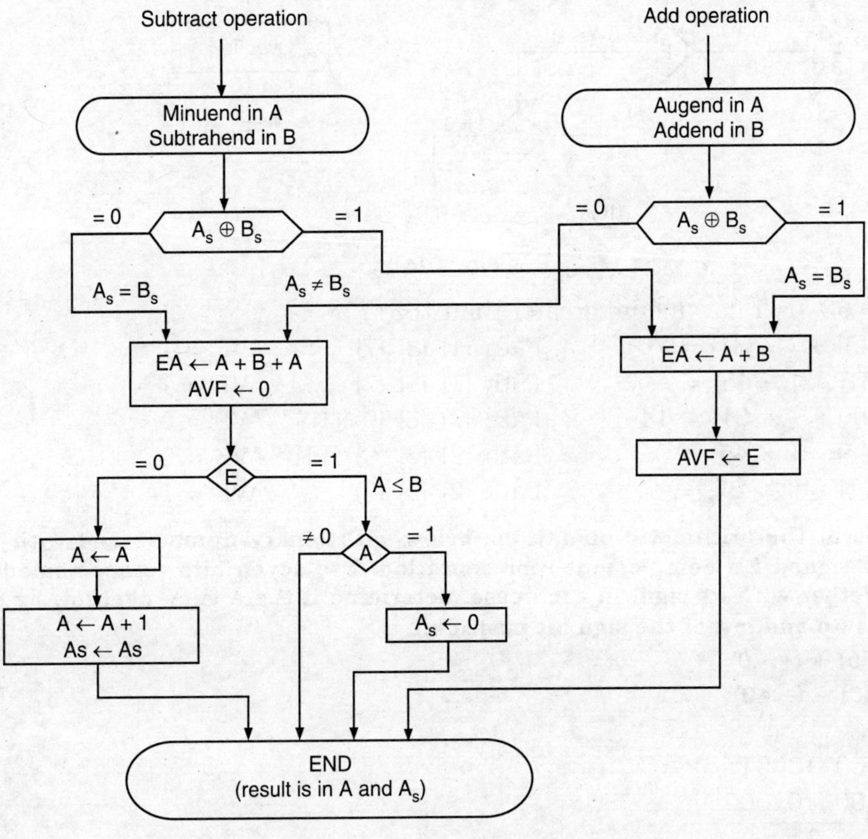

Fig. 4.33. Flow chart for add and subtract operations

Solution:

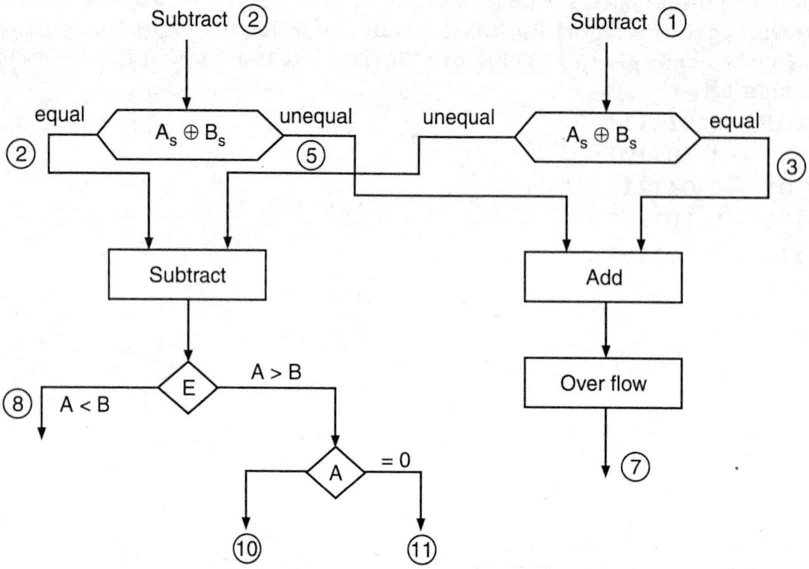

Fig. 4.34.

$2^6 - 1 = 63$, over flow if sum greater than $|63|$

(*a*) (+ 45) + (+ 31) = 76	Path (1) (3) (7)	AVF = 1
(*b*) (− 31) + (− 45) = − 76	Path (1) (3) (7)	AVF = 1
(*c*) (+ 45) − (+ 31) = 14	Path (2) (6) (9) (10)	AVF = 0
(*d*) (+ 45) − (+ 45) = 0	Path (2) (6) (9) (11)	AVF = 0
(*e*) (− 31) − (+ 45) = − 76	Path (2) (5) (7)	AVF = 1

Q.44. Perform the arithmetic operations below with binary numbers and with negative numbers in signed 2's complement representation. Use seven bits to accommodate each number together with its sign. In each case, determine if there is an overflow by checking the carries into and out of the sign bit position.

(*a*) (+ 35) + (+ 40)

(*b*) (− 35) + (− 40)

Solution:

(*a*) + 350 100011
+ 400 101000

+ 75 1 001011

Here $F = 0$ and $E = 1$

$F \oplus E = 1$; overflow

(*b*) + 35 1011101
− 40 1011000

− 70 0110101

Here $F = 1$ and $E = 0$

$F \oplus E = 1$; overflow

Q.45. Consider the binary numbers when they are in signed 2's complement representation. Each number has *n* bits: one for the and *k* = *n* − 1 for the magnitude. A negative number − X

is represented as $2^k + (2^k - X)$, where the first 2^k designates the sign bit and $(2^k - X)$ is the 2's complement of X. A positive number is represented as $0 + X$, where the 0 designates the sign bit, and X, the k-bit magnitude. Using these generalized symbols, prove that the sum $(\pm X) + (\pm Y)$ can be formed by adding the numbers including their sign bits and discarding the carry-out of the sign bit position. In other words, prove the algorithm for adding two binary numbers in signed – 2's complement representation.

Solution:

Case	*Operation in sign-magnitude*	*Operation in complement*	*Required result in sign 2's complement*
1.	$(+X) + (+Y)$	$(0 + X) + (0 + Y)$	$0 + (X + Y)$
2.	$(+X) + (-Y)$	$(0 + X) + 2^k + (2^k - Y)$	$0 + (X - Y)$ if X ³ Y $2^k + 2^k - (Y - X)$ if $X < Y$
3.	$(-X) + (+Y)$	$2^k + (2^k - X) + (0 + Y)$	$0 + (Y - X)$ if Y ³ X $2^k + 2^k - (X - Y)$ if $Y < X$
4.	$(-X) + (-Y)$	$(2^k + 2^k - X) + (2^k + 2^k - Y)$	$2^k + 2^k - (X + Y)$

It is necessary to show that the operations in column (*b*) produce the results listed in.

Case 1. Column (*b*) = column (*c*)

Case 2. If X ³ Y then $(X - Y)$ ³ 0 and consists of k bits.

Operation in column (*b*) gives: $2^{2k} + (X - Y)$.

Discard carry $2^k = 2^n$ to get $0 + (X - Y)$ as in column (*c*)

If $X < Y$, then $(Y - X) > 0$,

Operation gives $2^k = 2^k - (Y - X)$ as in column (*c*)

Case 3. Is the same azs case 2 with X and Y reversed.

Case 4. Operation in column (*b*) gives: $2^{2k} + 2^k - 2^k - (X - Y)$.

Discard carry $2^{2k} = 2^n$ to obtain result of (*c*):

$$2^k + (2^k - X - Y)$$

Q.46. Formulate a hardware procedure for detecting an overflow by comparing the sign of the sum with the signs of the augend and addend. The numbers are in signed – 2's complement representation.

Solution: Transfer augend sign into T_s

Then add: AC ¬ $AC + BR$

A_s will have sign of sum. Truth Table for combine circuit.

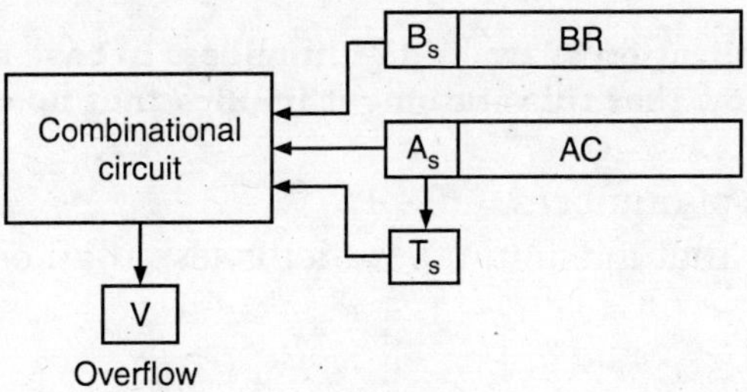

Fig. 4.35.

Boolean function for circuit:

$$V = T_s¢\ B_s¢\ A_s + T_s\ B_s\ A_s¢$$

Q.47. (a) Perform the operation (– 9) + (+ 6) = – 15 with binary numbers is signed – 1's complement representation using only five bits to represent each number (including the sign). Show that the overflow detection procedure of checking the inequality of the last two carries fails in this case.

(b) Suggest a modified procedure for detecting an overflow when signed – 1's complement numbers are used.

Solution: (a) – 910110 Add end around carry *F* as needed in signed – 1's complement addition

– 611001

– 1501111

01111

+ 1 = – 15

10000

$F = 1 \; E = 0 \leftarrow$ carries

$E \oplus F = 1$ but there should be no overflow since result is – 15.

(b) The procedure $V \leftarrow E \oplus F$ is valid for 1's complement numbers provided we check the result.

0 1111...11

As A When $V = 1$

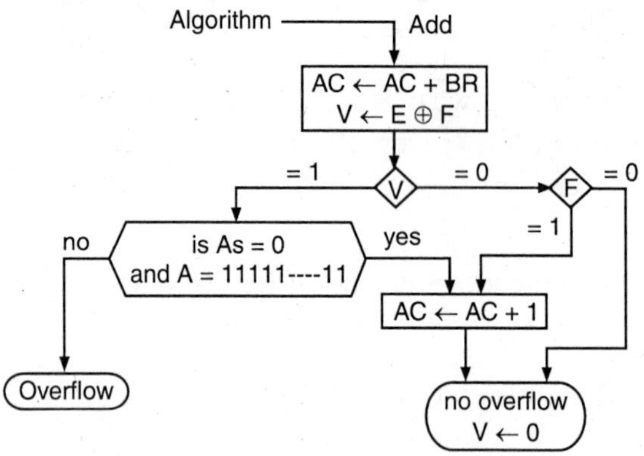

Fig. 4.36.

Q.48. Prove that the multiplication of two n-digit numbers in base r gives a product no more than 2n digits in length. Show that this statement implies that no overflow can occur in the multiplication operation.

Solution: Maximum value of numbers = $r^n - 1$

It is necessary to show that maximum product is less than or equal $r^{2n} - 1$.

$$\text{Maximum product} = (r^n - 1)(r^n - 1) = r^{2n} - 2r^n + 1$$
$$\leq r^{2n} - 1$$

which gives, $2 \leq 2r^n$

or $1 \leq r^n$

This is always true since $r \geq 2$ and $n \geq 1$.

Q.49. Show the contents of registers *E*, *A*, *Q* and *SC* (as in Table given below) during the process of multiplication of two binary numbers, 11111 (multiplicand) and 10101 (multiplier). The signs are not included.

Table 4.5. Numerical Example for Binary Multiplier

Multiplicand B = 10111	E	A	Q	SC
Multiplier in *Q*	0	0000	1011	101
Q_n = 1; add *B*	10111			
First partial product	0	10111		
Shift right *EAQ*	0	01011	11001	100
Q_n = 1; add *B*		10111		
Second partial product	1	00010		
Shift right *EAQ*	0	10001	01100	011
Q_n = 0; shift right *EAQ*	0	01000	10110	010
Q_n = 0; shift right *EAQ*		00100	01011	001
Q_n = 1; add *B*		10111		
Fifth partial product	0	11011		
Shift right *EAQ*	0	01101	10101	000

Solution: Multiplicand $B = 11111 = (31)_{10}$ $31 \times 21 = 651$

Multiplier in Q	E	A	Q	SC $Q = (21)_{10}$
	0	00000	10101	101
Q_n = 1, add B		11111		
	0	11111		
Shr EAQ		01111	11010	100
Q_n = 0, shr EAQ		00111	11101	011
Q_n = 1, add B		11111		
	1	00110		
Shr EAQ	0	10011	01110	010
Q_n = 0 shr EAQ		01001	10111	001
Q_n = 1, add B		11111		
	1	01000		
Shr EAQ		10100	01011	000
		(651)	10	

Q.50. Show the contents of registers *E*, *A*, *Q* and *SC* (as in figure given below) during the process of division of

(*a*) 10100011 by 1011;

(*b*) 00001111 by 0011. (Use of dividend of eight bits)

Divisor $B = 10001$, $\overline{B} + 1 = 01111$

	E	A	Q	SC
Dividend:		**01110**	**00000**	**5**
shl *EAQ*	**0**	**11100**	**0000**	
add $\overline{B}$ + 1		**01111**		
***E* = 1**	**1**	**01011**		
Set Q_n = 1	**0**	**01011**	**00001**	**4**
shl *EAQ*		**10110**	**00010**	
Add $\overline{B}$ + 1		**01111**		
***E* = 1**	**1**	**00101**		
Set Q_n = 1	**1**	**00101**	**00011**	**3**
shl *EAQ*	**0**	**01010**	**00110**	
Add $\overline{B}$ + 1		**01111**		
***E* = 0; leave Q_n = 0**	**0**	**11001**	**00110**	
Add *B*		**10001**		
Restore remainder		**01010**		**2**
shl *EAQ*	**0**	**10100**	**01100**	
Add *B* + 1		**01111**		
***E* = 1**	**1**	**00011**		
Set Q_n = 1	**1**	**00011**	**01101**	**1**
shl *EAQ*	**0**	**00110**	**11010**	
Add $\overline{B}$ + 1		**01111**		
***E* = 0; leave Q_n = 0**	**0**	**10101**	**11010**	
Add *B*		**10001**		
Restore remainder	**1**	**00110**	**11010**	**0**
Neglect *E*				
Remainder in *A*:		**00110**		
Quotient in *Q*:			**11010**	

Solution: (*a*) $\dfrac{10100011}{1011} = 1110 + \dfrac{1001}{1011} \quad \dfrac{163}{11} = 14 + \dfrac{9}{11}$

$$B = 1011 \quad \overline{B} + 1 = 0101 \quad \text{DVF} = 0$$

	E	A	Q	SC
Dividend in AQ	0	1010	0011	100
shl EAQ	1	0100	0110	
add $\overline{B}$ + 1, suppress carry			0101	
E = 1, set Q_n to 1	1	1001	0111	011
shl EAQ	1	0010	1110	
add $\overline{B}$ + 1, suppress carry		0101		
E = 1, set Q_n to 1	1	0111	1111	010
shl EAQ	0	1111	1110	

add $\overline{B}$ + 1, carry to E		0101		
E = 1, set Q_n to 1	1	0100	1111	001
shl EAQ	0	1001	1110	
add $\overline{B}$ + 1, carry to E		0101		
E = 0, leave Q_n = 0	0	1110	1110	
add B		1011		
restore remainder	1	1001	1110	000
	remainder		quotient	

(b) $\frac{1111}{0011}$ = 0101 B = 0011 $\overline{B}$ + 1 = 1101

	E	A	Q	SC
Dividend in Q, A = 0		0000	1111	100
shl EAQ	0	0001	1110	
add $\overline{B}$ + 1, suppress carry		1101	1110	
E = 0, leave Q_n = 0	0	1110		
add B		0011		
restore partial remainder	1	0001		011
shl EAQ	0	0011	1100	
add $\overline{B}$ + 1		1101		
E = 1, set Q_n to 1	1	0000	1101	010
shl EAQ	0	0001	1010	
add $\overline{B}$ + 1		1101		
E = 1, set Q_n to 1	1	0000	1101	010
shl EAQ	0	0001	1010	
add $\overline{B}$ + 1		1101		
E = 0, leave Q_n = 0	0	1110	1010	
add B		0011		
restore partial remainder	1	0001		001
shl EAQ	0	0011	0100	
and $\overline{B}$ + 1		1101		
E = 1, set Q_n to 1	1	0000	0101	000
	remainder		quotient	

Q.51. Show that adding B after the operation A + B + 1 restores the original value of A. What should be done with the end carry?

Solution: A + B + 1 performs: $A + 2^n - B = 2^n + A - B$

adding B: $(2^k + A - B) + B = 2^n + A$

remove end – carry 2^n to obtain A.

Q.52. Why should the sign of the remainder after a division be the same as the sign of the dividend?

Solution: To correspond with correct result. In general $\frac{A}{B} = Q + \frac{R}{B}$ where A is dividend, Q the quotient and R the remainder. Four possible signs for A and B:

$$\frac{+52}{+5} = +10 + \frac{+2}{+5} = +10.4 \qquad \frac{-52}{+5} = -10 + \frac{-2}{+5} = -10.4$$

$$\frac{+52}{-5} = -10 + \frac{+2}{-5} = -10.4 \qquad \frac{-52}{-5} = +10 + \frac{-2}{-5} = +10.4$$

The sign of the remainder (2) must be same as sign of dividend (52).

Q.53. Design an array multiplier that multiplies two 4-bit numbers. Use AND gates and binary address.

Solution: Add one more stage of figure given below with 4 AND gates and a 4-bit adder.

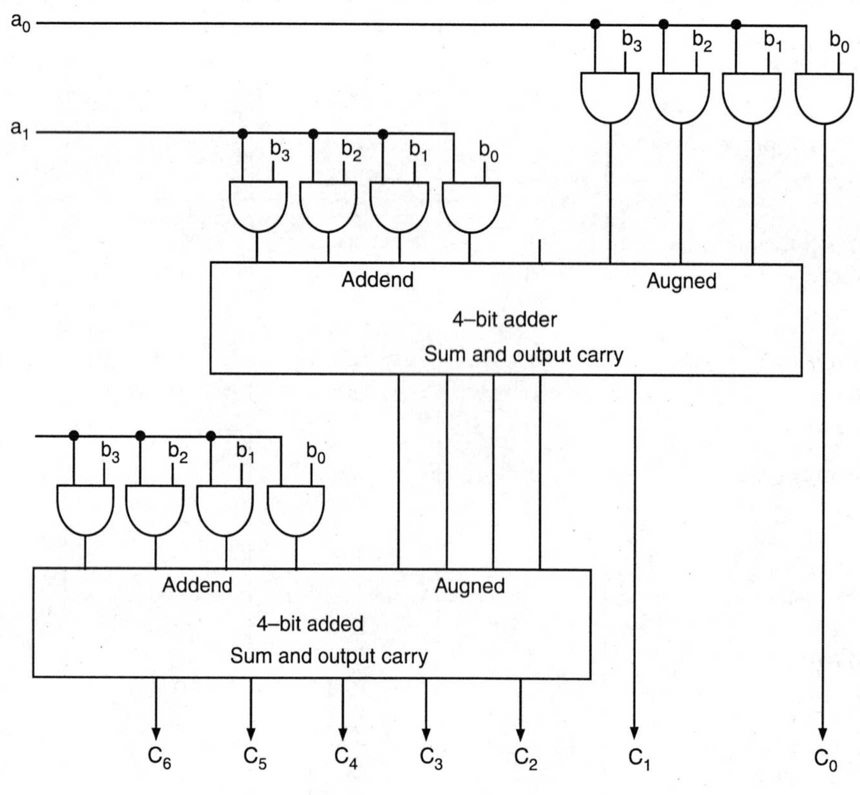

Fig. 4.37. 4-bit by 3 bit array multiplier

Q.54. Show the step by step multiplication process using Booth algorithm (as in Table given below) when the following binary numbers are multiplied. Assume 5-bit registers that hold signed numbers. The multiplicand in both cases is + 15.

(a) (+15) × (+ 13)

(b) (+15) × (– 13)

Table 4.6. Example of Multiplication with Booth Algorithm

Q_n	Q_{n+1}	$BR = 10111$ $\overline{BR} + 1 = 01001$	AC	QR	Q_{n+1}	SC
		Initial	**00000**	**10011**	**0**	**101**
1	**0**	**Subtract BR**	**01001**			
			01001			
		ashr	**00100**	**11001**	**1**	**100**
1	**1**	**ashr**	**00010**	**01100**	**1**	**011**
0	**1**	**Add BR**	**10111**			
			11001			
		ashr	**11100**	**10110**	**0**	**010**
0	**0**	**ashr**	**11110**	**01011**	**0**	**001**
1	**0**	**Subtract BR**	**01001**			
			00111			
		ashr	**00011**	**10101**	**1**	**000**

Solution:

(*a*) (+ 15) × (+ 13) = + 195 = $(0\ 011\ 000011)_2$

$BR = 0\ 1111\ (+15);$

$\overline{BR} + 1 = 10001\ (-15);$

$Q_R = 01101\ (+13)$

Q_n	Q_{n+1}		AC	QR	Q_{n+1}	SC
		Initial	00000	01101	0	101
1	0	Subtract BR	10001			
			10001			
		ashr	11000	10110	1	100
0	1	Add BR	01111			
			00111			
		ashr	00011	11011	0	011
1	0	Subtract BR	10001			
			10100			
		ashr	11010	01101	1	010
1	1	ashr	11101	00110	1	001
0	1	Add BR	01111			
			01100			
		ashr	00110	00011	0	000
			+ 195			

(*b*) (+ 15) × (– 13) = – 195 = (11 00111101)2¢ Complement

$BR = 011111\ (+15);$

$(\overline{BR} + 1 = 10001\ (-15);)$

$QR = 10011\ (-13)$

Q_n	Q_{n+1}		AC	QR	Q_{n+1}	SC
		Initial	00000	10011	0	101
1	0	Subtract BR	10001			
			10001			
		ashr	11000	11001	1	100
1	1	ashr	11100	01100	1	011
0	1	Add BR	01111			
			01011			
		ashr	00101	10110	0	010
0	0	ashr	00010	11011	0	001
1	0	Subtract BR	10001			
			10011			
		ashr	110001	11101	1	000
			– 195			

Q.55. Drive an algorithm in flowchart form the nonrestoring method of fixed-point binary division:

Solution:

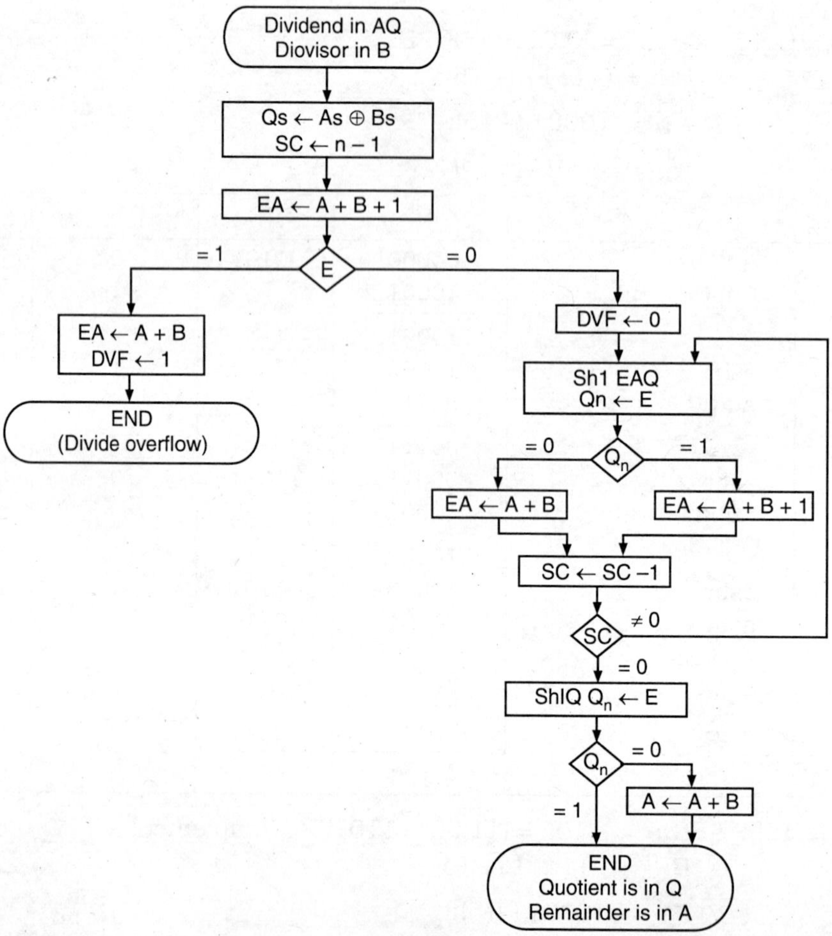

Fig. 4.38.

Q.56. Derive an algorithm for evaluating the square root of a binary fixed-point number.

Solution: The algorithm for square-root is similar to division with the radicand being equivalent to the divident and a "test value" being equivalent to the divisor.

Let A be the radicant, Q the square-root; and R the remainder such that $Q^2 + R = A$ or $\sqrt{A} = Q$ and a remainder.

General comments:

(i) For k bits in A (k even), Q will have $\frac{k}{2}$ bits: $Q = q_1\ q_2\ q_3 \ldots \frac{qk}{2}$

(ii) The first test value if 01
The second test value is $0_{q1}01$
The third test value is $00_{q1\ q2}01$
The fourth test value is $000_{q1q2q3}01$, etc.

(iii) Mark the bits of A in groups of two starting from left.

(iv) The procedure is similar to the division restoring method as shown inthe following example:

$q_1\ q_2\ q_3\ q_4$
1 1 0 1 = Q = 13

$\sqrt{10\ 10\ 10\ 01}$ = Q = 169

01 subtract first test value

01	answer positive; let $q_1 = 1$
01	answer positive; let $q1 = 1$
0110	bring down next pair
0101	subtract second test value $0q101$
0001	answer positive; let $q_2 = 1$
000110	bring down next pair
001101	subtract third test value $00q_1q_201$
negative	answer negative; let $q_3 = 0$
00110	restore partial remainder
00011001	bring down next pair
00011001	subtract fourth test value $000q_1q_2q_301$
Remainder = 00000	answer positive (zero); let $q_4 = 1$

Q.57. A binary floating-point number has seven bits for a biased exponent. The constant used for the bias is 64.

(a) List the biased representation of all exponents from – 64 to + 63.

(b) Show that a 7 bit- magnitude comparator can be used to compare the relative magnitude of the two exponents.

(c) Show that after the addition of two biased exponents it is necessary to subtract 64 in order to have a biased exponents sum. How would you subtract 64 by adding its 2's complement value?

(d) Show that after the subtraction of two biased exponents it is necessary to add 64 in order to have a biased exponent difference.

Solution: (*a*)

e = *exponent* *e*	*e* + 64	*e* + 64 = *biased exponent biased exponent*
– 64	– 64 + 64 = 0	0 000 000
– 63	– 63 + 64 = 1	0 000 001
– 62	– 62 + 64 = 2	0 000 010
– 1	– 1 + 64 = 63	0 111 111
– 0	0 + 64 = 64	1 000 000
+ 1	1 + 64 = 65	1 000 001
+ 62	62 + 64 = 126	1 111 110
+ 63	63 + 64 = 127	1 111 111

(*b*) The biased exponent follows the same algorithm as a magnitude comparator

(*c*) $(e_1 + 64) + (e_2 + 64) = (e_1 + e_2 + 64) + 64$
subtract 64 to obtain biased exponent sum.

(*d*) $(e_1 + 64) - (e_2 - 64) = e_1 + e_2$
Add 64 to obtain biased exponend difference.

Q.58. Device an algorithm in flowchart from the comparison of two signed biary numbers when negative numbers are in signed – 2's complement representation:

(*a*) By means of a subtraction operation with the signed – 2's complement numbers.

(*b*) By scanning and comparing pairs of bits from left to right.

Solution: (*a*)

$$AC = As\, A_1\, A_2\, A_3 \ldots A_n$$
$$BS = B_s\, B_1\, B_2\, B_3 \ldots B_n$$

If signs are unlike – The one iwht a 0 (plus) is larger.

If signs are alike – both numbers are either positive or negative.

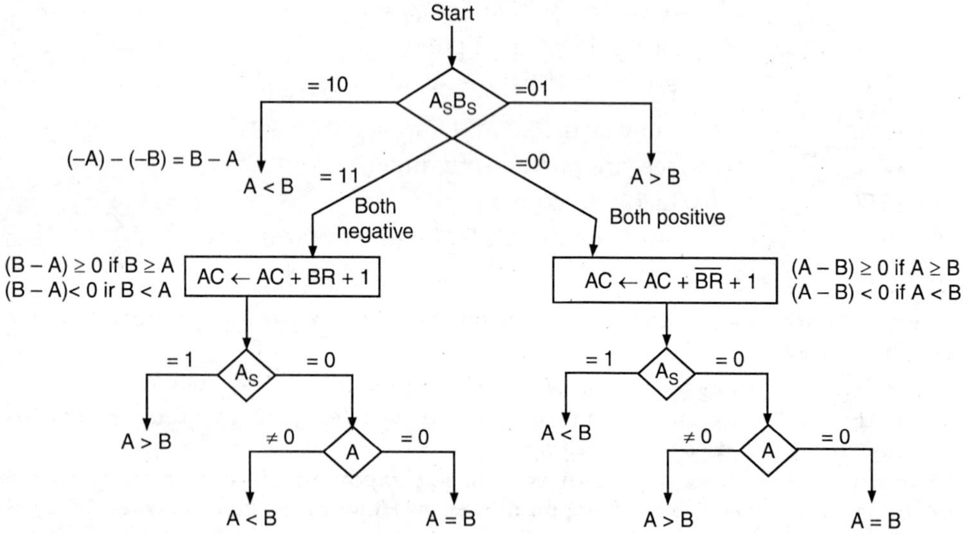

Fig. 4.39.

(*b*)

	A_s	A_1	$A_2 \ldots A_n$
+ 2	0	0	0 0 0 1 0
+ 1	0	0	0 0 0 0 1

0	0	0	00000
−1	1	1	11111
−2	1	1	11110
−3	1	1	11101

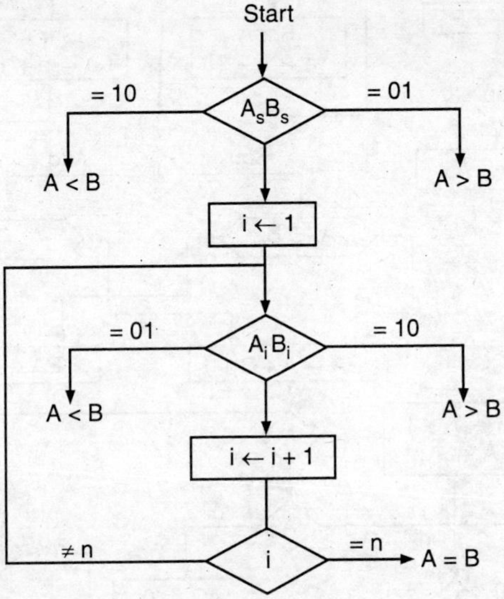

Fig. 4.40.

Q.59. Repeat Problem 22 for signed-magnitude binary numbers.

Solution:

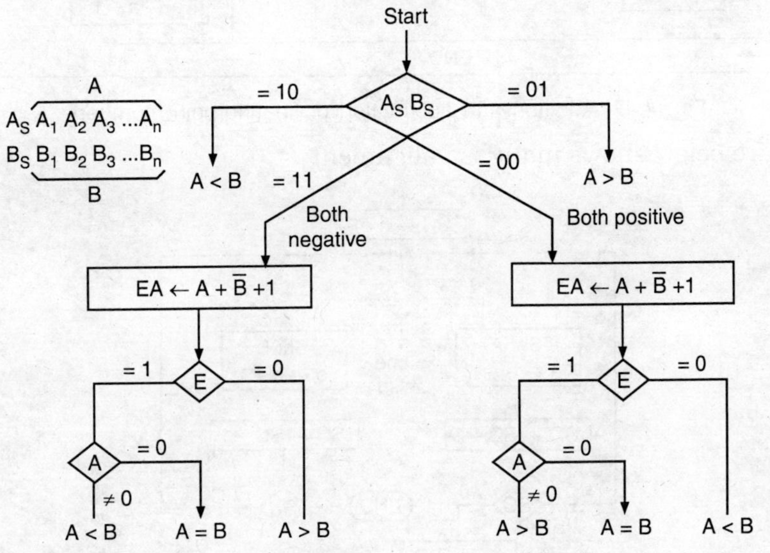

Fig. 4.41.

Q.60. Let n be the number of bits of the mantissa in binary floating point number. When the mantissas are aligned during the addition or subtraction the exponent difference may be greater than $n - 1$. If this occurs, the mantissa with the smaller exponent is shifted entirely out of the registor. Modify the mantissa alignment in figure given below by including a sequence counter. SC that couns the number of shifts. If the number of shifts is greater than $n - 1$,the large number is then used to determine the results.

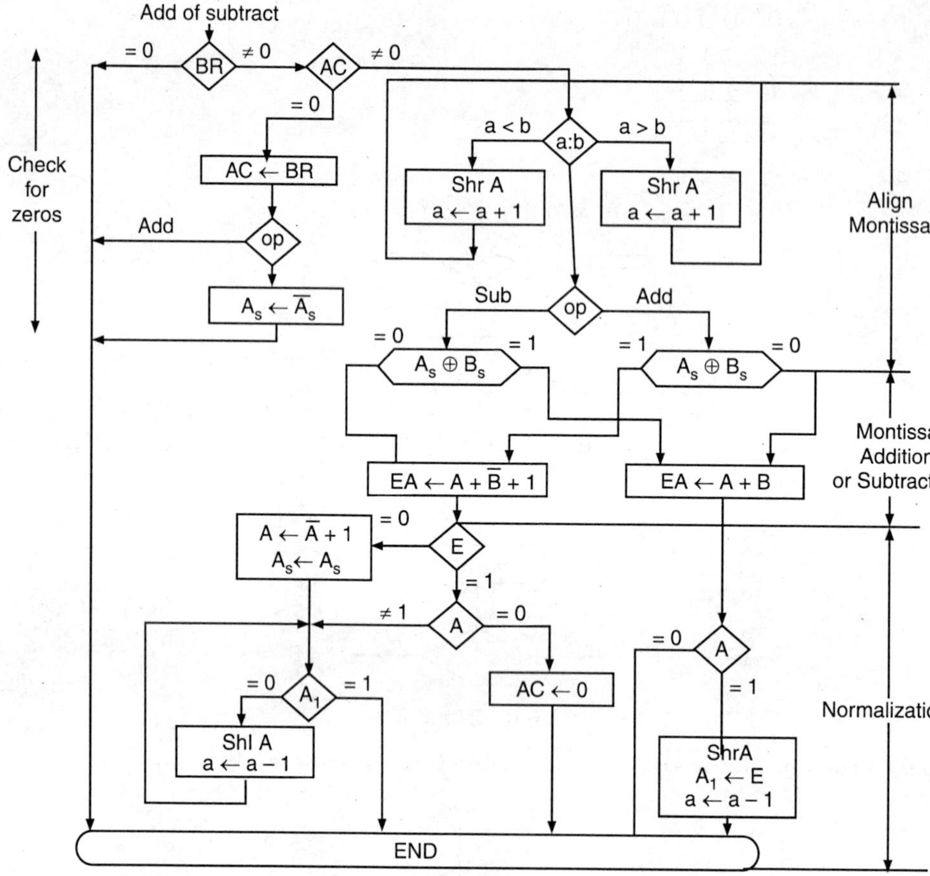

Fig. 4.42. Addition and subtraction of floating-point numbers

Solution: Figure below shows mantissa alignment

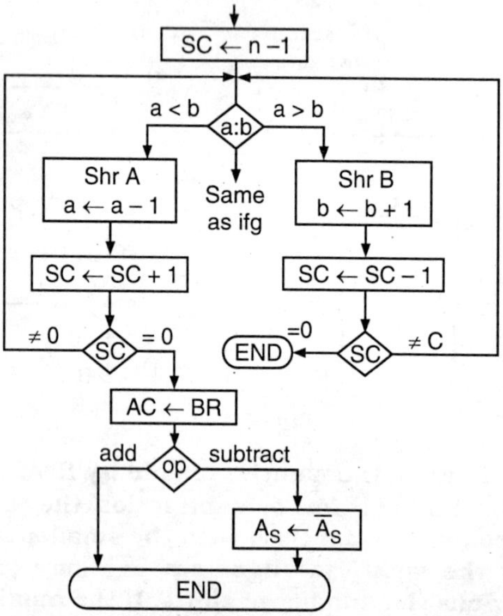

Fig. 4.43.

Q.61. The procedure for aligning mantissa during addition or subtraction of floating point numbers can be stated as follows:

Subtract the smaller exponent from the larger and shift right mantissa having the smaller exponent a number of places equal to the difference between the exponents. The exponent of the sum (or difference) is equal to larger exponents. Without using a magnitude comparator, assuming biased exponents, and taking into account that only the AC can be shifted, derive an algorithm in flowchart form for aligning the mantissa and placing the larger exponent in the AC.

Solution: Let "*e*" be a flip-flop that holds end carry after exponent addition.

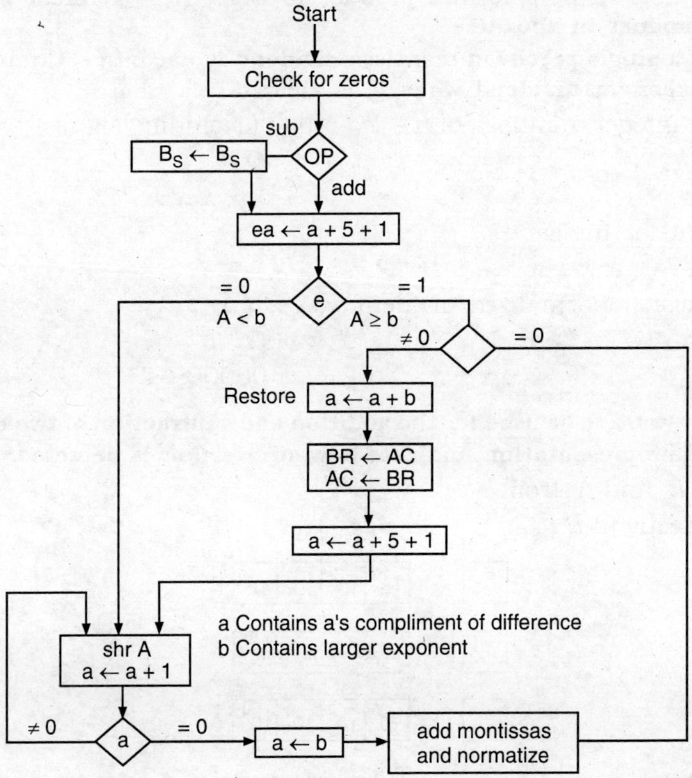

Fig. 4.44.

Q.62. Show that the division of two normalized floating-point numbers with fractional mantissas will always result in a normaliazed quotient provided a dividend alignment is carried out prior to the division operation.

Solution: Dividend A = 0.1 *xxxx* where x = 0.1 divisor *B* = 0.1 *xxxx*

(*a*) If $A < B$ then after shift we have A = *xxxx* and 1st quotient bit is 1.

(*b*) If $A \geq B$, dividend alignment results in A = 0.01 *xxxx*. Then after the left shift $A \geq B$ and first quotient bit = 1.

Q.63. The algorithms for the floating-point arithmetic operations in Sec. 10-5 neglect the possibility of exponent overflow or underflow.

(*a*) Go over the three flowcharts and find where an exponent overflow may occur.

(*b*) Repeat (*a*) for exponent underflow. An exponent underflow occurs if the exponent is more negative than the smallest number that can be accommodated in the register.

(c) Show an exponent overflow or underflow can be detected by the hardware.

Solution:

(a) When the exponents are added or incremented.

(b) When the exponents are subtracted or decremented.

(c) Check end-carry after addition and carry after increment or decrement.

Q.64. If we assume integer representation for the mantissa of floating-point numbers, we encounter certain scaling problems during multiplication and division. Let the number of bits in the magnitude part of the mantissa be $(n - 1)$. For integer representation.

(a) Show that if a single-precision product is used, $(n - 1)$ must be added to the exponent product in the *AC*.

(b) Show tht if a single-precision mantissa dividend is used, $(n - 1)$ must be subtracted from the exponent dividend when Q is cleared.

Solution: Assume integer mantissa of $n - 1 = 5$ bits (excluding sign)

(a) Product: *A* *Q*

$x\ x\ x\ x\ x$ $x\ x\ x\ x\ x.\ {}^*2^{z}$

binary-point for integer

Product in AC: $x\ x\ x\ x\ x\ .\ {}^*\ 2^{Z+5}$

(b) Single precision normalized divident: $x\ x\ x\ x\ x.\ {}^*\ 2^{z}$

Divident in AQ: *A* *Q*

$x\ x\ x\ x\ x$ $00000.{}^*\ 2^{z-5}$

Q.65. Show the hardware to be used for the addition and subtraction of two decimal numbers in signed-magnitude representation. Indicate how an overflow is detected.

Solution: Neglect B_e and A_e from

Apply carry directly to *E*.

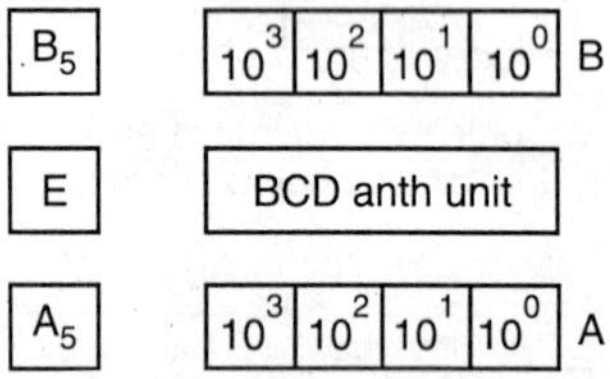

Q.66. Show that 673 – 356 can be computed by adding 673 to the 10's complement of 356 and discarding the end carry. Draw the block diagram of three-stage decimal arithmetic unit and show how this operation is implemented. List all input bits and output of the unit.

Solution: 673 673

\+

–356

317 10 s complement of 356 = 644

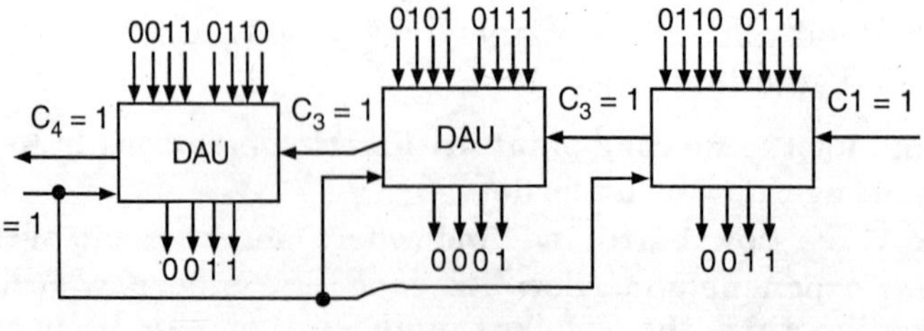

Fig. 4.45.

Q.67. Show that the lower 4 bit binary adder in figure given in the figure below can be replaced by one full-adder and two half-adders.

Solution:

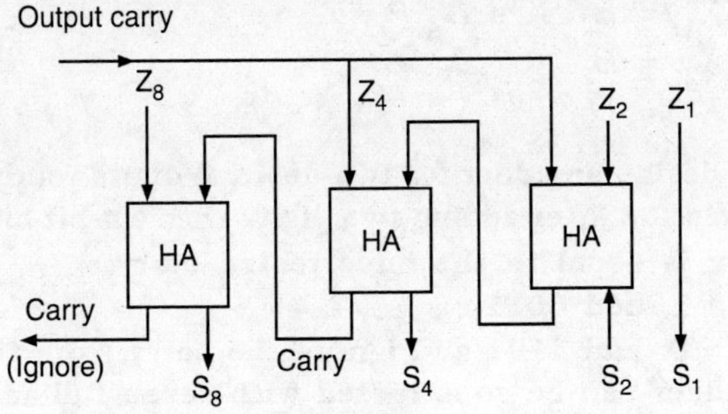

Fig. 4.46.

Q.68. Using combinational circuit design techniques, derivet he Boolean functions for the BCD 9's complementer of figure given below. Drasw the logic diagram.

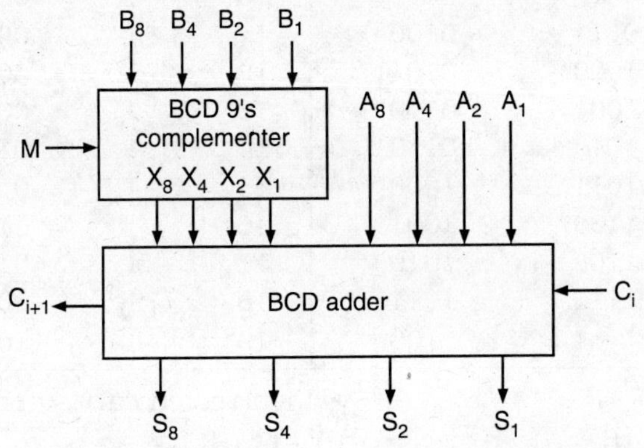

Fig. 4.47. One stage of a decimal arithmetic unit

Solution:

	Input				Output				
	B_8	B_4	B_2	B_1	X_8	X_4	X_2	X_1	
0	0	0	0	0	1	0	0	1	9
1	0	0	0	1	1	0	0	0	8
2	0	0	1	0	0	1	1	1	7
3	0	0	1	1	0	1	1	0	6
4	0	1	0	0	0	1	0	1	5
5	0	1	0	1	0	1	0	0	4
6	0	1	1	0	0	0	1	1	3
7	0	1	1	1	0	0	1	0	2
8	1	0	0	0	0	0	0	1	1
9	1	0	0	1	0	0	0	0	0

1	1		
X	X	X	X
		X	X

		1	1
1	1		
1	X	X	X
		X	X

		1	1
		1	1
X	X	X	X
		X	X

1			1
1			1
X	X	X	X
1		X	X

$$d(B_8\ B_2\ B_1) = \Sigma\ (10, 11, 12, 13, 14, 15) \text{ are don't-care conditions}$$
$$X_8 = B'_8\ B'_4\ B'_2$$
$$X_4 = B_4\ B'_2 + B'_4\ B'_2$$
$$X_2 = B_2$$
$$X_1 = B'_2$$

Q.69. It is necessary to design an adder for two decimal digits represented in the excess –3 code. Show that the correction after adding two digits with a 4-bit binary adder is as follows:

(a) The output carry is equal to the uncorrected carry.

(b) If output carry = 1, add 0011.

(c) If out put carry = 0, add 1101 and ignore the carry from this addition. Show that the excess – 3 adder can be constructed with seven full-adders and two inverters.

Solution:

	Z	*Y*			*Z*		*Y*
dec	**uncorrected**	**corrected**	**dec**		**uncorrected**		**corrected**
0	0110	0011	10	1	0000	1	0011
1	0111	0100	11	1	0001	1	0100
2	1000	0101	12	1	0010	1	0101
3	1001	0110	13	1	0011	1	0110
4	1010	0110	14	1	0100	1	0111
5	1011	1000	15	1	0101	1	1000
6	1100	1001	16	1	0110	1	1000
7	1101	1010	17	1	0111	1	1010
8	1110	1011	18	1	1001	1	1011
9	1111	1100	19	1	1001	1	1100

No output carry

$Y = Z - 3 = Z + 13 - 16$

ignore carry

Uncorrected carry = output carry

$Y - Z + 3$

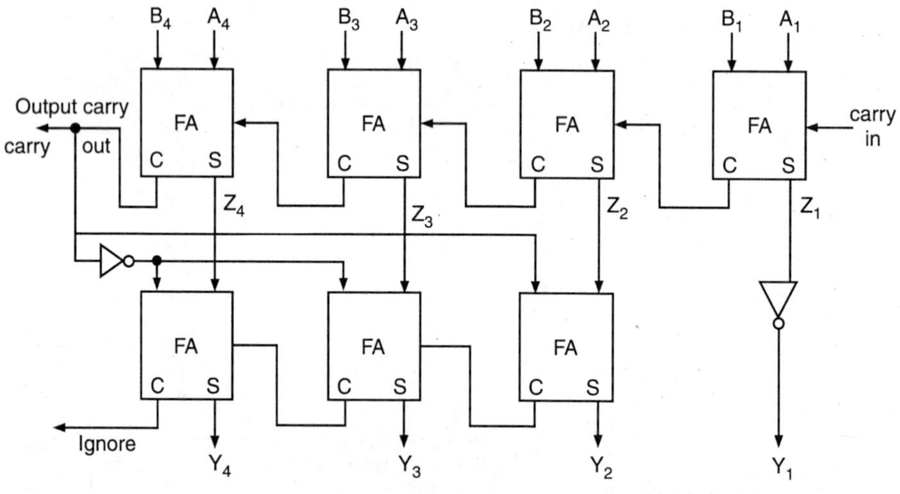

Fig. 4.48.

Q.70. Derive the circuit of a 9's complementer when decimal digits are represented in the excess-3 code. A mode control input determines whether the digit is complemented or not. What is the advantage of using this code over BCD?

Solution: The excess-3 code is self complementing code. Therefore, to get 9's complement we need to complement each bit.

$M = 0$ for $x = B$

$M = 0$ for $x =$ 9's complement of B

M	B_i	$X_i = B_i \oplus M$
0	0	0 $X_i = B_i$
0	1	1
1	0	1 $X¢_i = B¢_i$
1	1	0

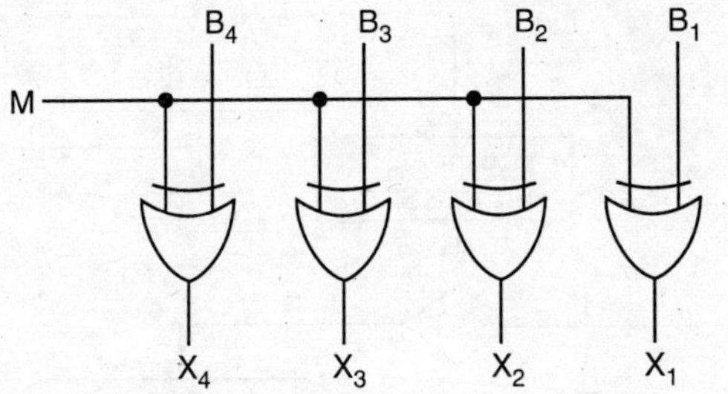

Fig. 4.49.

Q.71. Show the hardware to be used for the addition and subtraction of two decimal numbers with negative numbers insigned – 10's complement representation. Indicate how an overflou is detected. Derive the flowchart algorithm and try a few numbers to convince yourself that the algorithm produces correct results.

Solution:

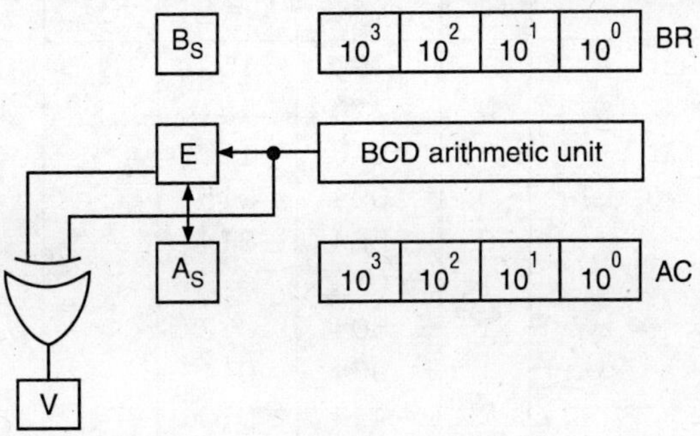

Fig. 4.50.

Q.72. Show the content of registers A, B, Q and SC during the decimal multiplication (Figure) of

(*a*) 470 × 152 and

(*b*) 999 × 199.

Assume three-digit registers and take the second number as the multiplier.

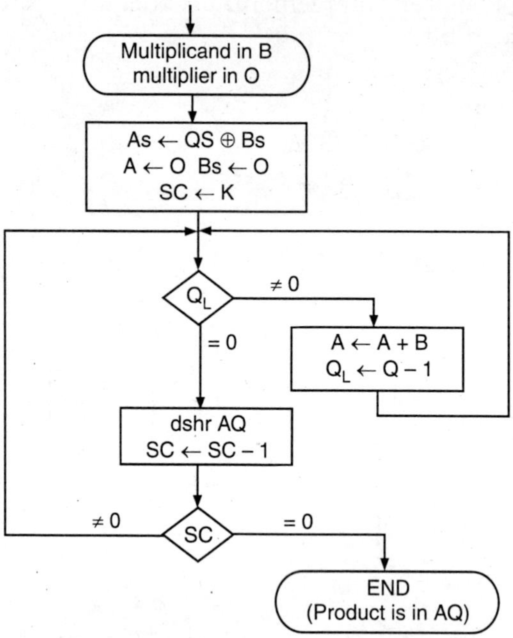

Fig. 4.51. Flowchart for decimal multiplication

Solution: (*a*) $B = 740$

	Ae	*A*	$Q = Q_1$	SC
Initial	0	000	152	3
$Q_L\ ^{1}\ 0$	0	470	151	
$Q_L\ ^{1}\ 0$	0	940	150	
$Q_L = 0$, dshr	0	094	015	2
	0	564	014	
	1	034	013	
$Q_L\ ^{1}\ 0$	1	504	012	
	1	974	011	
	2	444	010	
$Q_L\ ^{1}\ 0$, dshr	0	244	401	1
$Q_L\ ^{1}\ 0$	0	714	400	
$Q_L = 0$, dshr	0	071	440	0
		Product		

(*b*)

999	
× 199	
8991	first partial product $A_e = 8$
+ 89910	
98901	second partial product $A_e = 9$
+ 99900	
198801	final product $A_e = 1$

Q.73. Show the content of registers A, E, Q and SC during the decimal division (figure) of 1680/32. Assume two digit registers.

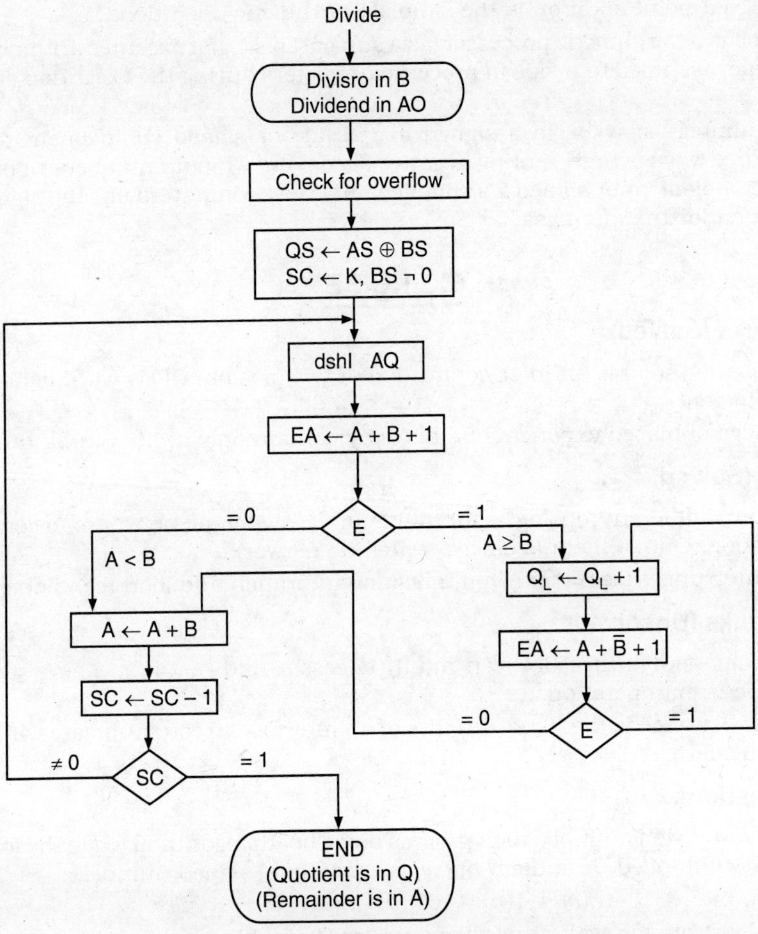

Fig. 4.52. Flowchart for decimal division

SUMMARY

1. The four basic arithmetic operations are addition, subtraction, multiplication and division.
2. Multiplication of two fixed-point binary numbers in signed magnitude representation is done with paper and pencil by a process of successive shift and add operations.
3. Booth algorithm gives a procedure for multiplying binary integers in signed-2's complement representation.
4. Division of two fixed-point binary numbers in signed magnitude representation is done with paper and pencil by a process of successive compare, shift and subtract operations.
5. The division operation may result in a quotient with an overflow.
6. Representation for floating point numbers is known as the IEEE 754 floating point standard.
7. The IEEE 754 standard has two main representations:
   (*i*) single precision, (*ii*) double precision

8. Arithmetic processing unit in central processing unit is used to execute all these operations. The arithmetic instructions are performed generally on binary or decimal data. Fixed-point numbers are used to represent integers or fractions. We can have signed or unsigned negative numbers. Fixed-point addition is the simplest arithmetic operation.
9. A processor has an arithmetic processor (as a sub part of it) that executes arithmetic operations. The data type, assumed to reside in processor, registers during the execution of an arithmetic instruction.
10. Negative numbers may be in a signed magnitude or signed complement representation. There are three ways of representing negative fixed point – binary numbers signed magnitude, signed 1's complement or signed 2's complement. Most computers use the signed magnitude representation for the mantissa.

EXERCISE

(I) Fill in the Blanks (Solved)

1. A .... system is an inter connection of two or more CPUs with memory and input-output equipment.
2. The advantage of biased exponents is that they contain only .... numbers.

(II) True or False (Solved)

1. There are many different topologies for multistorage switching networks to control processor communication. One of them is omega switching network.
2. A priority interrupt system is a communication of branch and loop techniques.

(III) Fill in the Blanks (Unsolved)

1. There are some similarities between multi processor and .... system since both support concurrent operations.
2. The .... needs the usual circuits of an interface to communicate with the CPU and input/output device.

(IV) Detailed Questions

1. Show the step-by-step multiplication process using booth algorithm, when the following binary numbers are multiplied. Assume 5-bit registers to hold signed numbers.

   (*a*) $+ 10 \times + 14$ (*b*) $+ 10 \times - 14$
2. Derive an algorithm for evaluating the square root of a binary fixed-point number.
3. Why should the sign of the remainder after a division be the same as the sign of the dividend?
4. Show that there can be no mantissa overflow after a multiplication operation.
5. Show the hardware to be used for the addition and subtraction.
6. List four peripheral devices that produce an acceptable output for a person to understand.
7. What is the difference between isolated I/O and memory mapped I/O? What are the advantages and disadvantages of each?
8. Give at least six status conditions for the setting of individual bits in the status register of an asynchronous communication interface.
9. What programming steps are required to check when a source interrupts the computer while it is still being serviced by a previous interrupt request from the same source?
10. Why does DMA have priority over the CPU when both request a memory transfer?
11. What is the advantage of using interrupt – initiated data transfer over transfer under program control without an interrupt?
12. The following transfer statements specify a memory. Explain the memory operation in each case:

   (*i*) $R_2 \leftarrow M[AR]$ (*ii*) $M\ (AR) \leftarrow R_3$

   (*iii*) $R_5 \leftarrow M\ (R_5)$.

13. Register A holds the 8 bit binary 11011001. Determine the B-operand and the logic micro-operation to be performed in order to change the value in A to:
    (*i*) 01101101
    (*ii*) 11111101.
14. An 8 bit register contains binary value 11001010. What is the register value after a circular shift right.
15. Starting from an initial value of R = 11011101, determine the sequence of binary values in R after a logical shift-left, followed by a circular shift-right followed by a logical shift right and a circular shift left.
16. Draw the diagram of bus system using 4 × 1 multiplex.

Answers

(I) Fill in the Blanks (Solved)

1. computer
2. positive

(II) True or False (Solved)

1. True
2. False

❐❐❐

Chapter

5

Processor Organization

Chapter Outline

- Introduction
- Parts of Processors
- Blocks Diagram of an Accumulators Based CPU
- Function of Each unit of Accumulator Based CPU
- General Register Organization
- Stack Organization
- Converting From Infix Notation
- Instruction Format
- Addressing Modes

5.0 INTRODUCTION

The part of the computer that performs the bulk of data-processing operations is called CPU. The CPU is made up of three major parts, as shown in Fig. 5.1. The register set stores intermediate data used during the execution of the instructions. The ALU performs the required micro-operations for executing the instructions. The control unit supervises the transfer of information among the registers and instructs the ALU as to which operation to perform.

5.1 PARTS OF PROCESSOR

Arithmetic/Logic Unit

- performs all mathematical and logical operations within the CPU.

Control Unit

- executes each control signal in the sequence determined by the decoder
- the signals generated depend on the instruction and external inputs.

Registers

- high speed memory locations used to store important information during CPU operations.

Program Counter

- holds the address of the next instruction to be executed.

• incremented every time a new instruction is fetched.

Instruction Register

• holds a copy of the currently executing instruction.

Buffer Registers

• isolate the external bus of the CPU from its internal bus.

Status Register

• stores the program status word, which consists of flag bits and control bits
• flag bits = set automatically by certain events during arithmetic and logic operations
• control bits = set by the program in order to enable certain modes in CPU operation

Stack Pointer

• points to the top of the stack
• stack = a set of contiguous memory locations where information items are added or removed in a stack-like fashion

General Purpose Registers

• programmable registers for various use
• "scratch pad"

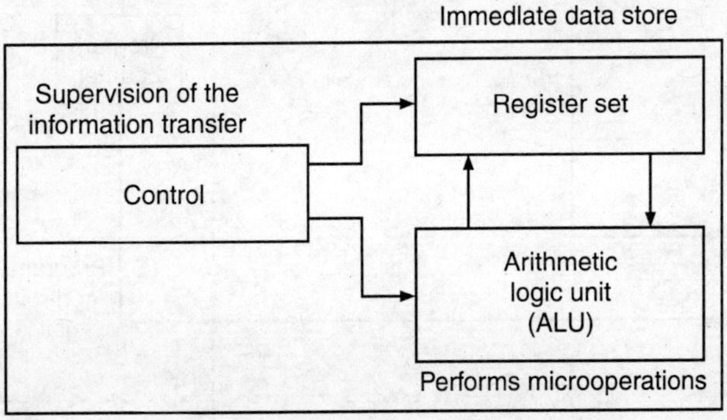

Fig. 5.1. Major components of CPU

Instruction set provides the specification for the design of the CPU. The design of CPU involves choosing the hardware for implementing the machine instructions.

In the following two sections we present the two main CPU organizations.

A programmer using assembly language has to be aware of the register set, the memory structure, the type of data supported by the instructions, and the function of each.

Most computers fall into one of three types of CPU organizations:

1. Single accumulator organization (uses **one** address field).
2. General register organization (uses **two or three** address fields).
3. Stack organization (uses **zero** address field).

5.2 BLOCK DIAGRAM OF AN ACCUMULATOR BASED CPU

The figure below shows the simple accumulator based CPU given by Neumann which is the basis for almost all CPUs. It consists minimum number of registers and necessary circuits. In this CPU one register called the accumulator is used to store an input data or an output result in the execution of most instructions. Hence it is known as accumulator based CPU.

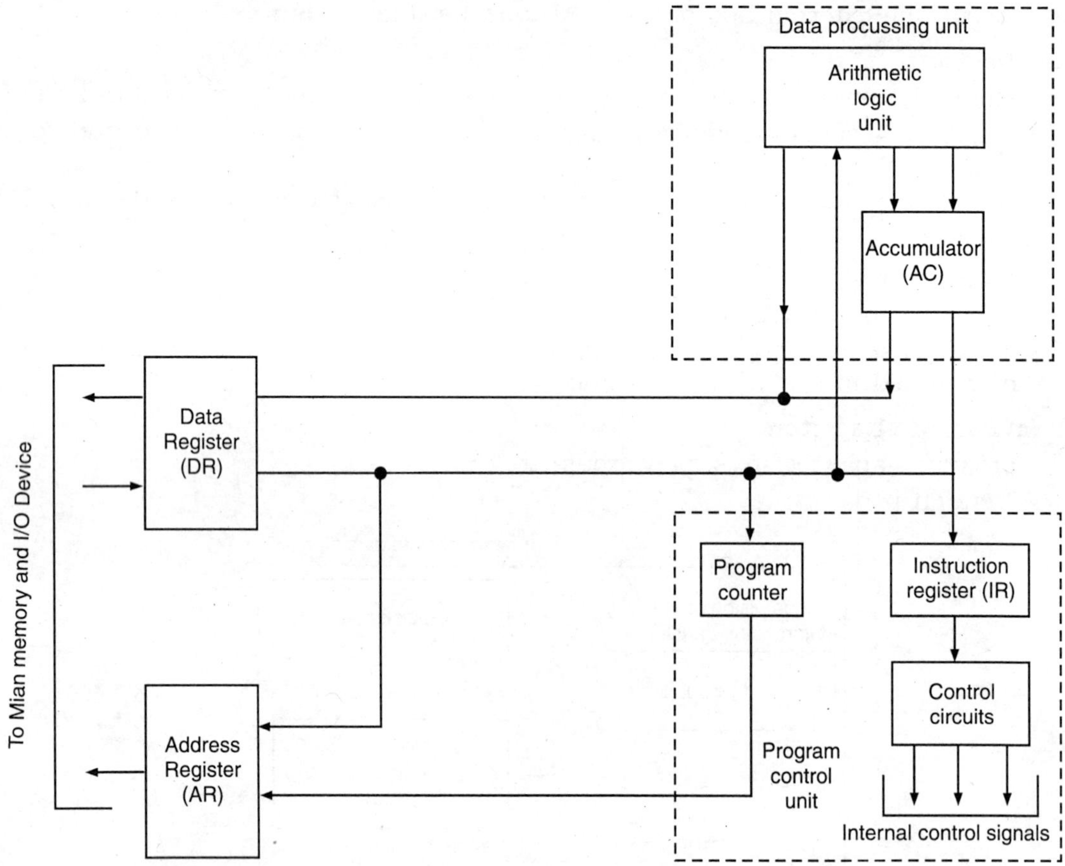

Fig. 5.2.

The accumulator is the main register of the arithmetic logic unit. The data register acts as buffer between the CPU and main memory. The other major registers are the program counter PC which stores the address of the next instruction which is to be executed. Instruction register IR which holds the opcode of the current instruction which is to be executed and the memory address register AR which holds the address of the current instruction.

5.3 FUNCTION OF EACH UNIT OF ACCUMULATOR BASED CPU

The accumulator based CPU consists the following units:

1. Data Processing Unit
2. Program Control Unit
3. Memory and input/output interface Unit

1. **Data Processing Unit:** This is the unit where the data is processed to get the output, *i.e.* some results. This unit can be sub-divided into:

(i) Arithmetic Logic Unit ALU
(ii) Accumulator AC

(i) **Arithmetic Logic Unit:** The arithmetic logic unit performs all arithmetic and logical operations. The necessary data is supplied by the accumulator and other registers. It performs the required operation on the operands contained in the instruction whenever it is activated by control unit. The results of all the arithmetic and logical operations are stored back in accumulator.

(ii) **Accumulator:** The accumulator is the main operand register of the arithmetic logic unit. The data is first loaded into the accumulator before it is transferred to the arithmetic logic unit for any operation. In performing these operations generally one of the operand is held in the accumulator. The result after performing the operation is stored again in accumulator. Accumulator is also referred as primary register. It is the most complicated register in the computer because it serves as a processor register. A simple accumulator based CPU consists only one Accumulator. It is also possible to have two or more accumulators if the processor is of the type multiprocessor.

2. **Program Control Unit:** This is the subsystem of CPU which helps in making various parts of the CPU to work in a proper sequence, i.e. it controls various parts of CPU. It provides the necessary timing and control signals to all operations depending upon the instructions in the program. It also controls the movement of data within the CPU registers, I/O devices and memory. This unit basically consists of:

(i) Program Counter
(ii) Instruction Register
(iii) Control Circuits.

(i) **Program Counter:** It is also one of the registers of CPU. It is used to store the address of the next instruction which is to be executed, i.e. it holds the address of the next instruction to be read from memory after the current instruction is executed. This register goes through a counting sequence and causes the computer to read sequential instructions previously stored in memory unless or otherwise a branch instruction comes into the picture. The address part of a branch instruction is transferred to PC to become the address of the next instruction.

(ii) **Instruction Register:** This register which holds the opcode of the current instruction is one of the high speed registers under the program control unit of the CPU. Generally this register is not directly available to the programmer.

The operation code (Opcode) of an instruction is a group of bits that define such operations as add, subtract, multiply, division, shift, and compliment, etc.

(iii) **Control Circuits:** This is the most important part of the program control unit. It is responsible for all-most every operation that takes place in CPU. These circuits provide the necessary control signals for fetching instructions decoding the opcodes. In most cases the output of the instruction register is given to these circuits. The internal control signals are the output of these circuits.

3. **Memory and input/output interface Unit:** To this unit input/output devices and memory are connected. This unit consists of:
   (*i*) Data register and
   (*ii*) Address registers.
   (*i*) **Data Registers:** This register DR acts as a buffer between the CPU and main memory. It is used as input register to store an operand during the execution of an instruction.
   (*ii*) **Address Register:** This contains the address of the present instruction which is obtained from the program control unit. In most of the cases the address is contained in the memory hence it is also known as Memory Address Register MAR.

In addition to this, the accumulator based CPU consists special registers such as stack pointer and flag registers. The stack pointer is used to hold the address of the most recently entered item into the stack. It always shows the top of the stack. Some times temporarily it holds the contents of program counter. The operations performed with in the stack are very fast.

The flag registers indicates the status. Depending upon the result the flags are set. Most commonly used flags in the flag register are zero flag and carry flag.

5.4 GENERAL REGISTER ORGANIZATION

Storing pointers, counters, return addresses, temporary results, and partial products into the memory is not efficient: referring to memory locations is time consuming, it is more efficient to use registers. If CPU has a large number of registers, a common bus is used to connect the registers. Arithmetic Logic Unit (ALU) is used to perform various micro-operations. A bus organization of seven CPU registers is shown in Fig. 5.3.

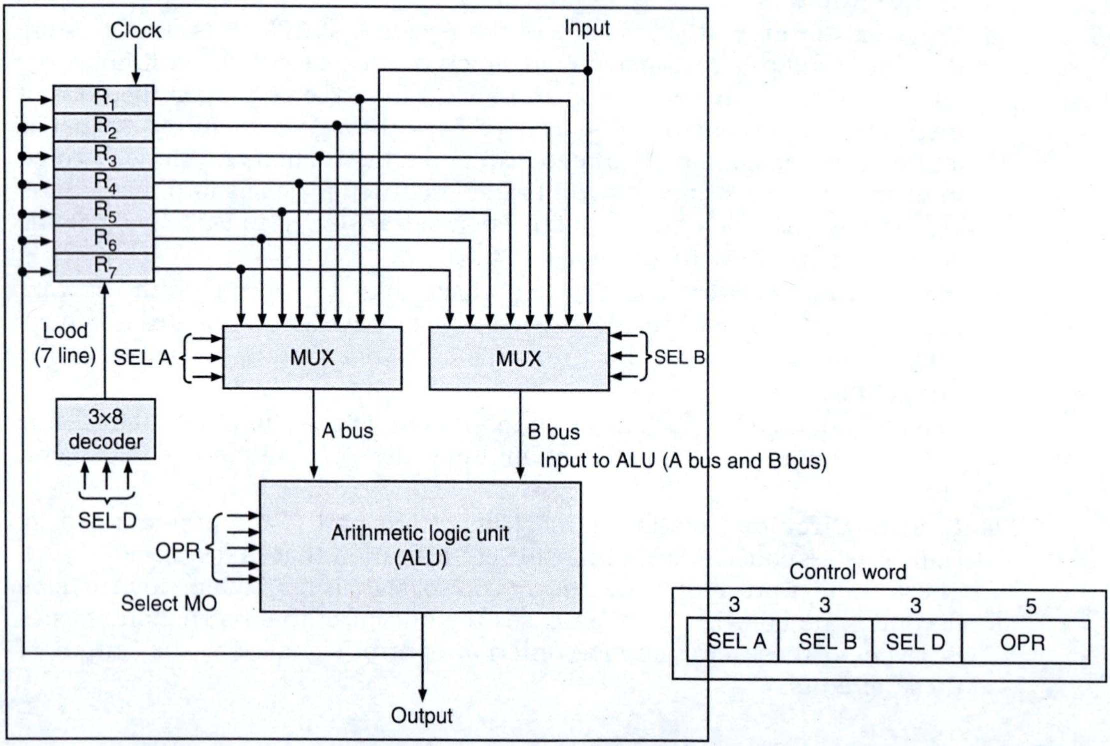

Fig. 5.3. Register set with common ALU

Control unit operates the CPU bus system. For example:

$R_1 \quad R_2 + R_3$

1. SELA is used to place R_2 into bus A.
2. SELB is used to place R_3 into bus B.
3. OPR selects the arithmetic addition.
4. SELD is used to transfer the result into R_1.

The buses are implemented with multiplexers of 3-state gates. The state of 14 binary selection inputs specifies a control word. The 14-bit control word (when applied to the selection inputs specify a micro-operation). The encoding of register selections in specified in Table 5.1.

Table 5.1 Encoding of register selection fields

Binary Code	SELA	SELB	SELD
000	Input	Input	No e
001	R1	R1	R1
001	R2	R2	R2
010	R3	R3	R3
100	R4	R4	R4
101	R5	R5	R5
110	R6	R6	R6
111	R7	R7	R7

Input = external input

If SELD = 000 then the content of the output in bus is available in the external output

ALU provides arithmetic and logic operations. The function table of the ALU (presented here) is listed in Table 5.2.

Table 5.2. Encoding of ALU operations

OPR	***Operation***	***Symbol***
Select	Transfer A	TSFA
00000	Increment A	INCA
00001	Add $A + B$	ADD
00101	Subtract $A - B$	SUB
00110	Decrement A	DECA
01000	AND A and B	AND
01010	OR A and B	OR
01100	XOR A and B	XOR
01110	Complement A	COMA
10000	Shift right A	SHRA
11000	Shift left A	SHLA

Examples: There are some examples of micro-operations and corresponding control words:

Table 5.3. Examples of micro-operations for the CPU

Symbolic Designation					
Microoperation	SEl A	SEl B	SEl D	OPR	Control Word
R1 ← R2 ← R3	R2	R3	R1	SUB	010 011 001 00101
R4 ← R4 V R5	R4	R5	R4	OR	100 101 100 01010
6 ← R6 + 1	R6	—	R6	INCA	110 000 110 00001
R7 ← R1	R1	—	R7	TSFA	001 000 111 00000
Output ← R2	R2	—	None	TSFA	010 000 000 00000
Output ← Input	Input	—	None	TSFA	000 000 000 00000
R4 ← shl R4	R4	—	R4	SHLA	100 000 100 11000
R5 ← 0	R5	R5	R5	XOR	101 101 101 01100

(XOR (x,x) = 0 (XOR (x,x) = 0

The most efficient way to generate control words with a large number of bits is to store them in a memory unit, which is referred to as control memory (control store). By reading consecutive control words from memory, it is possible to initiate the desired sequence of micro-operations for the CPU, micro-programmed control.

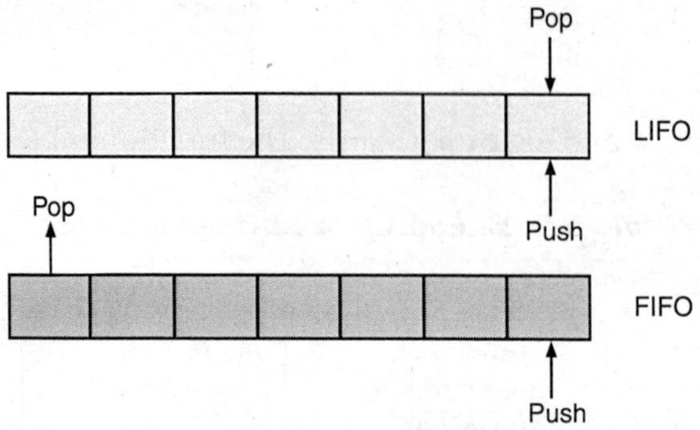

Fig. 5.4.

In computer science, a stack is a data structure that works on the principle of Last In First Out (LIFO). This means that the last item put on the stack is the first item that can be taken off, like a physical stack of plates. A stack-based computer system is one that is based on the use of stacks, rather than being register based.

5.5 STACK ORGANIZATION

5.5.1 Register Stack

Stack can reside in a portion of a large memory unit or it can be organized as a collection of a finite number of (fast) registers. Fig. 5.5. organization of 64-word register stack.

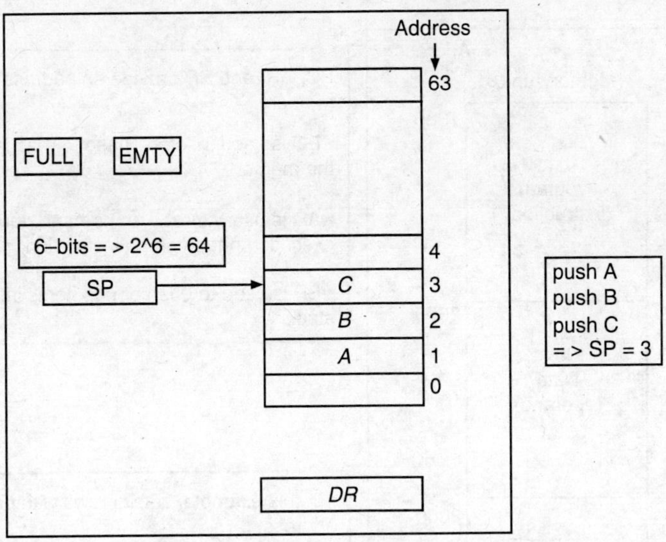

Fig. 5.5. Block diagram of a 64-word stack0

1. Push (performed if stack is not full, i.e. if FULL = 0):

$SP \leftarrow SP + 1$	Increment stack pointer
$M[SP] \leftarrow DR$	Write item on top of the stack
If $(SP = 0)$ then (Full $\leftarrow$ 1)	Check if stack is full
EMPTY $\leftarrow$ 0	Mark the stack not empty

2. Pop (performed if stack is not empty, i.e. if EMTY = 0):

$DR \leftarrow M[SP]$	Read item from the top of stack
$SP \leftarrow SP + 1$	Decrement stack pointer
If $(SP = 0)$ then (EMPTY $\leftarrow$ 1)	Check if stack is empty
FULL $\leftarrow$ 0	Mark the stack not full

5.5.2. Memory Stack

Stack can also be implemented with RAM attached to a CPU: A portion of memory is assigned to a stack operation, a processor register is used as a stack pointer. Fig. 5.6 shows how a portion of memory partitioned into three segments: program, data, and stack. Most computer do not provide hardware for checking stack overflow or underflow if registers are used to store the upper limit (*e.g.* 3000) and the lower limit (*e.g.* 4000), then after push SP can be compared against the upper limit register, and after pop against the lower limit register. The advantage of the memory stack is that CPU can refer it without having to specify an address: the address in always in SP and automatically updated during a push or pop instruction.

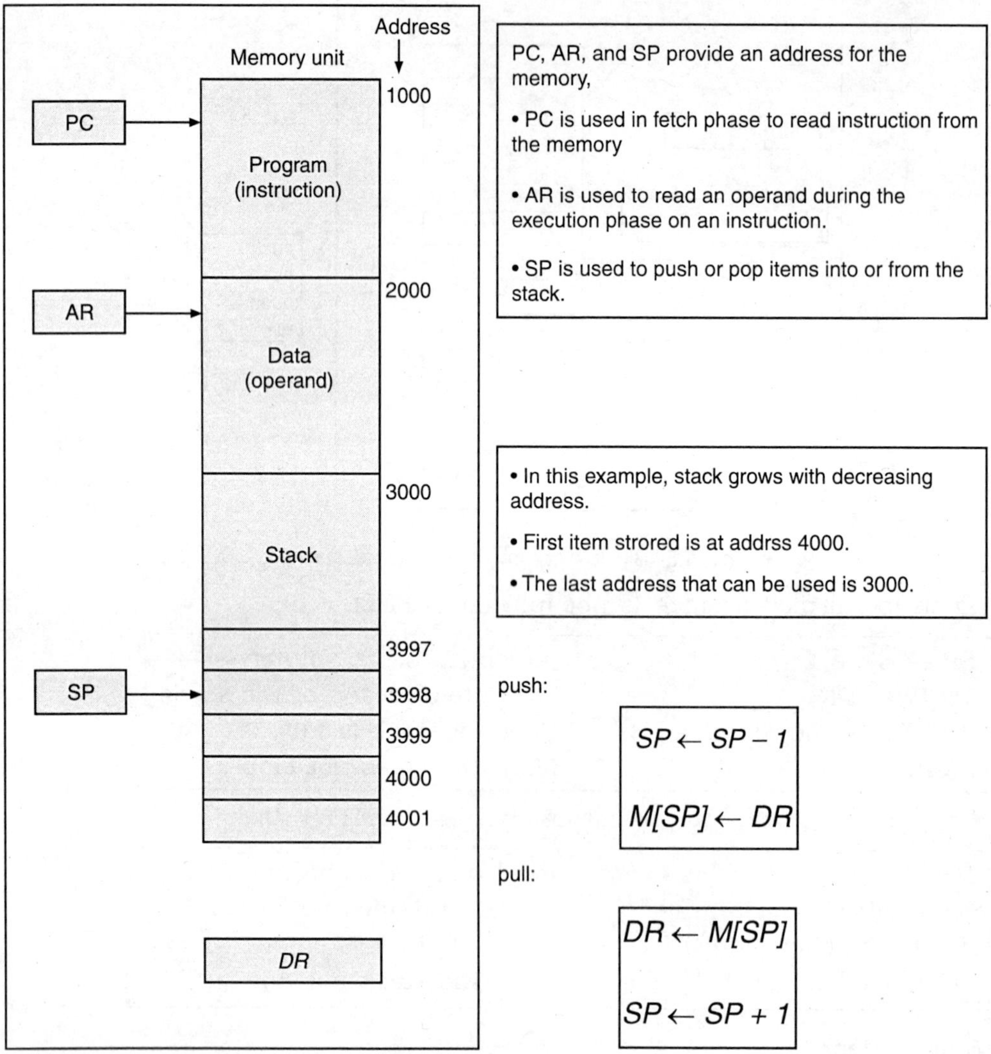

Fig. 5.6. Computer memory with program, data and stack segments

5.5.3. Reverse Polish Notation

A stack is effective for evaluating arithmetic expressions. Arithmetic operations are usually written in infix notation: each operator resides between the operands,

e.g.: $(A * B) + (C * D)$,

where * denotes multiplication: $A * B$ and $C * D$ has to be computed and stored. after the two products, sum $(A * B) + (C * D)$ is computed, there is no straight forward way to determine the next operation that is performed.

Arithmetic expressions can be presented in prefix notation (also referred to as Polish notation by Polish mathematician Lukasiewicz): operators are placed before the operands. The postfix notation (Reverse Polish Notation (RPN)) places the operator after the operand.

e.g.: $A + B$, infix notation

+ *AB*, prefix notation

AB +, postfix notation (RPN)

Reverse Polish Notation (RPN) , also known as postfix notation, is an arithmetic formula notation, derived from the Polish notation introduced in 1920 by the Polish mathematician Jan Lukasiewicz. RPN was invented by Australian philosopher and computer scientist Charles Hamblin in the mid-1950s, to enable zero-address memory stores.

As a user interface for calculation the notation was first used in Hewlett-Packard's desktop calculators from the late 1960s and then in the HP-35 handheld scientific calculator launched in 1972. In RPN the operands precede the operator, thus dispensing with the need for parentheses. For example, the expression 3 * (4 + 7) would be written as 3 4 7 + *, and done on an RPN calculator as "3", "Enter", "4", "Enter", "7", "+", "*". (Alternatively, and more-compactly, it could also be re-ordered and written as 4 7 + 3 *, and done on an RPN calculator as "4", "Enter", "7", "+", "3", "*".)

Implementations of RPN are stack-based; that is, operands are popped from a stack, and calculation results are pushed back onto it. Although this concept may seem obscure at first, RPN has the advantage of being extremely easy, and therefore fast, for a computer to analyze.

Practical Implications

Calculations proceed from left to right.

There are no brackets or parentheses, as they are unnecessary.

Operands precede operator. They are removed as the operation is evaluated.

When an operation is made, the result becomes an operand itself (for later operators).

There is no hidden state. No need to wonder if you hit an operator or not.

Example: The calculation: ((1 + 2) * 4) + 3 can be written down like this in RPN:

1 2 + 4 * 3 +

The expression is evaluated in the following way (the Stack is displayed after Operation has taken place):

Input	*Stack*	*Operation*
1	1	Push operand
2	1, 2	Push operand
+	3	Addition
4	3, 4	Push operand
*	12	Multiplication
3	12, 3	Push operand
+	15	Addition

The final result, 15, lies on the top of the stack at the end of the calculation.

5.6 CONVERTING FROM INFIX NOTATION

Like the evaluation of RPN, conversion from infix notation to RPN is stack-based. Infix expressions are the form of math most people are used to, for instance 3 + 4 or 3 + 4 * (2 – 1). For the conversion there are 2 text variables (strings), the input and the output. There is also a stack holding operators not yet added to the output stack. To convert, the program reads each letter in order and does something based on that letter.

A simple conversion

Input: 3 + 4

Add 3 to the output queue (whenever a number is read it is #added to the output)
Add 4 to the output queue
Push + (or its ID) onto the operator stack
After reading expression pop the operators off the stack and #add them to the output.
In this case there is only one, "+".
Output 3 4 +

This already shows a couple of rules:

- All numbers are added to the output when they are read.
- At the end of reading the expression, pop all operators off the stack and onto the output.

The algorithm in detail:

- While there are tokens to be read:

  Read a token.

  - If the token is a number, then add it to the output queue.
  - If the token is an operator, o1, then:
    1. while there is an operator, o2, at the top of the stack, and either o1 is left-associative and its precedence is less than or equal to that of o2, or o1 is right-associative and its precedence is less than that of o2, pop o2 off the stack, onto the output queue;
    2. push o1 onto the operator stack.
  - If the token is a left parenthesis, then push it onto the stack.
  - If the token is a right parenthesis, then pop operators off the stack, onto the output queue, until the token at the top of the stack is a left parenthesis, at which point it is popped off the stack but not added to the output queue. If the stack runs out without finding a left parenthesis, then there are mismatched parentheses.
  - When there are no more tokens to read, pop all the tokens, if any, off the stack, add each to the output as it is popped out and exit. (These must only be operators; if a left parenthesis is popped, then there are mismatched parentheses.)

Example: Consider the arithmetic expression:

$$(3 * 4) + (5 * 6)$$

In RPN, it is expressed as

$$34 * 56 * +$$

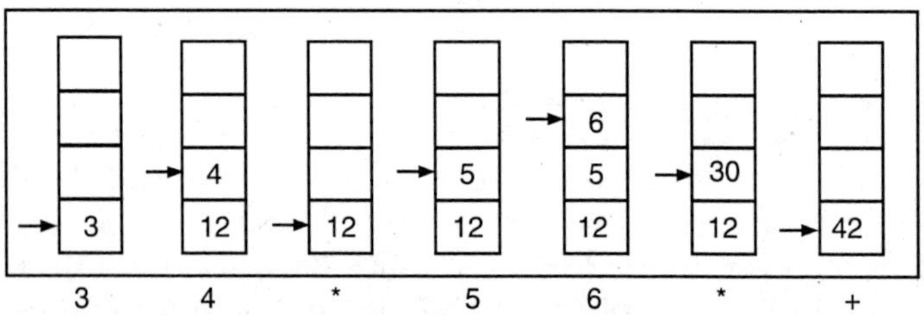

Fig. 5.7. Stack operations to evaluate 3 * 4 + 5 * 6

Example: Consider the expression

$$X = (A + B) * (C + D)$$

Convert the expression to the reverse polish notation and show the stack operations for evaluating the result.

The conversion: $(A + B) * (C + D)$ à $AB + CD + *$

Instructions to be executed in stack:

PUSH	A
PUSH	B
ADD	
PUSH	C
PUSH	D
ADD	
MUL	
POP	X

The stack operations will be as follows:

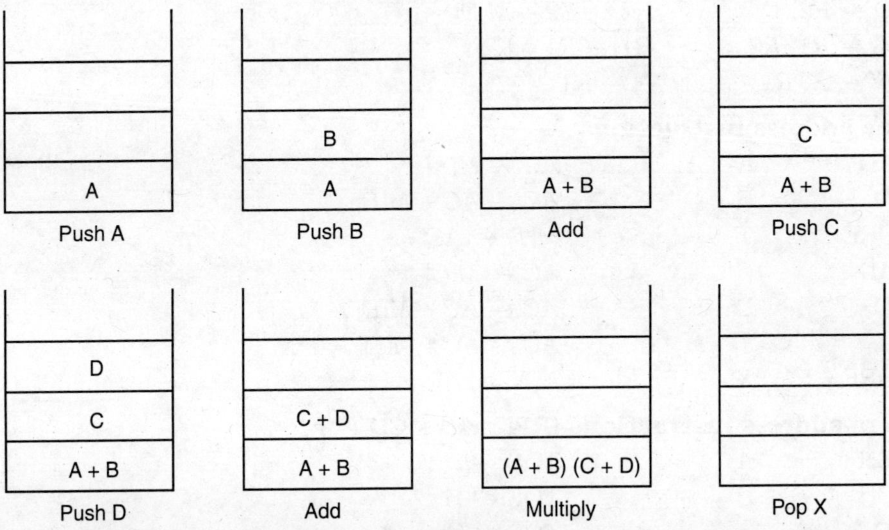

Fig. 5.8.

5.7 INSTRUCTION FORMAT

The format of an instruction is usually depicted in a rectangular box symbolizing the bits of the instruction as they appear in memory words or in a control register. The bits of the instruction are divided into groups called fields. The most common fields found in instruction formats are:

1. Operation code specifying the operation: add, subtract, complement, etc.
2. Address field designating a memory address or a register.
3. Mode field for specifying the way for determining the effective address of an operand.

The number of address fields in the instruction format of a computer depends on the internal organization of its registers. E.g.: MIPS (a RISC microprocessor architecture developed by MIPS Computer Systems Inc.).

Computers may have instructions of several different lengths containing varying number of addresses. The number of address fields in the instruction format of a computer depends on the internal organization of its register.

Example: To illustrate the influence of the number of addresses on computer programs, we will evaluate the arithmetic statement

$$X = (A + B) * (C + D)$$

using zero, one, two, or three address instructions.

1. Using three address instructions

ADD	$R1, A, B$	$R1 \leftarrow M[A] + M[B]$
ADD	$R2, C, D$	$R2 \leftarrow M[C] + M[D]$
MUL	$X, R1, R2$	$X \leftarrow R1 * R$

2. Using two address instructions

MOV $\leftarrow R1, A$	$R1 \leftarrow$ M[A]
ADD $\leftarrow R1, B$	$R1 \leftarrow R1 + M[B]$
MOV $\leftarrow R2, C$	$R2 \leftarrow$ M[C]
ADD $\leftarrow R2, D$	$R2 \leftarrow R2 + M[D]$
MUL $\leftarrow R1, R2$	$R1 \leftarrow R1 * R2$
MOV $\leftarrow X, R1$	$X \leftarrow R1$

3. Using one address instructions

LOAD	A	$AC \leftarrow M[A]$
ADD	B	$AC \leftarrow AC + M[B]$
STORE	T	$M[T] \leftarrow AC$
LOAD	C	$AC \leftarrow M[C]$
ADD	D	$AC \leftarrow AC + M[D]$
MUL	T	$AC \leftarrow AC * M[T]$
STORE	X	$M[X] \leftarrow AC$

4. Using zero address instructions (RPN: *AB* + *CD* + *)

PUSH	A	TOS $\leftarrow A$
PUSH	B	TOS $\leftarrow B$
ADD		TOS $\leftarrow (A + B)$
PUSH	C	TOS $\leftarrow C$
PUSH	D	TOS $\leftarrow D$
ADD		TOS $\leftarrow (C + D)$
MUL		TOS $\leftarrow (C + D) * (A + B)$
POP	X	$M[X] \leftarrow$ TOS

The MIPS ISA instructions fall into three categories: *R*-type, *I*-type, and *J*-type. Not all ISAs divide their instructions this neatly. This is one reason to study MIPS as a first assembly language. The format is simple.

R-type

R-type instructions refer to register type instructions. Of the three formats, the R-type is the most complex.

B31-26	***B25-21***	***B20-16***	***B15-11***	***B10-6***	***B5-0***
opcode	register s	register t	register d	Shift amount	function

This is the format of the R-type instruction, when it is encoded in machine code.

The prototypical R-type instruction is:

Add $rd, $rs, $rt

where $rd refers to some register d (d is shown as a variable, however, to use the instruction, you must put a number between 0 and 31, inclusive for d). $rs, $rt are also registers.

The semantics of the instruction are ;

$$R[d] = R[s] + R[t]$$

where the addition is signed addition.

You will notice that the order of the registers in the instruction is the destination register ($rd), followed by the two source registers ($rs and $rt).

However, the actual binary format (shown in the table above) stores the two source registers first, then the destination register. Thus, how the assembly language programmer uses the instruction, and how the instruction is stored in binary, do not always have to match.

Let's explain each of the fields of the R-type instruction.

- **opcode (B31-26):** Opcode is short for "operation code". The opcode is a binary encoding for the instruction. Opcodes are seen in all ISAs. In MIPS, there is an opcode for add.
  The opcode in MIPS ISA is only 6 bits. Ordinarily, this means there are only 64 possible instructions. Even for a RISC ISA, which typically has few instructions, 64 is quite small. For R-type instructions, an additional 6 bits are used (B5-0) called the function. Thus, the 6 bits of the opcode and the 6 bits of the function specify the kind of instruction for R-type instructions.
- **rd (B25-21):** This is the destination register. The destination register is the register where the result of the operation is stored.
- **rs (B20-16):** This is the first source register. The source register is the register that holds one of the arguments of the operation.
- **rt (B15-11):** This is the second source register.
- **shift amount (B10-6):** The amount of bits to shift. Used in shift instructions.
- **function (B5-0):** An additional 6 bits used to specify the operation, in addition to the opcode.

I-type instructions

I-type is short for "immediate type". The format of an I-type instuction looks like:

B31-26	*B25-21*	*B20-16*	*B15-0*
opcode	Registers	registert	immediate

The prototypical I-type instruction looks like:

Add $rt, $rs, immed

In this case, $rt is the destination register, and $rs is the only source register. It is unusual that $rd is not used, and that $rd does not appear in bit positions B25-21 for both R-type and I-type instructions. Presumably, the designers of the MIPS ISA had their reasons for not making the destination register at a particular location for R-type and I-type.

The semantics of the instruction are:

$$R[t] = R[s] + (IR15)16\ IR15\text{-}0$$

where IR refers to the instruction register, the register where the current instruction is stored. (IR15)16 means that bit *B*15 of the instruction register (which is the sign bit of the immediate value) is repeated 16 times. This is then followed by IR15-0, which is the 16 bits of the immediate value.

Basically, the semantics says to sign-extend the immediate value to 32 bits, add it (using signed addition) to register R[s], and store the result in register $rt.

J-type instructions

J-type is short for "jump type". The format of an J-type instuction looks like:

B31-26	***B25-0***
opcode	target

The prototypical I-type instruction looks like:

***j* target**

The semantics of the j instruction (j means jump) are:

$$PC\ (PC31-28\ IR25-0\ 00)$$

where *PC* is the program counter, which stores the current address of the instruction being executed. You update the *PC* by using the upper 4 bits of the program counter, followed by the 26 bits of the target (which is the lower 26 bits of the instruction register), followed by two 0's, which creates a 32 bit address. The jump instruction will be explained in more detail in a future set of notes.

5.8 ADDRESSING MODES

The addressing mode specifies a rule for interpreting or modifying the address field of the instruction before the operand is actually referenced. Computers use addressing mode techniques for the purpose of accommodating one or both of the following provisions:

1. To provide programming versatility for the user: pointers to memory, counters for loop control, indexing data, etc.
2. To reduce the number of bits in the addressing field of the instruction.

The decoding phase of an instruction cycle determines the addressing mode(s) and the locations (registers and/or memory locations) of operands. Depending on the CPU, an instruction can have more than one address field, and each address field may be associated with its own particular addressing mode.

Classifying Instructions According to Number of Operands

Possible to have anywhere from 0 to 4 or more operands. Fewer operands make for shorter but less flexible instructions since some of the operands are implicit.

Zero Address: any stack architecture - *e.g.*, IJVM Mic -1

One Address: accumulator is implicit operand - *e.g.*, PDP-8, IAS

Two Address: one operand is both source and destination - *e.g.*, Intel 80 × 86

Three Address: two source operands and one destination operand

VAX has 3 address instructions (a good example of a CISC architecture)

ARC has 3 register instructions (a good example of RISC architecture)

Four Address: 4th operand is either address of next instruction (obsolete technique) or for some reason the instruction requires more than three operands (*e.g.*, integer division produces quotient and remainder).

The "Standard" Addressing Modes – A Functional View

Level refers to the number of times memory is referenced to obtain the operand. "Half-levels" (non-canonical) refer to register look up.

EAddr denotes "effective address"; C( ) denotes "contents of" or "contents at".

Binding time refers to the latest time that the effective address can be fixed.

The value of the Operand Field of an instruction can be considered as either an Address or a Displacement value

Immediate Mode (level 0): operand is part of instruction; i.e. operand contained in Operand Field - binding time is at assembly time: ARC, IJVM, and Intel 80 × 86.

Register Addressing (level 1/2?): operand contained in register - binding time is assembly time: ARC, Intel 80 × 86.

Direct (level 1): effective address is part of instruction or directly obtained (calculated) from instruction - binding time is assembly tiime: PDP-8, ARC, IJVM, Intel 80 × 86.

Indirect (level 2): sometimes called "deferred"; operand field of instruction contains address of the address of the operand. This is like a "pointer" which can be modified - binding time is run time.

Register Indirect Addressing: effective address of operand contained in a register EAddr = C(Reg) ARC, Intel 80 × 86.

Memory Indirect Addressing: effective address of operand contained in a word in memoty Operand Field contains an Address so EAddr = C(Address), PDP-8.

Indexed or Base (or Index)) + Offset (Indirect) or Register + Offset (Indirect): Level 1 1/2- effective address obtained by adding value of operand field or offset to contents of register - EAddr = C(Reg) + Offset - binding time is run time Sometimes called displacement addressing.

Array Type: Fixed Base (address) + Variable Register Offset Operand Field contains a base Address so EAddr = base Address + C(Reg) used to implement arrays.

Record Type: Fixed Register + Variable Offset (address) Operand Field contains a Displacement so EAddr = C(Reg) + Displacement used to implement records and stack frames (local variables for subroutines).

Program Counter Based: PC + offset: relative addressing Operand Field contains a Displacement so EAddr = C(PC) + Displacement Near and Short Jump instructions on the Intel 80 × 86 use this.

Base + Index (Indirect) or Double Indexed (Indirect): Level 1 1/2 - effective address is obtained by adding the contents of two registers - EAddr = C(Reg1) + C(Reg2) - binding time is run time: ARC, Intel 80 × 86.

Base + Index + Offset (Indirect) - Level 1 1/2 - effective address obtained by adding contents ot two registers plus an offset - EAddr = C(Reg1) + C(Reg2) + offset - binding time is run time: Intel 80 × 86.

Indexed With Scaling: Base + Register Offset * Scaling Factor - binding time is run time variant of Indexing useful for array calculations where size of component is multiple bytes long.

Operand Field contains base Address then EAddr = base Address + C(Reg) * Scaling Factor.

Auto-indexing: Register Indirect or Indexing with auto-increment /decrement of register.

Auto-Increment: EAddr = C(Reg); Reg = C(Reg) + 1. Note indirect-increment order!

Auto-Decrement: Reg = C(Reg) - 1; EAddr = C(Reg). Note decrement-indirect order!

PDP-8 auto-indexing mode increments then uses indirection: IReg = C(IReg) + 1; EAddr = C(IReg)

Stack Addressing: push and pop - a variant of register indirect with auto-increment/decrement using the SP register implicitly.

Push: EAddr = C(SP). SP = C(SP) -1 ; C(C(SP)) = Source auto-decrement!

Pop: EAddr = C(SP); Destination = C(C(SP)); SP = C(SP) + 1; auto-increment.

Some "Not-So-Standard" Addressing Modes Auto-increment/Auto-decrement Indirect

EAddr = C(C(Reg)); Reg = C(Reg) + 1; PDP-11 Mode 3 Addressing.

The PDP-11 used the PC as the register to implement direct addressing this way!

Reg = C(Reg) - 1; EAddr = C(C(Reg)); PDP-11 Mode 5 Addressing.

Index 80 × 86 Supports this using Base Registers (BX, BP) and Index Registers (SI, DI).

Indexing Indirect: Indexing coupled with indirection (rare?)

postindexing: indexing after indirection: EAddr = C(Address) + C(Reg)

preindexing: indirection after indexing: EAddr = C(Address + C(Reg))

Multiple Indirection: requires indirect bit as part of address field (PDP-10 could do this)

In this chapter, we will cover the following modes:

1. **Implied mode:** The operands are implicitly specified by the instruction, e.g.: "complement accumulator".
2. **Immediate mode:** The operand is specified in the instruction. Can be used e.g. to initialize register to constant value (immediate operand).
3. **Register mode:** Operands are in registers that reside within the CPU. The particular register is selected with the register field of the instruction.
4. **Register indirect mode:** The content of a register specifies the address of the operand in memory.
5. **Autoincrement or autodecrement mode:** Similar to register indirect mode but the content of the register is automatically incremented/decremented after data access.
6. **Direct address mode:** The effective address is equal to the address part of the instruction. The operand resides in this address.
7. **Indirect address mode:** The address field gives the address where the effective address is stored in memory.
8. **Relative address mode:** The content of the program counter (PC) is added to the address part of the instruction in order to obtain the effective address, i.e. the effective address is relative to the address of the next instruction. This address mode can be used in branch type instructions when the branch address is in the area surrounding the instruction word.
9. **Indexed addressing mode:** Content of an index register (special CPU register) is added to the address part of the instruction to obtain the effective address. The index register can be incremented for accessing consecutive operands.
10. **Base register addressing mode:** The content of a base register in added to the address part of the instruction to obtain the effective address. The address part of the instruction gives the displacement relative to the base address.

Numerical Example

For each possible mode, calculate the effective address and the operand that must be loaded into AC. Use the instruction defined in Fig. 5.9. Table 5.4 show the tabular list of the numerical example.

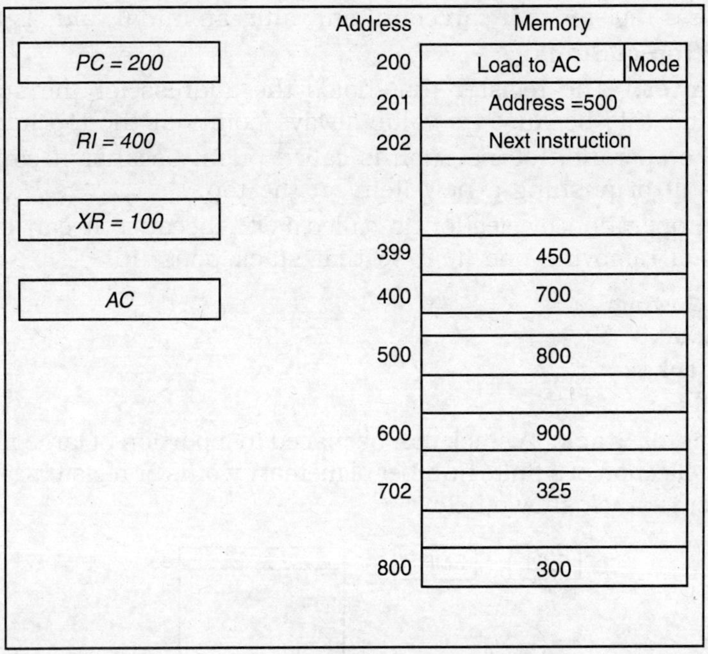

Fig. 5.9. Numerical example for addressing modes

Table 5.4. Tabular list of the numerical example

Addressing mode	Effective Address	Content of AC
Direct address	500	800
Immediate operand	201	500
Indirect address	800	300
Relative address	702	325
Indexed address	600	900
Register	—	400
Register indirect	400	700
Autoincrement	400	700
Autodecrement	399	450

Decrements R1 prior the execution (→ Autodecrement)

SOLVED PROBLEMS

Q.1. What is the stack organization? Describe all functionality regarding to stack?

Sol. Stack is very useful in the CPU of most computers. Stack is also called last in first out (LIFO). A stack is a storage device that stores information in such a manner that the item last is the first item retrieved. The stack in digital computers is essentially a memory unit with an address register that can count only after an initial value is loaded into it.

Various Function of Stack:

(*a*) **Stack Pointer:** The register that holds the address for the stack is called a stack pointer (SP) because its value always points at the top item in the stack.

(*b*) **PUSH:** The operation of insertion is called push, because it can be thought of as the result of pushing a new item on the top.

(*c*) **POP:** The operation of election is called POP., because it can be though of as the result of removing one item so that stack pops up.

Q.2. Define the following:

(*a*) Register stack

(*b*) Memory stack

(*c*) Stack limits

Solution: (*a*) **Register Stack:** A stack can be placed in a portion of large memory or it can be organized as a collection of a finite number of memory words or registers. The organization of 64-word register is stack shown below:

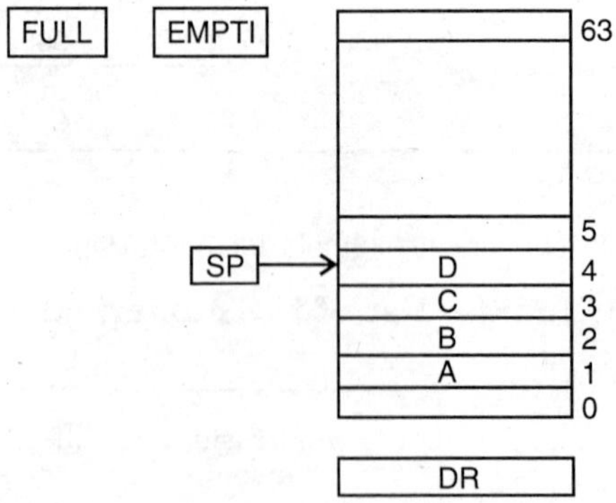

Fig. 5.10. Block diagram of 64-word stack

The stack point register SP contains a binary numbers whose value is equal to the address of the word is currently on top of the stack. So, in above block diagram, the content of SP is now 4. To remove the top item, the stack is popped by reading the memory word at address 3 and determination the content of SP. If stack is not full then use PUSH.

PUSH operation:

SP ←	SP + 1	increment stack pointer
M[SP] ←	DR	write item on top of the stack

If (SP = 0) then (Full ← 1), check if stack is full

EMPTY ←	0	mark the stack not Empty.

Pop Operation: If Empty = 0

DR ←	M[SP]	Read item from the top of stack
SP ←	SP – 1	Decrement stack pointer

It (SP = 0) then (empty ← 1) check if stack is empty.

Full ←	0	Mark the stack not full.

Memory Stack: A stack can exist as a stand alone unit or can be implemented in a random access memory attached to a CPU. The implementation of a stack in the CPU is done by assigning a portion of memory to a stack operation and using a processor register as a stack pointer. Portion of computer memory partitioned into three segments.

Program

Data and

Stack

The program counter PC points at the address of next instrument in the program.

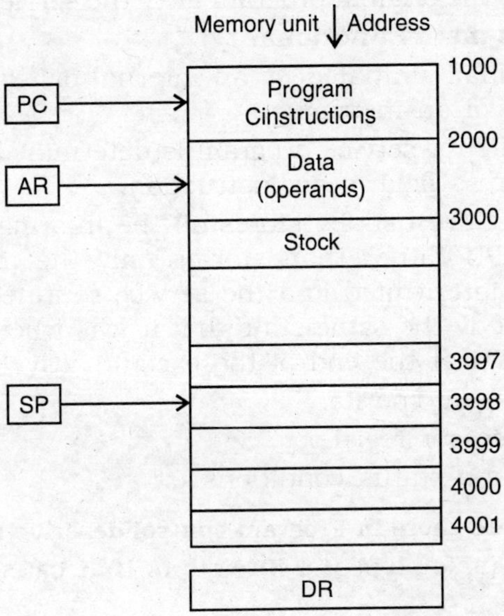

Fig. 5.11.

The address register array points at an array of data the stack points SP points at top of stack.

Stack Limits: Most computers do not provide hardware to check for stack overflow (Full stack) underflow (empty stack). The stack limits can be by using two processor register are to hold the upper limit and other to hold the lower limit. After a push operation, SP is compared with upper limit register and after a operation, SP is compared with the lower limit register.

Q.3. What is Reverse Publish notation? Change the following in reverse polish notation?

(a) A + B × C

(b) (A + B) × C

(c) A$ B × C – D + E/F/(G + H)

(d) A – B (C× DBE)

Solution: A stack organization is very effective for evaluating arithmetic expressions. The post fix notation referred to as reverse polish notation (RPN). Places the operator after the operands

e.g., A + B ← in fix notation

+ AB ← Prefix of polish notation

AB + ← Post fix or reverse polish notation

The reverse polish notation is in a form suitable for stack manipulation. The expression A × B + C × D is written in RPN AB × CD +

(a) ABC × +

(*b*) AB + C^x

(*c*) AB $ C × D EF/GH + / +

(*d*) ABC DE $ × –

Q.4. What is Program interrupt? Describe all functionality of the interrupt.

Solution: Program interrupt refers to the transfer of a program control from a currently running program to another service program as a result of external or internal generated request. Control returns to the original program after the service program is executed.

Interrupt Procedure is main Functionality:

(*a*) The interrupt is usually initiated by an internal or external signal rather than from the execution of an instruction

(*b*) The address of interrupt service program is determined by the hardware rather than from the address field of an instruction.

(*c*) The interrupt procedure usually stores all the information necessary to define the state of the CPU rather than storing only the program counter. After a program has been interrupted and the service routine been executed the CPU must return to exactly the same state that it was when the interrupt occurred. The state of the CPU at the end of the execute. Cycle is determined from:

(*i*) The content of program counter

(*ii*) The content of processor register

(*iii*) The content of certain status conditions

Q.5. How many interrupts are there in Program control describe briefly?

Solution: There are three major types of interrupts that cause a break in the normal execution of a program:

(*a*) External interrupts

(*b*) Internal interrupts

(*c*) Software interrupts

(*a*) External interrupts: External interrupts come from input out (I/O) devices, from a timing devices from a circuit monitoring the power supply or from any other external source.

(*b*) Internal interrupts: Internal interrupts arise from illegal erroveous use of an instruction or data. Internal interrupts are also called traps. Examples of internal interrupt caused by internal error conditions arc.

Register overflow, attempt to divide by zero, an invalid operation code, stack over flow and protection violation. These error condition usually occur as a result of a permeative termination of the instruction execution.

(*c*) Software interrupt: A software interrupt is initiated by executing an instruction. Software interrupt is a special call instruction that behaves like an interrupt rather than a subroutine call. It can be used by the programmer to initiate an interrupt procedure at any desired point in the program. The most common use of software interrupt is associated with a supervisor call instruction. This instruction provides means for switching from CPU user mode to supervisor mode.

Q.6. What are the differences between external and internal interrupts?

Solution: The difference between internal and external interrupt is that

(*i*) The internal interrupt is initiated by some exceptional condition caused by the program itself rather than by an external event.

(*ii*) Internal interrupts are synchronous with the program while external interrupts are asynchronous.

(*iii*) If the program is rerun, the internal interrupts will occur in the same place each time.

(*iv*) External interrupts depend on external conditions that are independent of the program being executed at the time.

(*v*) External or internal interrupts are initiated from signals that occur in the hardware of the CPU.

Q.7. What is the difference between user mode and supervisor mode?

Solution: A program written by a user must run in the user mode when an input or output transfer is required, the supervisor mode is requested by means of a supervisor call instruction. This instruction causes a software interrupt that stores the old CPU state and brings in a new PSW that belongs to the supervisor mode.

Q.8. Define the following terms:

(*a*) Program status word (PSW)

(*b*) RISC

(*c*) CISC

Solution: (*a*) PSW: (Program Status Word) The collection of all status bit conditions in the CPU is sometimes called a program status word or PSW. The PSW is stored in a separate hardware register and contains the status information that characterizes the state of the CPU.

(*b*) RISC: (Reduced Instruction Set Computer) A computer with instructions with simple constructs, so they can be executed much faster within the CPU without having to use memory as often. This type of computer is classified as a reduced instruction set computer or RISC.

RISC Characteristics:

1. Relatively few instructions
2. Relatively few addressing modes
3. Memory access limited to load and store instruction,
4. All operations done within the registers of the CPU.
5. Fixed length, easily decoded instruction formed.
6. Single cycle instruction execution.
7. Hardware rather than micro Programmed control.

CISC: A computer with large number of instruction is classified as a complex instruction set computer.

Characteristics of CISC:

1. A large number of instructions typically from 100 to 250 instructions.
2. Some instructions that perform specialized tasks and are used infrequently.
3. A large variety of addressing mode typically from 5 to 20 different modes.
4. Variable length instruction format.
5. Instruction that manipulate operands in memory.

Q.9. A bus-organized CPU similar to figure given below has 16 registers with 32 bits in each, an ALU and a destination decoder.

(*a*) How many multiplexers are there in the bus A and what is the size of each multiplexer.

(*b*) How many selection inputs are needed for B MUX and MUXB.

(c) How many inputs and outputs are there in the decoder?

(d) How many inputs and outputs are there in the ALU for data, including input and output carries?

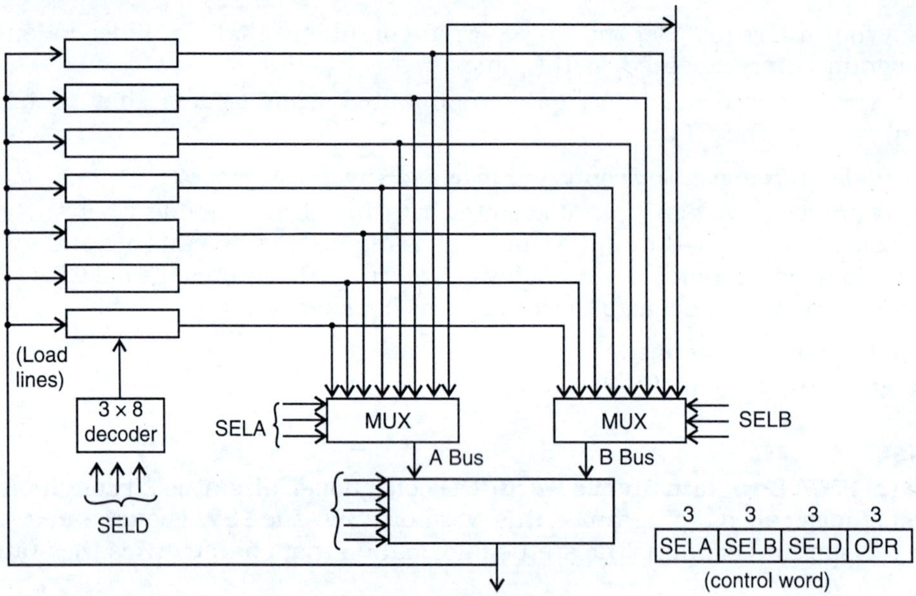

Fig. 5.12.

(e) Formulate a control word for the system assuming that the ALU has 35 operations.

Solution:

(a) 32 multiplexers, each of size 16 × 1

(b) 4 inputs each, to select one of 16 registers.

(c) 4–to-16 line decoder.

(d) 32 + 32 + 1 = 65 data input lines 32 + 1 = 33 data output lines

(e) 4 4 4 6 = 18 bits

SEL A	SEL B	SEL D	OPR

Q.10. The bus system of figure given in question 8. Has the following propagation delay times:

30 ns for the signals to propagate through the multiplexer

80 ns to perform the ADD operation in the ALU

20 ns delay in the destination decoder, and

10 ns to clock the data into the destination register

What is the minimum cycle time that can be used for the clock?

Solution: 30 + 80 + 10 = 120 n sec (The recorder signals propagate at the same as the mixes)

Q.11. Specify the control word that must be applied to the processor of figure given in Q. 8 implement the following micro operations.

(a) $R_1 \leftarrow R_2 + R_3$

(b) $R_4 \leftarrow R_4$

(c) $R_5 \leftarrow R_5 - 1$

(*d*) $R_6 \leftarrow$ **Shl** R_1

(*e*) $R_7 \leftarrow$ **input**

Solution:

	SEL A	SEL B	SEL D	OPR	Control word
(*a*) $R_1 \leftarrow R_2 + R_3$	R_2	R_3	R_1	ADD	01001100100010
(*b*) $R_4 \leftarrow R_4$	R_4	—	R_4	COMA	100×××10001110
(*c*) $R_5 \leftarrow R_5 - 1$	R_5	—	R_6	DECA	101×××10100110
(*d*) $R_6 \leftarrow$ Shl R_1	R_1	—	R_6	SHLA	001×××11011000
(*e*) $R_7 \leftarrow$ input	Input	—	R_7	TSFA	000×××11100000

Q.12. Determine the micro-operations that will be execute in the process of figure given in Q. 8. When the following 14-bit control words arc applied.

(*a*) **00101001100101**

(*b*) **00000000000000**

(*c*) **01001001001100**

(*d*) **00000100000010**

(*d*) **11110001110000**

Solution:

Control world	SEL A	SEL B	SEL D	OPR	Micro operation
(*a*) 00101001100101	R_1	R_2	R_3	SUB	$R_3 \leftarrow R_1 - R_2$
(*b*) 00000000000000	Input	Input	None	TSFA	Output $\leftarrow$ Input
(*c*) 01001001001100	R_2	R_2	R_2	XOR	$R_2 \leftarrow R_2 + R_2$
(*d*) 00000100000010	Input	R_1	None	ADD	Output $\leftarrow$ Input + R_1
(*d*) 11110001110000	R_7	R_4	R_3	SHRA	$R_3 \leftarrow$ Shr R_7

Q.13. Let SP = 000000 in the stack of figure given below. How many items are there in stack it

(*a*) **FULL = 1 and EMPTY = 0?**

(**b**) **FULL = 0 and EMPTY = 1?**

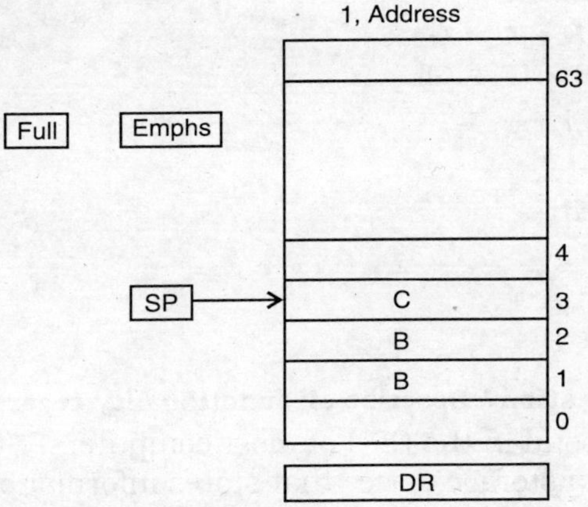

Fig. 5.13.

Solution: (*a*) Stack full with 64 items

(*b*) Stack Empty.

Q.14. A stack organized such that SP always points at the next empty location on the stack. This means that SP can be initiated to 4000 in figure given below and first item in the stack is storeding location 4000. List the micro-operations for the Push and Pop operations.

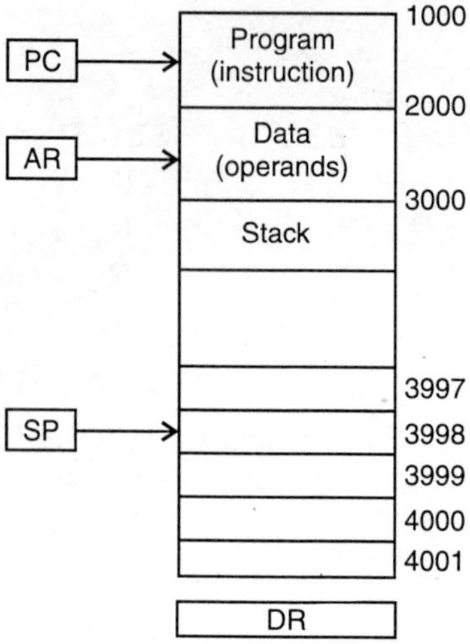

Fig. 5.14. Computer memory with Program, data dn stack segments

Solution: PUSH : M[SP] ← DR

SP ← SP – 1

POP : SPH

DR ← M[SP]

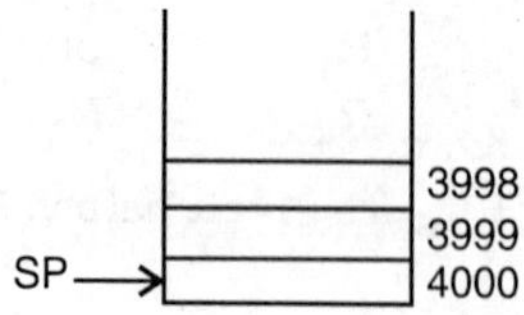

Q.15. Convert the following arithmetic expressions from infix to reverse polish notation.

(*a*) $A \times B + C \times D + E \times F$

(*b*) $A \times B + A \times CB \times D + C \times G$

(*c*) $A + B \times [C \times D + E \times (F + G)]$

(*d*) $\dfrac{A \times [B + C \times (D + E)]}{F \times (G + H)}$

Solution: (*a*) AB * CD * EX* ++

(*b*) AB * ABD* CE* + * +

(*c*) FG + E* CD* + B* A +

(*d*) ABCDE + α + * FGH ≠ */

Q.16. What is stack organization? Describe all functionality regarding to stack?

Solution: Stack is very useful in the CPU of most computers. Stack is also called last in first out (LIFO). A stack is a storage device that stores information in such a manner that the item stored last is the first item retrieved. The stack in digital computers is essentially a memory unit with an address register that can count only after an initial value is loaded into it.

Various function of stack

(*a*) **Stack pointer.** The register that holds the address for the stack is called a *stack pointer* (SP) because its value always points at the top item in the stack.

(*b*) **PUSH.** The operation of insertion is called push, because it can be thought of as the result of pushing a new item on top.

(*c*) **POP.** The operation of deletion is called POP, because it can be thought of as the result of removing one item so that the stack pops up.

Q.17. Define the following:

(*a*) **Register stack**

(*b*) **Memory stack**

(*c*) **Stack limits.**

Solution: (*a*) Register Stack.

A stack can be placed in a portion of large memory or it can be organized as a collection of a finite number of memory words or registers. The organization of a 64-word register is stack shown below:

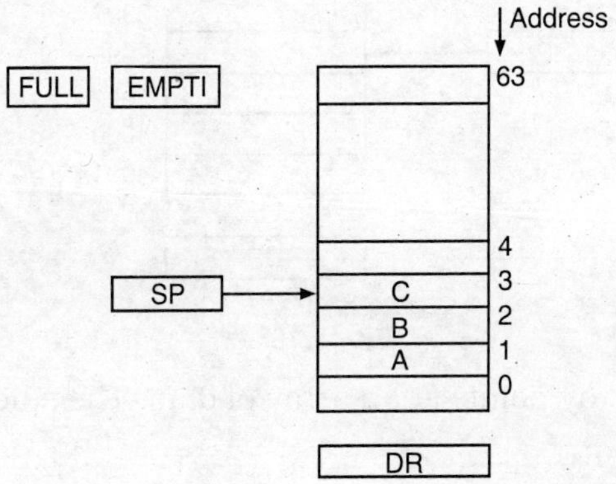

Fig. 5.15. Block diagram of 64-word stack

The stack point register SP contains a binary numbers, whose value is equal to the address of the word that is currently on top of the stack. So, in above block diagram, the content of SP is now 3. To remove the top item. The stack is popped by reading the memory word at address 3 and decrementing the conent of SP. If tack is not full then use PUSH.

Push Operation

SP ← SP + 1 Increment stack pointer

M[SP] ← DR Write item on top of the stack

If (SP = 0) then (Full ← 1), check if stack is full

EMPTY ← 0 Mark the stack not empty.

POP OPERATION: If Impty = 0

DR ← M[SP] **Read item from the top of stack**

SP ← SP – 1 **Decrement stack pointer**

If (SP = 0) then (empty ← 1) **Check if stack is empty**

FULL ← 0 **Mark the stack not full**

(*B*) Memory Stack

A stack can exist as a stand alone unit or can be implemented in a random access memory attached to a CPU. The implementation of a stack in the CPU is done by assigning a portion of memory to a stack operation and using a processor register as a stack pointer. Portion of computer memory partitioned into three segments.

Program,
Data and
Stack.

The program counter PC points at the address of next instruction in the program.

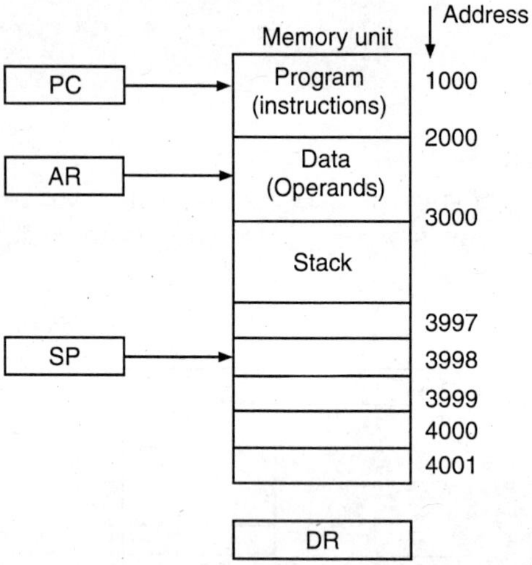

Fig. 5.16.

The address register array points at an array of data. The stack point SP points at top o stack.

(c) Stack Limits

Most computers do not provide hardware to check for stack overflow (full stack) or underflow (empty stack). The stack limits can be checked by using two processor register are to hold the upper limit and other to hold the lower limit. After a push operation, SP is compared with upper limit register and after a pop operation, SP is compared with the lower limit register.

Q.18. What is Reverse polish notation? Change the following in reverse polish notation?

(a) A + B * C)
(b) (A + B) * C
(c) A $ B * C – D + E/F/(G + H)
(d) NA – B(C * D $ E).

Solution: A stack organization is very effective for evaluating arithmetic expressions. The post fix notation, referred to as reverse polish notation (RPN), places the operator after the operands.

e.g.,

$A + B$ — In fix notation
$+ AB$ — Prefix or polish notation
$AB +$ — Postfix or reverse polish notation.

The reverse polish notation is in a form suitable for stack manipulation. The expression $A * B + C * D$ is written in RPN $AB * CD * +$:

(a) $ABC * +$
(b) $AB + C *$

(*c*) $AB \$ C * D - EF/GH + / +$

(*d*) $ABCDE \$ * -$

Q.19. What is program interrupt? Describe all functionality of the interrupt.

Solution: Program interrupt refers to the transfer of a program control from a currently running program to another service program as a result of an external or internal generated request. Control returns to the original program after the service program is executed.

Interrupt procedure is main functionality:

(*a*) The interrupt is usually initiated by an internal or external signal rather than from the execution of an instruction.

(*b*) The address of interrupt service program is determined by the hardware rather than from the address field of an instruction.

(*c*) The interrupt procedure usually stores all the information necessary to define the state of the CPU rather than storing only the program counter.

After a program has been interrupted and the service routine been executed, the CPU must return to exactly the same state that it was when the interrupt occured. The state of the CPU at the end of the execute cycle is determined from:

(*i*) The content of the program counter.

(*ii*) The content of all processor register.

(*iii*) The content of certain status conditions.

Q.20. How many interrupts are there in program control, describe briefly?

Solution: There are three major types of interrupts that cause a break in the normal execution of a program:

(*a*) External interrupts.

(*b*) Internal interrupts.

(*c*) Software interrupts.

(*a*) External Interrupts

External interrupts come from input-output (input/output) devices. from a timing devices, from a circuit monitoring the power supply or from any other external source.

(*b*) Internal Interrupts

Internal interrupts arise from illegal or erroneous use of an instruction or data. Internal interrupts are also called **traps**. Examples of interrupt caused by internal error conditions are:

Register over flow, attempt to divide by zero, an invalid operation code, stack over flow and protection violation. These error condition usually occur as a result of a premature termination of the instruction execution.

(*c*) Software Interrupts

A software interrupt is initiated by executing an instruction. Software interrupt is a special call instruction that behaves like an interrupt rather than a subroutine call. It can be used by the programmer to initiate an interrupt procedure at any desired point in the program. The most common use of software interrupt is associated with a supervisor call instruction. This instruction provides means for switching from CPU user mode to supervisor mode.

Q.21. What are the differences between external and internal interrupts?

Solution: The difference between internal and external interrupts is that:

(*i*) The internal interrupt is initiated by some exceptional condition caused by the program itself rather than by an external event.

(*ii*) Internal interrupts are synchronous with the program while external interrupts are asynchronous.

(*iii*) If the program is rerun, the internal interrupts will occur in the same place each time.

(*iv*) External interrupts depend on external conditions that are independent of the program being executed at the time.

(*v*) External or internal interrupts are initiated from signals that occur in the hardware of the CPU.

Q.22. What is the difference between user mode and supervisor mode?

Solution: A program written by a user must run in the user mode. When an input or output transfer is required, the supervisor mode is requested by means of a supervisor call instruction. This instruction causes a software interrupt that stores the old CPU state and brings in a new PSW that belongs to the supervisor mode.

Q.23. Define following terms:

(*a*) Program status word

(*b*) RISC

(*c*) CISC.

Solution: (*a*) PSW (Program Status Word)

The collection of all status bit conditions in the CPU is sometimes called a program status word or PSW. The PSW is stored in a seprate hardware register and contains the status information that characterizes the state of the CPU.

(*b*) RISC (Reduced Instruction Set Computer)

A computer with fewer instructs with simple constructs, so they can be executed much faster within the CPU without having to use memory as often. This type of computer is classified as a reduced instruction set computer or RISC.

RISC characteristics:

1. Relatively few instructions.
2. Relatively few addressing modes.
3. Memory access limited to load and store instructions.
4. All operations done within the registers of the CPU.
5. Fixed length, easily decoded instruction format.
6. Single cycle instruction execution.
7. Hardware rather than micro programed control.

(c) CISC

A computer with large number of instructions is classified as a complex instruction set computer.

Characteristics of CISC

1. A large number of instructions typically from 100 to 250 instructions.
2. Some instructions that perform specialized tasks and are used infrequently.
3. A large variety of addressing mode typically from 5 to 20 different modes.
4. Variable length instruction format.
5. Instructions that manipulate operands in memory.

Q.24. A bus-organized CPU similar to figure given below has 16 registers with 32 bits in each, an ALU, and a destination decoder.

- **(a) How meiny multiplexers are there in the A bus, and what is the size of each multimplexer?**
- **(b) How many selection inputs are needed for MUX A and MUX B?**
- **(c) How many inputs and outputs are there in the decoder?**
- **(d) How many inputs and outputs are there in the ALU for data, including input and output carries?**
- **(e) Formulate a control word for the system assuming that the ALU has 35 operations?**

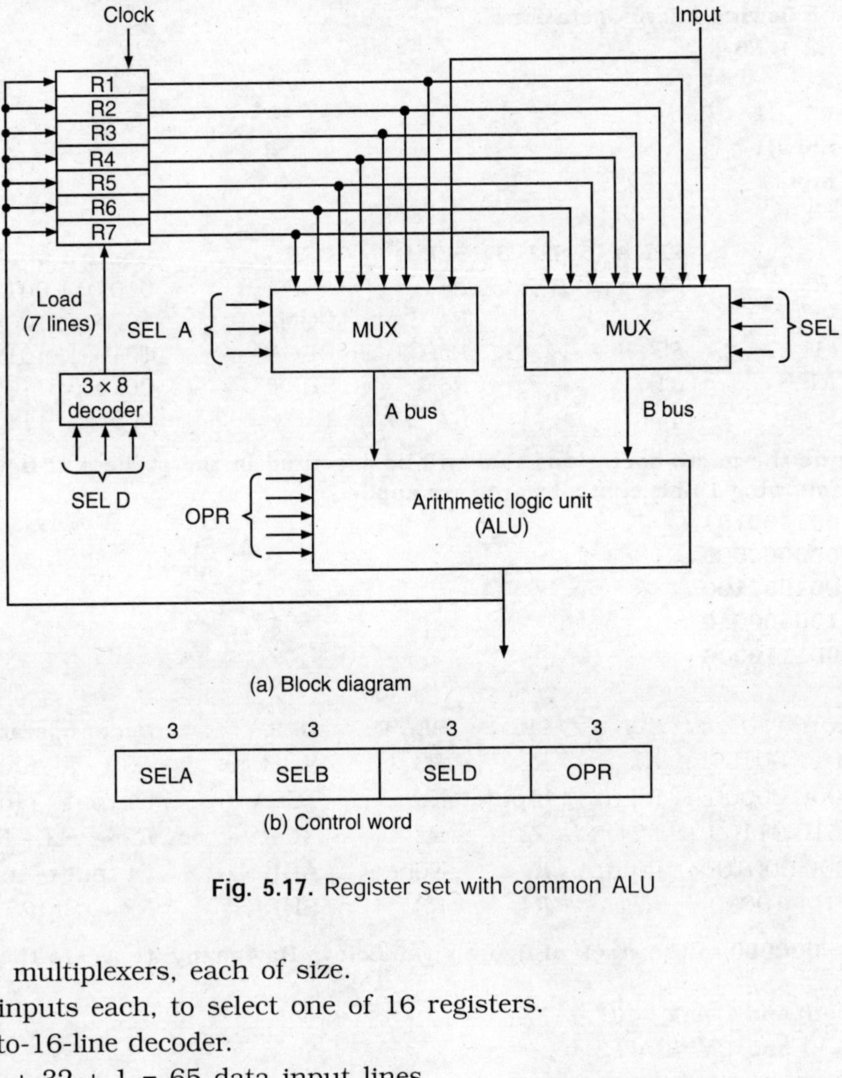

Fig. 5.17. Register set with common ALU

Solution:

(a) 32 multiplexers, each of size.

(b) 5 inputs each, to select one of 16 registers.

(c) 4-to-16-line decoder.

(d) 32 + 32 + 1 = 65 data input lines.

32 + 1 = 33 data output lines.

(e)

5	5	4	6	= 20 bits
SEL A	SEL B	SEL D	OPR	

Q.25. The bus system of figure given in Q. 9 has the following propagation delay times:

30 ns for the signals to propagate through the multiplexers,

80 ns to perform the ADD operation in the ALU,

20 ns delay in the destination decoder, and

10 ns to clock the data into the destination register.

What is the minimum cycle time that can be used for the clock?

Solution: 30 + 80 + 10 = 20 nsec.

(The decoder signals propagate at the same as the muxes).

Q.26. Specify the control word that must be applied to the processor of figure given in Q.9 implement the following micro-operations.

(a) R1 ← R2 + R3

(b) R4 ← R4

(c) R5 ← R5 – 1

(d) R6 ← Shl R1

(e) R7 ← input

Solution:

	SEL A	SEL B	SEL D	OPR	Control word
(a) R1 ← R2 + R3	R2	R3	R1	ADD	010 011 001 00010
(b) R4 ← R4	R4	—	R4	COMA	100 xxx 100 00000
(c) R5 ← R5 – 1	R5	—	R5	DECA	101 xxx 101 00110
(d) R6 ← Shi R1	R1	—	R6	SHUT	001 xxx 110 11000
(e) R7 ← Input	Input	—	R7	TSFA	000 xxx 111 00000

Q.27. Determine the micro-operations that will be executed in the process of figure given Q.9 when the following 14-bit control words are applied.

(a) 00101001100101

(b) 00000000000000

(c) 01001001001100

(d) 00000100000010

(e) 11110001110000

Solution:

	Control word	SEL A	SEL B	SEL D	OPR	Micro-operation
(a)	001 010 011 00101	R1	R2	R3	SUB	R3 ← R1 – R2
(b)	000 000 000 00000	Input	Input	None	TSFA	Output ← Input
(c)	010 010 010 01100	R2	R2	R2	XOR	R2 ← R2 + R2
(d)	000 001000 00010	Input	R1	None	ADD	Output ← Input + R1
(e)	111 100 011 10000	R7	R4	R3	SHRA	R3 ← Shr27

Q.28. Let SP = 000000 in the stack of figure given below. Haw many items are there in the stack if:

(a) FULL = 1 and EMTY = 0?

(b) FULL = 0 and EMTY = 1?

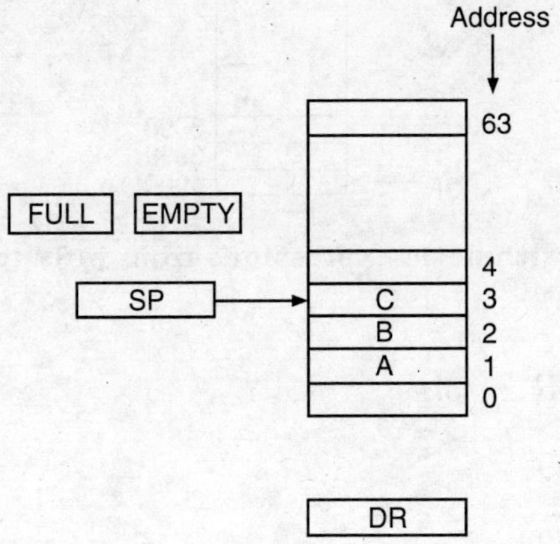

Fig. 5.18. Block diagram of a 64-word stack

Solution:

(*a*) Stack full with 64 items.

(*b*) Stack empty.

Q.29. A stack is organized such that SP always points at the next empty location on the stack. This means that SP can be initialized to 4000 in figure given below and the first item in the stack is stored in location 4000. List the microoperations for the push and pop operations.

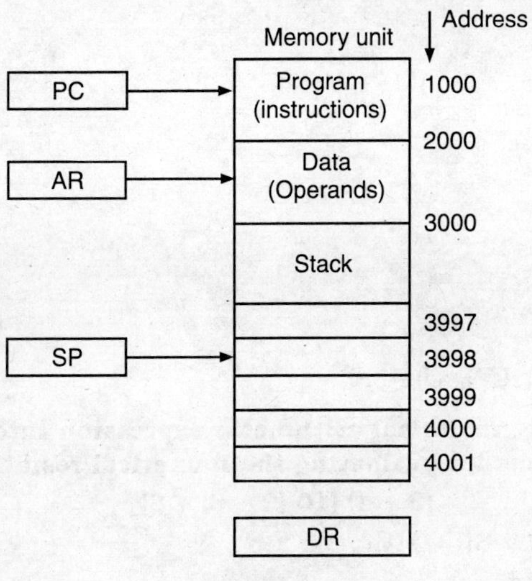

Fig. 5.19. Computer memory with program, data and stack segments

Solution: Push : $M[SP] \leftarrow DR$

$SP \leftarrow SP - 1$

POP : $SP \leftarrow SP + 1$

$DR \leftarrow M[SP]$

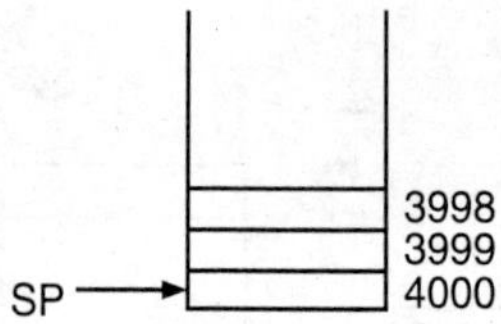

Q.30. Convert the following arithmetic expressions from infix to reverse Polish notation.

(a) A * B + C * D + E * F

(b) A * B + A * (B * D + C * E)

(c) A + B * [C * D + E * (F + G)]

(d) $\dfrac{A * [B + C * (D + E)]}{F * (G + H)}$

Solution:

(a) AB * CD * E * + +

(b) AB * ABD * CE * + * +

(c) FG + E * CD * + B * A +

(d) ABCDE + * + * FGH + * /

Q.31. Convert the following arithmetic expressions from reverse Polish notation to infix notation.

(a) ABCDE + * – /

(b) ABCDE * / – +

(c) ABC * /D – EF / +

(d) ABCDEFG + * + * + *

Solution:

(a) $\dfrac{A}{B - (D + E) * C}$

(b) $A + B - \dfrac{C}{D * E}$

(c) $\dfrac{A}{B * C} - D + \dfrac{E}{F}$

(d) (((F + G) * E + D) * C + B) * A

Q.32. Convert the following numerical arithmetic expression into reverse Polish notation and show the stack operations for evaluating the numerical result.

(3 + 4) [10 (2 + 6) + 8]

Solution: (3 + 4) [10 (2 + 6) + 8] = 616

RPN: 34 + 26 + 10 * 8 + *

				6		10		8		
	4		2	2	8	8	80	80	88	
3	3	7	7	7	7	7	7	7	7	616
3	4	+	2	6	+	10	*	8	+	*

Q.33. A first-in, first-out (FIFO) has a memory organization that stores information in such a manner that the item that is stored is the first item that is retrieved. Show how a FIFO memory operates with three counters. A write counter holds the address for writing into

memory. A read counter RC holds the address for reading from memory. An availables storage counter ASC indicataaes the number of words stored in FIFO. ASC is incremented for every word stored and decremented for every item that is retrieved.

Solution:

WRITE (if not full):

$M[WC] \leftarrow DR$

$WC \leftarrow WC + 1$

$ASC \leftarrow ASC + 1$

READ: (if not empty)

$DR \leftarrow M[RC]$

$RC \leftarrow RC + 1$

$ASC \leftarrow ASC - 1$

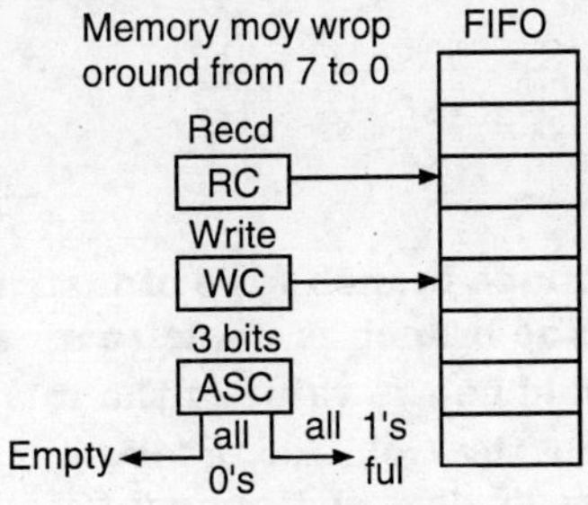

Fig. 5.20.

Q.34. A computer has 32-bit instructions and 12-bit addresses. If there are 250 two address instructions, how many one-address instructions can be formulated?

Solution:

8	12	12
Opcode	Address	Address

= 32 bits

(Two address instruction)

$$2^8 = 256 \text{ combinations.}$$

$$256 - 250 = \text{S combinations can be used for one address.}$$

Opcode	Address

6×2^{12} (one address instructions)

Maximum number of one address instruction = $6 \times 2^{12} = 24{,}576$

Q.35. The memory unit of a computer has 256K words of 32 bits each. The compter has an instruction format with four fields: an operation code field, a mode field to specify one of seven addressing modes, a register address field to specify one of 60 processor register, and a memory address. Specify the instruction format and the number of bits in each field if the in instruction is in one memory word.

Solution: $256\text{ K} = 2^8 \times 2^{10} = 2^{18}$ Address = 18 bits

Mode = 3 bits

Register = 6 bits

27 bits

Opcode 5

32 bits

Opcode	Mode	Register	Address
5	3	6	18

Q.36. A two-word instruction is stored in memory at an address designated by the symbol *W*. The address field of the instruction (stored at *W* + 1) is designated by the symbol *Y*. The operand used during the execution of instruction is stored at an address symbolized by *Z*. An index register contains the value of *X*. State how *Z* is calculated from the other addresses if the addressing mode of the instruction is:

(*a*) direct

(*b*) indirect

(*c*) relative

(*d*) indexed

Solution: $Z =$ Effective address

(*a*) Direct: $Z = Y$

(*b*) Indirect: $Z = M\ (Y)$

(*c*) Relative: $Z = Y + W + 2$

(*d*) Indexed: $Z = Y + X$

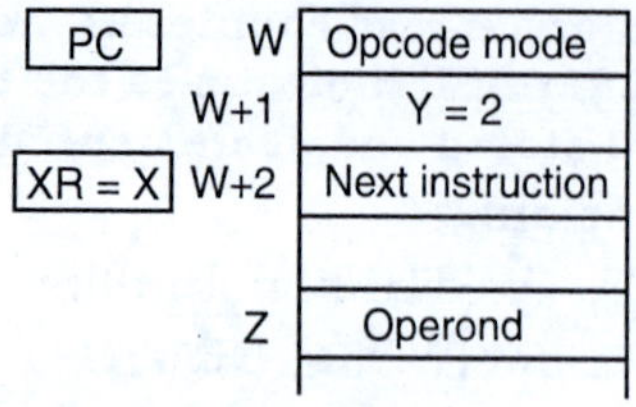

Fig. 5.21.

Q.37. A relative mode branch type of instruction stored in memory at an address equivalent to decimal 750. The branch is made to an address equivalent to decimal 500.

(*a*) What should be the value of the relative address field of the instruction (in decimal)?

(*b*) Determine the relative address value in binary using 12 bits. (Why must the number be in 2'8 complement?)

(*c*) Determine the binary value in PC after the fetch phase and calculate the binary value of 500. Then show that the binary value in PC plus the relative address calculated in part (*b*) is equal to the binary value of 500.

Solution:

(*a*) Relative address = 500 – 751 = – 251

(*b*) 251 = 000011111011; – 251 = 111100000101

(*c*) PC = 751 = 001011101111; 500 = 000111110100

PC = 751 = 001011101111

RA = – 251 = + 111100000101

EA = 500 = 000111110100

Q.38. How many times does the control unit refer to memory when it fetches and executes an indirect addressing modes instruction if the instruction is

(*a*) a computational type requiring and operand from memory;

(*b*) a branch type.

Solution: Assuming one word per instruction or operard.

(*a*) Computational type

Fetch instruction.

Fetch effective address.

Fetch operand

3 memory references

(*b*) Branch Type

Fetch instruction

Fetch effective addresses and transfer to *PC*

2 memory references

Q.39. What must the address field of an indexed addressing mode instruction be to make it the same as a register indirect mode instruction?

Solution: The address part of the indexed mode instruction must be set to zero,

Q.40. An instruction is stored at location 300 with its address field at location 301. The address field has the value 400. A processor register *R*1 contains the number 200. Evaluate the effective address if the addressing mode of the instruction is

(*a*) direct;

(*b*) immediate;

(*c*) relative;

(*d*) register indirect;

(*e*) index with *R*l as the index register.

Solution: Effective address

(*a*) Direct: 400

(*b*) Immediate: 301

(*c*) Relative: 302 + 400 = 702

(*d*) Register indirect: 200

(*e*) Indexed: 200 + 400 = 600

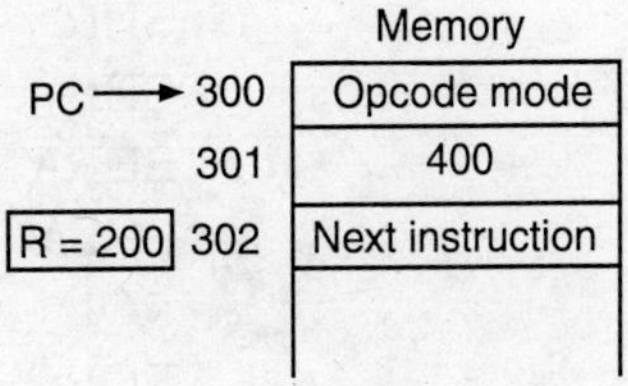

Fig. 5.22.

Q.41. Assuming an 8-bit computer, show the multiple precision addition of the two 32-bit unsigned numbers listed below using the add with carry instruction. Each byte is expressed as a two-digit hexadecimal number.

(6*E*, *C*3 56 7*A*) + (13 55 6*B* 8*F*)

Solution:	1 = *C*	0 = *C*	1 = *C*	0 = Reset initial carry
	6*E*	*C*3	56	7*A*
	??	??	??	??

Q.42. Perform the logic AND, OR and XOR with the two binary strings 10011100 and 10101010.

Solution:	10011100	10011100	10011100
	10101010 AND	10101010 OR	10101010 XOR
	10001000	11111110	00110110

Q.43. Given the 16-bit value 1001101011001101, what operation must be performed in order to:

(*a*) clear to 0 the first eight bits?

(*b*) set to 1 the last eight bits?

(*c*) complement the middle eight bits?

Solution:

(*a*) AND with: 0000000111111111

(*b*) OR with: 0000000011111111

(*c*) XOR with: 0000111111110000

Q.44. An 8- bit register contains the value 01111011 and the carry bit is equal to 1. Perform the eight shift operations given by the instructions listed in Table given below. Each time, start from the initial value given above,

Table 5.5. Typical Shift Instructions

Name	**Mnemonic**
Logical shift right	**SHR**
Logical shift left	**SHL**
Arithmetic shift right	**SHRA**
Arithmetic shift left	**SHLA**
Rotate right	**ROR**
Rotate left	**ROL**
Rotate right through carry	**RORC**
Rotate left through carr	**ROLC**

Solution:

Initial	:	01111011	C = 1
SHR	:	00111101	
SHL	:	11110110	
SHRA	:	00111101	
SHLA	:	11110110	(Overflow)
ROL	:	11110110	
RORC	:	10111101	
ROLC	:	11110111	

Q.45. Represent the following signed numbers in binary using eight bits + 83, – 83, + 68, – 68.

(*a*) Perform the addition (– 83) + (+ 68) in binary and interpret the result obtained.

(*b*) Perfrom the subtraction (– 68) – (+ 83) in binary and indicate if there is an.

(*c*) Shift binary – 68 once to the right and give the value of the shifted number in decimal.

(*d*) Shift binary – 83 once to the left and indicate if there is an overflow.

Solution: + 83 = 01010011– 83 = 10101101

+ 68 = 01000100 – 68 = 10111100

```
              10 carries
   - 83        10101101
(a) + 68     + 01000100
   -----     ----------
   - 15        11110001
   - 68        10111100
(b) - 83     + 10101101
   -----     ----------
   - 151       01101001
(c) - 68 = 10111100
    - 34 = 11011110
(d) - 83 = 10101101
    - 166 ≠ 01011010 (overflow)
```

Q.46. Show that the circuit labeled 'check for zero output'in figure given below is an 8-bot NOR gate.

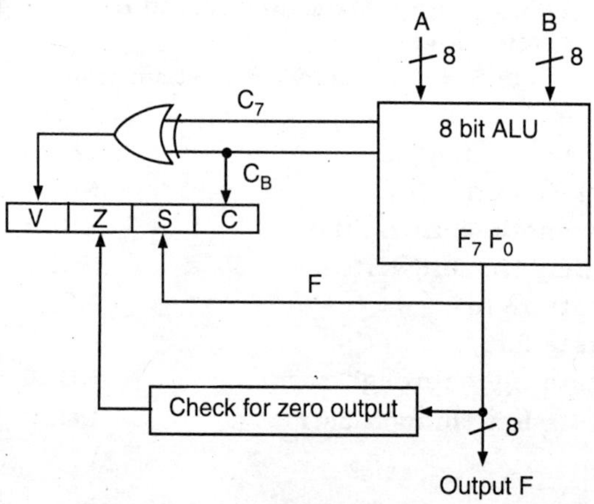

Fig. 5.23.

Solution: $Z = F_0' F_1' F_2' F_3' F_4' F_5' F_6' F_7' = (F_0 + F_1 + F_3 + F_4 + F_5 + F_6 + F_7)'$.

Q.47. An 8-bit computer has a register *R*. Determine the values of status bits *C*, *S*, *Z*, and *V* figure given; in Q. 30, after each of the following instructions. The initial value of register *R* in each case is hexadecimal 72. The numbers below are also in hexadecimal.

(*a*) Add immediate operand *C*6 to *R*.

(*b*) Add immediate operand *IE* to *R*.

(*c*) Subtract immediate operand 9*A* from *R*.

(*d*) AND immediate operand 8*D* to *R*.

(*e*) Exclusive-*OR* *R* with *R*.

Solution:

(*a*) 72 01110010
*C*6 11000110
———————
138 00111000
$C = 1$ $S = 0$ $Z = 0$ $V = 0$

(*b*) 72 01110010
IE 00011110
———————
138 10010000
$C = 0$ $S = 1$ $Z = 0$ $V = 1$

(*c*) 9*A* = 10011010
01100110 2's complement
72 01110010
———————
*D*8 11011000
$C = 0$ $S = 1$ $Z = 0$ $V = 0$ (Borrow = 1)

(*d*) 72 01110010
8*D* 10001101
———————
00 00000000
$C = 0$ $S = 0$ $Z = 1$ $V = 0$

(*e*) $C = 0$ $S = 0$ $Z = 1$ $V = 0$

Q.48. Two unsigned numbers *A* and *B* are compared by subtracting *A* – *B*. The carry status bit is considered as a borrow bit after a compare instruction in most commercial computers, so that *C* = 1 if *A* < *B*. Show that the relative magnitude of A and *B* ean be determined from inspection of status bits *C* and *Z* as specified in the tables.

Relation	Condition of Status Bits
$A > B$	$C = 0$ and $Z = 0$
$A \geq B$	$C = 0$
$A < B$	$C = 1$
$A \leq B$	$C = 1$ or $Z = 1$
$A = B$	$Z = 1$
$A \neq B$	$Z = 0$

TABLE for Conditional Branch Instruct

Mnemonic	Branch condition	Tested condition
BZ	Branch if zero	$Z = 1$
BNZ	Branch if not zero	$Z = 0$
BC	Branch if carry	$C = 1$
BNC	Branch if no carry	$C = 0$
BP	Branch if plus	$S = 0$
BM	Branch if minus	$S = 1$
BV	Branch if overflow	$V = 1$
BNV	Branch if not overflow	$V = 0$
	Unsigned compare conditions ($A - B$)	
BHI	Branch if higher	$A > B$
BHE	Branch if higher or equal	$A \leq B$
BLO	Branch if lower	$A < B$
BLOE	Branch if lower or equal	$A \leq B$
BE	Branch if equal	$A = B$
BNE	Branch if not equal	$A \neq B$
	Signed Compare Conditions ($A - B$)	
BGT	Branch if greater that	$A > B$
BGE	Branch if greater or equal	$A \geq B$
BLT	Branch if less than	$A < B$
BLE	Branch if less or equal	$A \leq B$
BE	Branch if equal	$A = B$
BNE	Branch if not equal	$A \neq B$

Solution: $C = 1$ if $A < B$, therefore $C = 0$ if $A \geq B$

$Z = 1$ if $A = B$, therefore $Z = 1$if $A \neq B$

For $A > B$ we must have $A \leq B$ provided $A \neq B$

or

For $A \leq B$ we must have $A < B$ or $A = B$

or $C = 1$ or $Z = 1$ $(C + Z) = 1$

Q.49. Two signed numbers *A* and *B* represented in signed – 2's complement form are compared by subtracting *A* – *B*. Status bits S, *Z* and *V* are set or cleared depending on the result of the operation, (Note that there is a sign reversal if an overflow occurs). Show that the relative magnitude of *A* and *B* can be determinded from inspection of the status bits as specified below. (See also Table for conditional branch instruction given in Q. 32).

Relation	Condition of Status Bits
$A > B$	$(S \oplus V) = 0$ and $Z = 0$
$A > B$	$(S \oplus V) = 0$
$A < B$	$(S \oplus V) = 1$
$A < B$	$(S \oplus V) = 1$ or $Z = 1$
$A = B$	$Z = 1$
$A \neq B$	$Z = 0$

Solution: $A > B$ implies that $A - B \geq 0$ (positive or zero)

Sign $S = 0$ if no overflow (positive)

or $S = 1$ if overflow (sign reversal)

Boolean expression:

$S'V' + SV = 1$

or $(S + V) = 0$

$A < B$ is the complement of $A \geq B$ $(A - B)$ negative

then $S = 1$ if $V = 0$

or $S = 0$ if $V = 1 (S \oplus V) = 1$

$A > B$ implies $A \geq B$ but not $A = B$

$(S \oplus V) = 0$ and $Z = 0$

$A \leq B$ implies $A < B$ or $A = B$

$(S \oplus V) = 1$ or $Z = 1$

Q.50. It is necessary to design a digital circuit with four inputs *C*, *S*, *Z* and *V* and 10 outputs, one for each of the branch conditions listed in Probs. 32 and 33. (The equal and unequal conditions are common to both tables). Draw the logic diagram of the circuit using two OR gates, one XOR gate, and five inverters.

Solution:

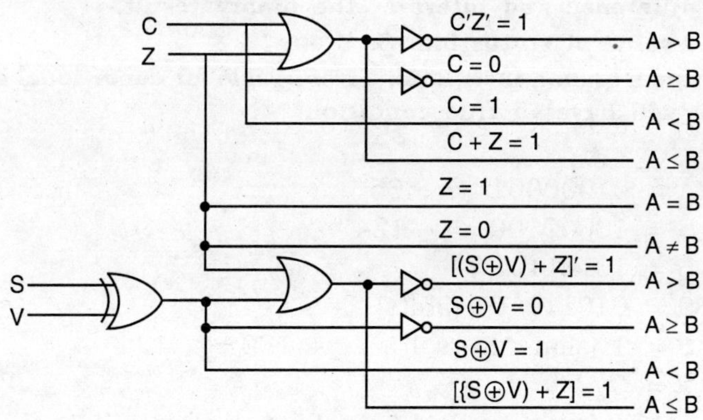

Fig. 5.24.

Q.51. Considser the two 8-bit numbers *A* = 01000001 and *B* = 10000100

(*a*) Give the decimal equivalent of each number assuming that

(*i*) they unsignsed, and

(*ii*) they are signed.

(*b*) Add the two binary numbers and interpret the sum assuming that the numbers are:

(*i*) unsigned, and

(*ii*) signed.

(*c*) Determine the values of the *C*, *Z*, *S* and *V* status bits after the addition.

(*d*) List the conditional branch instructions from Table for conditional branch instruction in Q.32.

Solution:

		Unsigned	Signed
(*a*)	$A = 01000001$	65	+ 65
	$B = 10000100$	132	– 124
(*b*)	$A + B = 11000101$	197	– 59

(*c*) $C = 0$ $Z = 0$ $S = 1$ $V = 0$

(*d*) BNC BNZ BM BNV

Q.52. The program in a computer compares two unsigned numbers *A* and *B* by performing a subtaction *A* – *B* and updating the status bits. Let *A* = 01000001 and *B* = 10000100.

(*a*) Evaluate the difference and intepret the binary result.

(*b*) Determine the values of status bits C (borrow) and *Z*.

(*c*) List the conditional branch instructions from Table for conditional branch instrucution in Q. 8.26 that a will have a true condition.

Solution:

(*a*) $A = 01000001 \quad = 65$

$B = 10000100 \quad = 132$

$A - B = 10111101 \quad = -67$ (2's complement of 01000011)

(*b*) C (borrow) = 1, $Z = 0$ $\quad 65 < 132$

$A < B$

(*c*) BL, BLE, BNE

Q.53. The program in a computer compares two signed numbers *A* and *B* by performing the subtraction *A* – *B* and updating the status bits. Let *A* = 01000001 and *B* = 10000100.

(*a*) Evaluate the difference and interpret the binary result.

(*b*) Determine the value of status bits S, *Z* and *V*.

(*c*) List the conditional branch instructions from Table for conditional branch instruction in Q. 32. that will haves a true condition.

Solution:

(*a*) $A = 01000001 \quad = +65$

$B = 10000100 \quad = -124$

$A - B = 10111101$

$+189 = 010111101$ (9bits)

(*b*) $S = 1$ (sign reversal) $\quad +189 > -124$

$Z = 0$

$V = 1$ (overflow) $\quad 65 > -124$

$A > B$

(*c*) BGT, BGE, BNE

Q.54. The content of the top of a memory stack is 5320. The content of the stack pointer SP is 3560. A two-word call subroutine instruction is located in memory at address 1120 followed by the address field of 6720 at location 1121. What are the content of PC, SP and the top of the stack:

(*a*) Before the call instruction is fetched from memory?

(*b*) After the call instruction is executed?

(*c*) After the return from subroutine?

Solution:

	PC	SP	Top of Stack
Initial	1120	3560	5320
Afer CALL	6720	3559	1122
After RETURN	1122	3560	5320

Q.55. What are the basic differences between a branch instruction, a call subroutine instruction, and program interrupt?

Solution: Branch instruction – Branch without being able to return

Subroutine call – Branch to subroutine and then return to calling program.

Program intesrrupt – Hardware initiated branch with possibility to return

Q.56. A computer responds to an interrupt request signal by pushing onto the stack the contents of PC and the current 'PSW' (program status word). It reads a new????? from memory from a location given by an interrupt address symbolized by?????. The first address of the service program is taken from memory at location IAD + 1.

(*a*) List the sequence of microoperations for the interrupt cycle.

(*b*) List the sequence of microoperations for the return from interrupt instructions.

Solution:

(*a*) $SP \leftarrow SP - 1$

$M(CP) \leftarrow PSW$

$SP \leftarrow SP - 1$

$M(SP) \leftarrow PC$

$TR \leftarrow IAD$ (TR is a temporary register)

$PSW \leftarrow M(TR)$

$TR \leftarrow TR + 1$

$PC \leftarrow M(TR)$

Go to fetch phase

(*b*) $PC \leftarrow M(SP)$

$SP \leftarrow SP + 1$

$PSW \leftarrow M(SP)$

$SP \leftarrow SP + 1$

Q.57. Three computers use register windows with the following characteristics. Determine the window size and the total number of registers in each computer.

	Computer 1	Computer 2	Computer 3
Global registers	10	8	16
Local registers	10	8	16
Common registers	6	8	16
Number of windows	8	4	16

Solution:

Window Size = $L + 2C + G$

Computer 1: $10 + 12 + 10 = 32$

Computer 2: $8 + 16 + 8 = 32$

Computer 3: $16 + 32 + 16 = 64$

Register file = $(L + C)\ W + G$

Computer 1: $(10 + 6)\ 8 + 10 = 16 \times 8 + 10 = 138$

Computer 2: $(8 + 8)\ 4 + 8 = 16 \times 4 + 8 = 72$

Computer 3: $(16 + 16)\ 16 + 16 = 32 \times 16 + 16 = 528$

Q.58. Give an example of a RISC *I* instructions that will perfrom the following operations.

(*a*) Decrement a register

(*b*) Complement a register

(*c*) Negate a register

(*d*) Clear a register to 0

(*e*) Divide a signed number by 4

(*f*) No operation

Solution:

(*a*)	SUB *R*22, # 1, *R*22	*R*22 ← *R*22 – 1 (Subtract 1)
(*b*)	XOR *R*22, # – 1, *R*22	*R*22 ← *R*22 ⊕ all 1's (*x*? 1 = *x*')
(*c*)	SUB *R*0, *R*22, *R*22	*R*22 ← 0 – *R*22
(*d*)	ADD *R*0, *R*0, *R*22	*R*22 ← 0 + 0
(*e*)	SRA *R*22, # 2, *R*22	Arithmetic shift right twice
(*f*)	OR *R*1, *R*1, *R*1	*R*1 ← *R*1 v *R*1
or	ADD *R*1, *R*0, *R*1	*R*1 ← *R*1 + 0
or	SLL *R*1, # 0, *R*1	shift left 0 times

SUMMARY

1. (*a*) A CPU can be defined as a general purpose instruction set processor responsible for program execution.
   (*b*) A CPU consists of address bus, data bus and control bus.
   (*c*) A computer with one CPU is called microprocessor and with more than one CPU in called multiprocessor.
   (*d*) The address bus in used to transfer addresses from the CPU to main memory or to I/O devices.
   (*e*) Data bus is the main path by which information is transferred to and from the CPU
   (*f*) Control bus is used by CPU to control various devices connected and to synchronise their operations with those of the CPU.
2. (*a*) A control unit take the instructions one by one to execute. It takes data from input devices and store it in memory. And also sends data from memory to the output device.
   (*b*) All arithmetic and logical operations are carried out by Arithmetic Logical Unit
   (*c*) A control unit and the arithmetic logical unit together is known as CPU
3. (*a*) The accumulator is the main register of the ALU.
   (*b*) In execution of the most of the instructions the accumulator is used to store a input data or output result.
   (*c*) Instructions register holds the opcode of the current instruction.
   (*d*) Memory address register holds the address of the current instructions.
4. A accumulator based CPU consists of (*a*) data processing unit (*b*) program control unit and (*c*) memory and input/output interface unit.
   (*a*) (*i*) In the data processing unit, data is processed to get some results.
   (*ii*) the accumulator in the main operand register of the ALU.
   (*b*) (*i*) Program control unit controls various parts of CPU.
   (*ii*) Program counter holds the address of the next instructions to be read from memory after the current instruction is executed.
   (*iii*) Instruction register holds the opcode of the current instruction.
   (*iv*) Control circuits hold the responsibility of every operation of the CPU.
   (*c*) (*i*) Data registers of the memory and input/output interface unit acts as a buffer between the CPU and main memory.
   (*ii*) Address register contains the address of the present instructions obtained from the program control unit.
5. (*a*) Stack pointer and flag register are special registers of CPU.
   (*b*) Stack pointers holds the address of the most recently entered item into the stack.
   (*c*) Flag registers indicates the status which depends on the results of the operation.
6. (*a*) Micro operations are the operations executed on data stored in registers.
   (*b*) A set of micro operations specified by an instruction is known as macro operation.

7. (*a*) A sequence of operations involved in processing an instruction constitutes an instruction cycle.
   (*b*) Fetch cycle in defined as the time required for getting the instruction code from main memory to CPU
   (*c*) Executed cycle is the time required to decode and execute an instruction.
   (*d*) Fetch cycle requires a fixed time slot and execute cycle requires variable time slot.
8. (*a*) a word length indicates the number of bits the CPU can process at a time.
   (*b*) A memory size indicates the total storage capacity of the CPU.
   (*c*) Word length is the indication of bit length of each register.
9. A computer is said to be operated based on stored program concept if it stores the instructions as well as data of a program in main memory when they are waiting to execute.
10. A stored program in main memory is executed instruction after instruction in a successive method by using program counter.

EXERCISE

(I) Fill in the Blanks (Solved)

1. SIMD stands for .... instruction stream, multiple data stream.
2. .... control is faster than microprogrammed control.

(II) True or False (Solved)

1. Indexing is one of the addressing mode.
2. MOV X is used to upload instruction in the processor.

(III) Fill in the Blanks (Unsolved)

1. In stack the values are stored in .... form.
2. .... is a visualisation of processing segments collection.

(IV) True or False (Unsolved)

1. Stack organized processor does not use address field for instructor ADD or MUL.
2. PSW stands for program stack word.

(V) Detailed Questions

1. A relative mode branch type of instruction is stored in memory at an address equivalent to decimal 750. The branch is made to an address equivalent to decimal 500.
   (*a*) What should be the value of the relative address field of the instruction in decimal?
   (*b*) Determine the relative address value in binary using 12 bits.
2. Write a program to evaluate the arithmetic statement
   (*a*) Using a general purpose register with three address instructions.
   (*b*) Using an accumulator type computer with one address instruction.
3. Perform the logic AND, OR , and XOR with the two binary strings 11011101 and 11001111.

Answers

(I) Fill in the Blanks (Solved)

1. single
2. Hardwired

(II) True or False (Solved)

1. True
2. False

□□□

6.2 INSTRUCTION SETS

Computers provide an extensive set of instructions to give the user the flexibility to carry out various computational tasks. The instruction set of different computers differ from each other mostly in the operands are determined from the address and mode field. The basic set of operations available in a typical computer is covered in this and in the next section.

Most of computer instructions can be classified into three categories:

- 1 - Data transfer instructions.
- 2 - Data manipulation instructions.
- 3 - Program control instructions.

In the next three sections we will briefly describe those types and list some sample instructions based on the instruction format of the ATMEL Microcontroller instruction set.

Fig. 6.2. ATMEL AVR Microcontroller

6.2.1 Data transfer instructions

Data transfer instructions move data in a computer from one place to another without changing the data content. The most common transfers are between memory and processor registers, between processor registers and input or output, and between the processor registers themselves. The following is a list of some data transfer instructions used in many computers.

Table 6.1. Typical Data Transfer instructions

Name	*Mnemonic*
Load	LD
Store	ST
Move	MOV
Exchange	XCH
Input	IN
Output	OUT
Push	PUSH
Pop	POP

Selected ATMEL Microcontroller Data Transfer instructions are listed below:

- LD – Load Indirect from Data Space to Register
- **Example:**
- clr r27 ; Clear X high byte
- ldi r26,$60 ; Set X low byte to $60
- **SPM –** Store Program Memory
- **ST (STD) –** Store Indirect From Register to Data Space
- **Example:**
- clr r27 ; Clear X high byte
- ldi r26,$60 ; Set X low byte to $60
- st X+,r0 ; Store r0 in data ;space loc. ;$60(X post ;inc)
- st X,r1 ; Store r1 in data ;space loc. $61
- ldi r26,$63 ; Set X low byte to $63
- st X,r2 ; Store r2 in data ;space loc. $63
- st -X,r3 ; Store r3 in data ;space loc.
- ;$62(X pre dec)
- **MOV –** Copy Register
- **MOVW –** Copy RegisterWord
- **IN –** Load an I/O Location to Register
- **OUT –** Store Register to I/O Location
- **PUSH –** Push Register on Stack
- **POP –** Pop Register from Stack

Data Manipulation Instructions

Data manipulation instructions perform operations on data and provide the computational capabilities for the computer. They are usually divided into three basic types:

- 1 - Arithmetic instructions.
- 2 - Logical and bit manipulation instructions.
- 3 - Shift instructions.

6.2.2 Arithmetic instructions

The four basic arithmetic operations are addition, subtraction, multiplication and division. A list of typical arithmetic instructions is given in Table 6.2. The increment instruction adds one to the value stored in a register or memory word. One common characteristic of the increment operations is that a binary number of all 1's when incremented produces a result of all 0's. The decrement instruction subtracts 1 from a value stored in a register or memory word. A number of all 0's when decremented produces a result of all ones.

- **Table 6.2. Typical Arithmetic instructions**

Name	***Mnemonic***
Increment	INC
Decrement	DEC
Add	ADD
Subtract	SUB
Multiply	MUL
Divide	DIV
Add with carry	ADDC
Subtract with borrow	SUBB
Negate (2's Complement)	NEG

Selected ATMEL Microcontroller Arithmetic instructions are listed below:

- **INC –** Increment
- **Example:**
- clr r22 ; clear r22
- inc r22 ; increment r22
- **DEC –** Decrement
- **Example:**
- ldi r17,$10 ; Load constant in r17
- add r1,r2 ; Add r2 to r1
- dec r17 ; Decrement r17
- **ADC –** Add with Carry
- **ADD –** Add without Carry
- **Example:**
- add r1,r2 ; Add r2 to r1 (r1=r1+r2)
- add r28,r28 ; Add r28 to itself (r28=r28+r28)
- **ADIW –** Add Immediate to Word
- **SUB –** Subtract without Carry
- **Example:**
- sub r13,r12 ; Subtract r12 from r13
- **SUBI –** Subtract Immediate
- **MUL –** Multiply Unsigned
- **Example:**
- mul r5,r4 ; Multiply unsigned r5 and r4
- movw r4,r0 ; Copy result back in r5:r4
- **MULS –** Multiply Signed
- **NEG –** Two's Complement

6.2.3 Logical and bit manipulation instructions

Logical instructions perform binary operations on strings of bits stored in registers. Some typical logical and bit manipulation instructions are listed in Table 6.3.

Table 6.3. Typical Logical and bit manipulation instructions

Name	***Mnemonic***
Clear	CLR
Complement	COM
AND	AND
OR	OR
Exclusive-OR	XOR
Clear carry	CLRC
Set Carry	SETC
Complement Carry	COMC
Enable interrupt	EI
Disable interrupt	DI

Selected ATMEL Microcontroller Logical and bit manipulation instructions are listed below:

- **CBI –** Clear Bit in I/O Register
- **Example:**
- cbi $12,7 ; Clear bit 7 in Port D
- **CBR –** Clear Bits in Register
- **COM –** One's Complement
- **Example:**
- com r4 ; Take one's complement of r4
- **AND –** Logical AND
- **Example:**
- and r2,r3 ; Bitwise and r2 and r3, result in ;r2
- **ANDI –** Logical AND with Immediate
- **OR –** Logical OR
- **Example:**
- or r15,r16 ; Do bitwise or between registers
- **ORI –** Logical OR with Immediate
- **EOR –** Exclusive OR
- **CLC –** Clear Carry Flag
- **Example:**
- add r0,r0 ; Add r0 to itself
- clc ; Clear Carry Flag
- **SEC –** Set Carry Flag
- **SEI –** Set Global Interrupt Flag
- **CLI –** Clear Global Interrupt Flag

6.2.4 Shift Instructions

Shifts are operations in which the bits of a word are moved to the left or right. Shift instructions may specify either logical shifts, arithmetic shifts, or rotate-type operations. Table 6.4. lists four types of shift instructions and four types of rotate instructions.

Table 6.4. Typical Shift and Rotate instructions

Name	*Mnemonic*
Logical shift right	SHR
Logical shift left	SHL
Arithmetic shift right	SHRA
Arithmetic shift left	SHLA
Rotate right	ROR
Rotate left	ROL
Rotate right through carry	RORC
Rotate left through carry	ROLC

Selected ATMEL Microcontroller Shift and Rotate instructions are listed below:

- **LSL –** Logical Shift Left
- **Example:**
- add r0,r4 ; Add r4 to r0
- lsl r0 ; Multiply r0 by 2
- **LSR –** Logical Shift Right

- **ASR –** Arithmetic Shift Right
- **Example:**
- ldi r16,$10 ; Load decimal 16 into r16
- asr r16 ; r16=r16 / 2
- ldi r17,$FC ; Load -4 in r17
- asr r17 ; r17=r17/2
- **ROL –** Rotate Left trough Carry
- **ROR –** Rotate Right through Carry
- **Example:**
- lsr r19 ; Divide r19:r18 by two
- ror r18 ; r19:r18 is an unsigned two-byte ;integer

Program Control Instructions

Program flow can be altered by instructions that modify the value of the program counter: important feature of a digital computer provides a control over the program flow and capability for branching to different program segments. Typical program control instructions are shown in Table 6.5.

Table 6.5. Typical program control instructions

Name	*Mnemonic*
Branch	BR
Jump	JMP
Skip	SKP
Call	CALL
Return	RET
Compare (by subtraction)	CMP
Test (by ANDing)	

Branch and jump instructions may be conditional or unconditional.

An unconditional branch instruction causes a branch to the specific address without any conditions, e.g.:

```
jmp   DisplayGreetinh
```

The conditional branch specifies a condition, e.g. branch if zero: only when the condition is met, the program counter is loaded with the branch address, e.g.:

Compare and test instructions can be used in setting conditions for subsequent conditional branch instructions.

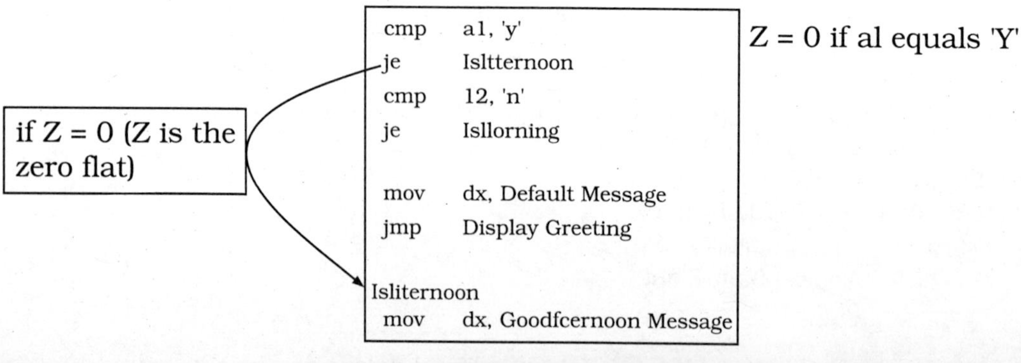

Compare performs an arithmetic subtraction: result is not saved - only status bit conditions are set as a result of operation. Similarly test performs logical AND of two operands and updates certain status bits.

Selected ATMEL Microcontroller Shift and Rotate instructions are listed below:

- **BRCC –** Branch if Carry Cleared
- **Example:**
- add r22,r23 ; Add r23 to r22
- brcc nocarry ; Branch if carry cleared
- ...
- nocarry: nop ; Branch destination (do nothing)
- **BRCS –** Branch if Carry Set
- **BRID –** Branch if Global Interrupt is Disabled
- **BRIE –** Branch if Global Interrupt is Enabled
- **BRVC –** Branch if Overflow Cleared
- **Example:**
- add r3,r4 ; Add r4 to r3
- brvc noover ; Branch if no overflow
- ...
- noover: nop ; Branch destination (do nothing)
- **BRVS –** Branch if Overflow Set
- **JMP –** Jump
- **Example:**
- mov r1,r0 ; Copy r0 to r1
- jmp farplc ; Unconditional jump
- ...
- farplc: nop ; Jump destination (do nothing)
- **SBI –** Set Bit in I/O Register
- **SBIC –** Skip if Bit in I/O Register is Cleared
- **CALL –** Long Call to a Subroutine
- **RCALL –** Relative Call to Subroutine
- **RET –** Return from Subroutine
- **CP –** Compare
- **Example:**
- cp r4,r19 ; Compare r4 with r19
- brne noteq ; Branch if r4 <> r19
- ...
- noteq: nop ; Branch destination (do nothing)
- **TST –** Test for Zero or Minus

6.2.5 Status Bit Conditions

The status register stores the values of the status bits (status register is composed of the status bits). Bits of the status register are modified as a result of an operation performed in the ALU. Fig. 6.3. shows the block diagram of an 8-bit ALU with a 4-bit status register.

C (carry) is set to 1 if carry C8 is 1.
C is cleared to D if C8 is 0.

S (sign) is set to 1 the highest-order bit F7 is 1. S is set to 0 if F7 is 0.

Z (zero) is set to 1 if the output of ALU is 0 (all 0's). Z is set to 0 otherwise.

V (overflow) is set to 1 is XOR of the last two carries is equal to 1 and cleared to 0 otherwise.

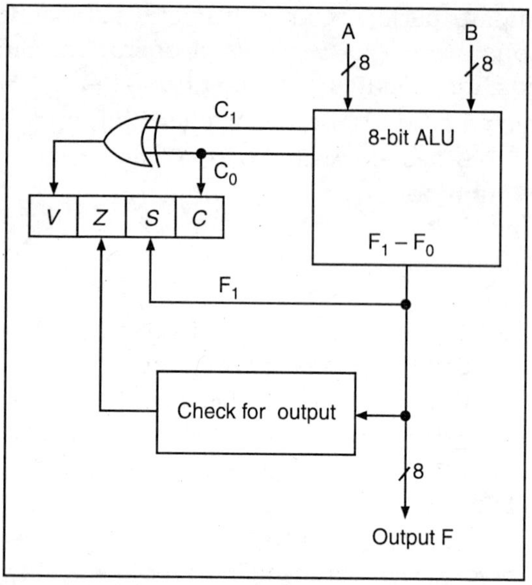

Fig. 6.3. Status register bits

Status bits can be checked after ALU operation to determine certain relationships that exist between the values of *A* and *B*. V indicates overflow *i.e.* for 8-bit ALU the result is greater than 127 or less than – 127. If Z is set, the result is zero: we can use, *e.g.* XOR operation to compare to numbers (the result is zero if *A* = *B*) and *Z* indicates the result of comparison. A single bit in *A* can be checked with a mask that contains 1 in that particular bit position (others being 0's) and by using AND operation.

6.2.6 Conditional Branch Instruction

Conditional branch instructions use the status bits for checking conditions for branching. Table 6.6. gives a list of the most common branch instruction.

Table 6.6. Conditional branch instructions

Mnemoice	Branch condition	Tested condition
BZ	Branch if zero	Z = 1
BNZ	Branch if not zero	Z = 0-
BC	Branch if carry	C = 1
BNC	Branch if no carry	C = 0
BP	Branch if plus	S = 0
BM	Branch if minus	S = 1
BV	Branch if overlflow	V = 1
BNV	Branch if no overflow	V = 0
Unsigned compare conditions (A – B)		
BHI	Branch if higher	A > B
BHE	Branch if higher or equal	A ≥ B
BLO	Branch if lower	A < B
BLOE	Branch if lower to equal	A ≤ B
BE	Branch if equal	A = B
BNE	Branch if not equal	A ≠ B
Signed compare conditions (A – B)		
BGT	Branch if greater than	A > B
BGE	Branch if greater or equal	A ≥ B
BLT	Branch if less than	A < B
BLE	Branch if less or equal	A ≤ B
BE	Branch if equal	A = B
BNE	Branch if not equal	A ¹ B

80×80 Z = 0 if al equals Y

```
cmp   al, 'y'
je    Isktner noon
cmp   al, 'x'
je    al, 'y'
mov   dx, Default Message
jmp
```

Numerical Example

- Let A = 11110000 and B = 00010100. Perform A - B, update the status bits, and list the conditional branch instructions.
- 1. The ALU takes the 2's complement of B and adds it to A.
- A: 11110000
- B + 1: 11101100
- A – B: 11011100
- 2. The status bits are
- C = 1, S = 1, V = 0, and Z = 0
- 3. For unsigned numbers:
- A = 240, and B = 20
- Then we have that A>B and A B, i.e. the instructions that will cause a branch are BHI, BHE, and BNE.
- 4. For signed numbers:
- A = - 16, and B = 20
- Then we have that A<B and A B, i.e. the instructions that will cause a branch are BLT, BLE, and BNE.

6.2.7 Subroutine Call and Return

For subroutine calls, different computers can use a different temporary location for storing the return address, some computers use the first memory location of the subroutine (like the "Basic Computer"), some store the return address in a fixed memory location, and some computers use a processor register. Stack memory is yet another possibility (the most efficient way): when a succession of subroutines is called (nested calls), the sequential return addresses can be pushed into the stack. The 'return from subroutine' instruction pops the return address (and assigns to program counter) from the top of the stack: we always have the return address for the last called subroutine.

- Subroutine call (stack based) microoperations:

$SP \leftarrow SP - 1$	Decrement stack pointer
$M[SP] \leftarrow PC$	Push content of PC onto the stack
$PC \leftarrow$ effective address	Transfer control to the subroutine

- Subroutine return microoperations:

$PC \leftarrow M[SP]$	Pop stack and transfer to PC
$SP \leftarrow SP + 1$	Increment stack pointer

By using subroutine stack each return address (in nested calls) can be pushed into the stack without destroying any previous values. In basic computer a recursive subroutine call would destroy the previous return address stored in the first memory location of the subroutine.

6.2.8 Program Interrupts

- In computer science, an interrupt is a signal from a device which typically results in a context switch: that is, the processor sets aside what it's doing and does something else.
- Digital computers usually provide a way to start software routines in response to asynchronous electronic events. These events are signaled to the processor via interrupt requests (IRQ). The processor and interrupt code make a context switch

into a specifically written piece of software to handle the interrupt. This software is called the interrupt service routine, or interrupt handler. The addresses of these handlers are termed interrupt vectors and are generally stored in a table in RAM, allowing them to be modified if required.

- Interrupts were originated to avoid wasting the computer's valuable time in software loops (called polling loops) waiting for electronic events. Instead, the computer was able to do other useful work while the event was pending. The interrupt would signal the computer when the event occurred, allowing efficient accommodation for slow mechanical devices.
- Interrupts allow modern computers to respond promptly to electronic events, while other work is being performed. Computer architectures also provide instructions to permit processes to initiate software interrupts or traps. These can be used, for instance, to implement co-operative multitasking.
- A well-designed interrupt mechanism arranges the design of the computer bus, software and interrupting device so that if some single part of the interrupt sequence fails, the interrupt restarts and runs to completion. Usually there is an electronic request, an electronic response, and a software operation to turn off the device's interrupt, to prevent another request.
- Program interrupt is similar to subroutine call, except:
    1. The interrupt is (usually) initiated by an internal of external signal rather than an execution of an instruction (software interrupts are exceptions).
    2. The address of the interrupt service program (routine) is determined by hardware rather than the address field of an instruction: the CPU must possess some form of HW procedure for selecting a branch address servicing the interrupt.
    3. Interrupt routine stores all the information (not just PC) necessary to recover the state of the CPU prior the return from the interrupt routine.
- After the interrupt routine the CPU must return exactly the same state that is was when the interrupt occurred. The state of the CPU at the end of the execute cycle (the interrupt is recognized in this phase) is determined from:
    1. The content of PC
    2. The content of all processor registers
    3. The content of status conditions
- Status bits (program status word PSW) stored in a separate status register.
- **Contains status information about the state of the CPU:** bits from ALU operation, interrupt enable bits, and CPU operation mode (system mode, user mode), for example.
- Some computer store only program counter (and PSW) prior entering to an interrupt routine. The interrupt routine must take care of storing and restoring the CPU status. CPU does not respond to an interrupt until the end of an instruction execution.

6.2.8.1 Interrupt cycle

- Contents of PC and PSW are pushed onto stack.
- The branch address is transferred to PC and new PSW is loaded into the status register.
- The interrupt routine can now be executed starting from the branch address (which may contain a branch instruction to a user defined service routine).

Return from interrupt

- The stack is popped to retrieve PWS to status register and return address to PC.
- CPU state is restored and the interrupted program can proceed like nothing had happen.

6.2.8.2 Interrupt types

There are three major types of interrupts that cause a break in the normal execution of a program. They can be classified as:

1. External interrupts

Come from input/output, timing, or any other external source.

e.g.: input/output device requesting new data, elapsed time of an event, power failure, etc.

2. Internal interrupts (traps)

Come from illegal or erroneous use of an instruction or data.

e.g.: overflow, division by zero, invalid operation code, stack overflow, and protection violation.

Usually occur as a result of a premature termination of the instruction execution: the service program determines the corrective measure to be taken (e.g. terminates the program)

3. Software interrupts

Initiated by an instruction (rather than HW signals).

A special call instruction that behaves like an interrupt.

- Can be used by a programmer to initiate an interrupt routine at any desired point in the program.
- Can be used for accessing operating system services.
- Practical Example
- Using INT-instruction (*i.e.* INT 21h) for invoking a DOS interrupt (see 80 × 86 example code shown earlier). The DOS interrupt example shown in Fig. 6.4.

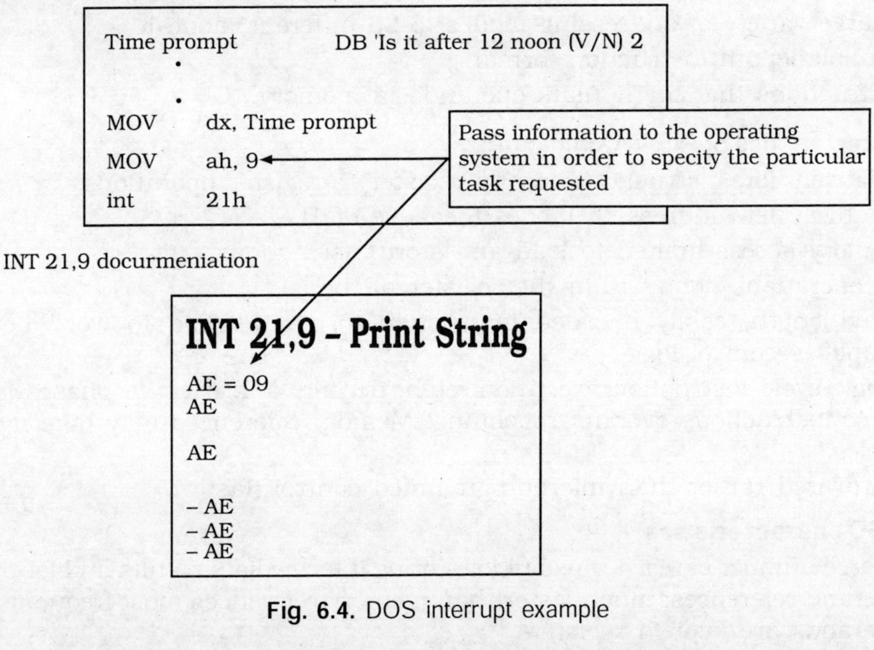

Fig. 6.4. DOS interrupt example

6.3 REDUCED INSTRUCTION SET COMPUTER (RISC)

- Instruction set determines the way that machine language programs are constructed. Early computers had small and simple instruction sets in order to minimize the (expensive) hardware needed for their implementation. Today many computers have instructions that include 100 to 200 instructions.
- Variety of data types.
- Large number of addressing mode.
- Complex Instruction Set Computer (CISC) has complex hardware and large instruction set: functions from software to hardware. In contrast, Reduced Instruction Set Computer (RISC) uses fewer and simpler instructions which can be executed faster within the CPU. RISC chips require fewer transistors (than CISC), which makes them cheaper to design and produce.
- There is still considerable controversy among experts about the ultimate value of RISC architectures.
- Its proponents argue that RISC machines are both cheaper and faster, and are therefore the machines of the future.
- Skeptics note that by making the hardware simpler, RISC architectures put a greater burden on the software. They argue that this is not worth the trouble because conventional microprocessors are becoming increasingly fast and cheap anyway.
- However, CISC and RISC implementations are becoming more and more alike. Many of today's RISC chips support as many instructions as yesterday's CISC chips. Today's CISC chips use many techniques formerly associated with RISC chips. One reason for the trend to provide a complex instruction set is to simplify the translation from high-level to machine language programs.

6.3.1 Characteristics of CISC Architecture

1. A large instruction set.
2. Instructions that perform special tasks and are used infrequently.
3. A large variety of addressing modes (5-20 different modes).
4. Variable-length instruction formats.
5. Instructions that manipulate operands in memory.

6.3.2 Characteristics of RISC Architecture

1. Relatively few instructions mostly register-to-register operations.
2. Relatively few addressing modes (because of 1).
3. Memory access limited to load and store instructions.
4. All operations done within the register of the CPU.
5. Fixed-length, easily decoded instruction format aligned to word boundaries simplifies control logic.
6. Single-cycle instruction execution fetch, decode, and execute phases for two to three instructions overlap: pipelining. Memory references may take more clock cycles.
7. Hardwired rather than microprogrammed control (faster).

Other RISC characteristics

1. A large number of register useful for storing intermediate results and for optimizing operand references: much faster than memory references most frequent accessed operands are kept in registers.

2. Use of overlapped register windows to speed-up procedure call and return.
3. Efficient instruction pipeline.
4. Compiler support for efficient translation of high-level language programs into machine language programs.

Overlapped Register Windows

A characteristic of some RISC processors is their use of overlapped register windows to provide the passing of parameters and avoid need for saving and restoring register values: speeds up procedure calls and returns. Each procedure call results in the allocation of a new window consisting of a set of registers current window pointer (CWP) is decremented: corresponds 'save' in Fig. 6.5.

- Each return statement increments the CWP: corresponds 'restore' in Fig. 6.4.
- Windows for adjacent procedures (nested calls) have overlapping registers that are shared to provide the passing of parameters and results.
- Local register can be used for local variables by using local registers there is no risk of corrupting data of another procedure (e.g. caller).
- Overlapped registers are used to pass parameters (in) and store results (out).
- Only one register windows is activated at any given time with a CWP.
- Each procedure call activates new register window by updating (decrementing) the CWP.

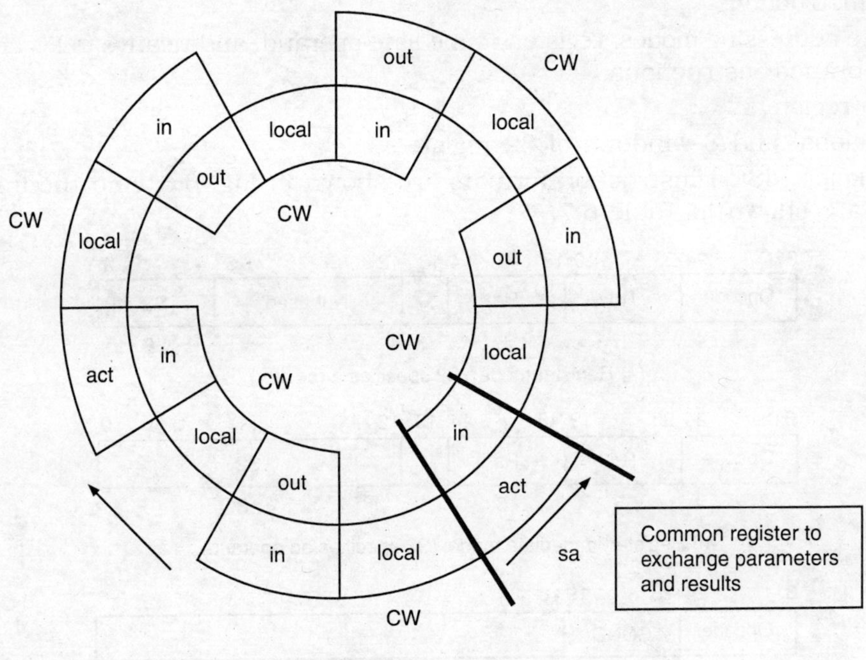

Fig. 6.5. Overlapped register windows

To summarize

- Register windows provide easy access to a large collection of registers and can reduce the need to save registers in memory.
- If you write a procedure with more parameters than common registers (reserved for input parameters), you will need to use the stack for any parameters beyond the amount of reserved number of registers.

- If your call sequence gets deeper than number of windows (as it probably will in most recursive procedures), you are again forced to use the stack.
- **In general**, the organization of register windows will have the following relationships :
    - Number of global registers = G
    - Number of local registers in each window = L
    - Number of registers common to two windows = C
    - Number of windows = W
    - Number of registers available for each window is calculated as follows:
        - window size = $L + 2C + G$
    - The total number of registers needed in the processor is:
        - Register file = $(L + C)W + G$
- **Berkeley RISC I**
- One of the first projects for showing the advantages of RISC architecture (in University of California, Berkeley).

  32-bit CPU

  32-bit addresses

  8 –, 16 –, or 32-bit data

  32-bit instruction format

  32 instructions

  three addressing modes: register, immediate operand, and relative to PC addressing for branch instructions.

  138 registers
- 10 global and 8 windows of 32 registers.
- Berkeley RISC I instruction formats are shown in Fig. 6.6. And the instruction set are shown in Table 6.7.

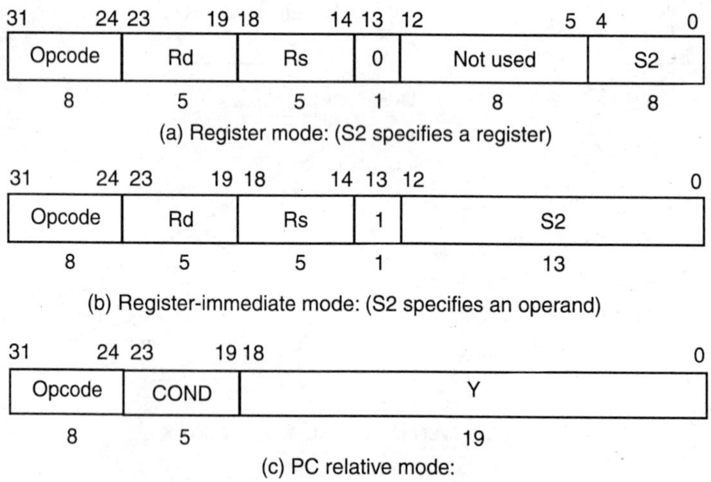

Fig. 6.6. Berkeley RISC I instruction formats

- Where:
    - **Rd field:** selects a destination register for result.
    - **Rs field:** one of the source registers.

- **S2 field:** second source, can specify a register or an immediate operand.
- ***Y* field:** 19-bit relative address.
- **COND field:** specify one of the possible branch conditions.

• From Table 6.6, register R0 contains all 0's, so it can be used in any field to specify a zero quantity. Number sign # used in an immediate operands.

Table 6.7. Instruction set of Berkeley RISC I

Opcode	*Operands*	*Register Transfer*	*Description*
Data manipolation instructions			
ADD	Rs. *S*2, *Rd*	*Rd* ← *Rs* + *S*2	Integer add
ADDC	Rs. *S*2, Rd	*Rd* ← *Rs* + *S*2 + carry	Add with carry
SUB	Rs. *S*2, *Rd*	*Rd* ← *Rs* – *S*2	Integer subtract
SUBC	Rs. *S*2, *Rd*	*Rd* ← *Rs* – *S*2 – carry	Subtract with carry
SUBR	Rs. *S*2, *Rd*	*Rd* ← *S*2 – *Rs*	Subtract reverse
SUBCR	Rs. *S*2, *Rd*	*Rd* ← *S*2 – *Rs* – carry	Subtract with carry
AND	Rs. *S*2, *Rd*	*Rd* ← *Rs* ∧ *S*2	AND
OR	Rs. *S*2, *Rd*	*Rd* ← *Rs* ∨ *S*2	OR
XOR	Rs. *S*2, *Rd*	*Rd* ← *Rs* ⊕ *S*2	Exclusive OR
SLL	Rs. *S*2, *Rd*	*Rd* ← *Rs* shifted by *S*2	Shift-left
SRL	Rs. *S*2, *Rd*	*Rd* ← *Rs* shifted by *S*2	Shift-right logical
SRA	Rs. *S*2, *Rd*	*Rd* ← *Rs* shifted by *S*2	Shift-right arithmetic
Data transfer instructions			
LDL	(Rs.) *S*2, *Rd*	*Rd* ← *M*[*Rs* + *S*2]	Load long
LDSU	(Rs.) *S*2, *Rd*	*Rd* ← *M*[*Rs* + *S*2]	Load short unsigned
LDSS	(Rs.) *S*2, *Rd*	*Rd* ← *M*[*Rs* + *S*2]	Load short signed
LDBU	(Rs.) *S*2, *Rd*	*Rd* ← *M*[*Rs* + *S*2]	Load by unsigned
LDBS	(Rs.) *S*2, *Rd*	*Rd* ← *M*[*Rs* + *S*2]	Load byte signed
LDHI	*Rd*, *Y*	*Rd* ← *Y*	Load immediate high
STL	*Rd*, (Rs.) *S*2	*M*[*Rs* + *S*2] ← *Rd*	Store long
STS	*Rd*, (Rs.) *S*2	*M*[*Rs* + *S*2] ← *Rd*	Store long
STB	*Rd*, (Rs.) *S*2	*M*[*Rs* + *S*2] ← *Rd*	Store byte
GETPSW	*Rd*	*Rd* ← *PSW*	Load states word
PUTPSE	*Rd*	*PSW* ← *Rd*	Set status word
Program control instructions			
JMP	CONND *S*2 (Rs.)	*PC* ← *Rs* + *S*2	Conditional jump
JMPR	COND, *Y*	*PC* ← *PC* + *Y*	Jump relative
CALL	*Rd*. *S*2 (Rs.)	*Rd* ← *PC* *PC* ← Rs. + *S*2 *CWP* ← *CWP* + 1	Call subroutine and change window
CALLR	*Rd*, *Y*	*Rd* ← *PC* *PC* ← *PC* + *Y* *CWP* ← *CWP* – 1	Call relative and change window
RET	*Rd*, *S*2	*PC* ← *Rd* + *S*2 *CWP* – *CWP* + 1	Return and change window
CALLINT	*Rd*	*Rd* ← *PC* *CWP* ← *CWP* – 1	Disable interrupts

RETINT	*Rd, S2*	$Pc \leftarrow Rd + S2$ $CWP \leftarrow CWP + 1$	Enable interrupts
GTLPC	*Rd*	$Rd \leftarrow PC$	Get last *PC*

6.4 REGISTERS

When the processor executes instructions, data is temporarily stored in small, local memory locations of 8, 16, 32 or 64 bits called **registers**. Depending on the type of processor, the overall number of registers can vary from about ten to many hundreds.

The main registers are:

- the **accumulator register** (*ACC*), which stores the results of arithmetic and logical operations;
- the **status register** (*PSW, Processor Status Word*), which holds system status indicators (carry digits, overflow, etc.);
- the **instruction register** (*RI*), which contains the current instruction being processed;
- the **ordinal counter** (*OC* or *PC* for *Program Counter*), which contains the address of the next instruction to process;
- the **buffer register**, which temporarily stores data from the memory.

6.4.1 Instruction Cycle, Fetch Cycle and Execution Cycle

The basic function of a CPU is to execute sequences of instructions stored in main memory which is external to the CPU. The sequence of operations involved in processing an instruction constitutes an instruction cycle. This can be subdivided into two major groups (*i*) Fetch cycle (FC) (*ii*) Execute cycle (EC). The behavior of the CPU during an instruction cycle can be explained by a sequence of micro operations. The time required for each micro-operation is known as CPU cycle time. It is generally denoted as tCPU-The reciprocal of cycle time is known as "Clock rate". Clock rate of CPU is measured in Mega hertz. The clock rate depends upon the type of the CPU.

Fetch and Execute cycles are performed in time periods of fixed or variable length. The fetch cycle is performed in a fixed time period whereas the execute cycle requires variable time periods. This length of time period depends upon the instruction to be executed. The total time required to complete the fetch cycle and execute cycle is known as instruction cycle time (IC) and IC = FC + EC.

Fetch Cycle: It is defined as the time required for getting the instruction code from main memory to CPU.

The instruction may also contain data or operand address. At the beginning of the fetch cycle the contents of program counter are sent to the memory. The memory responds by reading the contents of the received address and transferring them to the CPU. To transfer the address it requires one clock cycle, i.e. T1. In the next to clock cycles two operations are performed. The first of these is the reading of the instruction by the memory from the location whose address it has received in the previous clock cycle. This is performed during T2 time slot. In the next clock cycle, T3 the transfer of the instruction from the memory to the CPU takes place. This indicates that fetch cycle is a fixed time slot. This time is known as fetch1 cycle time.

Execute Cycle: The time required to decode and execute an instruction is known as execute cycle.

The time required for an execute cycle will depend on the instruction type. If the operands are in the CPU registers, then only the decoding and execute operations are performed. In most of the cases this is completed in just one clock cycle. If the data or the address of the. data are present then more cycles are required for fetching the data and executing from memory. This fetch operations is almost similar to fetch cycle. Hence execute cycle requires different time slots. This time is known as executive cycle time.

Instruction Cycle: We have seen that fetch cycle requires fixed time slot whereas execute cycle requires, variable width time slots. The total time required to perform fetch operation and execute operation is known as instruction cycle time.

The instruction cycle is the sum of fetch cycle and execute cycle:

Instruction cycle = Fetch cycle + Execute cycle

To control the operations of input/output devices CPU takes the help of interrupt. When interrupt is given by any input/output device the CPU suspends the execution of the program and gives service to the input/output device. Most of the input/output interrupts must be responded immediately. A test for the interrupts is carried out at the end of each instruction cycle. If there are no interrupt signals the fetch, execute and instruction cycles are carried out in a systematic order.

6.4.1.1 Instruction Format

The format of an instruction is usually depicted in a rectangular box symbolizing the bits of the instruction as they appear in memory words or in a control register. The bits of the instruction are divided into groups called fields. The most common fields found in instruction formats are:

1. Operation code specifying the operation: add, subtract, complement, etc.
2. Address field designating a memory address or a register.
3. Mode field for specifying the way for determining the effective address of an operand.

The number of address fields in the instruction format of a computer depends on the internal organization of its registers. E.g.: MIPS (a RISC microprocessor architecture developed by MIPS Computer Systems Inc.).

Computers may have instructions of several different lengths containing varying number of addresses. The number of address fields in the instruction format of a computer depends on the internal organization of its register. Most computers fall into one of three types of CPU organizations:

1. Single accumulator organization (uses **one** address field).
2. General register organization(uses two or **three** address fields).
3. Stack organization (uses **zero** address field).

Example: To illustrate the influence of the number of addresses on computer programs, we will evaluate the arithmetic statement

$$X = (A + B) * (C + D)$$

using zero, one, two, or three address instructions.

1. Using three address instructions

ADD	$R1, A, B$	$R1 \leftarrow M[A] + M[B]$
ADD	$R2, C, D$	$R2 \leftarrow M[C] + M[D]$
MUL	$X, R1, R2$	$X \leftarrow R1 * R$

2. Using two address instructions

MOV ← $R1, A$	$R1\ M[A]$
ADD ← $R1, B$	$R1 \leftarrow R1 + M[B]$
MOV ← $R2, C$	$R2 \leftarrow M[C]$
ADD ← $R2, D$	$R2 \leftarrow R2 + M[D]$
MUL ← $R1, R2$	$R1 \leftarrow R1 * R2$
MOV ← $X, R1$	$X \leftarrow R1$

3. Using one address instructions

LOAD	A	$AC \leftarrow M[A]$
ADD	B	$AC \leftarrow AC + M[B]$
STORE	T	$M[T] \leftarrow AC$
LOAD	C	$AC \leftarrow M[C]$
ADD	D	$AC \leftarrow AC + M[D]$
MUL	T	$AC \leftarrow AC * M[T]$
STORE	X	$M[X] \leftarrow AC$

4. Using zero address instructions (RPN: *AB* + *CD* + *)

PUSH	A	TOS ← A
PUSH	B	TOS ← B
ADD		TOS ← $(A + B)$
PUSH	C	TOS ← C
PUSH	D	TOS ← D
ADD		TOS ← $(C + D)$
MUL		TOS ← $(C + D) * (A + B)$
POP	X	$M[X]$ ← TOS

The MIPS ISA instructions fall into three categories: R-type, I-type, and J-type. Not all ISAs divide their instructions this neatly. This is one reason to study MIPS as a first assembly language. The format is simple.

R-type

R-type instructions refer to register type instructions. Of the three formats, the R-type is the most complex.

B31-26	*B25-21*	*B20-16*	*B15-11*	*B10-6*	*B5-0*
opcode	register s	register t	register d	Shift amount	function

This is the format of the R-type instruction, when it is encoded in machine code.

The prototypical R-type instruction is:

Add $rd, $rs, $rt

where $rd refers to some register d (d is shown as a variable, however, to use the instruction, you must put a number between 0 and 31, inclusive for d). $rs, $rt are also registers.

The semantics of the instruction are;

$$R[d] = R[s] + R[t]$$

where the addition is signed addition.

You will notice that the order of the registers in the instruction is the destination register ($rd), followed by the two source registers ($rs and $rt).

However, the actual binary format (shown in the table above) stores the two source registers first, then the destination register. Thus, how the assembly language programmer uses the instruction, and how the instruction is stored in binary, do not always have to match.

Let's explain each of the fields of the R-type instruction.

- **opcode (B31-26):** Opcode is short for "operation code". The opcode is a binary encoding for the instruction. Opcodes are seen in all ISAs. In MIPS, there is an opcode for add.
  The opcode in MIPS ISA is only 6 bits. Ordinarily, this means there are only 64 possible instructions. Even for a RISC ISA, which typically has few instructions, 64 is quite small. For R-type instructions, an additional 6 bits are used (B5-0) called the function. Thus, the 6 bits of the opcode and the 6 bits of the function specify the kind of instruction for R-type instructions.
- **rd (B25-21) ;** This is the destination register. The destination register is the register where the result of the operation is stored.
- **rs (B20-16):** This is the first source register. The source register is the register that holds one of the arguments of the operation.
- **rt (B15-11):** This is the second source register.
- **shift amount (B10-6):** The amount of bits to shift. Used in shift instructions.
- **function (B5-0):** An additional 6 bits used to specify the operation, in addition to the opcode.

I-type instructions

I-type is short for "immediate type". The format of an I-type instuction looks like:

B31-26	*B25-21*	*B20-16*	*B15-0*
opcode	Register s	register t	immediate

The prototypical I-type instruction looks like:

Add $rt, $rs, immed

In this case, $rt is the destination register, and $rs is the only source register. It is unusual that $rd is not used, and that $rd does not appear in bit positions B25-21 for both R-type and I-type instructions. Presumably, the designers of the MIPS ISA had their reasons for not making the destination register at a particular location for R-type and I-type.

The semantics of the instruction are;

$$R[t] = R[s] + (IR15)16\ IR15 - 0$$

where IR refers to the instruction register, the register where the current instruction is stored. (IR15)16 means that bit B15 of the instruction register (which is the sign bit of the immediate value) is repeated 16 times. This is then followed by IR15-0, which is the 16 bits of the immediate value.

Basically, the semantics says to sign-extend the immediate value to 32 bits, add it (using signed addition) to register R[s], and store the result in register $rt.

J-type instructions

J-type is short for "jump type". The format of an J-type instuction looks like:

B31-26	*B25-0*
opcode	target

The prototypical I-type instruction looks like:

j target

The semantics of the j instruction (j means jump) are:

$$PC \,(\, PC31 - 28 \; IR\, 25 - 0\; 00)$$

where PC is the program counter, which stores the current address of the instruction being executed. You update the PC by using the upper 4 bits of the program counter, followed by the 26 bits of the target (which is the lower 26 bits of the instruction register), followed by two 0's, which creates a 32 bit address. The jump instruction will be explained in more detail in a future set of notes.

6.4.2 Basic Computer Instructions

There are 25 no. of instructions chosen for basic computer.

Hexadecimal Code

Symbol	*I = 0*	*I = 1*	*Description*
AND	0xxx	8xxx	AND memory word to *AC*.
ADD	1xxx	9xxx	Add memory word to *AC*.
LDA	2xxx	Axxx	Load memory word to *AC*.
STA	3xxx	Bxxx	Store content of *AC* on memory.
BUN	4xxx	Cxxx	Branch unconditionally.
BSA	5xxx	Dxxx	Branch and save return address.
ISZ	6xxx	Exxx	Increment and skip if zero.
CLA	7800		Clear *AC*.
CLE	7400		Clear *E*.
CMA	7200		Complement *AC*.
CME	7100		Complement *E*.
CIR	7080		Circulate right *AC* and *E*.
CIL	7040		Circulate left *AC* and *E*.
INC	7020		Increment *AC*.
SPA	7010		Skip next instruction if *AC* positive.
SNA	7008		Skip next instruction if *AC* is negative.
SZA	7004		Skip next instruction if *AC* zero.
SZE	7002		Skip next instruction if *E* is 0.
HLT	7001		Halt computer.
INP	F800		Input character to *AC*.
OUT	F400		Output character from *AC*.
SKI	F200		Skip on input flag.
SKO	F100		Skip on output flag.
ION	F080		Interrupt on.
IOF	F040		Interrupt off.

6.4.3 Hardwired Control and Microprogrammed Control

There are two major types of control organization:

1. Hardwired control and
2. Microprogrammed control.

In the hardwired organization, the control logic is implemented with gates, flip-flops, decades and other digital circuits. It has the advantage that it can be optimized to produce a fast mode of operation.

In the microprogrammed organization, the control information is stored is a control memory. The control memory is programmed to initiate the required sequence of micro-operations. A hardwired control, as the name implies, require changes in the wiring among the various components if the design has to be modified or changed. In the microprogrammed control, any required changes or modifications can be done by updating the microprogram in control memory.

The block diagram of the control unit is shown in given Fig. 6.7.

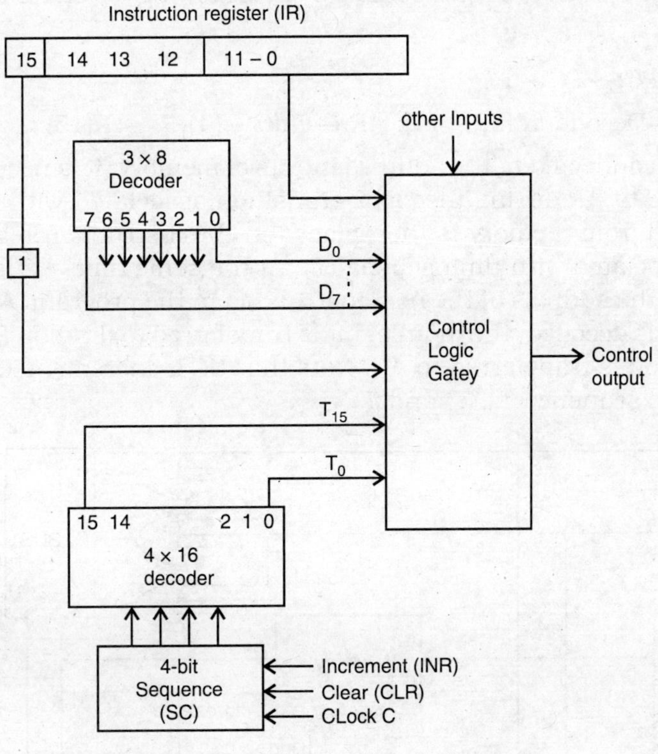

Fig. 6.7. Control Unit of Basic Computer

6.4.3.1 Instruction Cycle

A program in the memory unit of the computer consists of a sequence of instruction. The program is executed in the computer by going through a cycle for each instruction. Each instruction cycle in turn is sub-divided into a square of sub-cycle or phases. In the basic computer each instruction cycle consists of the following phases:

1. Fetch on instruction from memory.
2. Decode the instruction.
3. Read the effective address from memory if the instruction has an indirect address.
4. Execute the instruction.

Upon the completion of step 4, the control goes back to step 1 to fetch, decode and execute the next instruction. This process continues indefinitely unless a HALT instruction is encountered.

Notes:

Ist cycle: Insruction fetch.

IInd cycle: Instruction decode.

IIIrd cycle: Direct → Address fetch
Indirect → Operand fetch

1. Instruction Fetch and Decode: Initially, the program counter *PC* is paded with the address of the first instruction in the program. The sequence counter *SC* is cleared to 0, providing a decoded timing signal T_0. After each clock pulse, *SC* is incremented by one, so that the timing signals go through a sequence T_0, T_1, T_2 and so on. The micro-operations for the fetch and decode phases can be specified by the following register transfer statements.

T_0: $AR \rightarrow AC$

T_1: $IR \leftarrow M[AR]$, $PC \leftarrow PC + 1$

T_2: D_0, ..., $D_7 \leftarrow$ Decode IR (12 – 14), $AR \leftarrow IR(0 - 11)$, $I \leftarrow IR(15)$.

Since only *AR* is connected to the address inputs of memory, it is necessary to transfer the address from *PC* to *AR* during the clock transition associated with timing signal T_0. The instruction read from memory is then placed in the instruction register *IR* with the clock transition associated with timing signal T_1. At the same time, *PC* is incremented by one to prepare it for the address of the next instruction in the program. At the time T_2, the operation code in *IR* is decoded, the indirect bit is transferred to flip-flop *I*, and the address part of the instruction is transferred to *AR*. Note that *SC* is incremented after each clock pulse to produce the sequence T_0, T_1 and T_2.

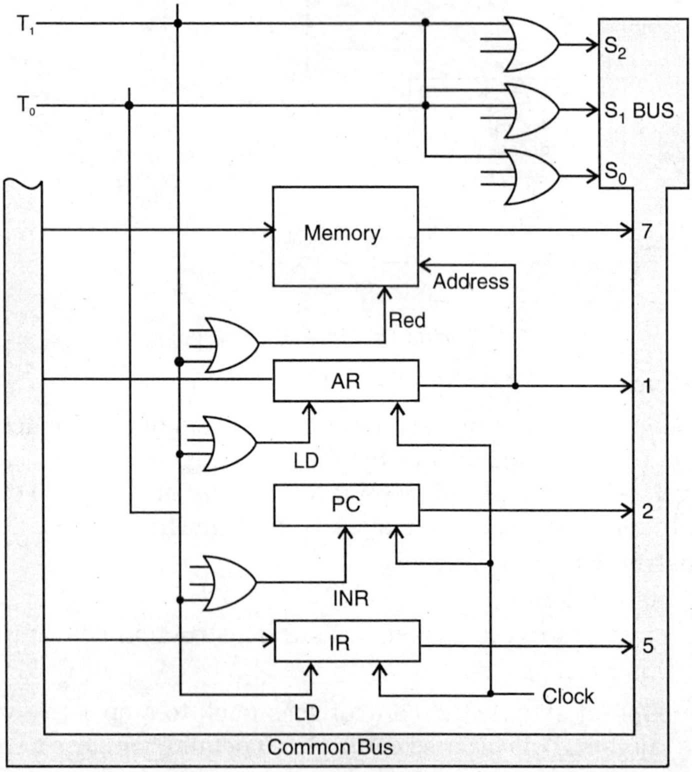

Fig. 6.8. Register transfer to the fetchphase

Given figure shows how the first two register transfer satements are implemented in the bus system. To provide the data path for the transfer of *PC* to *AR* we must apply timing signal T_0 to achieve the following connections:

1. Place the content of *PC* onto the bus by making the bus selection inputs $S_2S_1S_0$ equal to 010.
2. Transfer the content of the bus to *AR* by enabling the *LD* input of *AR*.

The next clock transition initiates the transfer from *PC* to *AR* since $T_0 = 1$. In order the implement the second statement.

T_1: $IR \leftarrow M[AR]$, $PC \leftarrow PC + 1$.

It is necessary to use timing signal T_1 to provide the following connections in the bus system.

1. Enable the read input of memory.
2. Place the content of memory onto the bus by making $S_2S_1S_0 = 111$.
3. Transfer the content of the bus to *IR* by enabling the *LD* input of *IR*.
4. Increment *PC* by enabling the *INR* input of *PC*.

The next clock transition initiates the read and increment operations since $T_1 = 1$.

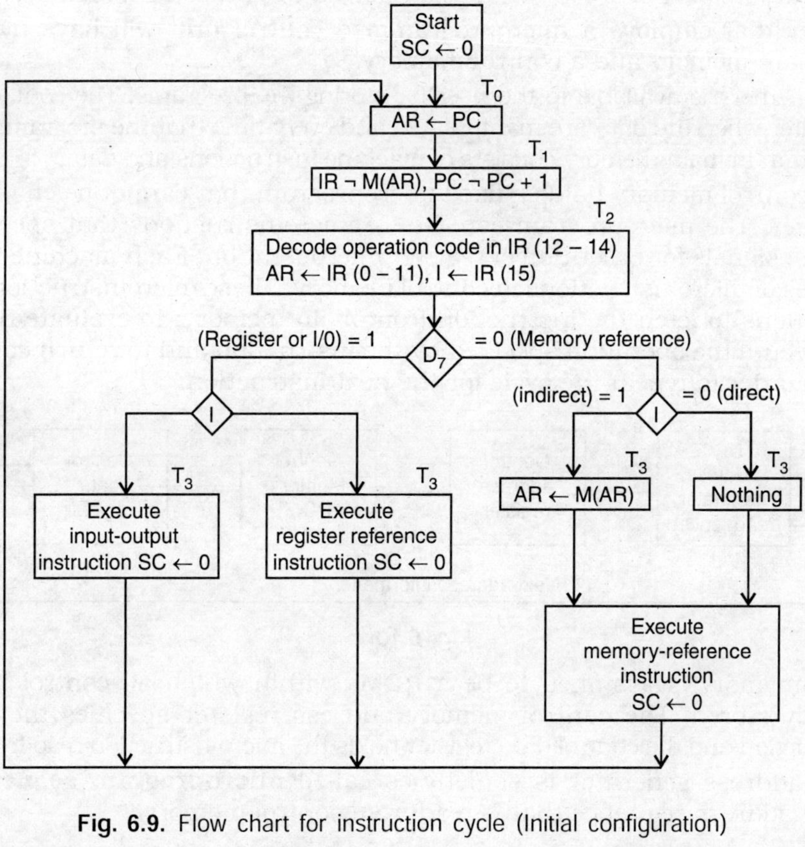

Fig. 6.9. Flow chart for instruction cycle (Initial configuration)

6.4.3.2 Memory—Reference Instruction

Symbol	*Operation Decoder*	*Symbolic Description*
AND	D_0	$AC \leftarrow AC \wedge M[AR]$
ADD	D_1	$AC \leftarrow AC + M[AR]$, $E \leftarrow C_{out}$

LDA	D_2	$AC \leftarrow M[AR]$
STA	D_3	$M[AR] \leftarrow AC$
BUN	D_4	$PC \leftarrow AR$
BSA	D_5	$M[AR] \leftarrow PC$, $PC \leftarrow AR + 1$
ISZ	D_6	$M[AR] \leftarrow M[AR] + 1$, If $M[AR] + 1 = 0$, then $PC \leftarrow PC + 1$

6.4.3.3 Microprogrammed Control

During any given time, certain micro-operations are to be initiated, while others remain idle. The control variables at any give time can be represented by a string of 1's and 0's called a control word.

As such, control words can be programmed to perform various operations on the components of the system. A control unit whose binary control variables are stored in memory is called a microprogrammed control unit. Each word in control memory contains within it a microinstruction. The microinstruction specifies one or more micro-operations for the system. A sequence of microinstructions constitutes a microprogram.

A memory that is part of a control unit is referred to as a control memory.

A computer that employs a microprogrammed control unit will have two separate memories: a main memory and a control memory.

The main memory is available to the user for storing the programs. The contents of main memory may alter when the data are manipulated and every time that the program is changed. The users program in main memory consists of machine instructions and data.

While the control memory holds a fixed microprogram that cannot be changed by the occassional user. The microprogram consists of microinstructions that specify various internal control signals for execution of register microperation. Each machine instruction initiates a series of microinstructions in control memory. These microinstructions generate the microperations to fetch the instruction from main memory; to evaluate the effective address, to execute the operation specified by the instruction and to return control to the fetch phase in order to repeat the cycle for the next instruction.

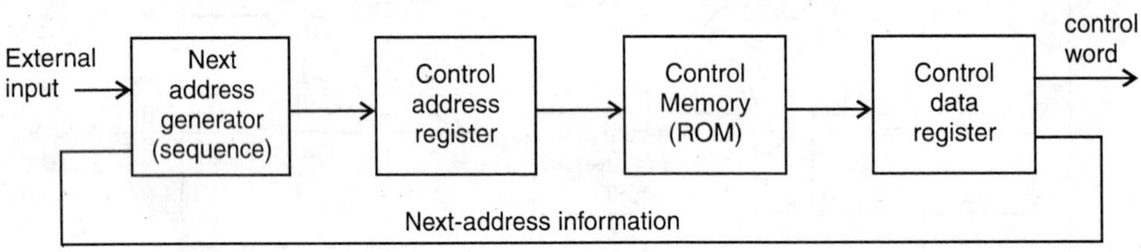

Fig. 6.10.

A control memory is assumed to be a ROM_X within which all control information is permanently stored. The control memory address register specifies the address of the microinstruction and the control data register holds the microinstruction read from memory.

The next address generator is sometimes called **microprogram sequencer,** as it determine the address sequence that is read from control memory.

Mapping of Instruction

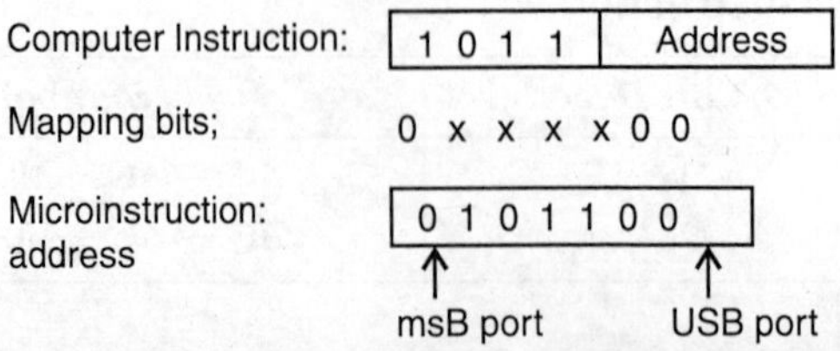

6.4.3.4 Microinstruction Format with next address field

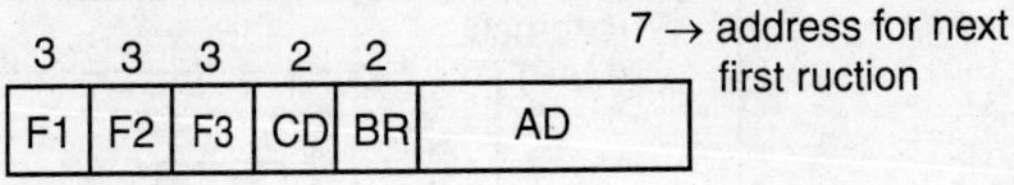

F_1, F_2, F_3: Microperation field.
CD: Condition for branching.
BR: Branch file.
AD: Address field.
Microinstruction code format (20 Bits).
Fetch Routine:

The microinstruction needed for the fetch routine are:

$AR \leftarrow PC$

$DR \leftarrow M[AR],\ PC \leftarrow PC + 1$

By using assembly language we can write the symbolic microprogram for the fetch routine as:

```
        ORG 64
FETCH:  PCTAR           U   JMP   NEXT
        READ, INCPC     U   JMP   NEXT
        DRTAR           U   MAP
```

The bit values of above program can be obtained as:

Binary Address	F_1	F_2	F_3	*CD*	*BR*	*Next*
1000000	110	000	000	00	00	1000001
1000001	000	100	101	00	00	1000010
1000010	101	000	000	00	11	0000000

6.4.3.5 Microprogram Address Sequencing and Microprogram Sequencer

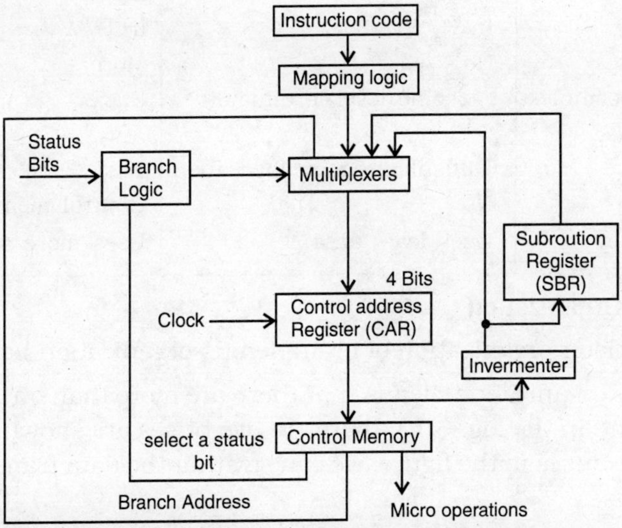

Fig. 6.11. Selection of address for control memory

Expanded Part of above Diagram: Microprogram sequencer for a control memory

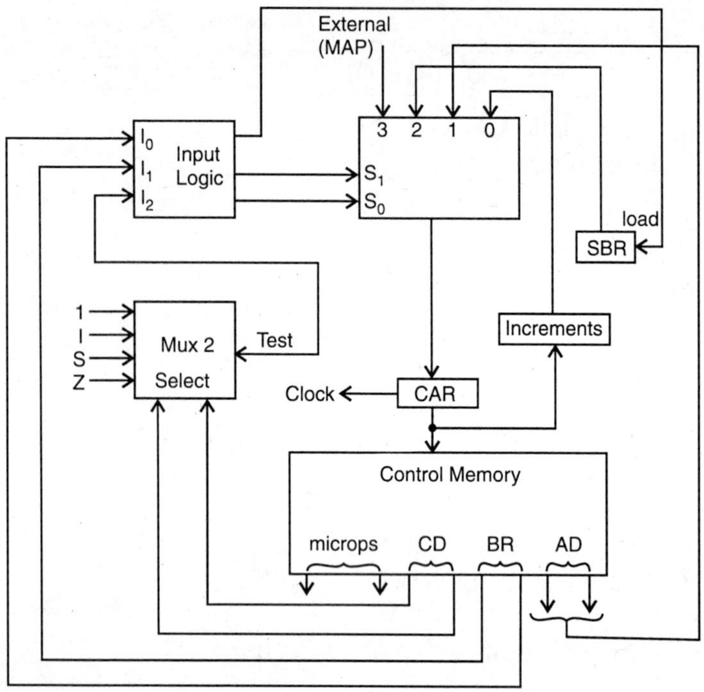

Fig. 6.12. Micropogam sequencer

6.4.3.6 Diffence between Hardwired and Microprogrammed

Characteristics	*Hardwired Control*	*Microprogrammed Control*
1. Speed	FAST	SLOW
2. Implementation	In H/W	In S/W
3. Flexibility	No	More
4. Ability to handle complex instruction sets	At some level is difficult	Easier
5. Design process	Difficult for more operation	Easy
6. Memory	No	Control memory (RAM or ROM)
7. Chip area efficiency	Uses least area	Uses more area

6.4.3.7 Multiple Bus Organization

Note: General register organization of multiple bus organization have same figure.

Here multiple bus organization means that there are more than one buses, which can work in a single circuit. In the Fig. 6.13, there are two buses are shown as bus A and Bus B and, there are more buses in the figure, we can say that the data transfer is not possible without the use of buses.

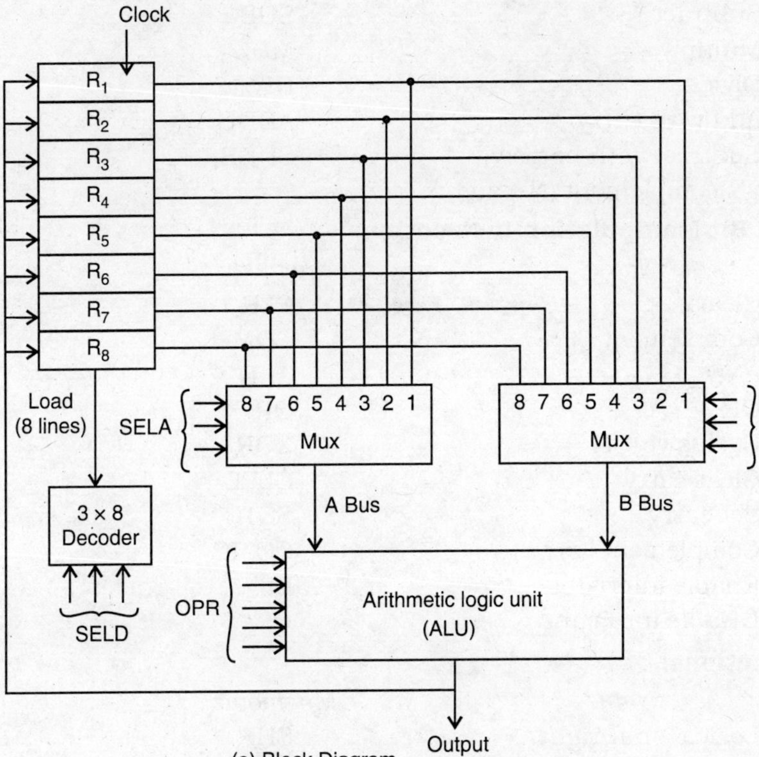

(a) Block Diagram

3	3	3	5
SELA	SELB	SELD	OPR

(b) Control word

Fig. 6.13. Multiple bus organization

6.4.3.8 Register Transfer Operations

1. Data Transfer: These are data transfer instructions

Name	*Mnemonic*
Load	LD
Stone	ST
Move	MOV
Exchange	XCH
Input	IN
Output	OUT
Push	Push
Pop	Pop

2. Data Manipulation: These are data manipulations instructions

(a) Arithmetic Instructions

Name	*Mnemonic*
Increment	INC
Decrement	DEC
Add	ADD

Subtract	SUB
Multiply	MUL
Divide	DIV
Add with carry	ADDC
Subtract with borrow	SUBB
2's complement (Negate)	

(b) Logical and Bit Manipulation Instructions:

Name	*Mnemonic*
Clear	CLR
Complement	COM
AND	AND
OR	OR
Exclusive-OR	XOR
Clear carry	CLRC
Set carry	SETC
Complement carry	COMC
Enable interrupt	EI
Disable interrupt	DI

(c) Shift Instructions:

Name	*Mnemonic*
Logical shift right	SHR
Logical shift left	SHL
Arithmetic shift right	SHRA
Arithmetic shift left	SHLA
Rotate right	ROR
Rotate left	ROL
Rotate right through carry	RORC
Rotate left through carry	ROLC

(3) Program Control Instructions:

Name	*Mnemonic*
Branch	BR
Jump	JMP
Skip	SKP
Call	CALL
Return	RET
Compare (by subtraction)	CMP
Test (by ANDING)	TST

6.4.3.9. Prefetching Microinstructions (Improving the Performance of CPU)

Performance improved greatly if the CPU can overlap the fetch and execute phases of instruction while one instruction is being executed, the next instruction can be prefetched from the memory. Modern processors include special unit often called the instruction unit whose main function is to fetch instruction and placed them into a queue ready for execution. In doing so if attempts to ensure that correct instructions are prefetched when a branch instruction is encountered.

Three fetching microinstructions presents some organizational difficults. Sometimes the status fetch and the results of the currently executed microinstructions are needed to determine the address of next microinstruction.

Thus, straight forward prefetching occassionally prefetches a wrong microinstruction. In these cases the disadvantages are minor and the prefetching technique are often used.

6.5 IMPLEMENTATION OF MICROPROGRAMMING CONTROL UNIT

6.5.1. Micro-programmed Control

The introduction and implementation of control part of a computer is explained in this chapter. Using two design approaches–award wired and micro-programmed. Hardwares control is discussed in brief because microprogramming is examined in detail. It is generally useful to separate a digital system into two parts: a data processing unit and control unit. The data processing unit is a network of functional units capable of performing certain operations on data. The purpose of control unit is to issue control signals or instructions to the data processing part. These control signals select the functions to be performed at specific times and route the data through the appropriate functional units. In other words, the data processing unit is logically recon figured by the control unit to perform certain set of micro operations. The sequence in which these micro-operations are performed is very in portent., so the control unit is intimately involved in the sequencing and timing of the data processing unit. Here we are concerned with the design of control units for digital systems The function of control unit is to fetch instructions from memory and interpret them to determine the control signals to be sent to the data processing units. Two central aspects of this process can be identified.

1. Instruction sequencing: That is, the methods by which instructions are selected for execution or the manner in which control of the processor is transferred one instruction to another.
2. Instruction Interpretation: The methods used for activating the control signals that cause the data processing unit to execute the instruction.

6.5.2 Microprogrammed Control Organization

Control units that use dynamic microprogramming employ a writable control memory. This type of memory can be used for writing but is used mostly for reading. A memory that is part of control unit is received to as control memory.

An advance development known as dynamic micro-programming permits a microprogram to be loaded initially from an auxiliary memory such as magnetic disk. Control units that use dynamic micro-programming use a writable control memory. This type of memory can be used for changing the micro-program.

A computer that uses a micro-programmed control unit usually has two memories–a main memory and control memory. The main memory is available to the user for storing their programs. The contents of main memory may change when the data are manipulated and every time the program is changed. The user's program in main memory consists of machine instructions and data. Where as, the control memory holds a fixed microprogram that cannot be altered by the user. The microprogram consists of micro instructions that specify various internal control signals for execution of register micro-operations. Each machine instruction initiates a series of micro instructions in control memory. These microinstructions generate the micro-operations to fetch the instruction from main memory, to evaluate the effective address, the execute the operation specified by the instruction, and to return control to the fetch phase in order to repeat the cycle for the next instruction.

A micro programmed control unit is shown in block diagram of Fig. 6.14. The control memory is usually a ROM, which stores all control information permanently. The control MAR specified the address of the microinstruction and the control data register holds the microinstruction read from memory. The microinstruction contains a control word that specified one or more micro-operations for the data processor. Once thee operations are executed the control must determine the next address. The location of the next microinstruction is generally the one next in sequence, otherwise, it may be located of the next microinstruction is generally the one next in sequence, otherwise, it may be located somewhere else in the control memory. For this reason it is necessary to use some bits of present address of the next micro instruction. The next address may also be a function of external in put conditions.

While the micro-operations are being executed the next address is computed in the next address generator circuit and then transferred into the control address register to read the next micro instruction. Hence a micro instruction contains bits for initiating micro-operations in the data processor part and bits that determine the address sequence of the control memory.

A micro program sequencer is the next address generator, as it determines the address sequence that is read from control memory. The address of the next microinstruction can be specified in several always depending on the sequencer inputs. The functions of a micro program sequencer are incriminating the control address register by one, loading into the control address register an address from control memory, transferring an external address, or loading an initial address to start the operations.

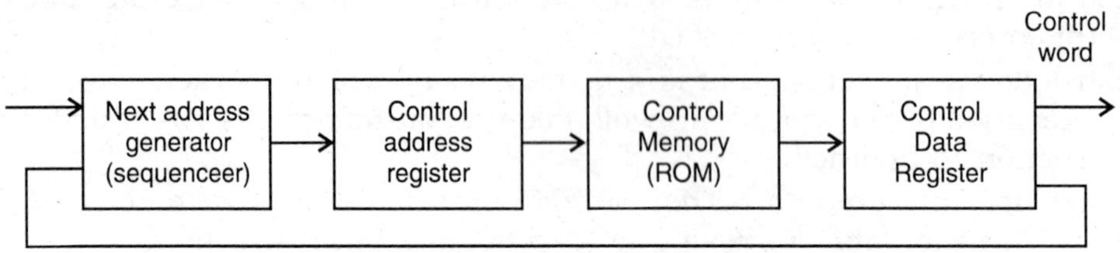

Fig. 6.14. Microprogramed control organization

The control DR (Data Register) stores the present microinstruction while the next address is computed and read from memory. The data Register is also called a pipeline register. It allows the execution of the micro-operations specified by the control word simultaneously, with the generation of the next micro instruction. This configuration requires a two phase clock, with one clock applied to the address register and the other to the data registers.

The main advantage of the micro-programmed control is that once the hardware configuration is built, there should be no need for further hardware or writing changes. If we want to make a different control sequence for the system, all we need to do is to specify a different set of micro instructions for control memory. The hardware configuration should not be changed for different operations. We have to change only the microprogram residing in control memory.

6.5.3. Microprogram Address Sequencing

Control memory stores the micro instructions in groups specifying a routine. Different computer instructions have different microprogram routine in control in control memory to generate the micro-operations that execute the instruction. The hardware must be capable of sequencing the microinstructions within a routine and be able to branch from one routine to another. To appreciate the address sequencing in microprogram control unit, let us enumerate the steps that the control must undergo during the execution of a single computer instruction.

When power is turned on in the computer, an initial address is loaded into the address register. This address is the address of the first micro instruct that activates the instruction fetch routine. The fetch routine may be sequenced by incriminating the control address register through the rest of its microinstruction. At the end of the fetch routine, the instruction is in the instruction register of the computer.

The control memory next mist go through the subroutine that decided the effective address of the operand. A machine instruction may have bits that specify various addressing modes, for example, indirect address and index registers. The effective address calculation routine in control memory can be reached through a branch microinstruction, which is conditioned on the status of the mode bits of the instruction. When the effective address computation routine is completed the address of the operand is available in the memory address register.

Now micro-operations are generated that execute the instruction fetched from memory. The operation code part of the instruction determined the micro-operations steps to be generated in processor registers. Each instruction has its own micro-program routine stored in a given location of control memory. The transformation from the instruction code bits to an address in control memory where the routine is located is referred to as a mapping process. A mapping procedure is a rule that transforms the instruction code into a control memory address. Once we reach at the required routine, the micro instructions that execute the instruction may be sequenced by incriminating the control address register, but sometimes the sequence of micro-operations will depend on values of certain status bits in processor register. Microprograms that employ will require an external register for storing the return address. Return addresses cannot be stored in ROM because the unit no writing capability.

After completion of instruction execution, control must return to the fetch routine. This is done by executing an unconditional branch microinstruction to the first address of the fetch routine. In the address sequencing capabilities required in a control memory are:

1. Incriminating of the control address register.
2. Unconditional branch or conditional branch, depending on status bit conditions.
3. A mapping process from the bits of the instruction to an address for the control memory.
4. A facility for subroutine call and return

Fig. 6.12 shows a block diagram of control memory and the associated hardware needed for selecting the next microinstruction address.

6.5.4 Microprogrammed Conditional Branching

The branch logic of Fig. 6.15 gives decision–making qualities in the control unit. The status conditions are special bits that give parameter information, for example the carry out of an adder, the sign bit of a number, the mode bits of an instruction and input or output status conditions. Information in these bits can be tested and actions initiated based on their condition whether their values is 1 or 0. The status bits together with the field in the microinstruction that specified a branch address, control the conditional branch decisions generated in the branch logic.

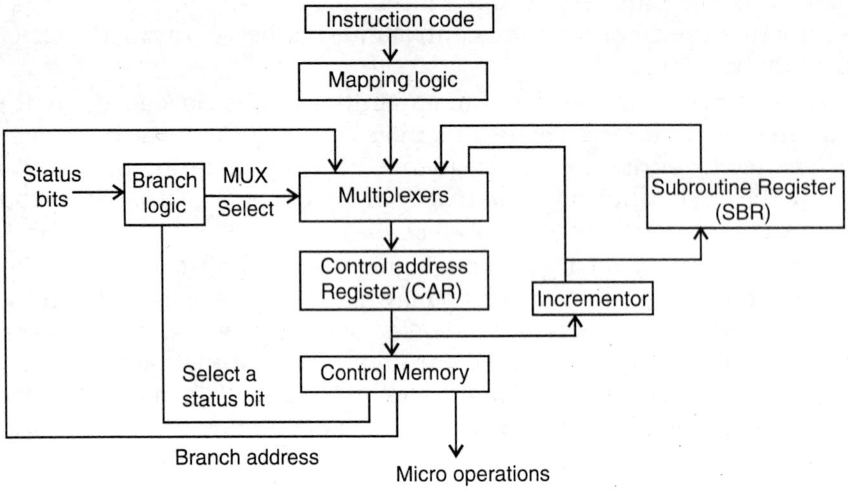

Fig. 6.15. Selection of address for control memory

We can implement the branch logic hardware in a variety of ways. The easiest way is to test the specified condition and branch to the indicated address if the conditions is met otherwise the address register is incremented.

A multiplexer can be used to implement it. Assume that there are eight status bit conditions in the system. Three bits in the microinstruction are used to specify any one of eight status bit conditions. These three bits give the selection variables for the multiplexer. If the selected status bit is in the 1 state, the output of the multiplexer is 1, otherwise it is o. A '1' output in the multiplexer generates a control signal to transfer the branch address from the micro instruction into the control address register. A 'O' output of the multiplexer causes the address register to be incremented.

We confinement an unconditional branch microinstruction by loading the branch address from control memory into the control address register. This can be done by fixing the value of one status bit at the input of the multiplexer, so that it is always bit select lines from control memory causes the branch address to be loaded into the control address register unconditionally.

6.5.5. Microprogrammed Mapping of Instruction

A special type of branch exists when a micro instruction specifies a branch to the first word in control memory where a microprogram routine for an instruction is located. The status bits for this type of branch are the bits in the operation code part of the instruction. For example, a computer with the simple instruction format as shown in Fig. 6.15.

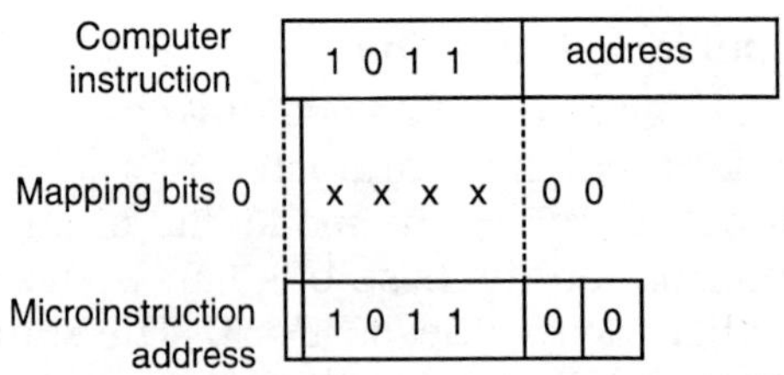

Fig. 6.16. Mapping from instruction code to the microinstruction address

An operation code of four bits which can specify upto 16 distinct instructions. Assume once again that the control memory has 128 words requiring the nine bits of the micro-operation fields will then be 000 100 101. It is important to realize that two or more conflicting micro-operations cannot be specified simultaneously. For example, a micro-operation field 010 001 000 has no meaning because it specified the operations to clear AC to 0 and subtract DR from AC at the same time.

Each micro-operation in Table 6.8 is defined with a register transfer statement and is assigned a symbol to use in a symbolic microprogram. All transfer–type micro-operation symbols use five letters. The first two letters designate the source register, the third letter is always a T, and the last two letters designate the destination register. For example, the micro-operation that specified the transfer AC ← DR (R_1 = 100) has the symbol DRTAC which stands for a transfer from DR to AC.

The CD (condition) field consists of two bits which are encoded to specify four status bit conditions assisted in Table 6.8. The first condition is always a 1 so that a reference to CD = 00 (or the symbol U) will always find the condition to the true. When this condition is used in conjunction with the BR (branch) field, it provides an unconditional branch operation. The indirect bit I is available from bit 15 of DR after an instruction is read from memory. The sign bit of AC provides the next status bit. The zero value, symbolized by z, is a binary variable whose value is equal to if all the bits in AC are equal to zero. We will use the symbols U, I, S and Z for the four status bits when we write microprograms in symbolic forms.

The BR (branch) field consists of two bits. It is sued, in conjunction with the address field AD, to chooses the address of the next microinstruction. As shown in Table 6.1, when Br = 00, the control from a jump (JMP) operation (which is similar to the branch) and when BR = 01, it performs a call to subroutine (CALL) operation. The two operations are identical except that a call microinstruction stores the return address, in the subroutine register SBR. The jump and call operations depend on the value of CD field.

If the status bit condition specified in the CD field is equal to 1, the next address in the AD field is transferred to the control address register CAR. Otherwise, CAR is incremented by 1. Address of seven bits for each operation code there exists a microprogram routine in control memory the executes the instruction. One simple mapping process the 4-bit operation code to 7-bit is shown in Fig. 6.8. The mapping consists of placing a 0 in the most significant bit of the address, transferring the four operation code bits and clearing the two least significant bits of the control address register. This provided for each computer instruction a microprogram routine with a capacity of four microinstructions. If the routine needs more than four microinstruction, it uses fewer than four microinstructions, the unused memory locations would be available for other routines.

6.5.6. Microinstruction Format

The microinstruction format for the control memory is shown in Fig. 6.17. The 20 bits of the microinstruction are divided into four functional parts. The three fields F_1 F_1 and F_3 specify micro-operation for the computer. The CD field selects status bit conditions. The BR field specified the type of branch to be used the AD field contains a branch address. The address field is seven bits wide, since the control memory has 128 = 27 words.

3	3	3	3	3	3
F_1	F_2	F_3	CD	BR	AD

F_1, F_2, F_3 — Micro-operation fields
C — Condition for branching

BR — Branch field
AD — Address field

Fig. 6.17. Microinstruction code Format (20 bits)

The micro-operations are subdivided into three fields of three bits each. The three bits in each field are encoded to specify seven distinct micro-operations, can be chosen for microinstruction one from. Each field. If fewer than three micro-operations can be chosen for a micro instruction one or more of the fields will use the binary code 000 for no operation. As shown in Table 6.1, a microinstruction can specify two simultaneous micro-operations from F_2 and F_3 and none

$$DR \leftarrow M[AR] \quad \text{with } F_2 = 100$$

and

$$PC \leftarrow PC + 1 \quad \text{with } F_3 = 101$$

6.5.7. Symbolic Microinstructions

The symbols defined in Table 6.8, can be used to specify microinstructions in symbolic form. A symbolic program can be translated into its binary equivalent by means of an assembler. A microprogram assembler is similar in concept to a conventional computer assembler as defined earlier. The simplest and most straight for ward way to formulate an assembly language for a microprogram is to define symbols for each field of the microinstruction and to give users the capability for defining their own symbolic addresses.

Each line of the assembly language microprogram defines a symbolic microinstruction. Each symbolic microinstruction is divided into five fields–label, micro-operation CD, BR and AD. The fields specify the following information

1. The label may be empty or it may specify a symbolic address. A label is terminated with colon (:)
2. The incorporations field consists of one, two or three symbols separated by commas, from those defined in Table 6.8. There may be no more than one symbol from each F field. The NOP symbol is used when the microinstruction has no micro operations. This will be translated by the assembler to nine zeros.
3. The CD field has one of the letters U, I,S or Z.
4. The BR field contains one of the four symbols defined in Table 6.8.
5. The AD field specified a value for the address field of the micro instruction in one of three possible ways.
   (*a*) With a symbolic address, which must also appear as a label.
   (*b*) With the symbol NEXT to designate the next address in sequence.
   (*c*) when the BR field contains a RET or MAP symbol, the AD field is left empty and is converted to seven zeros by the assembler.

Table 6.8. Symbols and Binary code for microinstruction

F_1	*Micro-operation*	*Symbol*
000	None	NDP
001	AC ← AC + DR	ADD
010	AC ← O	CLRAC
011	AC ← AC + 1	INCAC
100	AC ← DR	DRTAC
101	AR ← DR (0 – 10)	DRTAR
110	AR ← PC	PCTAR
111	M[AR] ← DR	WRITE

F_2	Micro-operation	Symbol
000	None	NDP
001	AC ← AC – DR	SUB
010	AC ← AC ∨ DR	OR
011	AC ← AC ∧ DR	AND
100	DR ← M[AR]	READ
101	DR ← AC	ACTDR
110	DR ← DR + 1	INCDR
111	DR(0 – 10) ← PC	PCTDR

F_3	Micro-operation	Symbol
000	None	NDP
001	AC ← AC ⊕ DR	XOR
010	AC ← AC	COM
011	AC ← Shl AC	SHL
100	AC ← Shr AC	SHR
101	PC ← PC + 1	INCPC
110	PC ← AR	ARTPC
111	Reserved	

CD	Conditions	Symbol	Comments
00	Always = 1	U	Unconditional branch
01	DR (15)	I	Indirect Address bit
10	AC (15)	S	Sign bit of AC
11	AC = 0	Z	Zero value in AC

CD	Conditions	Symbol Comments
00	JMP	CAR ← AD if condition = 1 CAR ← CAR + 1 if condition = 0
01	CALL	CAR ← AD SBR ← CAR + 1 if condition = 1 CAR ← CAR + 1 if condition = 0
10	RET	CAR ← SBR (Return from sub routine)
11	MAP	CAR (2–5) ← DR (11 – 14), CAR (0, 1, 6) ← 0

The Fetch Routine: The control memory has 28 words and each word contains 20 bits. To microprogram the control memory, if is necessary to determine the bit values of each of the 128 words. The first 64 words (addresses 0 to 63) are to be occupied by the routines for the 16 instructions. The last 64 words may be used for any other purpose. A convenient starting location for the Fetch routine is address 64. The micro instructions needed for the fetch routine are.

AR ← PC

DR ← M[AR], PC ← PC + 1

AR ← Dr (0 – 10), CAR (2 – 5) ← DR (11 – 14), CAR (0, 1, 6) ← 0

The fetch routine needs three microinstruction which are placed in control memory at addresses 64, 65 and 66. Using the assembly language conventions defined previously, we can write the symbolic microprogram for the fetch routine as follows:

```
            ORG 64
FETCH PCTAR U   JMP NEXT
READ, INCPC U   JMP NEXT
      DRTAR U   MAP
```

The translation of the symbolic microprogram to binary produces the following binary micro program. The bit values are obtained from Table 6.8.

Binary Address	F_1	F_2	F_3	*CD*	*BR*	*AD*
1000000	110	000	0000	00	00	1000001
1000001	000	100	101	00	00	1000010
1000010	101	000	000	00	11	0000000

The three microinstructions that constitute the fetch routine have been listed in three different representations the register transfer representation shows the internal register transfer operations that each microinstruction implements. The symbolic representation is useful for writing micro programs in an assembly language format the binary representation is the actual internal content that must be stored in control memory.

6.5.8. Microprogram Sequencer

The basic components of micro programmed control unit are the control memory and the circuit that select the next address. The address selection part is called a microprogram sequencer. A microprogram sequencer can be constructed with digital functions to suit a particular application.

The purpose of a microprogram sequencer is to present an address to the control memory so that a micro instruction may be read and executed. The next address logic of the sequencer determines the specific address source to be loaded into the control address register. The choice of the address source is guided by the next address information bits that the sequencer receives from the present Micro-operation.

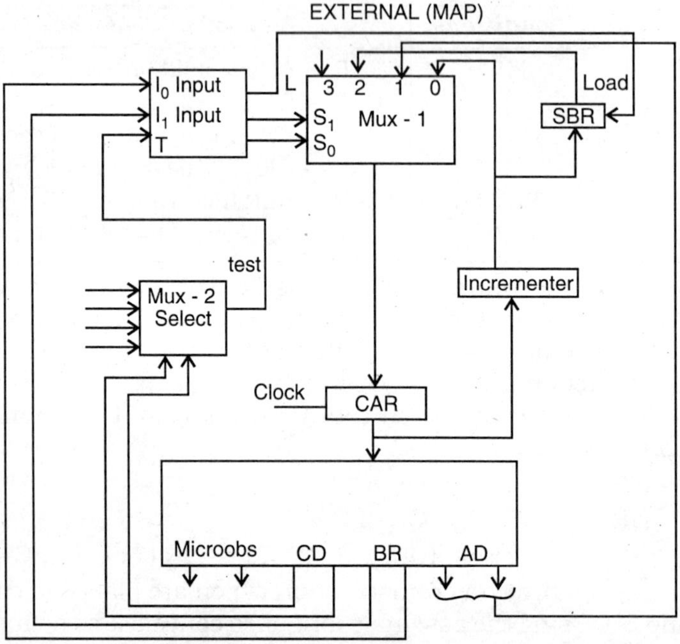

Fig. 6.18. Microprogram sequencer

To illustrate the internal structure of a typical microprogram sequencer we will show a particular unit that is suitable for use in the micro program computer example developed in the preceding section. Block diagram of the microprogram sequencer is shown in Fig. 6.10.

The control memory is included in the diagram to show the interaction between the sequencer and the memory attached to it. There are two multiplexers in the circuit. The first multiplexer selects an address from one of four sources and routes it into a control address register CAR. The second multiplexer tests the value of a selected status bit and the result of the test is applied to an input logic unit. The output from CAR provides the address for the control memory. The content of CAR is incremented and applied to one of the multiplexer inputs and to the subroutine register SBR. The other three inputs to multiplexer number 1 come from the address field of the present microinstruction, from that maps the instruction. Although the diagram shows a single subroutine register, a typical sequencer will have a register stack about four to eight levels deep In this way, a number of subroutines can be active at the same time. A push and pop operation in conjunction with a stack pointer. Stores and the return address during he call and return macroinstructions.

The CD (condition) field of the microinstruction selects one of the status bits in the second multiplexer if the bit selected is equal to 1, the T (test) variable is equal to 0. The T value together with the two bits from the BR (branch) field, go to an input logic circuit. The input logic in a particular sequencer will determine the type of operations that are available in the unit. Typical sequencer operations are -increment branch, or jump, call and return from subroutine load and external address. Push or pop the stack and other address sequencing operations. Some commercial sequencer have three or four inputs in addition to the T inputs and thus provide a wider range of operations.

6.6 CONCEPT OF HORIZONTAL AND VERTICAL MICROPROGRAMMING

Writing microprograms is quite similar to writing programs in machine language for a very primitive computer. It is as time consuming and error prone and requires in depth knowledge of the hardware. Just as for machine language programming, symbolic assembler languages have been developed to facilitate the writing of microprograms. In the early seventies some attempts to develop high level languages and compilers for microcode have even been made, but the evolution towards fully integrated processors with all their microcode in ROM has considerably reduced the interest in such high level languages, as the cost of highly optimized microcode, written by outstanding experts, can be shared by the millions of users of the processor some typical examples of microcode.

Table 6.9. A Example of part of a microprogram

label	next minstr.	destination	bits controlling all parts of the computer system
fetch	IF inter.		P > mem.add.; P + 1 > P
	THEN jump	intseq	
	ELSE next		
	inst.map ...		mem.dat.> IR
inst.n	next ...		*whatever is needed to execute instruction n.*
	jump ...	fetch	
intseq	next ...		*whatever is needed to identify the interrupt, to save the processor state and to fetch the first instruction of the interrupt handler*
	inst.map		

Horizontal and Vertical Microprogramming

This is often called "horizontal microprogramming". It allows to control separately every function in the computer and to perform as many things as possible simultaneously. It also results in extremely wide microinstructions (up to several hundreds of bits) which are difficult to master, even for very good computer designers. Horizontal microprogramming is mainly used in very high performance computers, where simultaneity is the key towards performance.

If one observes table A, it is obvious that the "Destination Address" field is used in very few microinstructions. As this field is very wide as it contains an address in the microstore, having it in every microinstruction is quite a waste of very expensive microstore space. In addition, one can imagine that many of the bits controlling parts of the computer are not really independent: for instance, it would make no sense to put. the contents of two different registers simultaneously on the same bus. Based on these observations, it would make sense to define several different formats of microinstructions in which each bit can have several different functions, depending on the format. Fig. 6.10(A) gives an example of such encoded microinstructions and Fig. 6.10(A) shows the block diagram of a control unit adapted to such encoded microinstructions. Encoding the microinstructions reduces, of course, the freedom of the instruction set designer, so that fewer operations can be performed simultaneously, but it also reduces considerably the size of the microstore and simplifies somewhat the task of writing microprograms. When the microinstructions are encoded, one uses the expression "vertical microprogramming". Vertical microprogramming is mostly used when cost is more important than performance.

Next µinstr. control	***µinstruction format***	***Bits with functions depending on the microinstruction format***
Next µinstr. control	***µinstruction format 1***	***Bits controlling all parts of computer system***
Next µinstr. control	***µinstruction format 2***	***Address in microstore of the next microinstruction to be fetched***

Fig. 6.19. (a) Vertical microinstructions

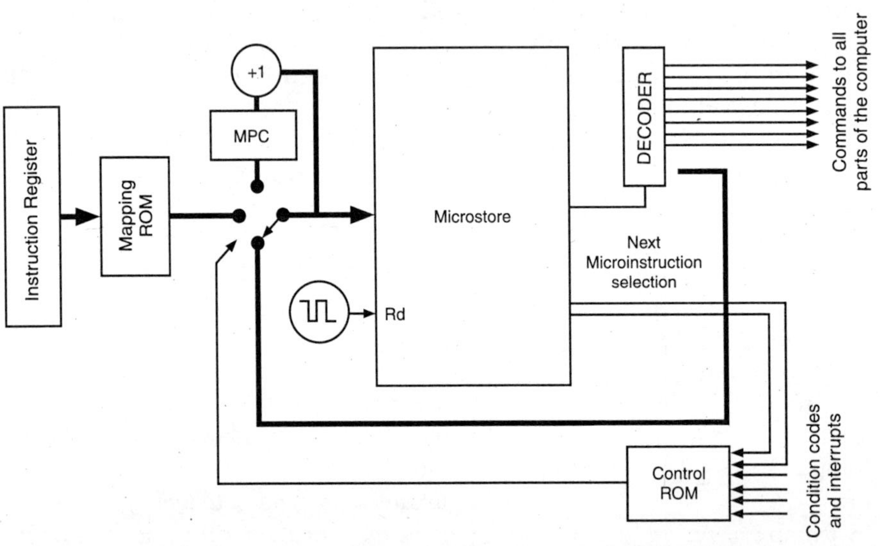

Fig. 6.19. (b) Control unit with vertical microprogramming

SOLVED PROBLEMS

Q.1. Define the following:

(a) Control memory

(b) Control word

(c) Sequencer

(d) Pipe line register.

Solution:

(a) Control Memory: Control units that use dynamic micro programming employ a writable control memory. This type of memory can be used for writting, but is used mostly for reading. A memory that is part of a control unit is received to as control memory.

(b) Control Word: The control variables at any given time can be represented by a string of 1's and 0's called *control word*.

(c) Sequencer: The next address generator is sometimes called a *microprogram sequencer*, as it determines the address sequence that is read from control memory. The address of the next micro instruction can be specified is several ways, depending on the sequencer inputs.

(d) Pipe line Register: The control data register holds the present micro instruction while the next address is computed and read from memory. The data register is some times called *pipe line register*.

Q.2. How many types of control organizations are there? Which one is most beneficial for computer system and why?

Solution: There are two major types of control organization:

(a) Hardware control

(b) Microprogram control.

Hardware control is more advantage as compared to microprogram control.

In the hardware organization, the control logic is implemented with gates, Flip-flops, decoders and other digital circuits. It has the advantage that it can be optimized to produce a fast mode of operation.

Mostly computer based on the reduced instruction set computer (RISC) architecture concept use hardwired control rather than a control memory with a microprogram. A hardwired control, as the name implies, requires changes in the wiring among the various components, if the design has to modified or changed.

In the microprogrammed organization, the control information is stored in a control memory. The control memory is programmed to initiate the required sequence of micro-operation. The main advantage of microprogrammed control is the fact that once the hardwired configuration is established, there should be no need for further hardware or wiring changes.

Q.3. Define the computer instruction format. What is possible/reference instructions, describe it?

Solution: In computer instruction format, there are three fields.

15	14	11	0
I	Opcode	Address	

It consists of 2 byte length [(0 – 15) bit].

(i) **Indirect addressing (I):** Its length is 1 bit field,

(*ii*) **Opcode:** It is 4 bit operation code.

(*iii*) **Address:** It is 11 bit address field.

There are 16 possible memory-reference instructions we have describe four of them.

Symbol	*Opcode*	*Description*
ADD	0000	AC ← AC + M[EA]
BRANCH	0001	If (AC < 0) then (PC ← EA)
STORE	0010	M[EA] ← AC
EXCHANGE	0011	AC ← M[EA], M[EA] ← AC

- The ADD instruction adds the content of the operand found in the effective address to the content OF AC.
- The **BRANCH** instruction causes a branch to the effective address if the operand in AC is negative. The program proceeds with the next consecutive instruction if AC is not negative. The AC is negative if its sign bit is a 1.
- **STORE** instruction transfers the content of AC into memory word specified by the effective address.
- **EXCHANGE.** The exchange instruction swaps the data between AC and the memory word specified by the effective address.

Each computer instruction must be microprogrammed.

Q.4. Describe the micro instruction format for the control memory.

Solution: Micro instruction format is 20 bits in length. It divided into four functional parts. The three fields F_1, F_2 and F_3 specify micro-operations for the computer. The CD field selects status bits condition:

- the BR field specifies the type of Branch to be used.
- the AD field contains a branch address. The address field is seven bits wide, since the control memory has $128 = 2^7$ words.

3	3	3	2	2	7
F_1	F_2	F_3			AD

F_1, F_2, F_3 : Micro-operation fields

CD : Condition for branching

BR : Franchfield

AD : Address field

- Micro-operations are sub-divided into three fields of three bits each. These bits in each field are encoded to specify seven distinct micro operations. This gives 21 micro-operations. If fewer than three micro operation are used, one or more of the helds will use the binary code 000 for no operation.

Symbols and Binary Code for Micro Instruction Fields

F_1	*Micro operation*	*Symbol*
000	None	NOP
001	AC ← AC + DR	ADD
010	AC ← 0	CLRAC
011	AC ← AC + 1	INCAC
100	AC ← DR	DRTAC
101	AR ← DR(0 – 10)	DRTAR

F_2	Micro operation	Symbol
110	AR ← PC	DCTAR
111	M [AR] ← DR	WRITE
000	None	NOP
001	AC ← AC – DR	SUB
010	AC ← AC ∪ DR	OR
011	AC ← AC ∩ DR	AND
100	DR ← M[AR]	READ
101	DR ← AC	ACTDR
110	DR ← DR + 1	INCDR
111	DR (0 – 10) ← PC	PCTDR

F_3	Micro operation	Symbol
000	None	NOP
001	AC ← AC ⊕ DR	XOR
010	AC ← $\overline{AC}$	COM
011	AC ← SNAC	SHL
100	AC ← SHRAC	SHR
101	PC ← PC + 1	INCPC
110	PC ← AR	ARTPC
111	Reserved	

CD	*Conditions*	*Symbol*	*Comments*
00	Always = 1	U	Unconditional branch
01	DR (15)	I	Indirect Address bit
10	AC (15)	S	Sign bit of AC
11	AC = 0	Z	Zero value in AC

CD	*Conditions*	*Symbol Comments*
00	JMP	CAR ← AD if condition = 1 CAR ← CAR + 1 if condition = 0
01	CALL	CAR ← AD SBR ← CAR + 1 if condition = 1 CAR ← CAR + 1 if condition = 0
10	RET	CAR ← SBR (Return from sub routine)
11	MAP	CAR (2–5) ← DR (11 – 14), CAR (0, 1, 6) ← 0

Q.5. A logic network has two data inputs *A* and *B*, and two control inputs *C0* and C_1. It implements the function *F* according to the following table:

C_1	C_2	*F*
0	**0**	$\overline{A+B}$
0	**1**	$A+B$
1	**0**	$A \oplus B$

Implement the circuit using one 4 to AB multiplexer, one 2 input exclusive OR gate, one 2 input AND gate, and 2 input OR gate and one inverter.

Solution:

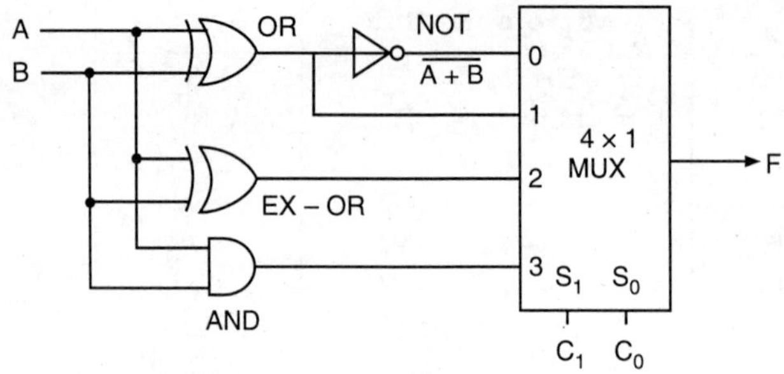

Fig. 6.20.

Q.6. Write the difference between a microprocessor and a microprogram? Is it possible to design a microprocessor without a microprogram? Are all microprogrammed computers also microprocessors?

Solution: A microprocessor is a small size CPU (computer on a chip). Microprogram is a program for a sequence of microoperations. The control unit of a microprocessor can be hardwired or microprogrammed, depending on the specific design. A microprogrammed computer does not have to be a microprocessor.

Q.7. Explain the difference between hardwired control and microprogrammed control. Is it possible to have a hardwired control associated with a control memory?

Solution: Hardwired control, by definition, does not contain a control memory.

Q.8. Define the following:

(*a*) Microoperation
(*b*) Microinstruction
(*c*) Microprogram
(*d*) Microcode

Solution: (*a*) Microoperation

It is an elementary digital computer operation.

(*b*) Microinstruction

It is an instruction stored in control memory.

(*c*) Microprogram

It is a sequence of microinstruction.

(*d*) Microcode

It is same as microprogram.

Q.9. The microprogrammed control organization shown in figure given below has the following propagation delay times. 40 ns to generate the next address, 10 ns to transfer the address into the control address register, 40 ns to across the control memory ROM, 10 ns to transfer the microinstruction into the control data register, and 40 ns to perform the required micro-operations specified by the control word. What is the maximum clock frequency that the control can use? What would the clock frequency be if the control data register is not used?

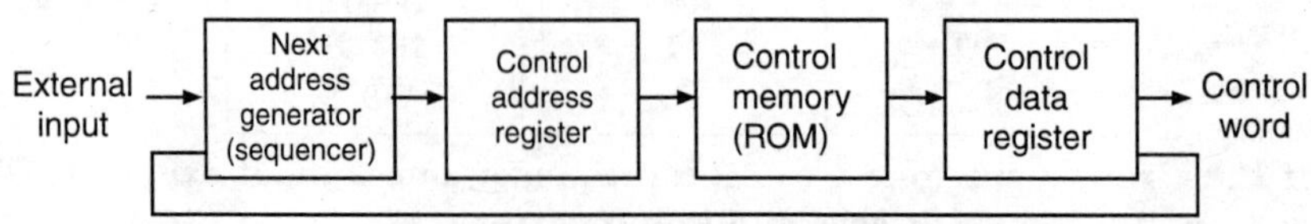

Fig. 6.21. Microprogrammed control organization

Solution:

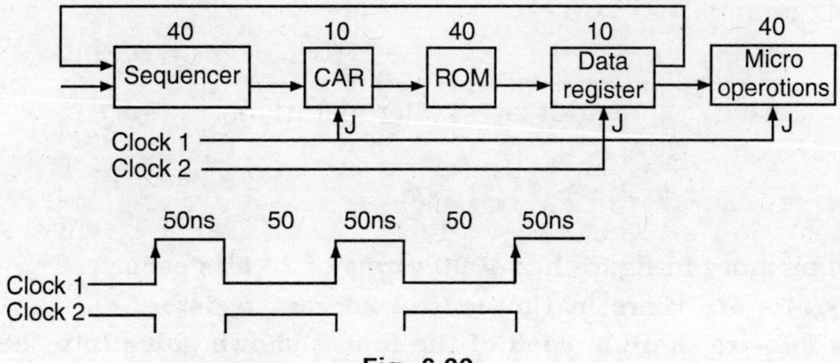

Fig. 6.22.

$$= \frac{1000}{100} \times 10^6 = 10 \text{ MHz}$$

If the data register is removed, we can use a single phase clock with a frequency

$$= \frac{1}{90 \times 10^{-9}} = 11.1 \text{ MHz}$$

Q.10. The system shown in figure given below uses a control memory of 1024 words of 32 bits each. The microinstruction has three fields as shown in the diagram. The microoperations field has 16 bits.

(*a*) How many bits are there in the branch address field and the select field?

(*b*) If there are 16 status bits in the system, how many bits of the branch logic are used to select a status bit?

(*c*) How may bits are left to select an input for the multiplexers?

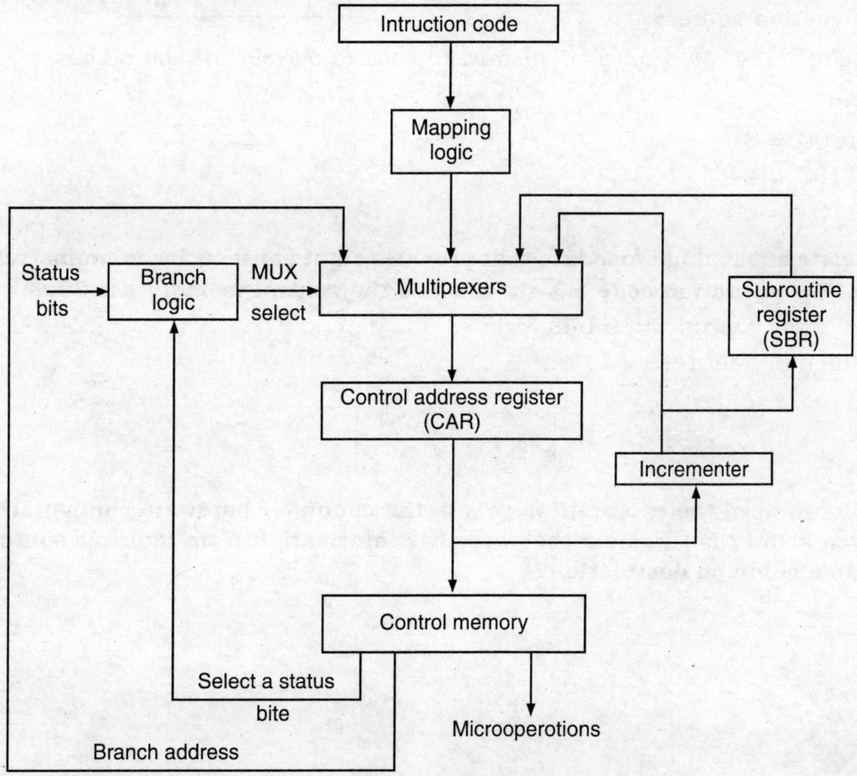

Fig. 6.23.

Solution: Control memory = $2^{10} \times 32$

(*a*)

6	10	16	= 32 bits
Select	Address	Microperations	

(*b*) 4 bite

(*c*) 2 bits

Q.11. The control memory in figure has 4096 words of 24 bits each.

(*a*) How many bits are there in the control address register?

(*b*) How many bits are there in each of the inputs shown going into the multiplexers?

(*c*) What are the number of inputs in each multiplexer and how many multiplexesrs are needed?

Solution: Control memory = $2^{12} \times 24$

(*a*) 12 bits

(*b*) 12 bits

(*c*) 12 multiplexers, each of size 4-to-1 line.

Q.12. Using the mapping procedure described in figure given below give the first micro instruction address for the following operation code:

(*a*) 0010

(*b*) 1011

(*c*) 1111

Computer Instruction:

Mapping bits:

Microinstruction address:

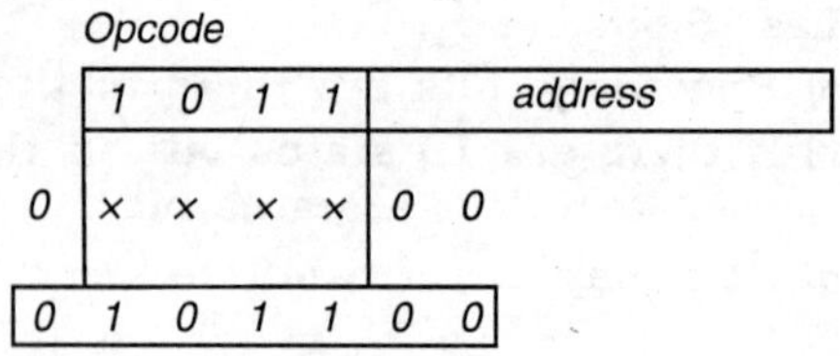

Fig. 6.24. *Mapping from instruction code to microinstruction address*

Solution:

(*a*) 0001000 = 8

(*b*) 0101100 = 44

(*c*) 0111100 = 00

Q.13. Formulate a mapping procedure that provides eight consecutive microinstructions for each routine. The operation code has six bits and the control memory has 2048 words.

Solution: Opcod = 6 bits

Control memory address = 11 bits

00 | × × × × × × | 000

| × × × × × × |

Q.14. Why do we need the two multiplexers in the computer hardware configuration shown in figure given below? Is there another way that information from multiple sources can be transferred to a common destination?

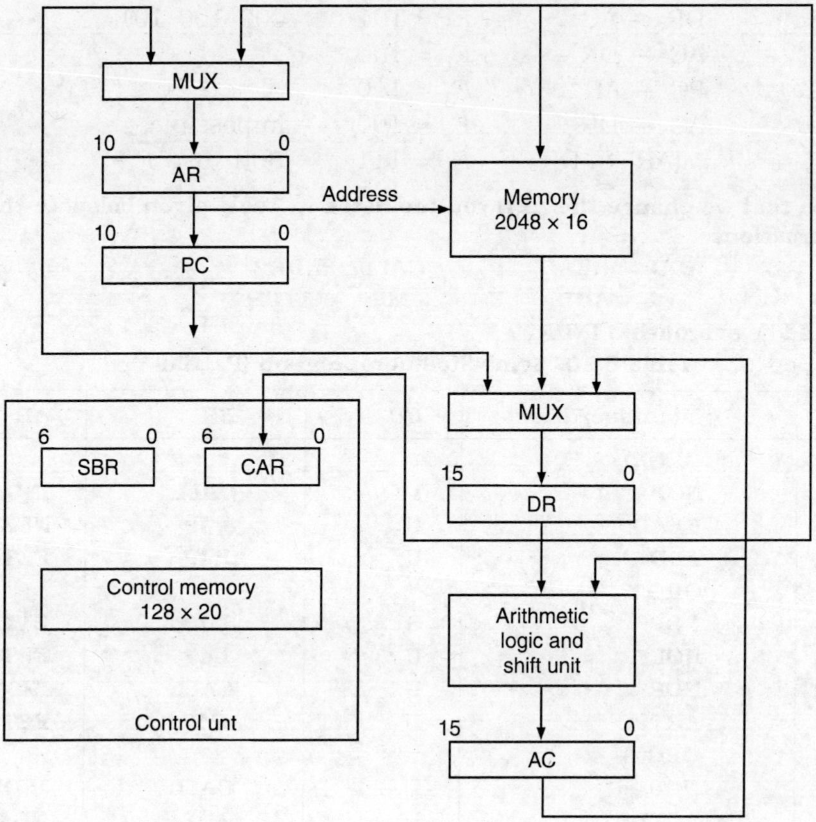

Fig. 6.25. Computer hardware configurtion

Solution: Either multiplexers, three state gates or gate logic (equivalent to a mux) are needed to transfer information from many sources to a common destination.

Q.15. Give the 9-bit Micro-operation field for the following microoperalions:

(a) AC ← AC + 1 DR ← DR + 1

(b) PC ← PC + 1 DR ← M[AR]

(c) DR ← AC AC ← DR

Solution:

	F_1	F_2	F_3			
(a)	011	110	000	INCAC	INCUR	NOP
(b)	000	100	101	NOP	READ	INCPC
(c)	100	101	000	DRTAC	ACTDR	NOP

Q.16. Convert the following symbolic micro-operations to register transfer statements and to binary.

(a) READ, INCPC

(b) ACTDR, DRTAC

(c) ARTPC, DRTAC, WRITE

Solution:

				Binary
(a)	READ	DR ← M[AR]	F_2 = 100	000 100 101
	DRTAC	AC ← DR	F_3 = 101	

(*b*) ACTDR	DR ← AC	F_2 = 101	000 100 101
DRTAC	AC ← DR	F_1 = 100	
(*c*) ARTPC	PC ← AR	F_3 = 110	
DRTAC	AC ← DR	F_1 = 100	Impossible,
WRITE	M[AR] ← DR	F_1 = 111	Both use F_1

Q.17. Suppose that we change the ADD routine listed in Table given below to the following two microinstructions.

```
ADD:  READ   I   CALL   INDR2
      ADD    U   JMP    FETCH
```

What should be subroutine INDR2?

Table 6.10. Symbolic Microprogram (Partial)

Label	*Microoperations*	*CD*	*BR*	*AD*
ADD:	ORG 0			
	NOP	I	CALL	INDIRECT
	READ	U	JMP	NEXT
	ADD	U	JMP	FETCH
	ORG 4			
BRANCH:	NOP	S	JMP	OVER
	NOP	U	JMP	FETCH
OVER:	NOP	I	CALL	INDIRECT
	ARTPC	U	JMP	FETCH
	ORG 8			
STORE:	NOP	I	CALL	INDIRECT
	ACTDR	U	JMP	NEXT
	WRITE	U	JMP	FETCH
	ORG 12			
EXHANGE:	NOP	I	CALL	INDIRECT
	READ	U	JMP	NEXT
	ACTDR, DRTAC	U	JMP	NEXT
	WRITE	U	JMP	FETCH
	ORG 64			
FETCH:	PCTAR	U	JMP	NEXT
	READ, INCPC	U	JMP	NEXT
	DRTAR	U	MAP	
INDRCT:	READ	U	JMP	NEXT
	DRTAR	U	RET	

Solution: If $I = 0$, the operand is read in the first microinstruction and added to *AC* in the second. If $I = 1$, the effective address is read into *DR* and control goes to INDR2. The subroutine must read the operand into *DR*.

```
INDR2:  DRTAR   U   JMP   NEXT
        READ    U   RET   —
```

Q.18. The following is a symbolic microprogram for an instruction in the computer

```
ORG40
NOP     S   JMP    FETCH
NOP     Z   JMP    FETCH
NOP     I   CALL   INDRCT
ARTPC   U   JMP    FETCH
```

(a) Specify the operation performed when the instruction is executed.

(b) Convert the four microinstructions into their equivalent binary form.

Solution:

(a) Branch if $S = 0$ and $Z = 0$ (Positive and non-zero AC) – See last instruction in problem 7 – 16.

(b)

40 :	000	000	000	10	00	1000000
41 :	000	000	000	11	00	1000000
42 :	000	000	000	01	01	1000011
43 :	000	000	110	00	00	1000000

Q.19. Add the following instructions to the computer (*EA* is the effective address). Write the symbolic microprogram for each routine as in Table given in Q. 17. (Note that *AC* must not change in value unless the instruction specifies a change in *AC*).

Symbol	***Opcode***	***Symbolic Function***	***Description***
AND	**0100**	**AC ← AC M[EA]**	**AND**
SUB	**0101**	**AC ← AC – M[EA]**	**Subtract**
ADM	**0110**	**M[EA] ← M[EA] + AC**	**Add to memory**
BTCL	**0111**	**AC ← AC M[EA]**	**Bit clear**
BZ	**1000**	**If (AC – O) then (PC ← EA)**	**Branch if AC zero**
SEQ	**1001**	**If (AC = M[EA]) then (PC ← PC + 1)**	**Skip if positive**
BPNZ	**1010**	**If (AC > 0) then (PC ← EA)**	**Branch if positive and non - zero**

Solution:

	0RG16			
AND:	NOP	1	CALL	INDRCT
READ	U	JMP	NEXT	
ANDOP:	AND	U	JMP	FETCH
	ORG 20			
SUB:	NOP	I	CALL	INDRCT
	READ	U	JMP	NEXT
	SUB	U	JMP	FETCH
	ORG 24			
ADM:	NOP	I	CALL	INDRCT
	READ	U	JMP	NEXT
	DRTACACTDR	U	JMP	NEXT
	ADD	U	JMP	EXCHANGE + 2
	ORG 28			
BTCL:	NOP	I	CALL	INDRCT
	READ	U	JMP	NEXT
	DRTACACTDR	U	JMP	NEXT
	COM	U	JMP	ANDOP
	ORG 32			
BZ:	NOP	Z	JMP	ZERO
	NOP	U	JMP	FETCH
ZERO:	NOP	I	CALL	INDRCT
	ARTPC	U	JMP	FETCH

	ORG 36			
SEQ:	NOP	I	CALL	INDRCT
	READ	U	JMP	NEXT
	DRTAC, ACTDR	U	JMP	NEXT
	XOR (OR SUB)	U	JMP	BEQ 1
	ORG 69			
BEQ1:	DRTAC, ACTDR	Z	JMP	EQUAL
	NOP	U	JMP	FETCH
EQUAL:	INCPC	I	JMP	FETCH
	ORG40			
BPNZ:	NOP	S	JMP	FETCH
	NOP	Z	JMP	FETCH
	NOP	I	CALL	INDRCT
	ARTPC	U	JMP	FETCH

Q.20. Write a symbolic microprogram routine for the ISZ (increment and skip if zero) instruction (Table given below). Note that DR = 0 status condition is not available in the CD field of the computer. However can exchange AC and DR and check if AC = 0 with the Z bit.

Table 6.11. Memory-Reference Instructions

Symbol	*Operation decoder*	*Symbolic description*
AND	D_0	**AC ← AC M[AR]**
ADD	D_1	**AC ← AC + M[AR], EC_{out}**
LDA	D_2	**AC ← M[AR]**
STA	D_3	**M[AR] ← AC**
BUN	D_4	**PC ← AR**
BSA	D_5	**M[AR] ← PC, PC ← AR + 1**
ISZ	D_6	**M[AR] ← M[AR] + 1**
		If M[AR] + 1 = 0 then PC ← PC + 1

Solution:

ISZ:	NOP	I	CALL	INDRECT
	READ	U	JMP	NEXT
	INCDR	U	JMP	NEXT
	DRTAC, ACTOR	U	JMP	NEXT (or past Indrct)
	DRTAC, ACTOR	Z	JMP	ZERO
	WRITE	U	JMP	FETCH
ZERO:	WRITE, INCPC	U	JMP	FETCH

Q.21. Write the symbolic microprogram routines for the BSA (branch and save address) instructions defined in (Table give in Q.20). Minimize the number of microinstructions.

Solution:

BSA:	NOP	I	Call	INDRCT
	PCTDR, ART PC	U	JMP	NEXT

Q.22. Show how outputs 5 and 6 of decoder F3 in figure given below are to be connected to the progran counter PC.

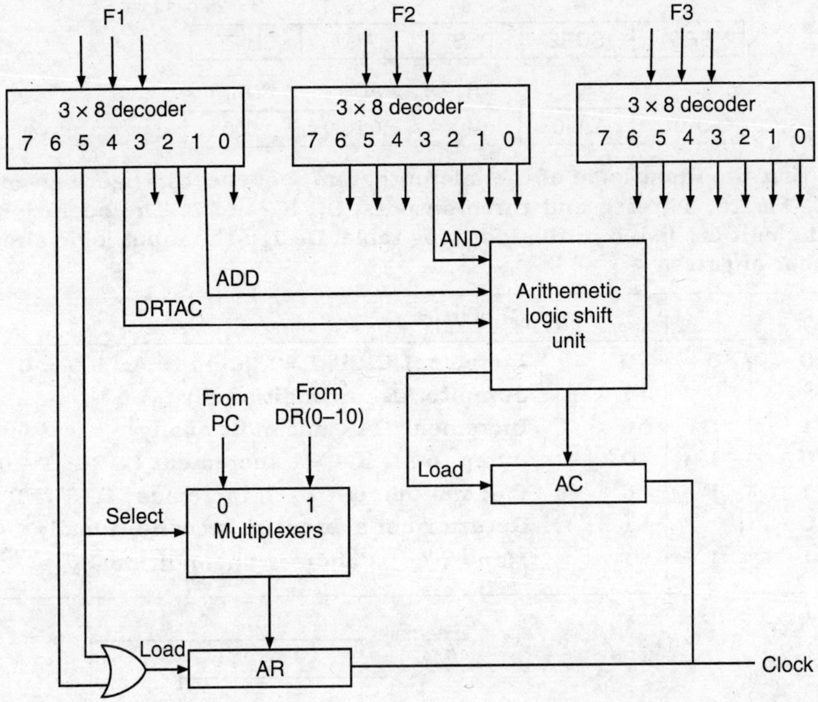

Fig. 6.26. Decoding of micro-operation field

Solution: F = 101 (5) PC ← PC + 1

From

F_3 output 6

F_3 output 5

Q.23. Show how a 9-bit micro-operation field in a microinstruction can be divided into sub fields to specify 46 micro-operations. How many micro-operations can specified in one microinstrucion?

Solution: A field of 5 bits can specify $2^5 - 1 = 31$ micro-operations.

A field of 4 bits can specify $2^4 - 1 = 15$ micro-operations

9 bits 46 micro-operations

Q.24. A computer has 16 registers, an ALU (arithmetic logic unit) with 32 operations, and a shifter with eight operations, all connected to a common bus system.

(a) Formulate a control word for a micro-operation.

(b) Specify the number of bits in each field of the control word and give a general encoding scheme.

(c) Show the bits of the control word that specify the microoperation $R_4 \leftarrow R_5 + R_6$.

Solution:

(a)

3	3	3	5
SELA	SELB	SELD	OPR

(b) 16 registers need 4 bits, ALU need 5 bits, and the shifter need 3 bits to encode all operations.

4	4	4	5	3 = 20 bits total
SRC1	SCR2	DEST	ALU	SHIFT

R_5	R_6	R_4	ADD	SHIFT
0101	0110	0100	00100	000

Q.25. Assume that the input logic of the microprogram sequencer of figure given below has four inputs 12, 11, 10, T (test), and three outputs, S_1, S_0 and *L*. The operations that are performed in the unit are listed in the following table. Design the input logic circuit using a minimum number of gates.

I_2	I_1	I_o	*Operation*
0	0	0	Increment CR if $T = 1$, jump to AD if $T = 0$
×	0	1	Jump to AD unconditionally
1	0	0	Increment CAR unconditionally
0	1	0	Jump to AD if $T = 1$, increment CAR if $T - 0$
1	1	0	Call subrouine if $T - 1$, increment CAR if $T = 0$
0	1	1	Return from subroutine unconditionally
1	1	1	Map external address unconditionally

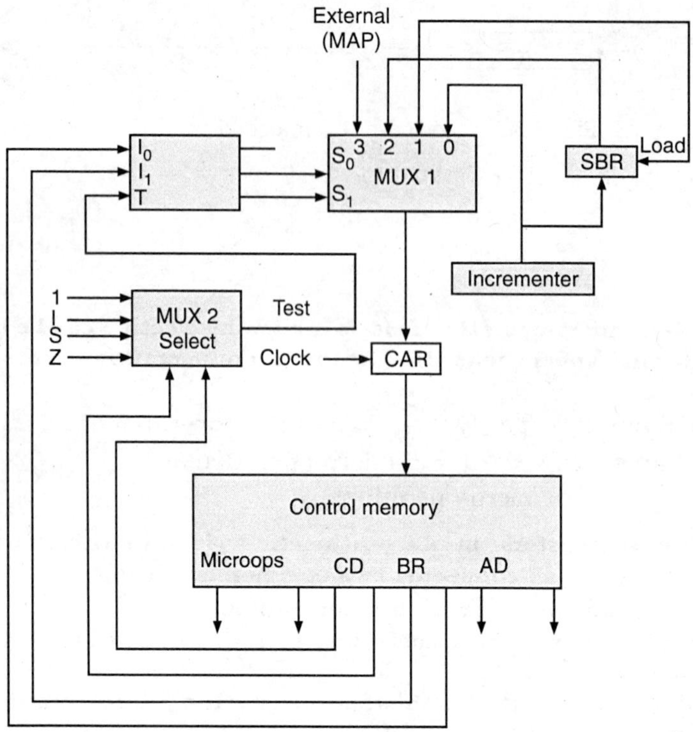

Fig. 6.27. Microprogram sequencer for a control memory

Solution:

I_2	I_1	I_0	T	S_1	S_0	L	
0	0	0	0	0	1	0	AD (1)
0	0	0	1	0	0	0	INC (0)
0	0	1	0	0	1	0	AD (1)
0	0	1	1	0	1	0	AD (1)
0	1	0	0	0	0	0	INC (0)
0	1	0	1	0	1	0	AD (1)
0	1	1	0	1	0	0	RET (2)
0	1	1	1	1	0	0	RET (2)
1	0	0	0	0	0	0	INC (0)
1	0	0	1	0	0	0	INC (0)
1	0	1	0	0	1	0	AD (1)
1	0	1	1	0	1	0	AD (1)
1	1	0	0	0	0	0	INC (0)
1	1	0	1	0	1	1	CALL (1)
1	1	1	0	1	1	0	MAP (3)
1	1	1	1	1	1	0	MAP (3)

I_2T_1 \ I_0T	00	01	11	10
00	1		1	1
01		1		
11		1	1	1
10			1	1

I_2T_1 \ I_0T	00	01	11	10
00				
01			1	1
11			1	1
10				

$$S_0 = I_2 I_1 + I_1 I_0 T + I_2 I_1 T$$
$$L = I_2 I_1 I_0 T$$

Q.26. Design a 7- bit combinational circuit incrementer for the microprogram sequencer of figure given in Q.25. (see figure given below). Modify the incrementer by including a control input *D*. When *D* = 0, the circuit increments by one, but when *D* = 1, the circuit increments by two.

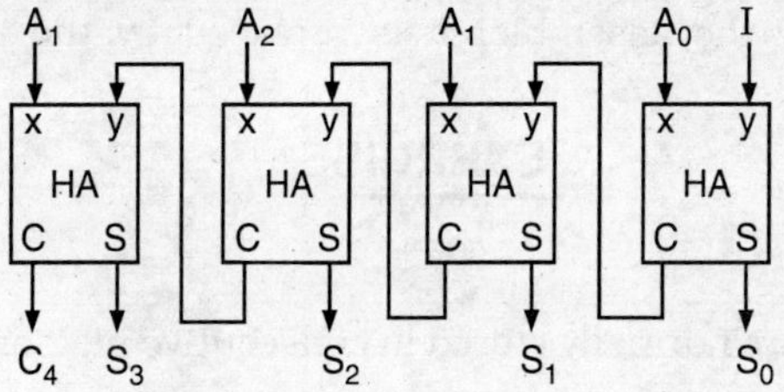

Fig. 6.28. 4-bit binary incrementer

Solution:

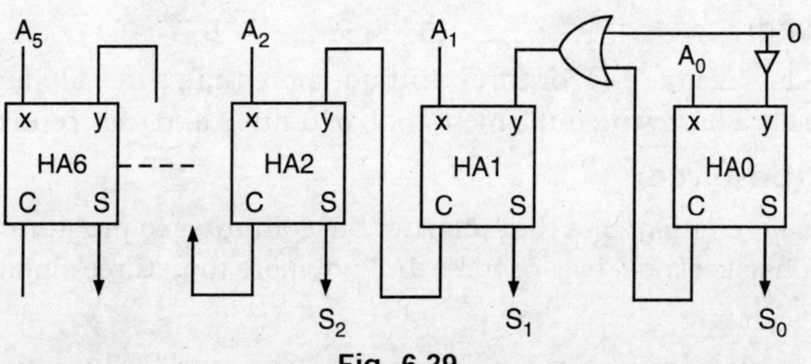

Fig. 6.29.

Q.27. Insesrt an exclusive–OR gate between MUX 2 and the input logic of figure given in Q.25. One input to the gate comes from the test output of the multiplexer. The other input to the gate comes from a bit labeled *P* (for polarity) in the microinstruction from control memory. The output of the gate goes to the input *T* of the input logic. What does the polarity control *P* accomplish?

Solution: *P* is used to determine the polarity of the selected status bit.

When $P = 0$, $T = G$ because $G \oplus = G$.

When $P = 1$, $T = G$ because $G \oplus 1 = G$.

When *G* is the value of the selected bit in MUX2.

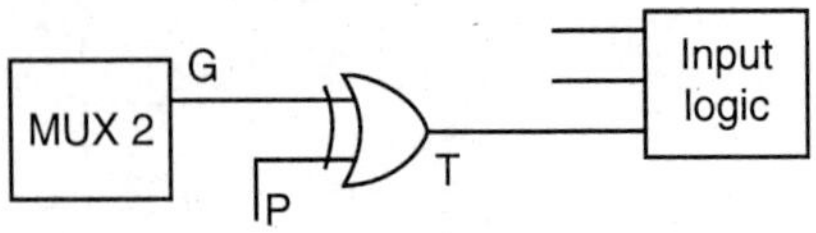

Fig. 6.30.

SUMMARY

1. Computer instructions are normally stored in consecutive memory locations and are executed sequentially one at a time.
2. The timing got all registers in the basic computer controlled by a clock.
3. An instruction is a group of bits that instruct the computer to program a specific operation.
4. A label consists of one, two or three, but not more than three alpha numeric characters.
5. A control memory is the unit which handles microinstructions.
6. Program resides in control memory is called microprogram, instruction is called microinstruction and operation is called micro-operation.
7. Microinstruction format is of 20 bits.
8. Micro programmed control unit will have two sperate memories– a main memory and a control memory.
9. The timing for all registers in the Basic computer is controlled by master clock generator.
10. Macros in an assembly language are used to repeat a sequence of instructions several times in a program. The instructions are stored in a macro body and whenever a macro name is encountered by the assembler it is replaced by the instruction.

EXERCISE

(I) Fill in the Blanks (Solved)

1. Computer .... are normally stored in consecutive memory locations and are executed sequentially one at a time.
2. The timing for all registers in the basic computer is controlled by a ....

(II) True or False (Solved)

1. A two phase consists of one, two, or three, but not more than three alphanumeric characters.
2. An interrupt cycle is a hardware implementation of branch and save return address operation.

(III) Fill in the Blanks (Unsolved)

1. An ....is a group of bits that instruct the computer to program a specific operation.
2. A .... consists of one, two, or three, but not more than three alphanumeric characters.

(IV) True or False (Unsolved)

1. The memory that is a part of a control unit is referred to as an accumulator.
2. The data register is sometimes called a buffer.

(V) Detailed Questions

1. What is the difference between a direct and an indirect address instruction? How many references to memory are needed for such type of instruction to bring an operand into a processor register?
2. Explain why each of the following micro-operations cannot be executed during a single clock pulse.
   (*a*) IR M[PC] (*b*) AC AC + TR (*c*) DR DR + AC (AC does not change)
3. Describe the control gates associated with the program counter PC in the basic computer.
4. Can the letter I be used as a symbolic address in the assembly language program defined for the basic computers? Justify the answer.
5. Write a program loop, using a pointer and a counter that clears to 0 the contents of hexadecimal location 500 through 5FF.
6. Is it possible to have a hardwired command associated with a control memory?
7. Define the following:
   (*a*) Micro-operation (*b*) Micro instruction (*c*) Micro program (*d*) Micro code
8. What is the difference between a microprocessor and a microprogram? Is it possible to design a microprocessor without a microprogram? Are all microprogrammed computers also microprocessors?

Answers

(I) Fill in the Blanks (Solved)

1. instructions
2. clock

(II) True or False (Solved)

1. False
2. True

❑❑❑

Chapter

7

Memory Organization

Chapter Outline

- Introduction
- Basic Concepts of Memory
- Memory Hierarchy
- Main Memory
- Block Diagram of Memory Device
- Ram and Rom Chips Format
- 2D and $2^{1/2}$D Memory Organizations
- Associative Memory
- Cache Memory Concept, Design Issues and Mapping Techniques
- Another Method of Address Mapping
- Replacemantal Algorithm
- Backup Storage
- Memory Management Hardware
- Virtual Memory
- Auxiliary Memory
- Address Space and Memory Space

7.0 INTRODUCTION

In computer's hardware language the peripheral devices can be defined as the devices that are connected to the computer in order to get most of the advantage out of it,if you in to detail description of peripheral devices it can be explained as the devices that are optional, and which are not required in principle. In earlier ages when computers were too much expensive and personal computer were very hard to afford, at that time the Motherboard, CPU (Central Processing Unit) and Memory RAM (Random Access Memory) & ROM (Read Only Memory)) were considered to be the main components of the computer and any other device additionally attached to the computer were considered as peripheral device so this means that the keyboard, mice, etc. at that time were considered as peripheral devices but now a days generally they are not considered peripheral devices.

The common peripheral devices are microphones, cameras, disk drives, scanners, etc.

There is a common understanding that among people that the internal devices such as Sound Cards etc are not peripheral because they are added in the computer case which is wrong following is the peripheral devices list for a general idea.

7.1 BASIC CONCEPTS OF MEMORY

The main memory is the central storage unit in a computer system. It is a relatively large and fast memory used to store programs and data during the computer operation. The principal technology used for the main memory is based on semiconductor integrated circuits. Integrated circuit RAM chips are available in two possible operating modes, *static* and *dynamic*.

The static RAM consists essentially of internal flip-flops that store the binary information. The stored information remains valid as long as power is applied to the unit.

The dynamic RAM stores the binary information in the form of electric charges that are applied to capacitors. The capacitors are provided inside the chip by MOS transistors. The stored charge on the capacitors tend to discharge with time and the capacitors must be periodically recharged by refreshing the dynamic memory. Refreshing is done by cycling through the words every few milliseconds to store the delying charge. The dynamic RAM offers reduced power consumption and larger read and write cycles.

The static RAM is used and has shorter read and write cycles.

Most of the main memory in a general-purpose computer is made up of RAM integrated circuit chips, but a portion of the memory may be constructed with ROM chips. Originally, RAM was used to refer to a random-access memory, but now it is used to designate a read/write memory to distinguish it from a read-only memory, although ROM is also random access. RAM is used for storing the bulk of the programs and data that are subject to change. ROM is used for storing programs that are permanently resident in the computer and for tables of constants that do not change in value once the production of the computer is completed.

The ROM portion of main memory is needed for storing an initial program called a *bootstrap loader*. The bootstrap loader is program whose function is to start the computer software operating when power is turned on. Since RAM is volatile, its contents are destroyed when power is turned off. The contents of ROM remain unchanged after power is turned off and on again. The startup of a computer consists of turning the power on and starting the execution of an initial program. Thus, when power is turned on, the hardware of the computer sets the program counter to the first address of the bootstrap loader. The bootstrap program loads a portion of the operating system from disk to main memory and control is

RAM and ROM chips are available in a variety of sizes. If the memory needed for the computer is larger than the capacity of one chips, it is necessary to continue a number of chips to form the required memory size.

An electronic circuitry that allows data to be store and required is known as memory. The memory unit that communicates directly with the CPU is called the main memory. The storage devices that provide backup storage are called auxiliary memory.

7.2 MEMORY HIERARCHY

The memory unit is an essential component in any digigtal computer since it is needed for storing programs and data. A very small computer with a limited application may be able to fulfil its intended task without the need of additional storage capacity. Most general purpose computers would run more efficiently if they were equipped with additional storage beyond the capacity of the main memory. There is just not enough space in one memory unit to accommodate all the programs used in a typical computer. Moreover, most computers users accumulated and continue to accumulate large amounts of a data-processing software. Not all accumulated information is needed by the processor at the same time. Therefore, it is more economical to use low-cost storage devices to serve as a backup for storing the information that is not currently used by the CPU. The memory unit that communicates directly with the CPU is called the *main memory.* Devices that provide back up storage are called *auxillary memory.* The most common auxiliary memory devices used in computer systems aer magnetic disks and tapes. They are used for storing system programs, large data files, and other backup information. Only progams and data currently needed by the processor reside in main memory. All other information is stored in auxiliary memory and transfered to main memory when needed.

The total memory capacity of a computer can be visualized as being a hierarchy of components. The memory hierarchy system consists of all storage devices employed in a computer system from the slow but high capacity auxiliary memory to a relatively faster main memory, to an even smaller and faster cache memory accessibles to the high-speed processing logic. Fig. 7.1. Shows the compontents in a typical memory hierarchy. At the bottom of the hierarchy, the relatively slow magnetic tapes used to store removable files. Next are the magnetic disks used as a backup storage. The main memory occupies a central position by being able to communicate directly with the CPU and with auxiliary memory devices through an I/O processor. When programs not residing in main memory are needed by the CPU, they are brought in from auxiliary memory. Programs not currently needed in main memory are transfered into auxiliary memory to provide space for currently used programs and data.

A special very-high speed memory called a *cache* is sometimes used to increase the speed of processing by making current programs and data available to the CPU at a rapid rate. The cache memory is employed in computer systems to compensate for the speed differential between main memory access time and processor logic. CPU logic is usually faster than main memory access time, with the result that processing speed is limited primarily by the speed of main memory. A technique used to compensate for the mismatch in operating speeds is to employ an extremely fast small cache between the CPU and main memory whose access time is close to processor logic clock cycle time. The cache is used for storing segments of programs currently being executed in the CPU and temporary data frequency needed in the present calculations.

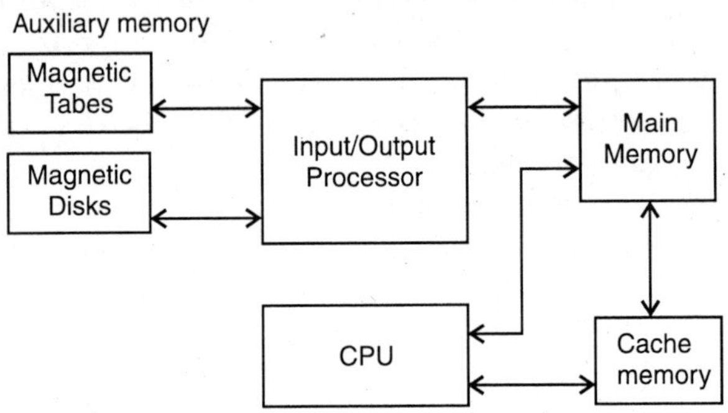

Fig. 7.1. Memory hierarchy in a computer system

By making programs and data available at a rapid rate, it is possible to increase the performance rate of the computer.

While the input/output processor manages data transfers between auxiliary memory and main memory, the cache organizations is concerned with the transfer of information between main memory and CPU. Thus, each is involved with a different level in the memory hierarchy system. The reason for having two or three levels of memory hierarchy is economics. As the storage capacity of the memory increases, the cost per bit for storing binary information decreases and the access time of the memory becomes longer. The auxiliary memory has a large storage capacity, is relatively inexpensive, but has low access speeds compared to main memory.

The cache memory is very small, relatively expensive, and has very high access speed. Thus, as the memory access speed increases, so does its relative cost. The overall goal of using a memory hierarchy is to obtain the highest-possible average access speed while minimizing the total cost of the entire memory system.

The cache memory is very small, relatively expensive, and has very high access speed. Thus, as the memory access speed increases, so does its relative cost. The overall goal of using a memory hierarchy is to obtain the highest-possible average access speed while minimizing the total cost of the entire memory system.

Auxiliary and cache memories are used for different purposes. The cache holds those parts of the program and data that are most heavily used, while the auxiliary memory holds those parts that are not presently used by the CPU. Moreover, the CPU has direct access to both cache and main memory but not to auxiliary memory. The transfer from auxiliary is usually done by means of direct memory access of large block of data. the typical access time ratio between cache and main memory is about 1 to 7. For example, a typical cache memory may have an access time of 100 ns, while main memory access time may be 700 ns. Auxiliary memory average access time is usually 1000 times that of main memory. Block size in auxiliary memory typically ranges from 265 to 2048 words, while cache block size is typically from 1 to 16 words.

Many operating systems are designed to enable the CPU to process a number of independent programs concurrently. This concept, called *multiprogramming*, refers to the existence of two or more programs in different parts of the memory hierarchy at the same time. In this way it is possible to keep all parts of the computer busy by working with several programms in sequence. For example, suppose that a program is being executed in the CPU and an input/output transfer is required. The CPU initiates the intput/output processor to start executing the transfer.

This leaves the CPU free to execute another program. In a multiprogramming system, when one program is waiting for input or output transfer, there is another program ready to utilize the CPU.

With multiprogramming, the need rises for running partial programs, for varying the amount of main memory in use by a given program, and for moving programs around the memory hierarchy. Computer programs are sometimes too long to be accommodated in the total space available in main memory. Moreover, a computer system uses many programs and all the programs cannot reside in main memory at all times. A program with its data normally resides in auxiliary memory. When the program or a segment of the program is to be executed, it is transferred to main memory to be executed by the CPU. Thus, one may think of auxillary memory as containing the totality of information stored in a computer system. It is the task of the operating system to maintain in main memory a portion of this information that is currently active. The part of the computer system that supervises the flow of information between auxiliary memory and main memory is called the *memory management system.*

In summarization, the memory hierarchy can be implemented as increasing speed, increasing size and increasing cost per bit, which is shown in Fig. 7.2 and 7.3.

(i) Memory Hierarchy in a computer system

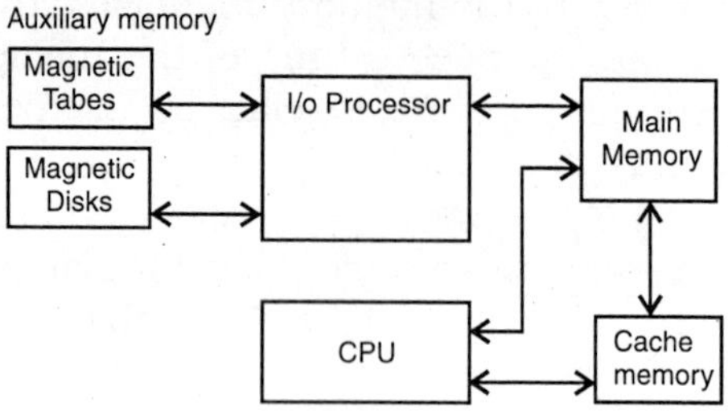

Fig. 7.2.

(ii) Memory Hierarchy

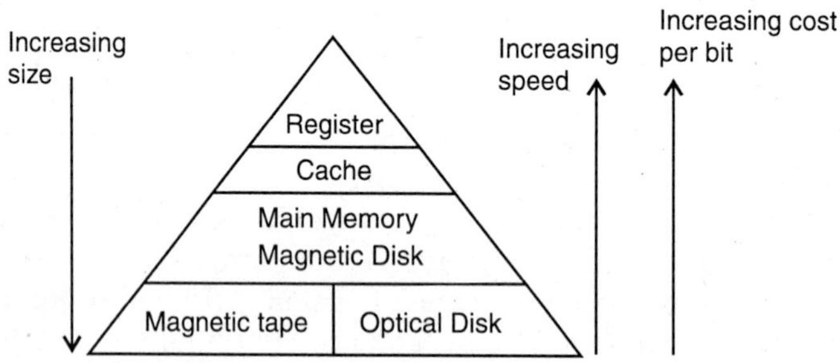

Fig. 7.3.

7.3 MAIN MEMORY

The memory unit that communicates directly with the CPU is called main memory. It is a relatively large and fast memory used to store programs and data during the computer operation. The main memory can be classified into two categories:

(i) RAM (Random Access Memory)

(ii) ROM (Read Only Memory)

7.3.1 RAM (Random Access Memory)

- RAM is a memory that can be read as well as written.
- RAM is a valuate memory so it losses the stored data when power is turned off.

There are two types of RAM:

(i) Static RAM or SRAM: Static RAM cells are basically flip-flops which can stay in a given time (*i.e.*, store a bit) as long as power to the circuit is not interrupt.

(ii) Dynamic RAM or DRAM: In the dynamic RAM, the data is stored in the form of charge on the capacitor. The capacitors are provided inside the chip by MOS transistors.

Comparison of SRAM and DRAM

Sr. No.	*Parameter*	*SRAM*	*DRAM*
1.	Circuit configuration	EachSRAM call is a flip flop consists	Each DRAM unit of one MOSFET and a capacity
2.	No. of components Per cell.	More	Less (only two)
3.	Memory cells/unit area	Less than DRAM	More than SRAM

4.	Refreshing	Not Required	Required
5.	Cost	More	Less
6.	Access time	Less so these are faster memory	More. So these are slower memory.

7.3.2 ROM (Read Only Memory)

- ROM is used for string programs that are permanently resident in the computer.
- It can be only read and cannot written
- Following advantages make the ROM an important part of many digital system.
  1. Low cost
  2. High speed
  3. Flexibility in system design
  4. ROM is a non volatile memory

Application of ROM

(*i*) For implementation of combinational circuits
(*ii*) For implementation of sequential circuits
(*iii*) Storage at microprocessor program
(*iv*) In character generation

The ROM portion of Main Memory is needed for storing initial program called a bootstrap loader.

Types of ROM:

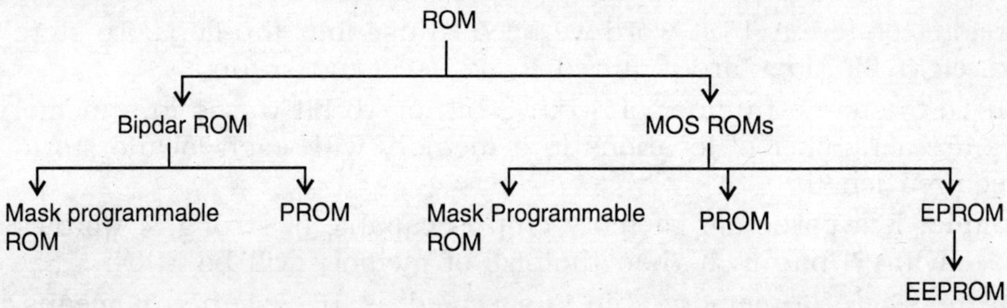

Fig. 7.4.

Characteristics of Memory

1. Memory capacity and its organization
2. Physical dimensions
3. Power consumption
4. Timing characteristic
5. Reliability
6. Cost

Comparison of RAM and ROM

Sr. No.	***Parameter***	***RAM***	***ROM***
1.	Operations involved	Reading and writing	Reading only
2.	Type of storage	Temporary	Permanent
3.	Types	SRAM, DRAM	RPOM, EPROM, EEPROM
4.	Application	Calculators, computers	Computers Microprocessors

7.3.3 Memory size

- The number of locations and no. of bits per word will very from memory.
- A flip-flop is equivalent to a single bit memory cell. Sl the basic element of a semiconductor memory is a flip-flop.

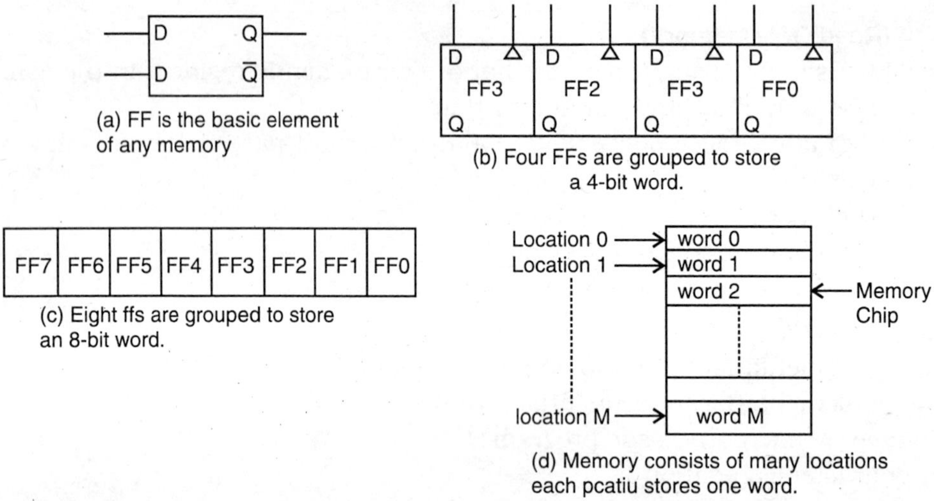

(a) FF is the basic element of any memory

(b) Four FFs are grouped to store a 4-bit word.

(c) Eight ffs are grouped to store an 8-bit word.

(d) Memory consists of many locations each pcatiu stores one word.

Fig. 7.5.

- Each flip-flop capable of staring on be binary bit, *i.e.* 0 or 1 only.
- Therefore to store a 4 bit word we need to use four flip-flops. To store an 8 bit word, eight flip flops are required to be used and so on.
- We have to store a number of 4 bit, 8 bit, or 16 bit words in a memory. Hence there are a number of locations in a memory with each locatio storing a word of required length.

  Example: It a particular memory chip is capable of strong M words with each word having N bits in it then the size of memory will be MXN.
- So if the size of a memory chip is specified as 16 × 4 then it means that this memory consists of 16 locations and each location can store one 4-bit word.
- The memory used values of number of words per cheap are 64, 256, 512, 1024, 2048, 4096, etc.
- And the common word size are 1, 4 and 8.
- Then what to do if we require to store 16 bit words? Memories with higher number of words and/or large word size can be formed by cascading a number of chips of smaller size.

7.4 BLOCK DIAGRAM OF MEMORY DEVICE

- There are three types of inputs to an MXN memory device:
  1. Address input lines
  2. Data input lines
  3. Control inputs
- And there are N number of data output lines.

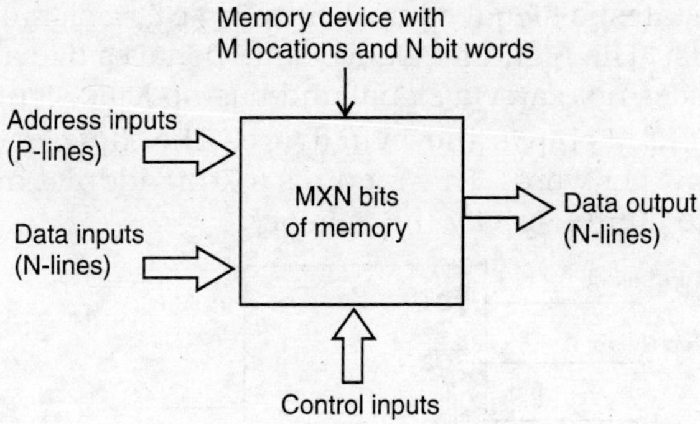

Fig. 7.6.

- **Data Inputs:** There are N-number of data input lines. The data to be stored is put on these lines word by word with each word N-bit longs. Data inputs are also called as Data bus.
- **Address Inputs:** There are "P" number of address input lines. These lines are used to specify the "address" of the required memory location, for reading the already stored data or writing a new data.
- To access any one of the M possible locations. We need P address lines such that
$$2^P = M$$
- For example, if M = 16 locations, then P = 4. So, we need 4 address lines to have system different combinations from 0000 to 1111.

Data Output lines

- The data available in the selected memory locations can be "read" on the data output lines.
- The number of data lines is N, i.e. equal to the number of bits per word. The output data lines are also called output data bus.

Power supply lines

In addition to the address, control and data lines, there are two more pins, required for the power supply and ground.

EXAMPLE 7.1. A ROM chip of 1024 × 8 bits has four select inputs and operates from a 5-volt power supply. How many publisher needed for the IC package? Draw a block diagram and bobble all input and output terminals in the ROM.

Solution: 10 pins for inputs ($\because 1024 = 2^{10}$)

4 for chip select 8 for output, and 2 for power supply

Total = 24 pins are needed.

7.5 RAM AND ROM (CHIP FORMAT)

Address Decoding and Selecting

RAM and ROM chips: A RAM chip is better suited for communication with the CPU if it has one or more control inputs that select the chip only when needed. Another common feature is bi-direction data bus that allows the transfer of data either. From memory to CPU during read operation, or from CPU to memory during write operation. A bidirectional bus can be constructed with three-state buffers. A three state buffer output can be placed in one of three possible states; a signal equivalent to logic 1, a signal equivalent to logic 0 are normal digital signals. The high impedance state behaves like an open circuit which means that the output does not carry a signal and has no logic significance.

The block diagram of RAM chip is shown in figure. The capacity of the memory is 128 words of eight bits (one by) per word. This requires a 7-bit address and g-bit bidirectional data bus. Read and write inputs specify the memory.

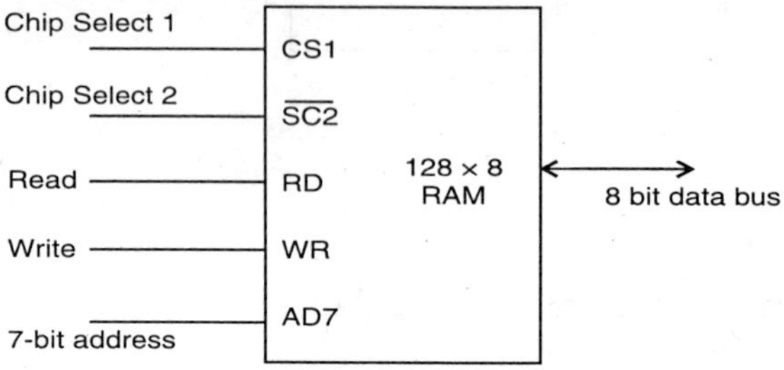

Fig. 7.7. Block diagram

Function Table 7.1.

CS1	*CS2*	*RD*	*WR*	*Memory function*	*State of data bus*
0	0	×	×	Inhibit	High Impedance
0	1	×	×	Inhibit	High impedance
1	0	0	0	Inhibit	High impedance
1	0	0	1	Write	Input data to RAM
1	0	1	×	Read	Output data from RAM
1	1	×	×	Inhibit	High impedance

Operation and two chip selects control inputs are for enabling the chip only when it is selected by the microprocessor. The availability of more than one control input to select the chip facilitates the decoding of the address lines when multiple chips are used in micro computer. The read and write inputs are sometimes combined into one line labelled R/w. When the chip is selected, the two binary states in this line specify the two operations of read for write.

The function table specified the operation of RAM chip. The unit is in operation only when CSI = 1 and CS2 = 0. The bar on the top of second variable indicates that this input is enabled when it is equal to 'O'. If the chip select inputs are not enabled or if they are enabled but the read write inputs are not enabled the memory is inhibited and its data bus is in a high impedance state when CS1 = 1 and CS2 = 0, the memory can be placed in a read or write mode. When WR input is enabled, the memory stores a byte from the data bus into a location specified by the address input lines. When the RD input is enabled, the content of the selected byte is placed into the data bus. The RD and WR signals control the memory operation as well as the bus buffers associated with the bidirectional data bus.

The ROM chip is organized in similar manner. However, since a ROM can only read. The data bus can only be in an output mode. The block diagram of ROM chip is shown in Figure. For same-size chip, it is possible to have more bits of ROM than of RAM, because the internal binary cells in ROM occupy less space than in RAM. For this reason, the diagram specified a 512 bytes-ROM while the RAM has only 128 bytes.

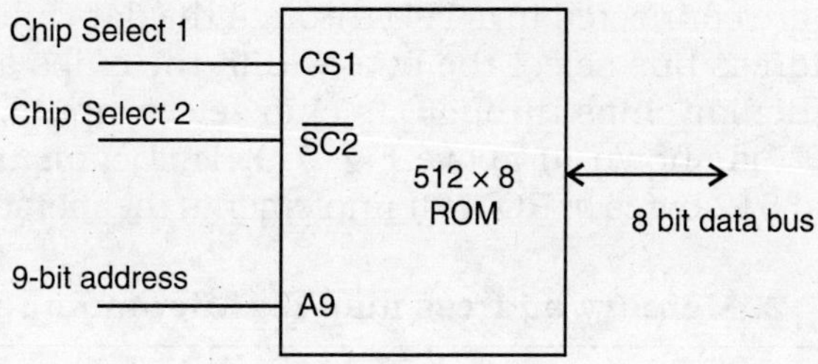

Fig. 7.8. Typical ROM Chip

The 9 address lines in ROM chip specify any one of 512 bytes stored in it. The two chip select inputs must be CS1 = 1 and CS2 = 0 for the unit to operate. Other wise the data bus is in high impedance state. There is no need for a read or write control because the unit can only read.

Memory connection to CPU

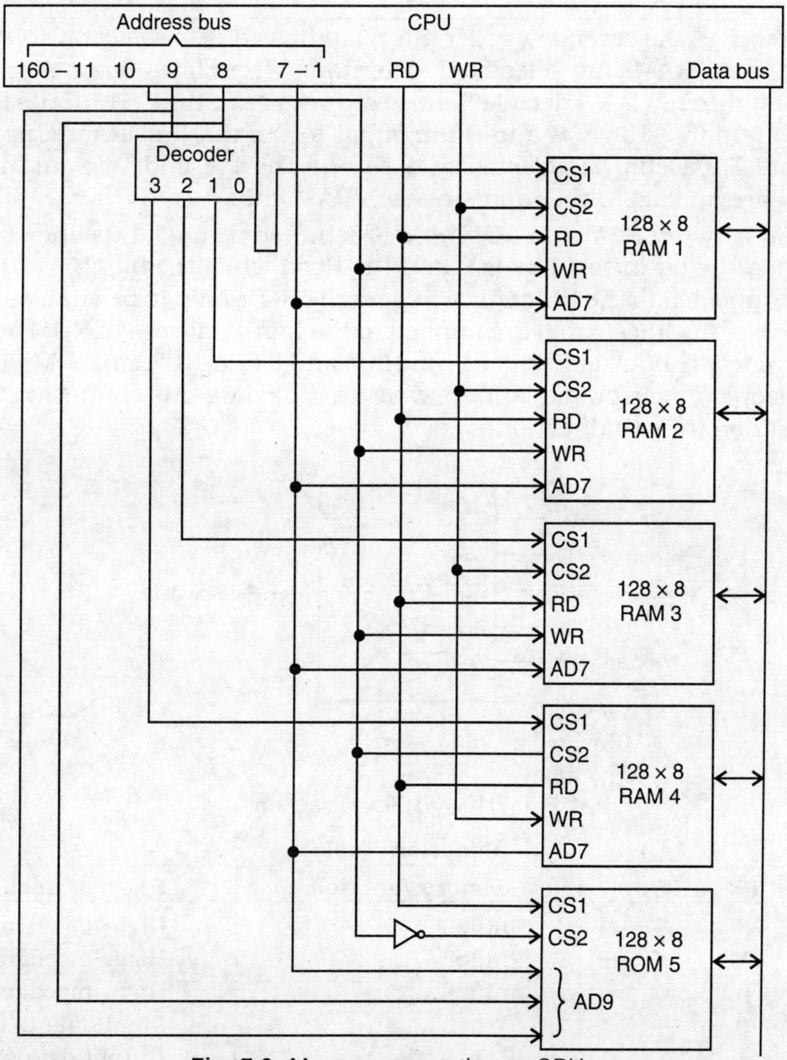

Fig. 7.9. Memory connection to CPU

RAM and ROM chips are connected to a CPU through the data and address buses. The low-order lines in the address bus select the byte within the chips and other lines in the address bus select a particular chips through its chip select inputs. The connection of - memory chips to the CPU is shown in above Fig. 7.9. This configuration gives the 512 bytes capacity of RAM and 512 bytes of ROM. It implements the memory main the following Table 7.2.

Table 7.2. Memory address map for Microcomputer

Component	Hexadecimal	Address bus									
	address	10	9	8	7	6	5	4	3	2	1
RAM 1	000–007F	0	0	0	×	×	×	×	×	×	×
RAM 2	0080–00FF	0	0	1	×	×	×	×	×	×	×
RAM 3	0100–017F	0	1	0	×	×	×	×	×	×	×
RAM 4	0180–01FF	0	1	1	×	×	×	×	×	×	×
ROM	0200–03FF	1	×	×	×	×	×	×	×	×	×

Each RAM receives the seven low order bits of address bus to select one of 128 possible bytes. The particular RAM chip selected is determined from lines 8 and 9 in the address bus. This is done through 2 × 4 decoder whose outputs go to the CS1 inputs in each RAM chip. Thus when address lines 8 and 9 are equal to 00, the first RAM chip is selected. When 01 second RAM chip is selected and so on. The PD and WR outputs from the microprocessor are applied to the inputs of each RAM chip.

The selection between RAM and ROM chip is achieved through bus line 10. The RAMs are selected when the bit in this line is 0 and the ROM when the bit is 1. The other chip select input is connected to the RD control line for the ROM chip to be enabled during read operation. Address bus lines 1 to 9 are applied to the input address of ROM without going through the decoder. This assigns 0 to 511 to RAM and 512 to 1023 to RDM. The data bus of the ROM has only an output capability, where as the data bus connected to RAM, can transfer information in both direction.

(*i*) RAM chips

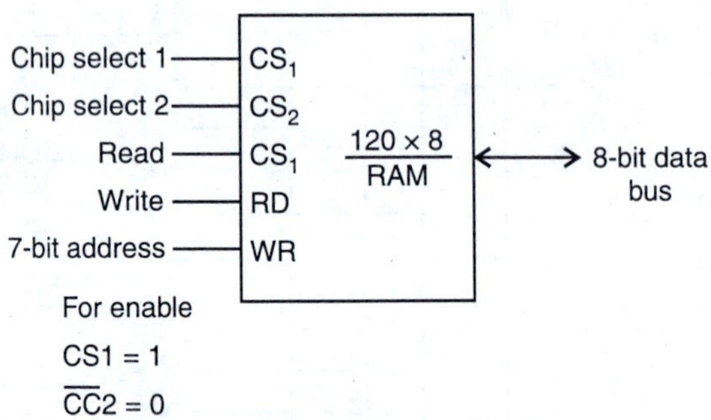

Fig. 7.10. (a) Block diagram

Function Table

CS1	*CS2*	*RD*	*WR*	*Memory function*	*State of data bus*
0	0	X	X	Inhibit	High–Impedance
0	1	X	X	Inhibit	High–Impedance
1	0	0	0	Inhibit	High–Impedance
1	0	0	1	Write	Input data to RAM
1	0	1	X	Read	Output data from RAM
1	1	X	X	Input	High-Impedance:

(*ii*) RAM Chip;

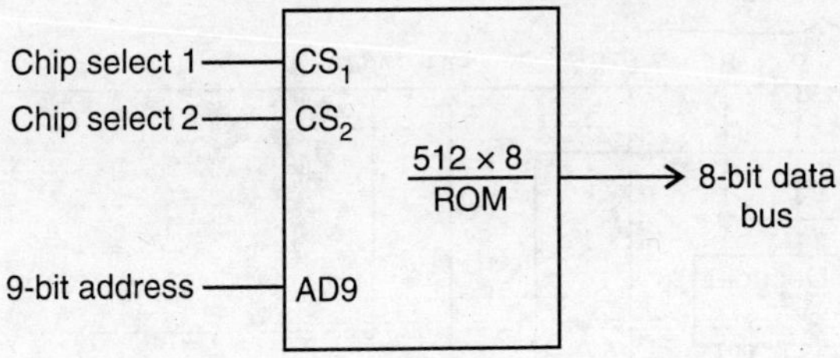

Fig. 7.10 (b) Block diagram of ROM Chip

Required size = M × N
Available size = m × n

$$\text{no. of chips required (K)} = \frac{M \times N}{m \times n}$$

EXAMPLE 7.2. A Computer system needs 512 bytes of RAM & 512 bytes of ROM. The capacity of RAM chip is 128 × 8 2 ROM is 512 × 8. Find out the no. of RAM chip required.

Solution: Required size = 512 bytes = 512 × 8.

Available size = 128 × 8

$$K = \frac{512 \times 8}{128 \times 8} = 4 . \Rightarrow K = 4$$

Memory address Map for Micro computer

Component	Hexadecimal address	Address bus									
		10	9	8	7	6	5	4	3	2	1
RAM 1	0000 – 007F	0	0	0	X	X	X	X	X	X	X
RAM 2	0080 – 00FF	0	0	1	X	X	X	X	X	X	X
RAM 3	0100 – 017F	0	1	0	X	X	X	X	X	X	X
RAM 4	0180 – 01FF	0	1	1	X	X	X	X	X	X	X
ROM	0200 – 03FF	1	X	X	X	X	X	X	X	X	X

Note: 1 Hexadecanoic Bit = 4 Bits of Boundary No.

0 0 7 F = 127
↓ ↓ ↓ ↓
0000 00000 0111 1111
0 0 8 0 = 128
↓ ↓ ↓ ↓
0000 0000 1000 0000

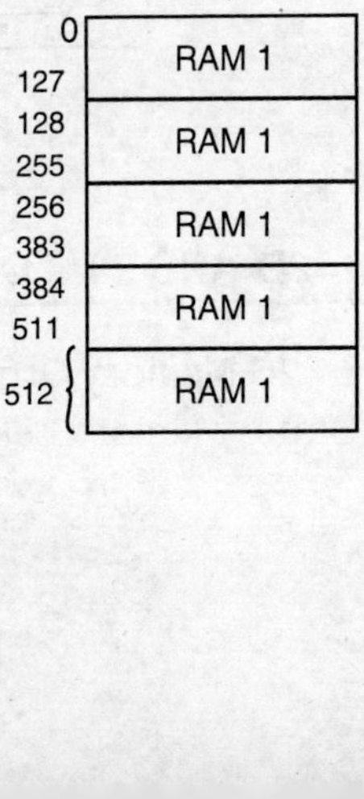

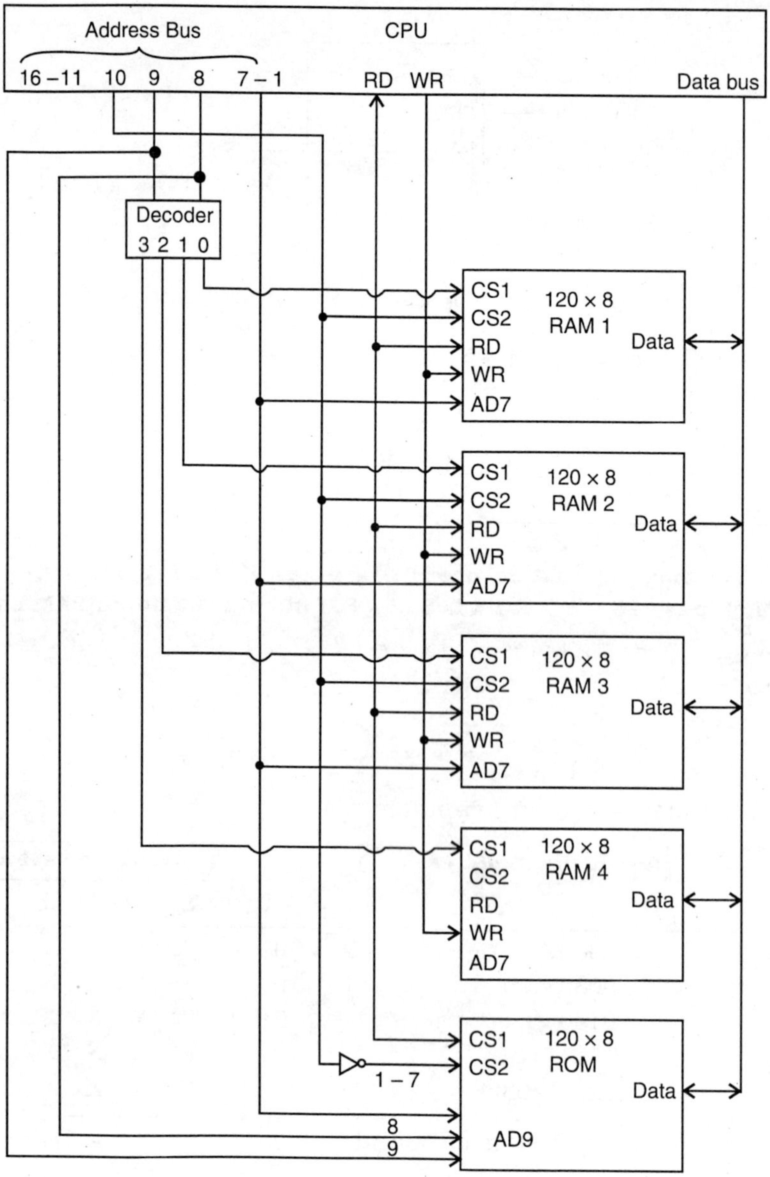

Fig. 7.11. Memory connection to the CPU

7. 6 2D AND 21/2 D MEMORY ORGANIZATIONS

7.6.1 2D Memory Organization

Static Memory cells can be organized into a 2D memory system a shown in Fig. 7.12.

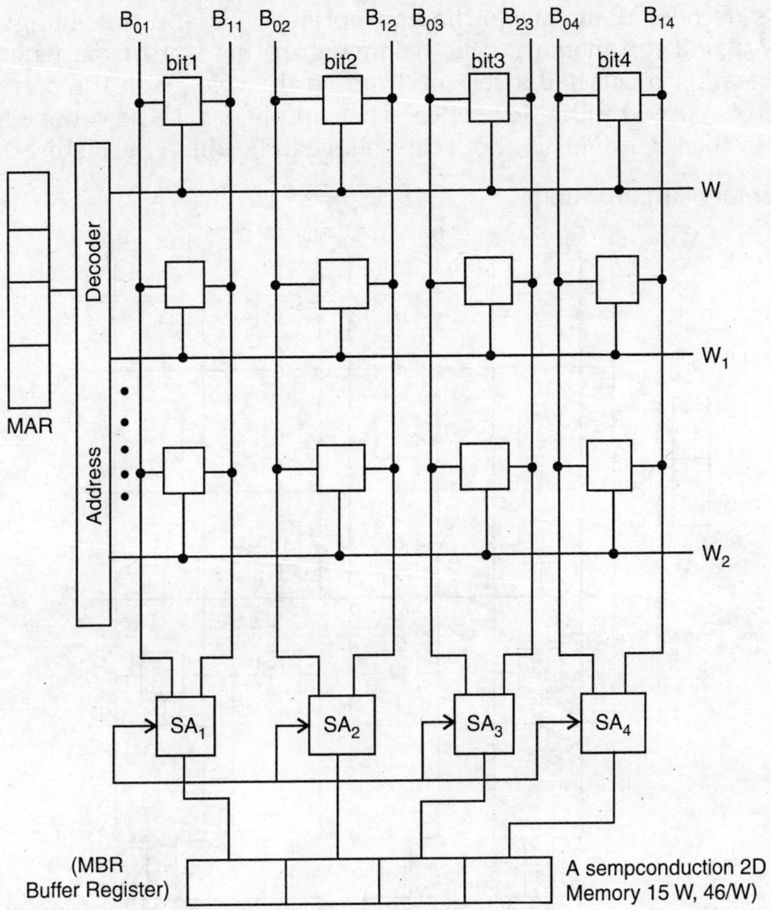

Fig. 7.12. 2D organizations

The figure illustrates the organization of 16 words 4 bit per word 2D Memory. There is an address decoder to decode the address in 4-bit MAR memory register), which select one out of 16 word wires V is the voltage applied to a selected word wire. A no. of memory cells equal to the bits per word, are connected to each word wire. For i = 1, 2, 3, 4, the bit wires Bot 8 Bit of the Bit i of all words are connected together & fed to the sense amplifier SAi as shown in Fig. 7.12.

Tow wire, by 1001 in word 2 (W2). The following operations are to be made. Word address 0010 to stored in MAR. When decoded it selects word W2 and voltage line on W2 is made V. The data 1001 is stored in MBR. SA1 brings voltage B11 $\rightarrow$ 1,

SA1 $\rightarrow$ B11 (for stored)

SA2 bring voltage on BO2 ($\rightarrow$ 0),

SA3 brings voltage on B_{03} for 0

SA4 brings voltage on B_{14} for 1

Thus Cell 1 of word 2 will store a 1

Cell 2 of word 2 will store a 0

Cell 3 of word 2 will store a 0

Cell 4 of word 2 will store a 1

To read the contents of a word. The address must be in MAR.

The address decoder then selects the appropriate wide and the voltage on the word wire is taken V/2. All the memory cells belonging to this word experience the voltage. Depending on weather a cell had a zero or 1 stared the voltage on the corresponding bit wire. Dip which is speed by the respective sense amplifiers. If the voltage in Bai Dip S_{Ai} will store Zero. In the Bit 1 of MBR and if the voltage in B_{1i} dip S_{ai} will store 1 bit i of MBR.

7.6.2 $2^{1/2}$ D memory Organization

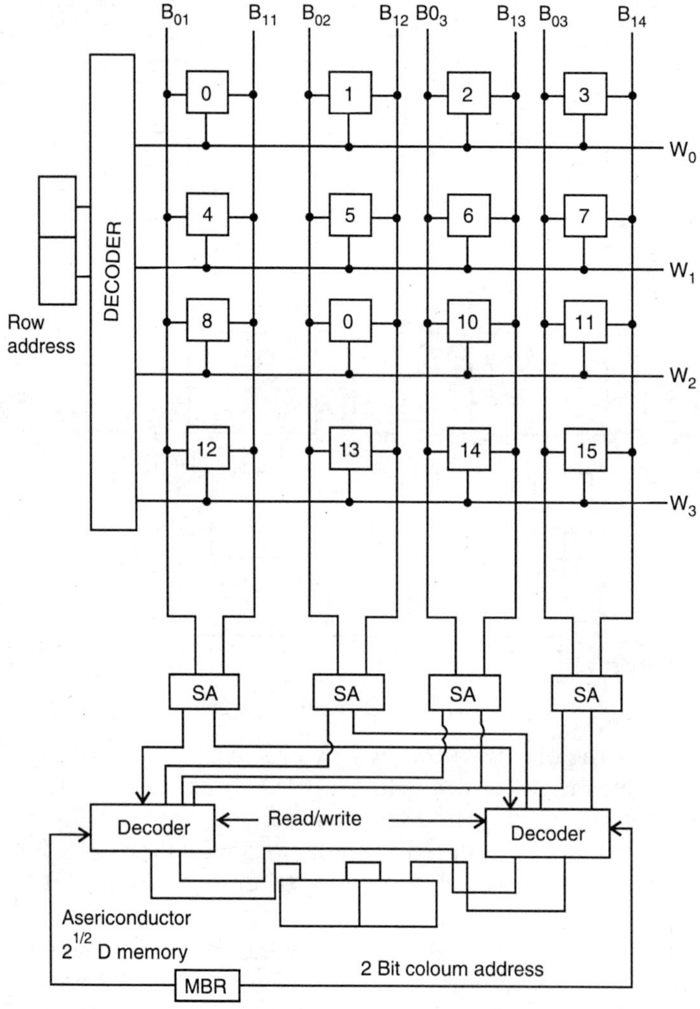

Fig. 7.13. $2^{1/2}$D organizations

Another organization is known as $2^{1/2}$ D organization uses only $\sqrt{W}$ words wires and allows the use of simple decoder. Fig. 7.13. shows the way, the cells, are organized on the memory. The four bit word address is split into 2-bit row address and 2-bit column address. Do read a bit from the memory two most significant bits of the, word address are entered into the rows register and two least significant bits of the word address entered into the coloumn register. The voltage of the selected row is lowered and del the 4 bits cells in this rows are read. The coloumn decoder routes only the bit in the selected coloumn as data O/p. To write a bit in the given cell the two most significant bits of the word address are entered into the low register and the two least significant of work address in the coloumn register. The bit to be written is entered into MBR. Appropriate row is selected by the decoder and the coloumn decoder selects the appropriate bit wire, Bit or B_{0i} and its voltage is lowered while writing the required information in the selected cell.

7.7. ASSOCIATIVE MEMORY

The auxiliary memory devices used in a computer system are magnetic disks, tapes, magnetic drums and more r ecently optical disks. The important features of any devices are its access mode, access time, transfer rate, capacity and cost.

The average time to reach a storage location in memory and obtain its contents is the access time. In electro-mechanical devices access time is equal to the sum of seek time (time required to position the read, write heads and on the location) and transfer time (the time required to transfer data to and from any device). Auxiliary storage is logically divided into records or blocks. A record/block consists of a number of words. Input/output from/on auxiliary memory is always done to enter blocks.

Many data processing applications require that the time required to find an item stored in memory be reduced to a minimum for efficiency. This can be done if stored data can be identified for access by content of the data rather than by its address. A memory unit addressed by its contents is caled associative memory or Content Addressable Memory (CAM). This type of memory can be accessed simultaneously in parallel on basis of data content. When a word is to be written into associative memory no address is specified, it is capable of finding an empty unused location to store the word. When a word is read from the memory, the content or part of the word is specified, the memory hardware locates all words which match the content and mark them for reading.

A Memory unit accessed by content is called an associative memory or Content Addressable Memory (CAM).

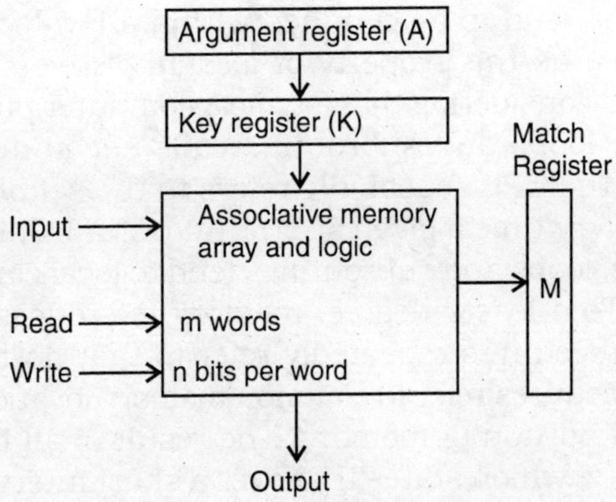

Fig. 7.14. Associative memory

A memory unit accessed by the content of the data it self is called on Associative Memory or Content Addressable Memory (CAM). This type of memory is accessed simultaneously and in parallel on the basic of data content rather than by specific address or location. When a word is written in associative memory, no address is given. The memory is capable of finding an empty unused location to store the word. When a word is to be read from an associative memory, the content of the word or part of the word, is specified.

An associative memory is more expensive than a random access memory because each cell must have storage capability as well as logic circuits for matching its contents with an external argument.

The Block diagram of associative memory is shown in Fig. 7.14. It consists of a memory array and logic for m words with n bits per word. The argument register a and key register K each have n bits, one for each bit of a word. The match register M has *m* bits, one for each memory word. Each word in memory is compared in parallel with the content of argument register. The words that match the bits of the argument register set a corresponding bit in the match register. After the matching process, those bits in the match register that have been set indicate the fact that their corresponding words have been matched.

The key register provides a mask for choosing particular field or key in the argument word. The entire argument is compared with each memory word if key register contains all 1s. Other wise, those bits in the argument that have 1's in their corresponding position of the key register are compared. Thus the key provides a mask or identifying piece of information which specified how the reference to memory is made.

To illustrate with a numerical example, suppose that the argument register A and the key register K have the bit organization shown below. Only the three left most bits of A are compared with memory words because K has 1's in these positions.

A	101	1111	00	
K	111	0000	00	
Word 1	100	1111	00	no match
Word 2	101	0000	01	match

Word 2 matches the unmasked arguments field because three left most bits of the argument and the word are equal.

7.8. CACHE MEMORY CONCEPT, DESIGN ISSUES AND MAPPING TECHNIQUES

Analysis of a large number of typical programs has shown that the references to memory at any given interval of time tend to be confined within a few localized aceas in memory. This phenomenon is known as the property of *locality* of *reference*. The reason for this property may be understood considering that a typical computer program flows in a straight-line fashion with program loops and subroutine calls encountered frequently. When a program loop is executed, the CPU repeatedly refers to the set of instructions in memory that constitute the loop. Every time a given subroutine is called, its set of instructions are fetched from memory. Thus, loops and subroutines tend to localize the references to memory for fetching instructions. To a lesser degree, memory references to data also tend to be localized. Table-loop-up procedures repeatedly refer to that portion in memory wher the table is stored. Iterative procedures refer to common memory locations and array of numbers are confined within a local portion of memory. The result of all these observations is the locality of reference property, which states that over a short interval of time, the addresses generated by a typical program refer to a few localized areas of memory repeatedly, while the remainder of memory is accessed relatively infrequently.

If the active portions of the program and data are placed in a fast small memory, the average memory access time can be reduced, thus reducing the total execution time of the program. Such a fast small memory is referred to as a *cache memory*. It is placed between the CPU and main memory as shown in Fig. 7.1. The cache memory access time is less than the access time of main memory by a factor of 5 to 10. The cache is the fastest component in the memory hierarchy and approaches the speed of CPU components.

The fundamental idea of cache organization is that by keeping the most frequency accessed instructions and data in the fast cache memory, the average age memory access time will approach the access time of the cache. Although the cache is only a small fraction of the size of main memory, a large fraction of memory requests will be found in the fast cache memory because of the locality of a reference property of programs.

The basic operations of the cache is as follows. When the CPU needs to access memory, the cache is examined. If the word is found in the cache, it is read from the fast memory. IF the word addressed by the CPU is not found in the cache, the main memory is accessed to read the word. A block of words containing the one just accessed is then transferred from main memory to cache memory. The block size may vary from one word (in one just accessed) to about 16 words adjacent to the one just accessed. In this manner, some data are transferred to the cache so that future references to memory find the required words in the fast cache memory.

The performance of cache memory is frequently measurd in terms of a quantity called *hit ratio*. When the CPU refers to memory and finds the word in cache, it is said to produce a *hit*. If the word is not found in cache, it is in main memory and it counts as a *miss*. The ratio of the number of hits divided by the total CPU references to memory (hits plus misses) is the hit ratio. The hit ratio is best measured experimentally by running representative programs in the computer and measuring the number of hits and misses during a given interval of time. Hit ratios of 0.9 and higher have been reported. This high ratio verified the validity of the locality of reference property.

The average memory access time of a computer system can be improved considerably by use of a cache. If the hit ratio is high enough so that most of the time the CPU accesses the cache instead of main memory, the average access time is closer to the access time of the fast cache memory. For example, a computer with cache access time of 100 ns, a main memory access time of 1000 ns, and a hit ratio of 0.9 produces an average access time of 20 ns. This is a considerable improvement over a similar computer without a cache memory, whose access time is 100 ns.

The basic characteristics of cache memory is its fast access time. Therefore, very little or no time must be wasted when searching for words in the cache. The transformation of data from main memory to cache memory is referred to as a *mapping* process. Three types of mapping procedures are of practical interest when considering the organization of cache memory:

1. Associative mapping
2. Direct mapping
3. Set-associative mapping

Cache memory is defined as very high speed memory, *i.e.* placed between the CPU and main memory.

When the CPU needs to access memory the cache is examined, If the word is found in the cache, it is read from the fat memory. If the word addressed by the CPU is found in the cache, the main memory is accused to read the word. A block of words contain the one but accessed is then transferred from main memory to cache memory.

The performance of cache memory is measured in terms of bit ratio.

"If the CPU refers to memory and finds the word in cache, it is called cache bit"

If the word not found in cache, it is calls as cache miss"

"The ratio of number of bits divided by the total no. of bits plus total no. of miss is called bit ratio.

Let to is the cache access time and h = bit ratio and tm is the memory access time then it is the average access time is given by relation.

$$\bar{t} = h_{tc} + (1 - h)\,(t_c + t_m)$$

the bit ratio h always lies is the closed interval of 0 and 1, [0, 1].

Suppose the ratio of main memory access time to cache memory access time is γ then the expression for the efficiency of the system that implies, a cache can devised a

Efficiency, $\boxed{E = \frac{t_c}{t}}$

After putting values of to from above from, we get.

$$F = \frac{1}{1 + \gamma\,(1 - h)}$$

where, $g = g = \frac{t_m}{t_c}$

If the active portions of the program and data are placed in a fast small memory, the average memory access time can be reduced thus reducing the total execution time of the program. Such a fast small memory is referred to as cache memory. It is placed between the CPU and main memory. The cache memory access time is less than the access time of main memory by a factor of 5 to 10. The cache is the fastest component in the memory hierarchy and approaches the speed of CPU components.

The basic operation of the cache is as follows: When the CPU needs to access memory, the cache is examined. If the word is found in the cache it is read from the fast memory. If the word addressed by the CPU is not found in the cache, the main memory is accessed to read the word. A block of words containing the one just accessed is then transferred from main memory to cache memory.

The performance of cache memory is frequently measured in terms of a quantity called hit ratio. When the CPU refers to memory and finds the word in cache, it is said to produce a hit. In the word is not found in cache, it is in main memory and counts as a miss. The ratio of the number of hits divided by the total CPU references to memory (hits plus misses) is the hit ratio. The hit ratio is the best measured experimentally by running representative programs in the computer and measuring the number of hits and misses during a given interval of time.

The basic characteristic of cache memory is its fastest access time. The transformation of data from main memory to cache memory is referred to as a mapping process.

Three types of mapping procedures are of Practical interest when considering the organization of chace memory:

7.8.1 Associative Mapping

7.8.2 Direct Mapping

7.8.3 Set-associative Mapping

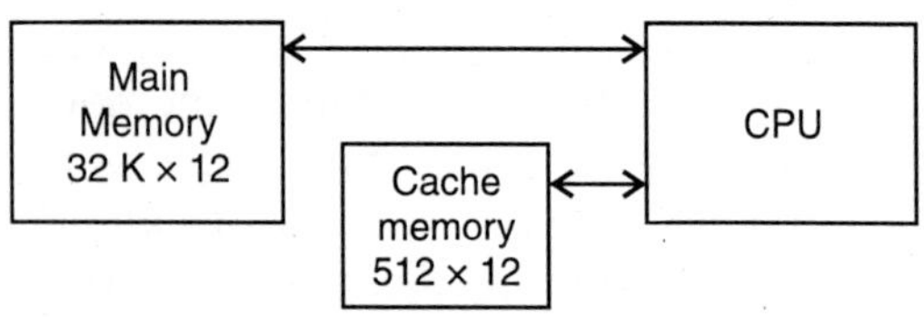

Fig. 7.15. Example of cache memory

The explanation of these three mapping procedures here is a specific example of a memory organization as shown in figure below. The main memory can store 32 k words of 12 bits each. The cache is capable of storing 512 of these words at any given time. For every words stored in the cache there is a duplicate copy in main memory. The CPU communicates with both of memories. It first sends a 15-bit address to cache. If there is a hit, the CPU accepts the 12 bit data from cache. If there is a miss, the CPU reads the word from main memory and the word is then transferred to cache.

Associative Mapping: The fastest and most flexible cache organization uses an associative memory. The associative memory stores both the address and consent (data) of he memory word. The address value of 15 bits is shown as a 5 digit octal number and its corresponding 12-bit word is shown as a four digit octal number. A CPU address of 15 bits is placed in the argument register and associative memory is searched for matching address. IF the address is found, the corresponding 12-bit data is read and sent to the CPU, if no match occurs, the main memory is accessed for the word. The address-data pair is then transferred to the associative cache memory. If the cache is full then address-data pair must be displaced using FIFO replacement policy.

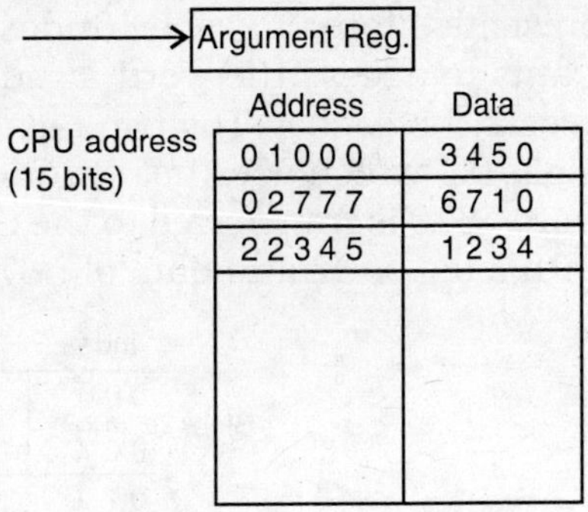

Fig. 7.16. Associative memory cache all nos. in octal

Direct Mapping: Associative memories are expensive compared to random-access memories because of the added logic associated with each cell. In this technique the CPU address of 15 bits is divided into two fields. The nine least significant bits constitute the index field and the remaining six bits form the tag field. The Fig. 7.18. shows that main memory needs an address that includes both tag and index bits. The number of bits in the index field is equal to the number of address bits required to access the cache memory.

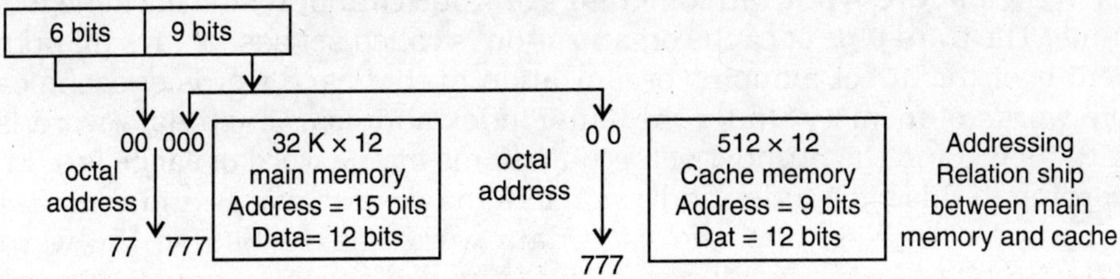

Fig. 7.17.

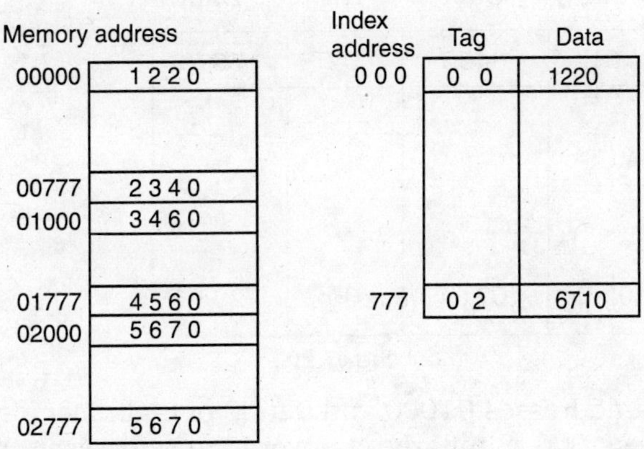

Fig. 7.18. (a) Main memory (b) cache memory

The disadvantage of direct mapping cache organization that hit can drop considerably if two or more words whose addresses have the same index but different tags are accessed repeatedly. However, this possibility is minimized by the fact that such words are relatively far apart in the address range consider the numerical example, the word at address zero is presently stored in cache (index 000, tag = 00, data = 1220). Suppose that the CPU now wants to access the word at address 02000. The index address is 000, so it is used to access the cache. The two tags are then compared. The cache tag is 00 but the address tag is 02, which does not produce a match. Therefore main memory is accessed and data word 5670 is transferred to the CPU. The cache word at index address 000 is then replaced with a tag of 02 and data of 5670.

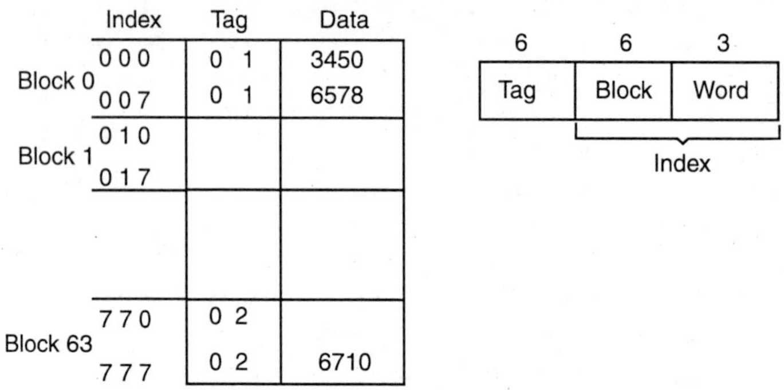

Fig. 7.19. Direct mapping cache with block size of 8 words

Associative Mapping: The disadvantage of direct mapping is that two words with the same index in their address but with different age value cannot reside in cache memory at the same time. The third type of cache organization is called set associative mapping, is an improvement over the direct mapping organization in that each words of cache can store two or more words of memory under the same index address. Each data word is stored together with its tag and the number of tag-data items in one word of cache is said to form a set. According to this each index address refers to two data words and their associated tags. Each tag requires six bits and each data word has 12 bits so the word length is 2 (6 + 12) = 36 bits. An index address of nine bits can accommodate 512 words. Thus the size of cache memory is 512 × 36

Index	Tag	Data	Tag	Data
000	01	3450	02	5670
77	02m	6710	00	2340

Fig. 7.20.

The words stored at addresses 01000 and 02000 of main memory are stored in cache memory at index address 000. Similarly, the words at addresses 02777 and 00777 are stored in cache at index address 777. When the CPU generates a memory request, the index value of the address is sued to access the cache. The tag field of the CPU address is then compared with both tags in the cache to determine if the match occurs, the comparison logic is done by an associative search of the tags in the set similar to an associative memory search. Thus the name set associative.

7.8.4 Writing into Cache

If the operation is a write there are two ways that the system can proceed.

Write through

The simplest and most commonly used procedure is to update main memory with every memory write operation, with cache memory being updated in parallel if it contains the word at the specified address. This is called the write through method. This method has the advantage that main memory always contains the same data as the cache.

Write Back

The second procedure is called the write back method. In this method only the cache location is updated during a write operation. The location is then marked by a flag so that later when the word is removed from the cache it is copied into main memory. The reason for the write back method is that during the time, a word residues in the cache, it may be updated several times.

EXAMPLE 7.3. Calculate average access time ($\bar{t}$), ratio of main memory access time to cache memory access time (g) and efficiency of a memory system whose parameters are

$$t_c = 160 \text{ ns.}$$
$$t_m = 960 \text{ ns}$$
$$h = 0.9$$

7.9 ANOTHER METHOD FOR ADDRESS MAPPING

There are three different mapping procedures while considering the organization of the cache memory. These mapping methods are:

1. Associative mapping
2. Direct mapping
3. Set-associative mapping

***(i)* Associative Mapping**

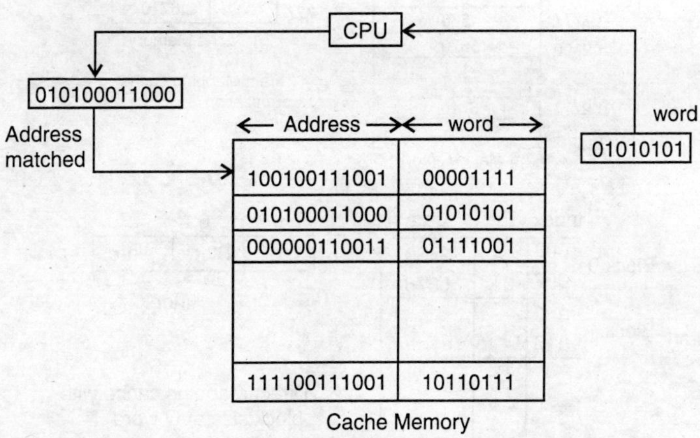

Fig. 7.21. Block diagram showing associative memory

***(ii)* Direct Mapping:** In the general case, there are 2^k words in cache memory ad 2^n words in main memory. The n-bit memory address and divided into two fields: k bits from the index field and n-k bits for the tag fields. The direct mapping cache organization uses the n-bit address to access the main memory and the k-bit index to access the cache

organization uses the n-bit address to access the main memory and the k-bit index to access

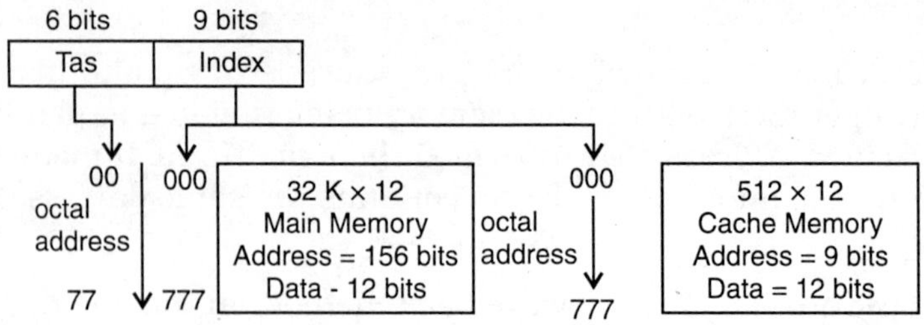

Fig. 7.22. Addressing relationships between main and cache memory

To see how the direct mapping organization operates, consider the numerical example. The word at address zero is presently stride in the cache (index = 000, tag = 00, data = 1220). suppose that the CPU now matix to access the word at address 00 000. The index address is 000, so it is used to access the cache.

The two tags are then compared. The cache tag is 00 but the address tag is 02, which absent produce a match. Therefore the main memory is accessed and the data word 5670 is transferred to the CPU. The cache word at index address 000 is then replaced with a tag of 02 and data of 5670.

The direct mapping example just describled uses a block size of one word. The same organization but using a block size of a 8 words is shown is IInd figure.

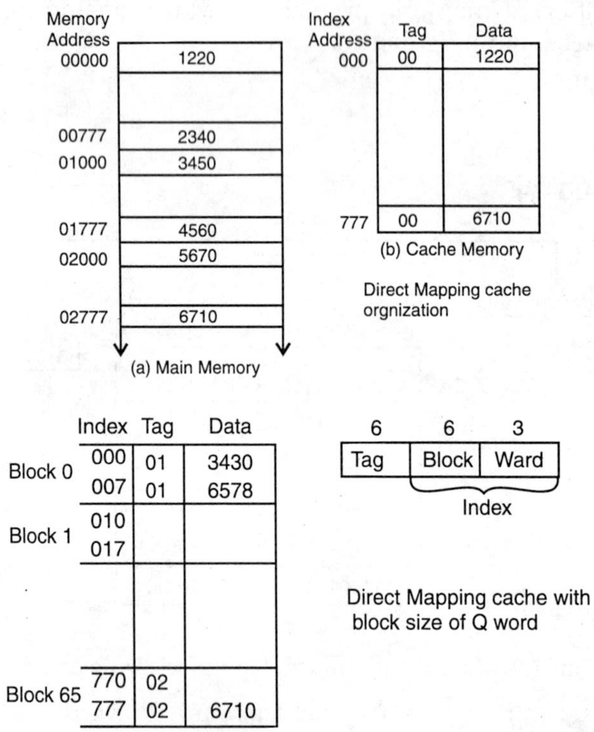

Fig. 7.23. Direct Mapping cache with block size of words

(3) Set-Associative Mapping: It was mentioned previously that the disadvantage of direct mapping is that two words with the same index in their address but with different tag values can not reside in cache memory at the dame time.

A third type of cache organization called set-associative mapping, is an improvement over the direct mapping organization in that, each word of cache can store two or more words of memory under the same index address. Each data word is stored together with its tag and the number of tag data times in one word of cache is said to form a set.

Example of set associative cache organization of for a set of two is shown in figure. Each index address refers to two data words and their associated tags . Each tag require six bits and each data word has 12 bits.

So, word length = 2(6 + 12) = 36 bits. An index address of nine bits can acuminated 512. Words So, size of cache memory is 512 × 36. It can accommodate 1024 words of main memory since each word of cache contains two data words.

Index	Tag	Data	Tag	Data
0000110010	00	10101010	01	10010110
00001111011	01	11000011	10	11000011

Two way set associative mapping

Note: 1 way mapping → Tag + Data

2 way mapping → 2 Tag + 2 Data

K way mapping → K Tag + k-Data

7.10. REPLACEMENT ALGORITHM

1. LRU (least Recently Used)
2. FIFO (First In First Out)
3. Most optimal Algorithm.
   → Updating method
   → Write Through Policy
   → Write Back Policy

1. Memory table for mapping rurality address virtual.

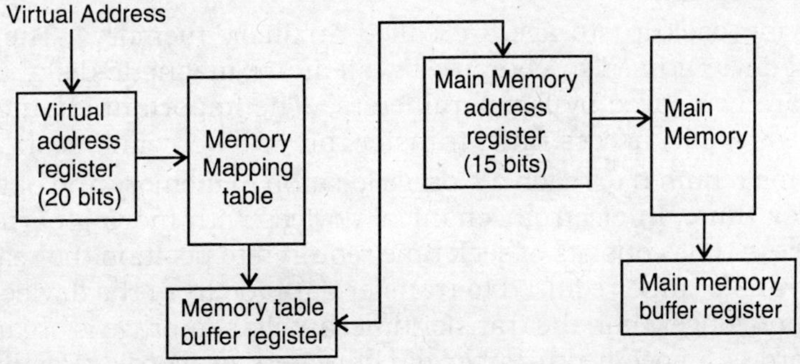

Fig. 7.24.

2. Address space and memory space fits into groups of 1 K words.

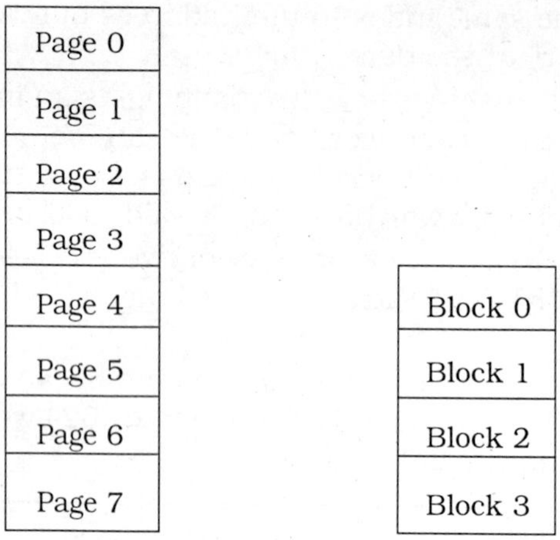

Address space $N = QK = 2^{13}$ Memory space $M = 4K = 2^{12}$

Address space and memory space splits into groups of 1 K words.

3.

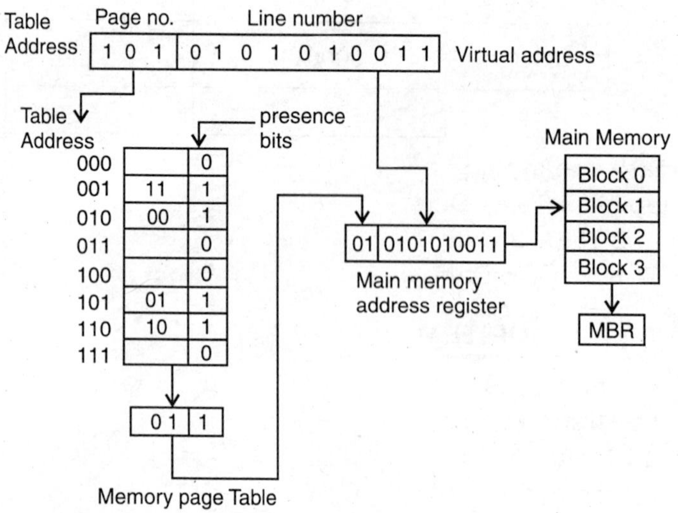

Fig. 7.25. Memory Table in a paged system

7.11 BACKUP STORAGE

Devices that provide backup storage are called auxiliary memory. The most common auxiliary memory devices used in computer system are magnetic disks and tapes. The physical devices are compared by few parameters. The important characteristics of any device are its access mode, access time, transfer rate, capacity and cost.

The average time required to reach a storage location in memory and obtain its contents is called the access time. In electromechanical devices with moving parts such as disks and tapes, the access time consists of seek time required to position the read-write head to a location and a transfer time required to transfer data too from the device. Because seek time is usually much longer than the transfer time, auxiliary storage is organized in records or blocks. A record is a specified number of characters or words. Reading or writing is done on entire records. The transfer rate is the number of characters or words that the device can transfer per second after it has been positioned at the running of the record.

The auxiliary memory devices used in a computer system are magnetic disks and tapes. The important features of any devices are its access mode, access time, transfer rate, capacity and cost.

The average time to reach storage location in memory and obtain its contents is called the access time.

In electromechanical devices

Access time = self time (time required to position on the location + Transfer time (time requires to transfer data to and from any device.

Let and bytes are stored per sector p sectors are there, per track, tracks per surface 8 m surfaces. Then the capacity of disk will be defined as →

$$\boxed{\text{Capacity} = m \times t \times p \times s \text{ bytes}}$$

If d = diameter of disk than dinette of recording is

$$\boxed{\text{Density} = \left(\frac{p \times s}{\pi \times d}\right) \text{bytes/inch}}$$

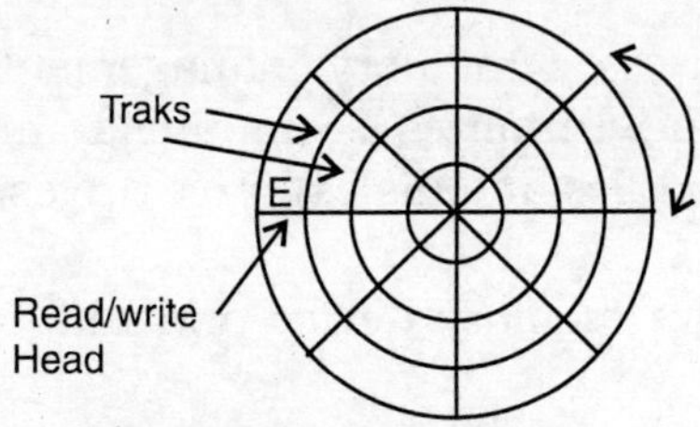

7.12 MEMORY MANAGEMENT HARDWARE

A Memory management system is a collection of hardware and software procedures for managing the various programs resitting in memory.

A segment is a set of logically related instructions Associated with a given name. The address generated by segmented program is called a logical address.

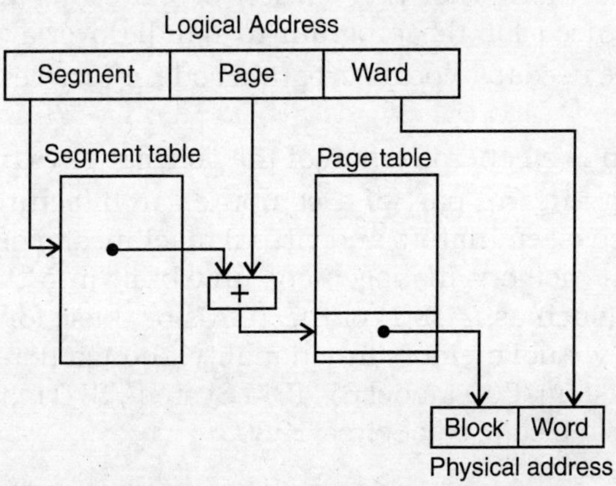

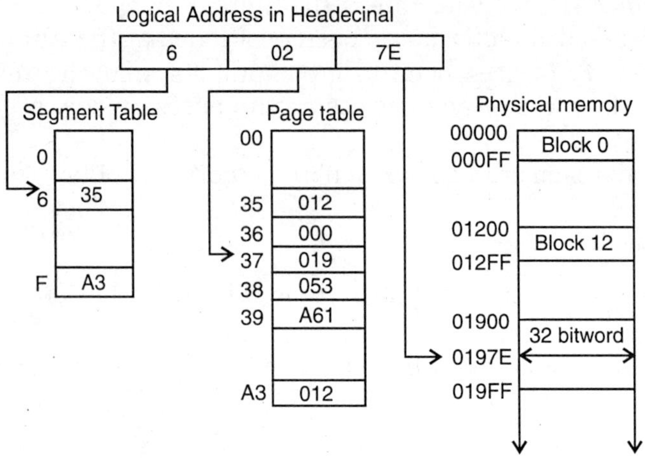

Fig. 7.26. Segment and page table mapping

7.13 VIRTUAL MEMORY

The program thinks it has a large range of contiguous addresses, but in reality the parts it is currently using are scattered around RAM, and the inactive parts are saved in a disk file.

In computing, virtual memory is a memory management technique developed for multitasking kernels. This technique virtualizes a computer architecture's various hardware memory devices (such as RAM modules and disk storage drives), allowing a program to be designed as though:

- there is only one hardware memory device and this "virtual" device acts like a RAM module.
- the program has, by default, sole access to this virtual RAM module as the basis for a contiguous working memory (an address space).

Systems that employ virtual memory:

- use hardware memory more efficiently than systems without virtual memory.
- make the programming of applications easier by:
  - hiding fragmentation.
  - delegating to the kernel the burden of managing the memory hierarchy; there is no need for the program to handle overlays explicitly.
  - obviating the need to relocate program code or to access memory with relative addressing.

Memory virtualization is a generalization of the concept of virtual memory.

Virtual memory is an integral part of a computer architecture; all implementations (excluding[dubious - discuss] emulators and virtual machines) require hardware support, typically in the form of a memory management unit built into the CPU. Consequently, older operating systems (such as DOS[1] of the 1980s or those for the mainframes of the 1960s) generally have no virtual memory functionality, though notable exceptions include the Atlas, B5000, IBM System/360 Model 67, IBM System/370 mainframe systems of the early 1970s, and the Apple Lisa project circa 1980.

Embedded systems and other special-purpose computer systems which require very fast and/or very consistent response times may opt not to use virtual memory due to decreased determinism; virtual memory systems trigger unpredictable interrupts that may produce unwanted "jitter" during input/output operations. This is because embedded hardware costs are often kept low by implementing all such operations with software (a technique called bit-banging) rather than with dedicated hardware. In any case, embedded systems usually have little use for multitasking features or complicated memory hierarchies.

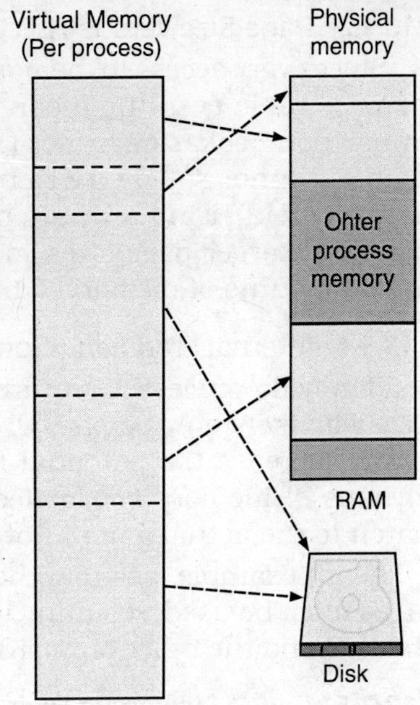

Fig. 7.27.

Virtual memory was introduced to the x86 architecture with the protected mode of the Intel 80286 processor; however, its segment swapping technique was discovered to scale poorly to larger segment sizes. The Intel 80386 introduced support for paging underneath the existing segmentation layer, and the page fault exception could be chained with other exceptions without causing a double fault. The 80386 did not have a Translation Lookaside Buffer (TLB), which made loading segment descriptors expensive and caused operating system designers to rely strictly on paging rather than a combination of paging and segmentation.

7.13.1 Paged Virtual Memory

In the context of computer virtual memory, a **page**, **memory page**, or **virtual page** is a fixed-length block of main memory that is contiguous in both physical memory addressing and virtual memory addressing. A page is usually the smallest unit of data for the following :

- memory allocation performed by the operating system for a program; and
- transfer between main memory and any other auxiliary store, such as hard disk drive.

Virtual memory abstraction allows a page that does not currently reside in main memory to be addressed and used. If a program tries to access a location in such page, an exception called a page fault is generated. The hardware or operating system is notified and loads the required page from auxiliary store automatically. A program addressing the memory has no knowledge of a page fault or a process following it. Thus a program can address more (virtual) RAM than physically exists in the computer.

A transfer of pages between main memory and an auxiliary store, such as hard disk drive, is referred to as paging or swapping.[1]

7.13.1.1. Page Size Trade-Off

Page size is usually determined by processor architecture. Traditionally, pages in a system had uniform size, for example 4096 bytes. However, processor designs often allow two or more, sometimes simultaneous, page sizes due to the benefits and penalties. There are several points that can factor into choosing the best page size.

7.13.1.2. Page Size Versus Page Table Size

A system with a smaller page size uses more pages, requiring a page table that occupies more space. For example, if a 2^{32} virtual address space is mapped to 4KB (2^{12} bytes) pages, the number of virtual pages is 2^{20} (= $2^{32}/2^{12}$). However, if the page size is increased to 32KB (2^{15} bytes), only 2^{17} pages are required.

7.13.1.3. Page Size Versus TLB Usage

Since every access to memory must be mapped from virtual to physical address, reading the page table every time can be quite costly. Therefore, a very fast kind of cache, the Translation Lookaside Buffer (TLB), is often used. The TLB is typically of limited size, and when it cannot satisfy a given request (a TLB miss) the page tables must be searched manually (either in hardware or software, depending on the architecture) for the correct mapping. Larger page sizes mean that a TLB cache of the same size can keep track of larger amounts of memory, which avoids the costly TLB misses.

7.13.1.4. Internal Fragmentation of Pages

Rarely do processes require the use of an exact number of pages. As a result, the last page will likely only be partially full, wasting some amount of memory. Larger page sizes clearly increase the potential for wasted memory this way, as more potentially unused portions of memory are loaded into main memory. Smaller page sizes ensure a closer match to the actual amount of memory required in an allocation.

As an example, assume the page size is 1024KB. If a process allocates 1025KB, two pages must be used, resulting in 1023KB of unused space (where one page fully consumes 1024KB and the other only 1KB).

7.13.1.5. Page Size Versus Disk Access

When transferring from disk, much of the delay is caused by seek time, the time it takes to correctly position the read/write heads above the disk platters. Because of this, large sequential transfers are more efficient than several smaller transfers. Transferring the same amount of data from disk to memory often requires less time with larger pages than with smaller pages.

7.13.1.6. Determining the Page Size in a Program

Most operating systems allow programs to discover the page size at runtime. This allows programs to use memory more efficiently by aligning allocations to this size and reducing overall internal fragmentation of pages.

Almost all implementations of virtual memory divide the virtual address space of an application program into pages; a page is a block of contiguous virtual memory addresses. Pages are usually at least 4 KiB (4×1024 bytes) in size, and systems with large virtual address ranges or large amounts of real memory generally use larger page sizes.

7.13.1.7. Page Tables

Almost all implementations use page tables to translate the virtual addresses seen by the application program into physical addresses (also referred to as "real addresses") used by the hardware to process instructions. Each entry in the page table contains a mapping for a virtual page to either the real memory address at which the page is stored, or an indicator that the page is currently held in a disk file. (Although most do, some systems may not support use of a disk file for virtual memory).

Systems can have one page table for the whole system, a separate page table for each application, a separate page table for each segment, a tree of page tables for large segments or some combination of these. If there is only one, different applications which are running at the same time share a single virtual address space, i.e. they use different parts of a single range of virtual addresses. Systems which use multiple page tables provide multiple virtual address spaces-concurrent applications think they are using the same range of virtual addresses, but their separate page tables redirect to different real addresses.

7.13.1.8. Dynamic Address Translation

If, while executing an instruction, a CPU fetches an instruction located at a particular virtual address, or fetches data from a specific virtual address or stores data to a particular virtual address, the virtual address must be translated to the corresponding physical address. This is done by a hardware component, sometimes called a memory management unit, which looks up the real address (from the page table) corresponding to a virtual address and passes the real address to the parts of the CPU which execute instructions. If the page tables indicate that the virtual memory page is not currently in real memory, the hardware raises a page fault exception (special internal signal) which invokes the paging supervisor component of the operating system (see below).

7.13.1.9. Paging Supervisor

This part of the operating system creates and manages page tables. If the address translation hardware raises a page fault exception, the paging supervisor accesses secondary storage, returns the page containing the required virtual address, updates the page tables to reflect the physical location of the virtual address and finally tells the dynamic address translation mechanism to restart the request. When all physical memory is already in use as is typical, the paging supervisor must free an area in primary storage to hold the swapped-in page. Freeing memory minimally requires updating the paging table to say that the page is in secondary storage. The supervisor saves time by not re-swapping pages that are already present in secondary storage.

Paging supervisors generally choose the page that has been least recently used, guessing that such pages are less likely to be requested. Every time the dynamic address translation hardware matches a virtual address with a real physical memory address, it time-stamps the page table entry for that virtual address.

7.13.1.10. Permanently Resident Pages

Operating systems have memory areas that are "pinned" - i.e., never swapped to secondary storage. For example:

- Interrupt mechanisms generally rely on an array of pointers to their handlers (input/output completion, timer event, program error, page fault, etc.). If the pages containing these pointers or the code that they invoke were pageable, interrupt-handling would become far more complex and time-consuming; particularly in the case of page fault interrupts.
- Page table pages are usually not pageable.[dubious - discuss]
- Data buffers that are accessed directly, for example by peripheral devices that use direct memory access (DMA) or by input/output channels. Usually such devices and the buses (connection paths) to which they are attached use physical memory addresses rather than virtual memory addresses. Even on buses with an IOMMU, which is a special memory management unit that can translate virtual addresses used on an input/output bus to physical addresses, the transfer cannot be stopped if a page fault occurs and then restarted when the page fault has been processed. So pages containing locations to which or from which a peripheral device is transferring data are either permanently pinned down or pinned down while the transfer is in progress.
- Timing-dependent kernel/application areas cannot tolerate the varying response time caused by paging. In particular the paging supervisor code or drivers for secondary storage devices must not be swapped out.

7.13.1.11. Virtual-Real Operation

In OS/VS1, OS/VS2 (SVS), MVS, z/OS, and similar OSes, some parts of systems memory are managed in virtual=real mode, where every virtual address corresponds to a real address. Those are:

- interrupt mechanisms
- paging supervisor and page tables, in older systems
- application programs that use non-standard methods of managing I/O, e.g., reading into an active channel program.

IBM's z/OS has 3 modes, V=V (virtual=virtual; fully pageable), V=R and V=F (virtual = fixed, i.e. "pinned down" but with DAT operating).[7][page needed].

7.13.2. Segmented virtual memory

Some systems, such as Burroughs B5500,[8] do not use paging to implement virtual memory. Instead, they use segmentation, that divide virtual address spaces into variable-length segments. A virtual address consists of a segment number and an offset within the segment.

Notably, the Intel 80286 supports a similar segmentation scheme as an option, but it is little used.

It is possible to combine segmentation and paging, dividing each segment into pages. In systems that combine them, such as Multics, IBM System/38 and IBM System i machines, virtual memory is usually implemented with paging, with segmentation providing memory protection.[9][10][11]

In the Intel 80386 and later IA-32 processors, the segments reside in a 32-bit linear, paged address space. Segments can be moved in and out of that space, and pages in that space can "page" in and out of main memory, providing two levels of virtual memory; however, few if any operating systems do so. Instead, they use only paging. Early x86 virtualization solutions (non-hardware-assisted) combined paging and segmentation because x86 paging offers only two protection domains, whereas a VMM/guest OS/guest applications stack (typically running on rings 0/1/3) needs three memory protection domains.[12]:22

The difference between paging and segmentation systems is not only about memory division. In Multics, System/38 and Prime machines, segmentation is visible to user processes, as part of memory model semantics. In other words, instead of memory that looks like a single large vector, memory is structured into multiple spaces.

This difference has important consequences; a segment isn't just a "page with a variable length", or a simple way to lengthen the address space (as in Intel 80286). In Multics, segmentation that can provide a single-level memory model, in which there is no differentiation between "process memory" and "file system". I.e., a process's active address space for both code and data consists of only a list of segments (files) which are mapped into the process's potential address space.[13]

This is not the same as the mechanisms provided in TENEX and TOPS-20 (PMAP), modern Unix-like systems (mmap), and Win32 systems (MapViewOfFile), because inter-file pointers don't work when mapping files into semi-arbitrary places. In Multics, a file (or a segment from a multi-segment file) is mapped into a segment in the address space, so files are always mapped at a segment boundary. A file's linkage section can contain pointers that are "unsnapped links"; an attempt to load the pointer into a register or make an indirect reference through it causes a trap (a field in the pointer can have a value that specifies that an attempt to dereference it should cause a trap). The unresolved pointer contains an indication of the name of the segment to which the pointer refers, and an offset within the segment; the handler for the trap will "snap the link" the pointer by mapping the segment into the address space, putting the segment number into the pointer, and changing the tag field in the pointer so that it no longer causes the trap), and return to the code where the trap occurred, re-executing the instruction that caused the trap.[14] This eliminates the need for a linker completely.[2] This also works when different processes map the same file into different places in their private address spaces.[15]

7.13.3. AvoidingThrashing

When paging is used a potential problem called "thrashing" can occur, in which the computer spends a disproportionate amount of its capacity swapping pages to and from a backing store and therefore performs useful work more slowly. Adding real memory is the simplest response, although improving application design, scheduling, and memory usage can help.

Virtual memory is a concept used some large computer systems that permit user to construct programs as though a large memory space were available, equal to the totality of auxiliary memory.

Address space → And ← Memory space →

An address used by a programmer will be called a virtual address and the set of such address called the address space. An address in main memory is called a location or physical address. The set of such locations is called the memory space.

Consider a computer with a main memory capacity of 32 K words (K = 1024). 14 bits as needed to specify a physical address in available auxiliary memory for storing 2^{20} = 1024 k words. Thus auxiliary memory has a capacity for storing information equivalent to the capacity of 32 main memories. If N is address space and M is memory address then for this example,

$$M = 1024 \text{ K, and } M = 32 \text{ K}$$

7.14 AUXILIARY MEMORIES

Data and programs are stored in main memory, as **random access memory (RAM)**, before and after processing by the **central processing unit (CPU)**. However, RAM is **volatile** -its contents disappear when a computer's power is turned off. So what does one do with data that will be reused, stored on a long-term basis, or is simply too large to fit into the main memory of a computer? Mechanical storage devices, called secondary storage or external storage, are used to store data externally. Secondary storage is non-volatile-that is, data and programs are permanent.

Types of Storage Devices

There are many secondary storage devices, including magnetic drums, magnetic tapes, magnetic disks, and optical disks. These devices vary with respect to their versatility, durability, capacity, and speed.

7.14.1. Magnetic Drums

These are very early high-speed, direct access storage devices used in the 1950s and 1960s. The magnetic drum is a metal cylinder coated with a sensitive magnetic material. The cylinder has tracks around its circumference. Each track has its own read/write head, and data are stored as magnetic spots on the tracks. Like the magnetic disk, which was introduced later, the read/write head can move quickly to any track, providing direct access to the data stored on the drum.

7.14.2. Magnetic Tape

This is one of the oldest secondary storage devices. It was first used for storing data in the early 1950s. At that time, the tape was made of a flexible metal and was stored on reels. The metal tape was plated with a thin film of iron, which allowed data to be stored as a series of small, magnetized spots. Although the tape provided a compact form of storage, it was extremely heavy and not universally accepted. It was not until a very thin, flexible material called **mylar** was developed that tape processing gained wide acceptance. This plastic mylar was coated with an iron oxide that could be magnetized to store data. In the 1950s and 1960s, magnetic tape was the primary means for storing large amounts of data.

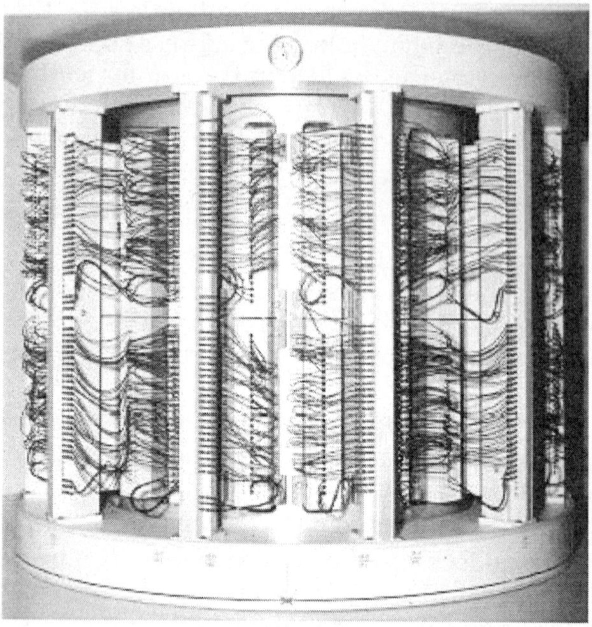

Fig. 7.28.

Data are stored onmagnetic tape in columns. Each **byte** of data (eight **bits**) utilizes one column across the width of the tape. Data are stored on the magnetic tape at different **densities**. Low density is 1,600 bytes per inch (bpi). Densities of 6,250 bpi and greater were common in the 1980s.

One type of computer memory storage device was the magnetic drum, popular in the 1950s and 1960s. As it quickly rotates, the drum records bits of electronic signal information.

The old reel-to-reel magnetic tapes are being replaced by tape cartridges, which are used to back up or archive data. A tape backup is a copy of the data used to restore lost data. If one loses the data on a hard disk, he can copy the data from the tape backup. Tape cartridges used on microcomputers are similar to audiocassettes and can hold up to 35 gigabytes (GB) of data. There are several different types of tape cartridges, including quarter-inch cartridge (QIC), digital audio tape (DAT), digital linear tape (DLT), Travan, and advanced digital recording (ADR).

Magnetic tapes are an inexpensive and relatively stable way to store large volumes of data. Tapes can be used repeatedly, although time and environmental factors such as heat and humidity do affect them. The principle disadvantage of magnetic tape is that it stores data sequentially-that is, one record is stored right after another. Retrieving data from the tape is slow since the tape must be read from the beginning up to the location of the desired record. Thus, magnetic tapes are not a good choice when one needs to find information rapidly.

7.14.3. Magnetic Disk

Currently, the most widely used secondary storage device is the magnetic disk. There are two kinds of magnetic disks: hard disks and floppy disks. Hard disks are thin, metallic platters developed in the 1950s. Each hard disk contains one or, more commonly, a series of platters that rotate on a shaft at a high rate of speed. The platters have a top and bottom surface where data can be recorded as magnetic spots. The platters have **concentric** circles called tracks where the data are actually stored. Access arms containing read/write heads move across the spinning platters to read or write data on the tracks. There is a read/write head for each platter surface. These read/write heads float on a cushion of air and do not actually touch the surface of the disk. When reading data from the disk, the read head senses the magnetic spots on the surface and transfers the data to main memory. When writing, the data are transferred from main memory and are stored as magnetic spots on the recording surface.

The speed of access to data on a disk is based on both the rotational speed of the disk and the speed of the access arm. The read/write heads must position themselves over the platter at the proper address and the disk pack must rotate until the proper data are located. The time it takes the read/write heads to position themselves to the correct address is referred to as access motion time. Rotational delay time is the time needed for the disk to rotate the data under the read/write head.

The read/write heads move horizontally (from left to right) to any of 200 positions, called cylinders, on the platter surface. The cylinder represents the circular tracks on the same vertical line within the disk pack. To find data on the disk, an address is used. This address consists of a cylinder number, the recording surface number, and the data record number. The computer's operating system figures out how to fit data into the fixed spaces on the disk.

The micro-computer utilizes a hard disk, but it also uses floppy disks, also called floppy diskettes. Floppy disks are the second kind of magnetic disk and are removable. The two common standard sizes of floppy disks are 8.9 centimeters (3.5 inches) and 13.3 centimeters (5.25 inches). It is difficult today to find a microcomputer that uses the 5.25-inch floppy disk. Floppy disks are made up of one platter constructed from polyester film rather than metal. This polyester film is covered with a magnetic coating. Data are stored on the floppy disk in much the same manner as on a hard disk-in the form of magnetic spots on tracks. However, floppy disks use sectors to store data rather than cylinders. In this method of storing data, the disk is divided into pie-shaped pieces called sectors. Each sector is assigned a unique address to locate data. Sectors are created on a floppy disk by formatting it. Floppy disks have a much lower data storage capacity (usually 1.44 MB for a 3.5-inch floppy disk) and a much slower data access rate than hard disks.

As software packages such as Microsoft Office became popular, the need for larger data capacity for items such as text and graphics in a portable form became necessary. The zip drive and zip disk were introduced in the 1990s as a relatively inexpensive large-storage-capacity floppy disk. The zip disk is 8.9 centimeters (3.5 inches), removable, and provides about 100 megabytes (MB) of data storage. In late 1998, a 250 MB version of the zip disk was introduced. While this zip disk has double the storage capacity of its predecessor, it is still only 8.9 centimeters (3.5 inches) in size!

The main advantages of using a magnetic disk as a secondary storage device are its speed and direct access capability. Data can be easily and rapidlyread, written, or accessed. Floppy disks provide the added advantage of portability. Disadvantages of using a magnetic disk as a secondary storage device include cost, environmental factors, user misuse and abuse, head crashes, and update problems.

The magnetic disk is more expensive than magnetic tape (DAT). The February 1998 CompUSA catalog reported that DAT cost $49.95 and stored 10,000 MB. This is a $.005 cost per megabyte. An 18 GB hard drive cost $230.00 or $.08 per megabyte. A 3.5 inch diskette cost $.50 and stored 1.4MB. This is a cost of $.35 per megabyte. A Zip Plus Drive cost $100.00 and stored 100MG to 250MG. This is a cost of $1.00 per megabyte. It was reported by Dataquest, a research firm, that the average cost of data storage across various media was 15 cents per megabyte in the year 2000. This cost decreased to 3 cents per megabyte in 2002, as noted by Effy Oz in Management Information Systems.

Magnetic disks, both hard and floppy, are also susceptible to environmental factors such as dust, dirt, and smoke. Any of these factors will cause a hard disk to fail. Because of this, hard disks are sealed. Floppy disks are also vulnerable to environmental factors. Also, because of their portability, the floppy disk is vulnerable to misuse or abuse by users.

Head crashes can occur with any magnetic disk technology. This is when the read/write head touches the surface of the disk platter, destroying it and all of the stored data. Head crashes are normally caused by misalignment of the platter and the read/write head. A head crash renders a magnetic disk unusable. Another disadvantage of any magnetic disk is that when updating data, the old data are written over, destroying them instantly and permanently. Unless proper precautions are taken, data may be written over by mistake.

7.14.4. Optical Disks

These are the newest secondary storage devices. Originally optical disks were called optical disk-read-only memory (OD-ROM) and are now called compact disk-read only memory (CD-ROM). Data are not recorded on optical disks magnetically, but with a laser device that burns microscopic holes on the surface of the disk. Binary information (0s and 1s) is encoded by the length of these bumps and the space between them. Optical disks can store much more data than floppy disks. Data can be stored in the form of text as well as pictures, sound, and full-motion video. The disks are not as sensitive to dust, dirt, and smoke as magnetic disks are, and they are portable.

CD-ROM is read-only storage. No new data can be written to it. Once recorded, the CD can be read an indefinite number of times. This is commonly referred to as write once/read many (WORM). CD-ROM has been used for storage of large financial, legal, educational, or demographic databases, and by the music industry. Encyclopedias and multimedia applications are also stored using CD-ROM technology.

Individuals and organizations can now record their own CD-ROMs using compact disk-recordable (CD-R) technology. CD-ReWritable (CDRW) is a newer technology that has been developed to allow users to create rewritable optical disks. Currently CD-RW has slower access speeds and is more expensive than magnetic storage devices.

Another optical storage medium is the digital video disk or digital versatile disk (DVD). This optical disk is the same size as a CD-ROM, but has much higher storage capacity. DVDs can store large amounts of data, video, graphics, digitized text, and audio, and are portable. It is likely that DVD will eventually replace the CD-ROM.

7.15 ADDRESS SPACE AND MEMORY SPACE

An address used by a programmer will be called a virtual address and set of such addresses the address space. An address in main memory is called a location or physical address. The set of such locations is called the memory space. Thus the address space is the set of addresses generated by programs as they reference instructions and data the address space 1s allowed to be larger that memory space in computers with virtual memory.

EXAMPLE 7.4. Supposse a cache is 10 times faster than main memory, and suppose that the cache can be used 90% of the time. How much speed up do we gain by using the cache?

Solution: Speed up = $\dfrac{1}{(1 - \text{\% of time cache can be used}) + \left[\dfrac{\text{\% of time cache can be used}}{\text{speed up using cache}}\right]}$

$$= \frac{1}{(1-0.9)+\dfrac{0.9}{10}} = \frac{1}{0.19} = 5.3.$$

Hence, we get a speed up from the cache of about 5.3 times.

EXAMPLE 7.5. Assume we have a machine where the CPI (cycles per instruction) is 2.0, when all memory accesses hit in the cache. The only data accesses aer loads aer stores, and these total 40% of the instuctions. If the miss penalty is 25 clock cycles and the miss rate is 2%, how much faster would the machine be if all instructions were cache hits?

Solution: First compute the performance for the machine that always hits:

$$\begin{aligned} \text{CPU execution}_{\text{time}} &= (\text{CPU clock cycles + Memory stall cycles}).\ \text{clock cycles} \\ &= (\text{IC} \times \text{CPI} + 0) \times \text{clock cycle [where IC} \equiv \text{Instruction cycle]} \\ &= \text{IC} \times 2.0 \times \text{clock cycle} \end{aligned}$$

Now for the machine with the real cache, first we compute memory stall cycles:

$$\begin{aligned} \text{Memory stall cycles} &= \text{IC} \times \text{Mem reference per instruction miss rate} \times \text{Miss penalty} \\ &= \text{IC} \times (1 + 0.4) \times 0.2 \times 25 = \text{IC} \times 0.7 \end{aligned}$$

Where the middle term (1 to 4) represents one instruction access and 0.4 data acceses per instruction. The total performance is thus

$$\text{(CPU execution time) Cache} = (\text{IC} \times 2.0 + \text{IC} \times 0.7) \times \text{clock cycle} = 2.7\ .\ \text{IC}\ .\ \text{clock cycle}$$

$$\therefore \frac{(\text{CPU execution time})_{\text{Cache}}}{\text{CPU execution time}} = \frac{2.7 \times \text{IC} \times \text{clock cycle}}{2.0 \times \text{IC} \times \text{clock cycle}} = 1.35$$

Hence, machine with no cache misses is 1.35 times faster.

EXAMPLE 7.6. For a machine, assume the cache miss penalty is 50 clock cycles, and all instructions normally take 2.0 clock cycles. Assume the miss rate is 2%, and there is an average of 1.33 memory references per instruction. What is the impact on performance when behaviour of the cache is included?

Solution:

$$\text{CPU time} = \text{IC} \times \left(\text{CPI}_{\text{execution}} + \frac{\text{Memory clock cycle}}{\text{Instruction}}\right) \times \text{clock cycle time}$$

Therefore, the performance, including cache misses, is

$$\begin{aligned} \text{(CPU time) with}_{\text{cache}} &= \text{IC} \times (2.0 + (1.33 \times 2\% \times 50) \times \text{clock cycle time} \\ &= \text{IC} \times 3.33 \times \text{clock cycle time} \end{aligned}$$

The clock time and instruction count (IC) are the same, with or without a cache, so CPU time increases with CPI from 2.0 for a "perfect cache" to 3.33 with a cache that can miss. Hence, including the memory hierarchy in the CPI calculations stretches the CPU time by a factor of 1.67. Without any memory hierarchy at all the CPI would increase to (2.0 + 50.1.33) or 68.5.

EXAMPLE 7.7. What is the impact of two different cache organizations on the performance of a CPU? Assume that the CPI with a perfect cache is 2.0 and the clock cycle time is 2 ns, that there are 1.3 memory reference per instruction, and that the size of both cache is 64 KB and both have a block rize of 32 bytes. One cache is direct mapped and the other is two way set associative. Assume CPU clock cycle time with set associative cache is stretched 1.10 times than the other. The cache miss penalty is 70 ns for either cache organization. Calculate the average memory access time and CPU performance.

Assume the hit time is one clock cycle. Also the miss rate of a direct-mapped 64 KB cache is 1.4% and the miss rate for a two-way set-associative cache of the same size is 1.0%.

Solution: Average Memory access time, T_{av} = Hit time + Miss rate × Miss penalty

$$T_{av\ (1\ way)} = 2.0 + (0.014 \times 70) = 2.98 \text{ ns}$$
$$T_{av\ (2\ way)} = 2.0 \times 1.10 + (0.010 \times 70) = 2.90 \text{ ns}$$

$$\text{CPU time} = \text{IC} \times \left(\text{CPI}_{\text{execution}} + \frac{\text{Misses}}{\text{Instruction}} \times \text{Miss Penalty} \right) \times \text{clock cycle time}$$

$$= \text{IC} \times \left[\left(\text{CPI}_{\text{Ex}} \times \text{clock cycle time} + \left(\frac{\text{Memory Accesses}}{\text{Instruction}} \right) \times \text{Miss rate} \times \text{Miss penalty} \times \text{Clock cycle time} \right) \right]$$

Substituting 70 ns for (Miss Penalty × Clock cycle time), we get

$$\text{CPU time (1 – way)} = \text{IC} \times \{2.0 \times 2 + (1.3 \times 0.014 \times 70)\} = 5.27 \times \text{IC}$$
$$\text{CPU time (2 – way)} = \text{IC} \times \{2.0 \times 2 \times 1.10 + (1.31 \times 0.010 \times 70)\} = 5.31 \times \text{IC}$$

$$\therefore \text{Relative Performance} = \frac{\text{CPU time}_{2\ \text{way}}}{\text{CPU time}_{1\ \text{way}}} = \frac{5.31 \times \text{IC}}{5.27 \times \text{Instruction Count}} = \frac{5.31}{5.27} = \mathbf{1.01}.$$

SOLVED PROBLEMS

Q.1. Define Multiprogramming.

Solution: Multiprogramming.

Many operating systems are designed to enable the CPU to process a number of independent programs concurrently, this concept, called multiprogramming refers to the existance of two or more programs in different parts of the memory hierarchy at the same time.

Q.2. What is associate memory? Make the diagram of associative memory.

Solution: The time required to find an item stored in memory can be reduced considerably if stored data can be identified fo access by the content of the data itself rather than by an address. A memory unit accessed by content is caled an associate memory or content Addressable Memory (CAM). This type of memory is accessed simultaneously in parallel on the basis of data content rather than by specific address or location. When a world is written in an associative memory, no address is given. The memory is capable of finding an empty unused location to store the word. When a word is to be read from an associative memory, the content of the word, or part of the word, is specified.

An associative memory is more expensive than RAM, because each cell must have storage capability as well as logic circuits for matching its content wtih an external argument.

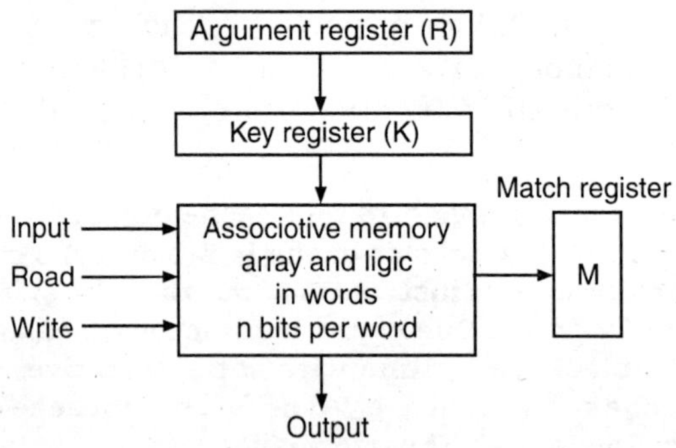

Fig. 7.29. Block diagram of associative memory

Q.3. Define:

(a) Locality of reference

(*b*) **Hit ratio**

(*c*) **Mapping.**

Solution: (*a*) Location of reference: Analysis of a large number of typical programs has shown that the references to memory at any given interval of time tend to be confined within a few localized areas in memory. This phenomenon is known as the property of locality of reference.

(*b*) Hit ratio: The performance of cache memory is frequently measured in terms of a quantity called hit ratio. When the CPU refers to memory and finds the words in cache, it is said to produce hit. If word is not found in cache, it is in main memory and it counts as a miss. The ratio of the number of hits divided by the total CPU references to memory (hits + misses) is the hit ratio. hit ratio of 0.9 and higher have been reported. This high ratio verifies the validity of the locality of reference property.

(*c*) Mapping: The transformation of data from main memory to cache memory is referred to as mapping process.

There are three types of mapping procedure are of partical interest when considering the organization of cache memory:

(*i*) Associative mapping

(*ii*) Direct mapping

(*iii*) Set associative mapping

Q.4. Describe the direct mapping.

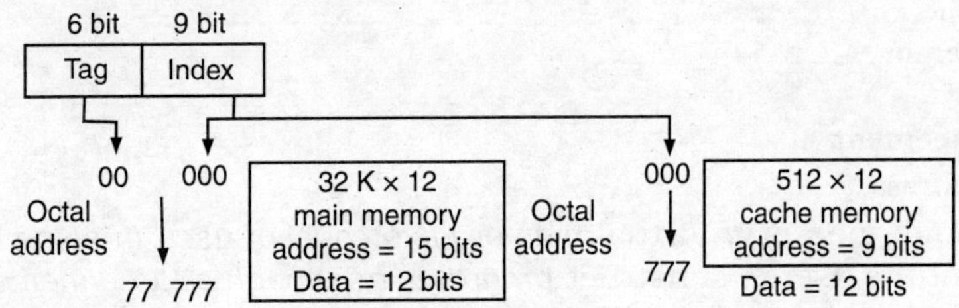

Fig. 7.30.

Solution: The CPU address of 15 bits is divided into two fields. The nine least significant bits constitutes the index field and remaining six bits from the tag field.

Figure shows that main memory needs an address that includes both the tag and the index bits. The number of 1 bit in the index field is equal to the number of address bits required to access the cache memory.

In the general case, there are 2^h words in cache memory and 2^n words i main memory.

The n bit memory address is divided into two fields:

k bits for the index field and nk bits for tage field,

The direct mapping cache organization uses the n bit address to access the main memory and the k-bit index to access the cache, when a new words is first brought into cache, the tag bits are stored along side the data bits. When the CPU generates the memory requests, the index field is used for the address to access the cache the tag field of the CPU address as compared with the tag in the word read from the cache. If two tages are matched, there is a hit and desired data world is in cache. If there is no match, there is a miss and then the required word is read from main memory. It is then stored in the cache together with the new tag replacing the previous value.

The word at address zero is presently stored in the cache (index = 000, tag, = 00, data = 1220). Suppose that the CPU now wants to access the word at address 02000. The index address is 000, so it is used to access the cache. If two tags are then compared, the cache tag is 00. The address tag is 02, which does not produce match. Therefore, the main memory is accessed and the data word 5670 is transferred to the CPU. The cache word at index address 000 is then replaced with a tag of 02 and data of 5670.

Direct mapping organization operates are shown below:

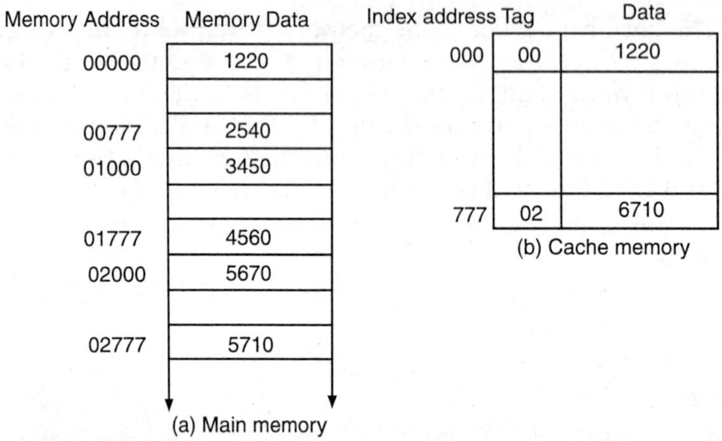

Fig. 7.31. Direct Mapping cache organization

Q.5. Define:

(a) Virtual memory

(b) Page fault

(c) Page replacement

(d) Logical address.

Solution: (a) Virtual memory: Virtual *memory is a concept used in some large computer system* that permit the user to construct *programs as though a large memory space were available,* equal to the totality of auxiliary memory. Each address that is referenced by the CPU goes through an *address mapping from the so-called virtual address to a physical address in main memory.*

A virtual memory system provides a mechanism for translating program generated *addresses into correct main memory locations. This is done dynamically, while programs are being executed in the CPU.*

(b) Page Replacement: A virtual memory system is a combination of shardware and software techniques. The memory management software system handles all the software operations for the efficient utilization of memory space. It must decide:

1. which page in main memory ought to be removed to make room for a new page.
2. which a new page is to be transferred from auxiliarly memory to main memory.
3. where the page is to be placed in main memory.

(c) Page Fault: When a program starts execution, one or more pages are transferred into main memory and the page table is set to indicate their position. The program is executed from main memory until it attempts to reference a page that is still in auxilary memory. This condition is called page fault when page fault occurs, the execution of the present program is suspended until the required page is brought into main memory.

(d) Logical address: The address generated by a segmented program is called a logical address. The logical address may be larger than the physical memory address as in virtual memory, but it may also be equal and sometimes even smaller than the length of the physical memory address.

Q.6. (*a*) How many 128 × 8 RAM chips are needed to provide a memory capacity of 2048 bytes?

(*b*) How many lines of the address bus must be used to access 2048 bytes of memory? How many of these lines will be common to all chips?

(*c*) How many lines must be decoded for chip and select? Specify the size of the decoders.

Solution:

(*a*) $\frac{2048}{128} = 16$ chips

(*b*) $2048 = 2^{11}$ 11 lines to address 2048 bytes

$128 = 2^7$ 7 lines to address each chip and 4 lines to decoder for selecting 16 chips

(*c*) 4 × 16 decoder

Q.7. A computer uses RAM chips of 1024 × 1 capacity.

(*a*) How many chips are needed, and how should their address lines be connected to provide a memory capacity of 1024 bytes?

(*b*) How many chips are needed to provide a memory capacity of 16K bytes? Explain in words how the chips are to be connected to the address bus.

Solution: (*a*) 8 chips are needed with address lines connected in parallel.

(*b*) 16 × 8 = 128 chips. Use 14 address lines ($16\text{ K} = 2^{14}$)

7 lines specify the chip address.

7 lines are decoded into chip-select inputs.

Q.8. A ROM chip of 1024 × 8 bits has four select inputs and operates from a 5-volt power supply. How many pins are needed for the IC package. Draw a block diagram and label all input and output terminate in the ROM.

Solution: 10 point for inputs, 4 for chip-select, 8 for outputs, and 2 for power. Total 24 point are needed.

Q.9. Extend the memory sytem of figure given below to 4906 byts of RAM and 4096 bytes of ROM. List the memory-address map and indicate what size decoders are needed?

Solution: $\frac{4096}{128} = 32$ RAM chips,

$\frac{4096}{512} = 8$ ROM chips.

$4096 = 2^{12}$,

There are 12 common address lines and 1 line to select between RAM and ROM.

Component	**Address**	**16 15 14 13**	**12 11 10 9**	**8 7 6 5**	**4 3 2 1**
RAM	0000 – 0FFF	0 0 0 0	5 × 32 decoder	x x x x	x x x x
ROM	1000 – 1FFF	0 0 0 1	3 × 8 decoder x	x x x x	x x x x

to $\overline{CS2}$ ↑

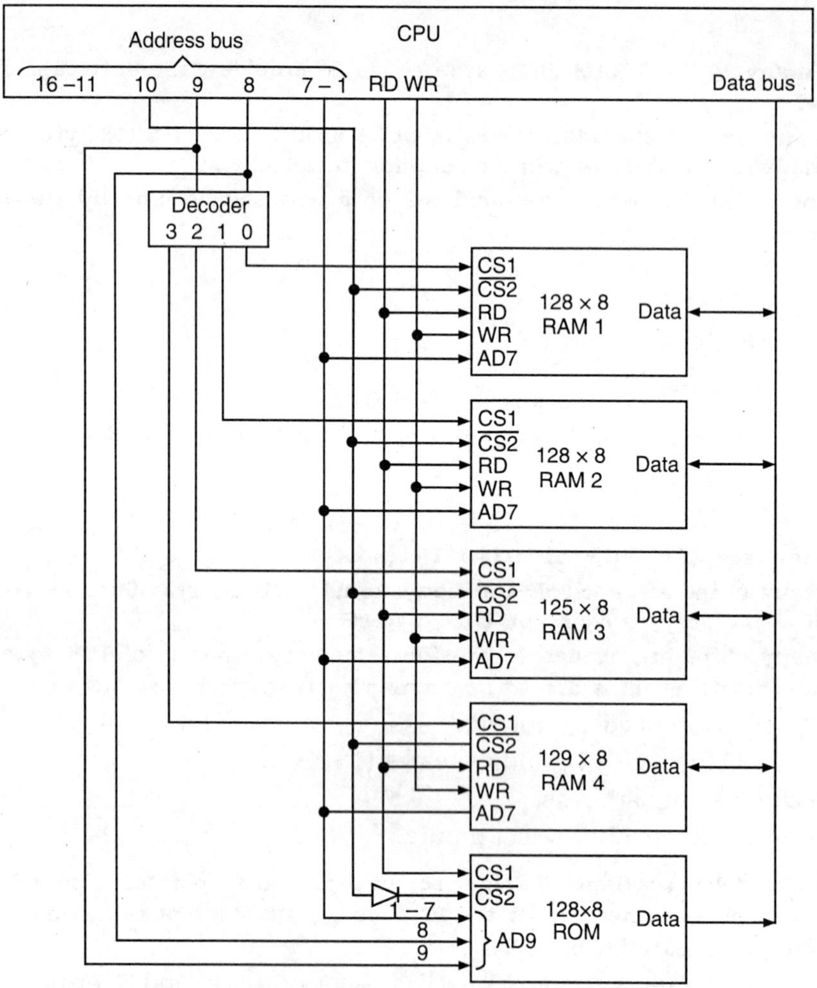

Fig. 7.32. Memory connection to the CPU

Q.10. A computer employs RAM chips of 256 × 8 and ROM chips of 1024 × 8. The computer system needs 2 K bytes of RAM, 4 K bytes of ROM and four interface units, each with four registers. A memory mapped I/O configuration is used. The two higest order bits of the address bus are assigned 00 for RAM, 01 for ROM, and 10 for interface registers.

(*a*) How many RAM and ROM chips are needed?

(*b*) How a memory-address map for the system?

(*c*) Give the address range in hexadecimal for RAM, ROM and interface.

Solution: RAM 2048/256 = 8 chips, $2048 = 2^{11}$, $256 = 2^8$

ROM 4096/1024 = 4 chips, $4096 = 2^{12}$, $1024 = 2^{10}$

Interface 4 × 4 = 16 registers; $16 = 2^4$

Component	Address	16 15 14 13	12 11 10 9	8 7 6 5	4 3 2 1
RAM	0000 – 07FF	0 0 0 0	0.3 × 8 decoder	x x x x	x x x x
ROM	4000 – 4FFF	0 0 0 1	2 × 4 dec. x x	x x x x	x x x x
Interface	8000 – 800F	1 0 0 0	0 0 0 0	0 0 0 0	x x x x

Q.11. An 8 bit computer has a 16-bit address bus. The first 15 lines of the address are used to select a bank of 32 K bytes of memory. The high order bit of the address is used to select a register which receives the contents of the data bus. Explain how this configuration can be used to extend the memory capacity of the system to eight banks of 32 K bytes each, for a total of 256 bytes of memory.

Solution: The processor selects the external register with an address 8000 hexadecimal. Each bank of 32 K bytes aer selected by adderss 0000.7FFF. The processor loads an 8-bit number into th register with a single 1 and 70'S. Each output of the register selects one of the 8 banks of 32 bytes through 2 chip-select input. A memory bank cn be changed by changing in number in the register.

Q.12. A magnetic dish system has the following parameters.

T_s = average time to position the magnetic head over a track
R = rotation speed of disk in revolutions per second
N_t = number of bits per track
N_s = number of bits per sector
Calcualte the average time T_α that it will take to read one sector.

Solution: Average time, T_α = Ts + time for half revolution + time to read a sector.

$$T_\alpha = T_s + \frac{1}{2R} + \frac{N_s}{N_t} \times \frac{1}{R}$$

Q.13. What is the transfer rate of an eight truck magnetic tape whose speed is 120 inches per second and whose density is 1600 bits per inch?

Solution: An eight-track tape reads 8 bits (one character) at the same time.

Transfer rate = 1600 × 120 = 192,000 character/sec.

Q.14. Obtain the complement function for the match logic of one world in an associative memory. in other words, show that M_is the sum of exclusive-OR functions. Draw the logic diagram for M_i and terminate it with an inverter to obtain M.

Solution:
$$M_i = \prod_{j=1}^{n} [(A_j \oplus F_{ij})' + k_j']$$

$$M_i = \sum_{j=1}^{n} (A_j \oplus F_{ij}) + k_j$$

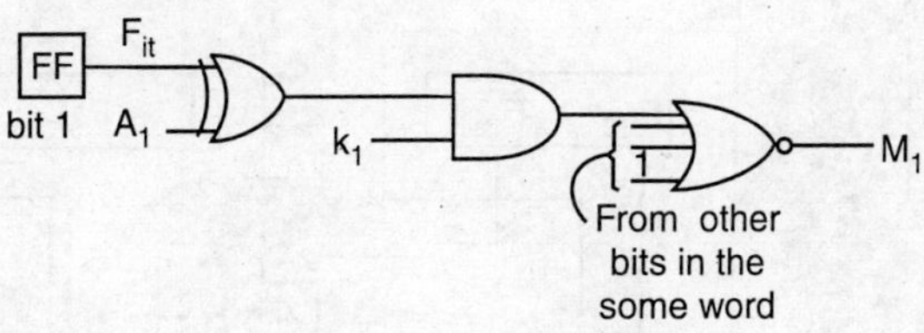

Q.15. Obtain the Boolean functions for the match logic of one word in an associative memory taking into consideration a tag bit that indicats whether the word is active or inactive.

Solution: A match occurs if $T_i = 1$

$$\text{match} = M_i T_i$$

Q.16. What additional logic is required to give a no-match result for a word in an associative memory when all key bits zeros?

Solution: $$M_i = M\left(\prod_{j=1}^{n} A_jF_{ij} + A_j'\ F_{ij}' + k_j'\right)(K_1 + K_2 + K_3 ... + K_n)$$

At least one key bit k_i must be equal to 1.

Q.17. Describe in words and by means of a block diagram how multiple matched words can be read out from an associative memory.

Solution:

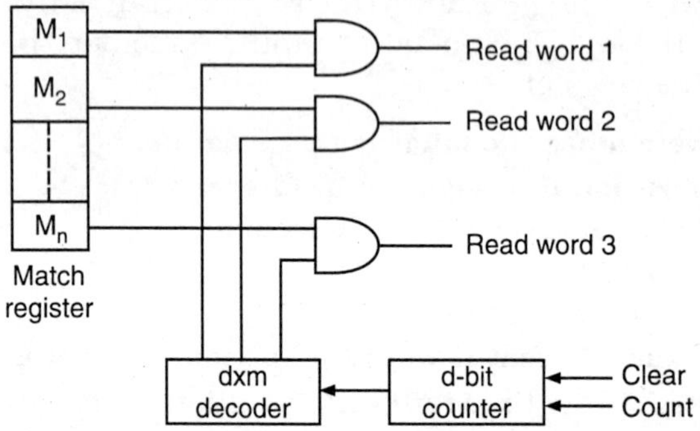

Fig. 7.33.

A d-bit counter drives a d-to-m line decoder where $2^d = m$
where, m = number of words in memory.

For each count, the M_i bit is checked and if 1, the corresponding read signal for work i is activated.

Q.18. Determine the logic of one cell and of an enture word for an associative memory that has an output indicator when the unmasked argument is greater than (but not equal to) the word in the associative memory.

Solution: Let $x_j = A_j\ F_{ij} + A_i\ F_{ij}$ (argument bit = memory word bit)

Output indicator $G_i = 1$ if:

$A_1\ F_{i1} = 1$ and K = 1 (First bit is $A = 1$ while $F_{i1} = 0$)

or if $x_1\ A_2\ F_{i2} = 1$ and $K_2 = 1$ (First pair of bits are equal and second bit is $A = 1$ while $F_{j2} = 0$)

$$G_i = (A_1\ F_{i1} + K_1)\ (x_1\ A_2\ F_{i2} + K_2)\ (x_1\ x_2\ F_{i3} + K_3)\ ...\ (x_1\ x_2\ ...\ x_{n-1}\ A_n\ F_{in} + k_n)$$

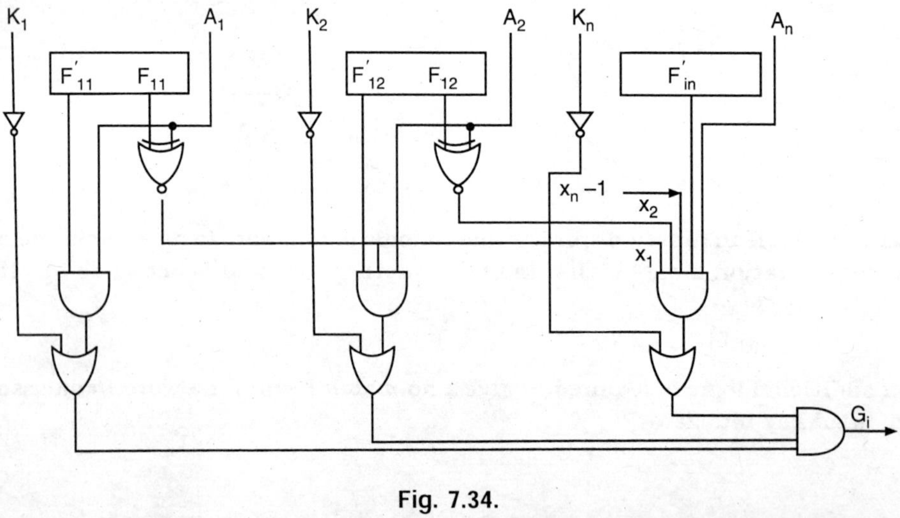

Fig. 7.34.

Q.19. A two way set associative cache memory uses blocks of four words. The cache can accomodate a total of 2048 words from main memory. The main memory size is 128 K × 32.

(a) Formulate all partinent information required to construct the cache memory.

(b) What is the size of the cache memory?

Solution: $128\ K = 2^{17}$

For a set size of 2, index address has 10 bits to accomodate $\frac{2048}{2} = 1024$ words of cache.

(a)

7 bits	10 bits
TAG	INDEX
← Block →	Word ←
8 bits	2 bits

(b)

	Tag1	Data 1	Tag 2	Data 2
	0 ← 7 →	← 32 bits →	← 7 →	← 32 bit3 →
1021				

Size of cache memory $= 1024 \times 2\ (7 + 32) = 1024 \times 78$

Q.20. The access of a cache memory is 100 ns and that of main memory 1000 ns. It is estimated that 80 percent of the memory requests are for read and the remaining 20 percent for write. The hit ratio for read accesses only is 0.9. A wirte through procedure is used.

(a) What is the average access time of the system considering only memory read cycles?

(b) What is the average access time of the system for both read and write requests?

(c) What is the ratio taking into consideration the write cycles?

Solution: (*a*) $0.9 \times 100 + 0.1 \times 11000 = 90 + 110 = 200$ nsec.

cache access memory access

(*b*) $0.2 \times 1000 + 0.8 \times 200 = 200 + 160 = 360$ nsec.

read access from (α)

(*c*) Hit ratio $= 0.8 \times 0.9 = 0.72$

Q.21. A four-way set associative cache memory has four words in each set. A replacement procedure base on the least recently used (LRU) algorithm is implemented by means of 2-bit counters associated with each word in the set. A value in the range 0 to 3 in thus recorded for each work. When a hit occurs, the counter associated with the referrenced world is set to 0, those counters with values originally lower than teh referenced one are incremented by 1, and all other remain unchanged. If a miss occurs, the word with counter is set to 0. The other three counters are incremented by 1. Show that this procedure works for the following sequence of word reference:

***A, B, C, D, B, E, D, A, C, E, C, E.* (Start with *A, B, C, D* as the initial four words, with word *A* being the least recently used).**

Solution: Sequence: *A B C D E D A C E C E*

LRU

Count value = 32 10

Initial words = *A B C D*

B is a hit	*A C D B*
E is a miss	*C D E F*
D is a hit	*C B E D*

A is a miss	$B\,E\,D\,A$
C is a miss	$E\,D\,A\,C$
E is a hit	$D\,A\,C\,E$
C is a hit	$D\,A\,E\,C$
E is a hit	$D\,A\,C\,E$

Q.22. A digital computer has a memory unit of 64 K × 16 and a cache memory of 1K words. The cache uses direct mapping with a block size of four words.

(a) How many bits are there in the tag, index, block, and word fields of the address format?

(b) How many bits are there in each word of cache, and how are they divided into functions? Include a valid bit.

(c) How many blocks can the cache accommodate?

Solution: 64 K × 16: 16 bit address; 16 bit data

(a) 6 8 2 = 16 bit address

TAG	BLOCK	WRD

Index = 10 bit cache address

(a) 1 6 16 = 23 bits in each word of cache

V	TAG	DATA

(c) 2^8 = 256 blocks of 4 words each

Q.23. An address space is specified by 24 bits and the corresponding memory space by 16 bits.

(a) How many words are there in the address space?

(b) How many words are there in the memory space?

(c) If a page consists of 2K words, how many pages and blocks are there in the system?

Solution: (a) Address phase = 24 bits 2^{24} = 164 K words

(b) Memory space = 16 bits 2^{16} = 24 K words

(c) $\frac{16\,K}{2\,K}$ = 8 K pages $\frac{64\,K}{2\,K}$ = 32 blocks

Q.24. A virtual memory has a page size of 1 K words. There are eight pages and four blocks. The associative memory page table cocntains the following entries:

Page	Block
0	3
1	1
4	2
6	0

Make a list of all virtual addresses (in decimal) that will cause a-page fault if used by the CPU.

Solution: The pages which are not in main memory are:

Page	Address	Address that will cause fault
2	2 *K*	2048 – 3071
3	3 *K*	3072 – 4095
5	5 *K*	5120 – 6143
7	7 *K*	7168 – 8191

Q.25. A vitual memory system has an address space of 8 K words, a memory space of 4 K, words and page block sizes of 1K words (see figure, given below). The following page reference changes occur during a given time iteval. (Only page changes are listed. F the same page is referenced again, it is not listed twice.)

4 2 0 1 2 6 1 4 0 1 0 2 3 5 7

Determine the four pages that are resident in main memory after each page reference change if the replacement algorithm used is

(*a*) FIFO

(*b*) LRU

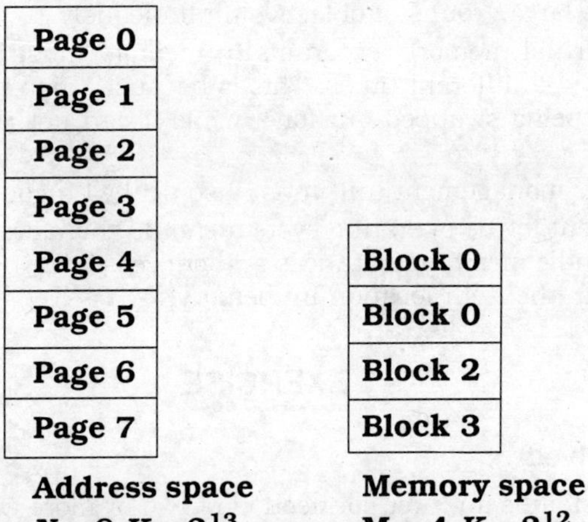

Fig. 7.18. *Address space and memory space split into groups of 1K words*

Solution: 4 2 0 1 2 6 1 4 0 10 2 3 5 7

Page reference	*Pages in main memory*	*Contents First in of FIFC*	*Pages in memory*	*LRU Most recently used*
initial	0124	4201	0124	4201
2	0124	4201	0124	4012
6	0126	2016	0126	0126
1	0126	2016	0126	0261
4	0146	0164	0246	2614
0	0146	0164	0146	6140
1	0146	0164	0146	6401
0	0146	0164	0146	6410
2	1246	1642	0124	4102
3	2346	6423	0123	1023
5	2345	4235	0235	0235
7	2357	2357	2357	2357

SUMMARY

1. Memory management is the act of managing computer memory. In its simpler forms, this involves providing ways to allocate portions of memory to programs at their request, and freeing it for reuse when no longer needed.
2. The management of main memory is critical to the computer system.
3. Virtual memory systems separate the memory addresses used by a process from actual physical addresses, allowing separation of processes and increasing the effectively available amount of RAM using disk swapping.
4. The quality of the virtual memory manager can have a big impact on overall system performance.
5. Garbage collection is the automated allocation and deallocation of computer memory resources for a program.
6. This is generally implemented at the programming language level and is in opposition to manual memory management, the explicit allocation and deallocation of computer memory resources.
7. Region-based memory management is an efficient variant of explicit memory management that can deallocate large groups of objects simultaneously.
8. In systems with virtual memory, programs in memory must be able to reside in different parts of the memory at different times. This is because when the program is swapped back into memory after being swapped out for a while it can not always be placed in the same location.
9. The virtual memory management unit must also deal with concurrency.
10. Memory management in the operating system should therefore be able to relocate programs in memory and handle memory references and addresses in the code of the program so that they always point to the right location in memory.

EXERCISE

(I) Fill in the Blanks (Solved)

1. A .... system is an inter connection of two or more CPUs with memory and input-output equipment.
2. The advantage of biased exponents is that they contain only .... numbers.

(II) True or False (Solved)

1. There are many different topologies for multistorage switching networks to control processor communication. One of them is omega switching network.
2. A priority interrupt system is a communication of branch and loop techniques.

(III) Fill in the Blanks (Unsolved)

1. There are some similarities between multi processor and .... system since both support concurrent operations.
2. The .... needs the usual circuits of an interface to communicate with the CPU and input/output device.

(IV) Detailed Questions

1. Show the step-by-step multiplication process using booth algorithm, when the following binary numbers are multiplied. Assume 5 - bit registers to hold signed numbers.

   (*a*) + 10 *x* + 14 (*b*) + 10 x - 14
2. Derive an algorithm for evaluating the square root of a binary fixed-point number.
3. Why should the sign of the remainder after a division be the same as the sign of the dividend?

4. Show that there can be no mantissa overflow after a multiplication operation.
5. Show the hardware to be used for the addition and subtraction.
6. List four peripheral devices that produce an acceptable output for a person to understand.
7. What is the difference between isolated I/O and memory mapped I/O? What are the advantages and disadvantages of each?
8. Give at least six status conditions for the setting of individual bits in the status register of an asynchronous communication interface.
9. What programming steps are required to check when a source interrupts the computer while it is still being serviced by a previous interrupt request from the same source?
10. Why does DMA have priority over the CPU when both request a memory transfer?
11. What is the advantage of using interrupt - initiated data transfer over transfer under program control without an interrupt?

Answers

(I) Fill in the Blanks (Solved)

1. computer
2. positive

(II) True or False (Solved)

1. True
2. False

Chapter

8

Input/Output Organization

8.0 INTRODUCTION

In computer's hardware language the peripheral devices can be defined as the devices that are connected to the computer in order to get most of the advantage out of it,if you in to detail description of peripheral devices it can be explained as the devices that are optional, and which are not required in principle. In earlier ages when computers were too much expensive and personal computer were very hard to afford, at that time the Motherboard, CPU (Central Processing Unit) and Memory (RAM (random access memory) & ROM (read only memory)) were considered to be the main components of the computer and any other device additionally attached to the computer were considered as peripheral device so this means that the keyboard, mice, etc. at that time were considered as peripheral devices but now a days generally they are not considered peripheral devices.

The common peripheral devices are microphones, cameras, disk drives, scanners, etc.

There is a common understanding that among people that the internal devices such as Sound Cards etc are not peripheral because they are added in the computer case which is wrong following is the peripheral devices list for a general idea.

Peripheral devise are the computer devices that are connected to the computer externally such as

Chapter Outline

printer, scanner, keyboard, mouse, tape device, microphone and external modem. They can be internal such as CD-ROM or internal modem and external as the scanner, printer and Zip drive. Peripheral devices can be classified according to their functions.

Input

Input devices are the type of the computer devices that are used to provide the control signals to the computer. Keyboard and the Mouse are the examples of the input devices.

Output

Output devices are the devices that are used to display the results. Printer, scanner, speaker and the monitor are the examples of the output devices.

Storage

A storage device is a device that is used to store the information such as hard disk drive, flash drive, floppy disk and the tape drive.

1. CD-ROM
2. CD-R
3. DVD-ROM
4. USB flash drive
5. Graphics Cards
6. Sound Cards
7. Modem
8. Network Card
9. Doking Station

Are some of the examples of peripheral devices.

Examples of Computer Peripheral Devices

A peripheral device is an external object that gives input and output for the computer. A few common input devices include:

1. keyboard
2. mouse
3. joystick
4. pen tablet
5. MIDI keyboard
6. scanner
7. digital camera
8. video camera
9. microphone

A few common output devices include:

1. monitor
2. projector
3. TV screen
4. printer
5. plotter
6. speakers

There are also devices that function as both input and output devices, for instance:

1. external hard drives

2. media card readers
3. digital camcorders
4. digital mixers
5. MIDI equipment

8.1 INPUT/OUTPUT PORTS

The picture below shows an internal view of some of the input/output ports on the right and a covered view on the left. Input/Output stands for input and output. The most common device for input is the keyboard. When you type, you are putting information into the computer, which is known as input. The most common device for output is the monitor. After the information has made its way through the computer, it is sent out to the monitor for us to see. This is known as output. On the back of computers are several input/output, (or input/output), ports. Above, on the very top are two PS/2 ports, normally used for mouse and keyboard connections. Below that are the USB, (or Universal Serial Bus), ports. Below those are two serial ports beside a long parallel port that is often used to connect to a printer. On the bottom right is a game port for joysticks or other game controllers. On the bottom left is a microphone hook up, a speaker hook up, and an additional hook up for another sound input device like a musical keyboard. Below is a picture of the input/output ports on a more recent computer.

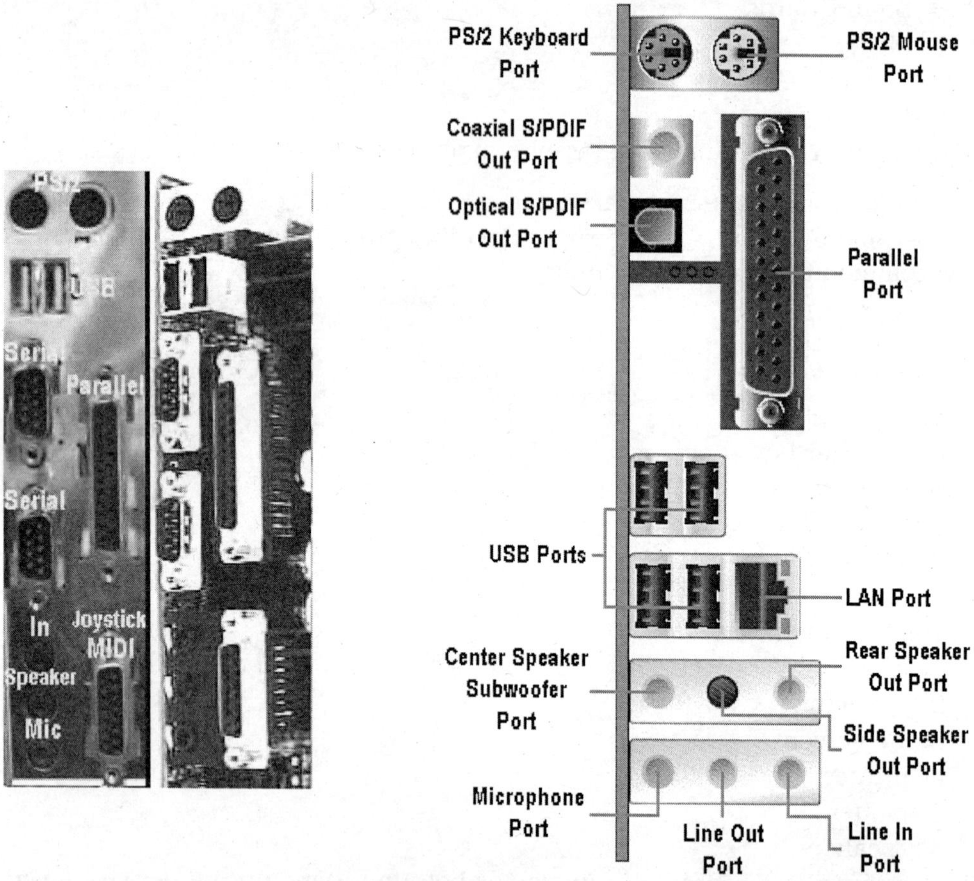

Fig. 8.1.

8.2. INTERRUPTS

A interrupt is a signal that stops the current program forcing it to execute another program immediately. The interrupt does this without waiting for the current program to finish. It is unconditional and immediate which is why it is called an interrupt.

The whole point of a interrupt is that the main program can perform a task without worrying about an external event.

Example

For example if you want to read a push button connected to one pin of an input port you could read it in one of two ways either by polling it or by using interrupts.

Polling

Polling is simply reading the button input regularly. Usually you need to do some other tasks as well, *e.g.* read an RS232 input so there will be a delay between reads of the button. As long as the delays are small compared to the speed of the input change then no button presses will be missed.

If however you had to do some long calculation then a keyboard press could be missed while the processor is busy. With polling you have to be more aware of the processor activity so that you allow enough time for each essential activity.

Hardware Interrupt

By using a hardware interrupt driven button reader the calculation could proceed with all button presses captured. With the interrupt running in the background you would not have to alter the calculation function to give up processing time for button reading.

The interrupt routine obviously takes processor time – but you do not have to worry about it while constructing the calculation function.

You do have to keep the interrupt routine small compared to the processing time of the other functions – its no good putting tons of operations into the ISR. Interrupt routines should be kept short and sweet so that the main part of the program executes correctly, *e.g.* for lots of interrupts you need the routine to finish quickly ready for the next one.

Hardware Interrupt or polling ?

The benefit of the hardware interrupt is that processor time is used efficiently and not wasted polling input ports. The benefit of polling is that it is easy to do.

Another benefit of using interrupts is that in some processors you can use a wake-from-sleep interrupt. This lets the processor go into a low power mode, where only the interrupt hardware is active, which is useful if the system is running on batteries.

Hardware Interrupt Common Terms

Terms you might hear associated with hardware interrupts are ISR, interrupt mask, non maskable interrupt, an asynchronous event, interrupt vector and context switching.

ISR	Interrupt Service Routine.
Interrupt vector	The address that holds the location of the ISR.
Interrupt mask	Controls which interrupts are active.
NMI	Non Maskable Interrupt – an interrupt that is always active.
Asynchronous event	An event that could happen at any time.
Context switching	Saving/restoring data before & after the ISR.

ISR

An ISR is simply another program function. It is no different to any other function except that it does context switching (saving/restoring processor registers) and at the end of the function it re-enables interrupts.

After the ISR finishes program execution returns to the original program and it continues exactly from where it was interrupted. The orriginal program will have no idea that this has happened.

Hardware Interrupt Vector

This is a fixed address that contains the location of your ISR - for a PIC micro it is usually address 4. In other micros there may be an interrupt vector for each vector - you have to make an interrupt routine for each one that you want to use. For PIC micros you have just one interrupt and you have to detect which interrupt triggered by examining the interrupt flag register(s).

You program the interrupt address with the address of your interrupt routine. Whenever the interrupt is triggered (and if the interrupt is unmasked) program operation jumps to the location of your interrupt routine.

Note: High level language compilers take care of all of this for you - in 'C' you just declare the function using the keyword **interrupt** (as the type returned from the function). It then puts the address of this routine in the interrupt vector.

A Basic Interrupt System

This is a general description of an interrupt system (biased slightly to PIC micros) which is true for any interrupt module and it is useful for understanding how to control and use interrupts.

Hardware Interrupt Block Diagram

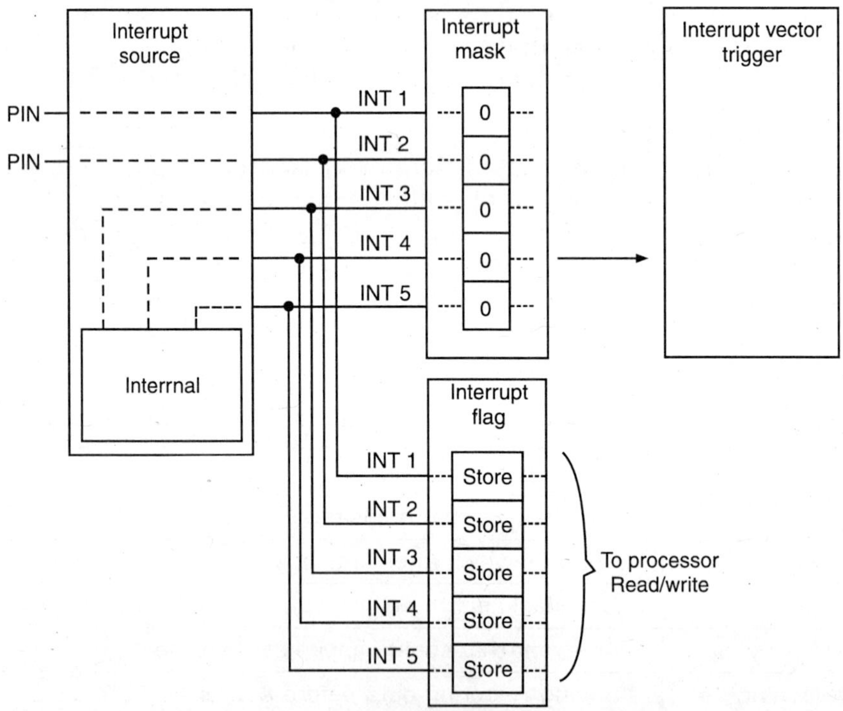

Fig. 8.2.

Input Sources

These are signals that start an interrupt. They can be external pins (where a rising or falling edge of a input triggers the interrupt) or internal peripheral interrupts, *e.g.* ADC completed, Serial Rx data received, timer overflow, etc. (The 16F877 has 15 interrupt sources).

> **Note:** you can ususlly control which edge (rising or falling) is used by setting bits in yet another register!

Interrupt Flags

Each hardware interrupt source has an associated interrupt flag - whenever the interrupt is triggered the corresponding interrupt flag is set. These flags are usually stored as bits within an interrupt register.

The processor can read from and write to the interrupt register, reading from it to find out which interrupts occurred and writing to it to clear the interrupt flags.

Interrupt Mask

The interrupt mask has a set of bits identical to those in the interrupt register. Setting any bit (or unmasking) lets the corresponding signal source generate an interrupt - causing the processor to execute the ISR.

> **Note:** When a bit in the mask is clear (masked) the ISR is not activated for that signal source but the interrupt register is still set by the signal source. So you can still detect what the hardware is doing by polling the interrupt register.

Hardware Interrupt Vector

The interrupt vector is a location in memory that you program with the address of your interrupt service routine (ISR). Whenever an unmasked interrupt occurs program execution starts from the address contained in the interrupt vector.

8.3 INTERRUPTS AND EXCEPTIONS

Much of the functionality embedded inside a personal computer is implemented by hardware devices other than the processor. For example, serial communication is handled by an Intel 8251A chip (or a 16550AF in newer computers), which is located on the computer's motherboard (usually inside an ASIC) while the keyboard is controlled by a 8042A microprocessor.

Since, each device operates at its own pace, a method is needed for synchronizing the operation of the processor with these devices. One solution is for the processor to sit in a tight loop, asking each device about its current state (a method known as polling). When data is available in one of the devices, the processor can then read and process the incoming bytes. While this method works, it has many disadvantages. First, it is very wasteful in terms of processing power, since the processor is constantly busy reading the status of the attached devices instead of executing some useful code. And second, when the rate of data transfer is extremely high, the processor might lose data bytes arriving from the hardware devices.

So, a different approach was devised. Instead of polling hardware devices to wait for their response, each device is responsible for notifying the processor about its current state. When a hardware device needs the processor's attention, it simply sends an electrical signal (hardware interrupt) through a dedicated pin in the interrupt controller chip (located on the computer's motherboard).

The interrupt controller serves as an intermediate between the hardware devices and the processor. Its responsibility is to alert the processor when one of the hardware devices needs its immediate attention. In this case, the processor stops its current activity and jumps to execute a function (interrupt handler) which was previously associated with the calling device (or more accurately, associated with the interrupt vector of the device).

While hardware interrupts have a significant importance, software interrupts also play a major role in the normal operation of the PC. These interrupts are synchronous rather than asynchronous, since they are generated by the processor itself when it encounters an INT instruction inside the code stream. Software interrupts are usually used by the operating system to publish its internal functions, thereby allowing applications to take advantage of these services.

Exceptions belong to a special type of software interrupts. They are generated by the processor itself whenever some unexpected critical event occurs. For instance, a page fault exception (interrupt 14) is triggered when the processor attempts to access a page, which is marked as not-present. The exception handler can then reload the page from disk (virtual memory) and restart the instruction which generated the exception.

Three types of exceptions can be generated by the processor: faults, traps and aborts. When a fault exception occurs, the CS and (E)IP registers which are pushed on the stack, point to the address of the instruction, which generated the exception. This gives the exception handler a chance to fix the condition which caused the exception to occur, before restarting the faulting instruction. Traps are similar to interrupts in the sense that they make the processor push the address of the next instruction to the stack, while aborts neglect to specify the location of the faulting instruction, since they are usually used to indicate severe errors (such as hardware errors or illegal system tables) which are not recoverable.

8.3.1. Software Interrupts

If you've ever programmed for DOS, you probably know that the INT 21h interface has a salient importance in linking applications with operating system services. For instance, calling an operating system function for opening a file, simply requires issuing an INT 21h interrupt with the correct parameters set in the processor registers.

The BIOS (Basic Input Output System) uses a similar software communication mechanism. The INT 10h, 13h and 16h are all interfaces to internal BIOS functions, which control the screen, the disk controller and the keyboard. Instead of accessing hardware devices directly, DOS uses the BIOS services to control the operation of the system. Modern operating systems such as Windows NT drop their reliance upon the BIOS in favor of faster mechanisms (device drivers) for accessing the PC's hardware.

When the processor bumps into an INT instruction, it pushes the address of the next instruction (CS and IP registers) and the contents of its flags register to the stack and then jumps to execute the interrupt handler. This ensures that when the handler code ends, the processor returns automatically to the original code stream (the code which was executing before the interrupt occured).

Software interrupts are still widely used by protected-mode operating systems in order to publish their internal services to the external world. However, most of these interfaces are now wrapped by functions and cannot be accessed directly (although it's possible, it's not recommended). For instance, Windows NT uses the undocumented interrupt 2Eh to allow control transfers between its user mode and kernel mode components. Whenever a non-privileged Windows application issues a call to an API function, the system routes the call until it reaches an INT 2Eh instruction. The IDT gate corresponding to this interrupt points to a handler inside ntoskrnl, which performs the required privileged operations on behalf of the application. Windows 95 and 98 use a similar technique to hop between user and kernel mode, but rely upon INT 30h instead of using INT 2Eh.

Here are the precise steps taken by the processor when it encounters an INT instruction :

1. Push Flags, CS and IP to the stack (in this order).
2. Multiply interrupt number by 4 and use the resulting number as an offset into the interrupt table (located at the beginning of the physical address space).
3. Get far address (CS and IP) of the interrupt handler from the table entry.
4. Disable interrupts (note that if you wish hardware interrupts to be serviced even when the processor executes inside an interrupt handler, you must issue a STI instruction at the beginning of the handler code).
5. Jump to execute the handler.
6. When the handler code ends (an IRET instruction is executed), pop CS, IP and the flags from the stack so that control returns to the currently active application.

Many applications (especially TSRs – Terminate and Stay Resident programs) hook software interrupts to supply additional services on top of those offered by DOS and the BIOS. One good example is Netware 3.1, which hooks INT 21h in order to add support for network file operations (such as reading a file located on a remote computer). Other pieces of code (mostly device drivers) hook hardware interrupts to monitor various devices and ensure their proper operation.

To hook an interrupt, a program need only to replace a specific interrupt table entry with the address of its own interrupt handler. Whenever this interrupt occurs, the handler is automatically invoked by the processor. A good programming practice is to place a call to the previous handler inside the interrupt handling function. This ensures that all previously installed hooks will get a chance to handle the incoming interrupt.

Note that when loading an application which uses software interrupts to call operating system services, there is no need to patch the application code with the actual memory addresses of the OS services. Therefore, the operating system can freely relocate its services in memory by simply updating the corresponding interrupt table entries with the new addresses.

8.3.2. Hardware Interrupts

Hardware interrupts are not very different in behavior. But still, when considering the path traveled by a hardware interrupt from the instant it leaves the hardware device until it reaches the processor, you must take into account the actions taken by the interrupt controller (or the 8259A if you prefer).

When the processor executes an INT instruction, it retrieves the interrupt code from the opcode itself. This code is later used by the processor to index the **IVT** (interrupt vector table) and find the address of the interrupt handler. On the other hand, when a hardware interrupt is detected, the interrupt controller sends the interrupt code to the processor via the data bus. When the code is finally acquired by the processor (either from the INT opcode or from the interrupt controller) the steps taken are equivalent to those discussed earlier.

Hardware interrupts in protected-mode behave in the same manner as described in the real-mode section. The only major difference is that when executing in protected-mode, the processor consults its IDT rather than searching the real-mode IVT for the address of the interrupt handler.

Table 8.1 presents the interrupt vectors occupied by the interrupt controller chips under different operating systems:

Table 8.1. Interrupt vectors occupied by the 8259A chips

OS	Vectors occupied by the master 8259A	Vectors occupied by the slave 8259A
DOS	8h - Fh	70h - 77h
Windows 95 / 98	50h - 57h	58h - 5Fh
Windows NT	30h - 37h	38h - 3Fh

8.3.3. Exceptions

Most of the information which was presented in the previous sections, applies to protected-mode as well. However, exception 13 (General Protection Fault) has a wider meaning in protected-mode and can indicate additional conditions than those defined for real-mode.

That's it. More information about the implementation of interrupts and exceptions in protected-mode can be obtained from the "Pentium Processor User's Manual" available from Intel.

When executing in real-mode, the following exceptions are supported:

Table 8.2. Real-Mode Exceptions

Exception	Vector	Condition
Division by Zero	0	Attempting to execute a DIV or an IDIV instruction with a divisor which equals zero.
Debug / Single Step	1	Used in conjunction with the debug registers to indicate a breakpoint hit. The processor also issues this interrupt after executing every instruction when the TRAP flag is set.
Breakpoint	3	The INT 3 instruction generates this exception. Since the opcode of this instruction is only one byte long, it is often used by debuggers to set a breakpoint in application code. The debugger needs only to replace the first byte of the instruction with the INT 3 opcode (11001100b) and wait for the breakpoint exception to occur. After the exception handler resumes execution and before returning to the debugged application, the debugger sets the INT 3 opcode to the previously overwritten byte and issues an IRET (interrupt return) instruction.
Overflow	4	When performing arithmetic instructions with signed operands, the processor set the OF flag to indicate an overflow. The INTO instruction tests this flag and if it is set - generates an exception.

Bounds Check	5	The BOUND instruction is used to verify that an array index does not exceed a certain limit. If the limit is exceeded, an exception is generated.
Invalid Opcode	6	Occurs when the processor executes one of the reserved opcodes or uses the LOCK prefix improperly. Can also indicate an invalid operand following an opcode.
Device Not Available	7	Attempting to execute a floating-point instruction when there is no coprocessor installed. The EM bit of the CR0 register can be cleared to disable this exception.
Double Fault	8	An interrupt occurs which has no corresponding entry in the IVT or a second exception is generated while the processor is executing a previously activated exception handler.
Stack Exception	12	Stack operation exceeds offset FFFFh or a selector pointing to a non-present segment is loaded into SS.
CS, DS, ES, FS, GS Segment Overrun	13	Word memory access at offset FFFFh or an attempt to execute past the end of the code segment.
Floating-Point Error	16	An error with the numeric coprocessor (Divide-by-Zero, Underflow, Overflow...).

Other exceptions exist only in **Virtual 8086 Mode**:

Table 8.3. Virtual-Mode exceptions

Exception	Vector	Condition
Invalid Task State Segment	10	Indicates that one of the checks made during a task switch failed.
Segment Not Present	11	Loading one of the segment registers with a selector to a segment marked as not-present.
Page Fault	14	Accessing a supervisor page from user privileged code or attempting to access a page which is marked as not-present.
Alignment Check	17	Memory access to an unaligned memory location (can only occur while executing ring 3 code).

8.3.4. Protected-Mode

Handling interrupts in protected-mode requires setting up a table in memory known as the IDT (Interrupt Descriptor Table). The **IDT** is somewhat different from the interrupt vector table that exists in real-mode. Instead of being "stuck" at physical address 0, the protected-mode IDT can float around in the linear address space with absolute freedom (although it is possible to change the address of the IVT while in real-mode, it is incompatible with the implementation of the 8086 processor). The linear address of the IDT is determined by a value set into the IDTR register using the LIDT instruction. Each entry in the IDT is 8-bytes long (as opposed to the 4-bytes entries of the IVT) and can contain one of three gate descriptors:

- A task gate – Causes a task switch to occur.
- An interrupt gate – Control is transferred to the interrupt handler with interrupts disabled.

- A trap gate – Control is transferred to the interrupt handler (the interrupt flag remains unchanged).

The following figure presents the internal structure of a trap gate entry:

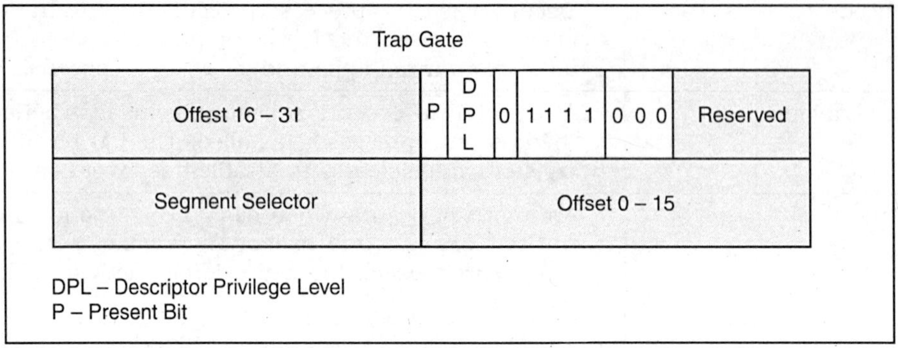

Fig. 8.3. Structure of a trap gate entry

Trap and interrupt gates have much in common since they are both used for control transfers between different privilege levels. The selector and offset fields specified in these gate descriptors are used to locate and execute the interrupt handler code. A control transfer is only permitted to a code residing in the same or a higher privilege level but never to a lower privileged one.

8.4 INPUT–OUTPUT INTERFACE

All Input–output devices, including hard disk drives, are much slow in access time and speed of transferring information as compared to CPU. Thus, an input output interface is required that provides a way of transferring information between internal storage and external I/O devices. This show in Fig. 8.4. All Input/Output devices have some form of controller/interface unit and the CPU communication with input/output divides via this controller. The purpose of inter face unit is to supervise and sysnchronize the input-output transfer there by resolving the differences between the computer system and the peripheral devices.

For example:

(*a*) A conversion of signal values may be required as most or all of the peripheral devices are electromagnetic or electromechanical in nature where as CPU and memory are electronic devices.

(*b*) The Peripheral devices are much slower as compared to CPU. Thus, there is a need for synchronization mechanism.

(*c*) The Peripheral devices and CPU/memory may differ in their word formats. In addition to interface unit, each peripheral may hare its own controller to super-wise its operations.

Data buffering is an important aspect of input/output controllers. Each controller has a memory area associated with it called as Buffer. Data buffering is a solution to the problem of mismatch in transfer rates between the memory and the CPU and peripheral devices. The CPU sends all output information to buffer for retransmission by the peripheral controller to the peripheral devices.

For example while sending text to a printer, the characters are transferred at a fast rate to the buffer associated with the printer controller which will transfer them on to the printer at the Printer's transfer rate. Similarly input from the devices is sent to the buffer from where it is transferred by the CPU to main memory at high speed.

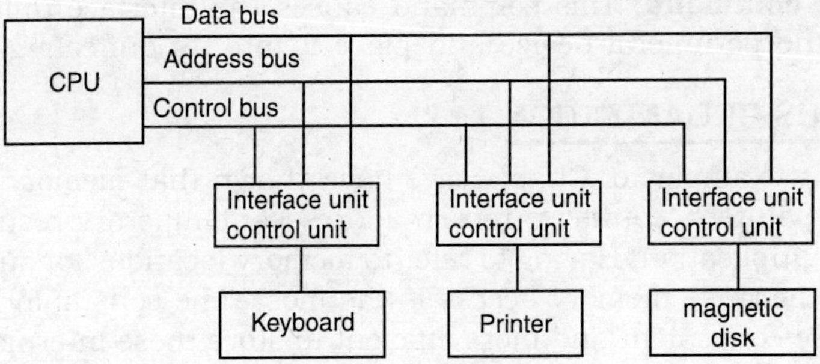

Fig. 8.4.

8.4.1 Input/Output Bus and Interface Module

All communication between the individual major components of the computer system takes place through the system bus. The bus is a cable, which is capable of carrying signals that represent data from one place to another. The bus used in a particular computer may be specific to that computer or may be an industry-standard bus. If an industry standard bus is used then it becomes easy to upgrade the computer by buying a component from an independent manufacturer. Most modern PCS use the PCI bus where as the older machine used the ISE bus.

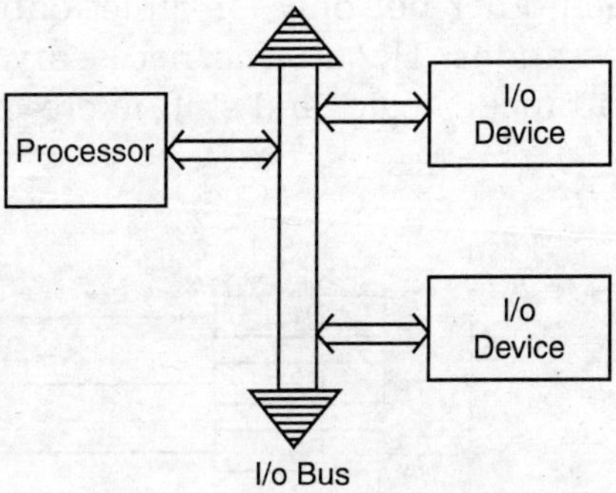

Fig. 8.5.

When the data is to be transferred from memory to a printer when it will be sent through the system bus. The control signals required to access memory and to activate the printer are also sent by the CPU via the system control bus.

In Fig. 8.5, it is shown that I/O bus consists of data bus, address bus and control bus. When CPU has to communicate with a particular device, it places the device address on the address bus together with the necessary control signals on the control bus. Each interface unit decodes the address the address and the control signals received from the I/O bus, interprets them and provides corresponding signals to the peripheral controller. All other peripheral devices whose device address does not match with the address placed by the Processor are disabled by their interfaces. The control information contained by the control bus is referred as an I/O command, which is actually an instruction and is executed in the interface and its attached peripheral unit. The I/O commands are classified into Four major types:

(*a*) **Control command:** It is used to activate the particular device and inform it about the operation to be performed. Each Peripheral has its own sequence of control commands depending on its mode of operation. For example magnetic tape may be instructed to start the tape moving inform ward motion or to rewind the tape etc. as desired.

(*b*) **Status command:** It is used by the Processor to check various status conditions in the interface unit and the Peripheral device. Basically, it is used to detect the errors that might occur during a transfer. These error conditions are indicated by setting corresponding bits in the status register which can be read by the Processor at certain intervals.

(*c*) **Data output command:** This command is issued to transfer the data contained in the data bus into one of the registers in the interface unit.

(*d*) **Data input command:** This command causes the interface unit to receive the data from the peripheral device and place it into its buffer register.

8.5 MULTIPLE BUS ORGANIZATION

In the programming examples of Chap. 6, we have shown that memory location as are needed for storing pointers, counters, return addresses, temporary results, and partial products during multiplication. Having to refer to memory locations for such applications is time consuming because memory access is the most time-consuming operation in a computer. It is more convenient and more efficient to store these intermediate values in processor registers. When a large number of registers are included in the CPU, it is most efficient to connect them through a common bus system. The registers communicate with each other not only for direct data transfers, but also while performing various micro-operations. Hence it is necessary to provide a common unit that can perform all the arithmetic, logic, and shift micro-operations in the processor.

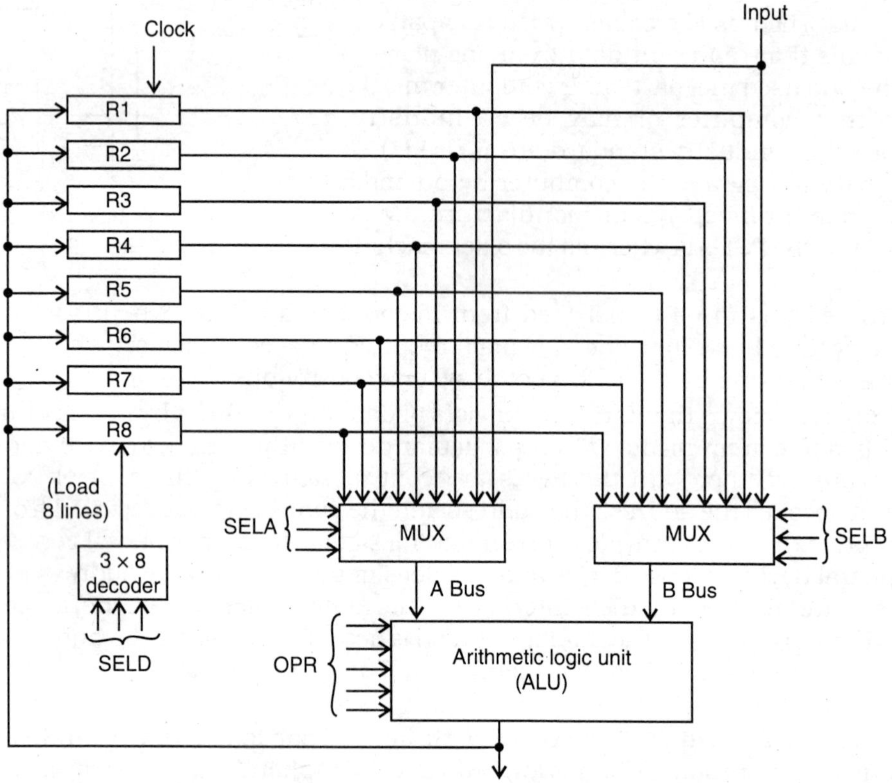

(a) Block diagram

3	3	3	5
SELA	SELB	SELD	OPR

(b) Control word

Fig. 8.6. Register set with common ALU

A bus organization for seven CPU registers is shown in Fig. 8.4. The output of each register is connected to two multiplexers (MUX) to form the two buses a and B. The selection lines in each multiplexer select one register or the input data for the particular bus. The a and B buses from the inputs to a common arithmetic logic unit (ALU). The operation selected in the ALU determines the arithmetic or logic micro-operation that is to be performed. The result of the micro-operation is available for output data and also goes into the inputs of all the registers. The register that receives the information from the output bus is selected by a decoder. The decoder activates one of the register load inputs, thus providing a transfer path between the data in the output bus and the inputs of the selected destination register.

The control unit that operates the CPU bus system directs the information flow through the registers and ALU by selecting the various components in the system. For example, to perform the operation the control must provide binary selection variables to the following selector inputs:

$$R1 \leftarrow R2 + R3$$

1. MUX A selector (SELA): to place the content of R2 into bus *A*.
2. MUX *B* selector (SELB): to place the content of R3 into bus *B*.
3. ALU operation selector (OPR): to provide the arithmetic addition $A + B$.
4. Decoder destination selector (SELD): to transfer the content of the output bus into *R*1.

The four control selection variables are generated in the control unit and must be available at the beginning of a clock cycle. The data from the two source registers propagate through the gates in the multiplexers and the ALU, to the output bus, and into the inputs of the destination register, all during the clock cycle interval. Then, when the next clock transition occurs, the binary information from the output bus is transferred into *R*1. To achieve a fast response time, the ALU is constructed with high-speed circuits. The buses are implemented with multiplexers or three-state gates, as shown in Sec. 4–3.

Control Word

There are 14 binary selection inputs in the unit, and their combined value specified a *control word*. The 14-bit control word is defined in Fig. 8.2 (*b*). It consists of four fields. Three fields contain three bits each, and one field has five bits. The three bits of SELA select a source register for the *A* input of the ALU. The three bits of SELB select a register for the *B* input of the ALU. The three bits of SELD select a destination register using the decoder and its seven load outputs. The five bits of OPR select one of the operations in the ALU. The 14-bit control word when applied to the selection inputs specify a particular micro-operation.

Table 8.4. Encoding of Register Selection Fields

Binary code	SELA	SELB	SELD
000	Input	Input	None
001	*R*1	*R*1	*R*1
010	*R*2	*R*2	*R*2
011	*R*3	*R*3	*R*3
100	*R*4	*R*4	*R*4
101	*R*5	*R*5	*R*5
110	*R*6	*R*6	*R*6
111	*R*7	*R*7	*R*7

binary code listed in the first column of the table specifies the binary code for each of the three fields. **The register selected by fields SELA, SELB, and SELD is the one whose decimal number is equivalent to the binary number in the code. When SELA or SELB is 000, the corresponding multiplexer selects the external input data. When SELD = 000, no destination register is selected but the contents of the output bus are available in the external output.**

The ALU provides arithmetic and logic operations. In addition, the CPU mist provide shift operations. The shifter may be placed in the input of the ALU to provide a per shift capability, or at the output of the ALU to provide post shifting capability. In some cases, the shift operations are included with the ALU. **The OPR field has five bits and each operation is designated with a symbolic name.**

Table 8.5. Encoding of ALU Operations

OPR code	*Operation*	*Symbol*
00000	Transfer *A*	TSFA
00001	Increment *A*	INCA
00010	Add *A* + *B*	ADD
00101	Subtract *A* – *B*	SUB
00110	Decrement *A*	DECA
01000	AND *A* and *B*	AND
01010	OR *A* and *B*	OR
01100	XOR *A* and *B*	XOR
01110	Complement *A*	COMA
10000	Shift right *A*	SHRA
11000	Shift left *A*	SHLA

Example of Micro-operations

A control word of 14-bits is needed to specify a micro-operation in the CPU. The control word for a given micro-operation can be derived from the selection variables. For example, the subtract micro-operation can be derived from the selection variables. For example, the subtract micro-operation given by the statement

$$R1 \leftarrow R2 - R3$$

specified *R*2 for the *A* input of the ALU, *R*3 for the *B* input of the ALU, *R*1 for the destination register, and an ALU operation to subtract *A* – *B*. Thus the control word is specified by the four fields and the corresponding binary value for each field is obtained from the encoding listed in Tables 8.4 and 8.5. The binary control word for the subtract micro-operation is 010 011 001 00101 and is obtained as follows:

Field:	SELA	SELB	SELD	OPR
Symbol:	R2	R3	R1	SUB
Control word:	010	011	001	00101

The control word for this micro-operation and a few others are listed in Table 8.3.

The increment and transfer Micro-operations do not use the *B* input of the ALU. For these cases, the B field is marked with a dash. We assign 000 to any unused field when formulating the binary control word, although any other binary number may be used. To place the content of a register into the output terminals we place the content of the register into the *a* input of the ALU, but none of the registers are selected to accept the data. The ALU operation TSFA places the data from the register, through the ALU, into the output terminals. The direct transfer from input to output is accomplished with a control word of all 0's (making the *B* field 000). A register can be cleared to 0 with an exclusive operation. This is because $x \oplus x = 0$.

Micro-operation	*SELA*	*SELB*	*SELD*	*OPR*	*Control Word*
		Symbolic Designation			
$R1 \leftarrow R2 - R3$	R2	R3	R1	SUB	010 011 001 00101
$R4 \leftarrow R4$ V $R5$	R4	R5	R4	OR	100 101 100 01010
$R6 \leftarrow$ R6 + 1	r6	—	R6	INCA	110 000 110 00001
$R7 \leftarrow R1$	R1	—	R7	TSFA	001 000 111 00000
Input $\leftarrow R2$	R2	—	None	TSFA	010 000 000 00000
Output $\leftarrow$ Input	Input	—	None	TSFA	000 000 000 00000
$R4 \leftarrow$ shl $R4$	R4	—	R4	SHLA	100 000 100 11000
$R5 \leftarrow 0$	R5	R5	R5	XOR	101 101 101 01100

It is apparent from these examples that many other micro-operations can be generated in the CPU. The most efficient way to generate control words with a large number of bits is to store them in a memory unit. A memory unit that stores control words is referred to as a control memory. By reading consecutive control words from memory, it is possible to initiate the desired sequence of micro-operations for the CPU. This type of control is referred to as micro-programmed control. A micro-programmed control unit is shown in Fig. 8.7. The binary control word for the CPU will come from the outputs of the control memory marked "micro-ops."

8.6 ASYNCHRONOUS DATA TRANSFER

In digital system, the internal operations are synchronized by using clock pulses generated by a common pulse generator. The clock pulses are applied to all the registers within a unit and thus all the data transfers between internal registers take place simultaneously on the occurrence of the clock pulse. CPU and I/O interface are two separate units designed independent of each other. Now, the data transfer among these two units is said to be synchronous if the interface registers and the CPU registers share a common clock. On the other hand, if the internal timing in each unit is independent of the other, *i.e.*, each unit uses its own clock. For internal registers, the transfer is asynchronous. Most communications b/w computers and I/O devices are asynchronous.

Asynchronous data transfer between two units requires to transmit control signals between he units in order to indicate the time at which data transmission is to start. This can be done in two ways.

8.6.1 Strobe Control

In this method, a strobe pulse is supplied by one of the units to the other to indicate when the transfer has to start.

Handshaking: Here, each data being transferred is accompanied by a control signal to indicate the presence of data on the bus. The unit that receives the data acknowledges the receipt of data by sending another control signal to the source of data.

Strobe Control

In strobe control method of data transfer s single control signal called as strobe is used to time each transfer. This strobe pulse can be activated either by the source or destination.

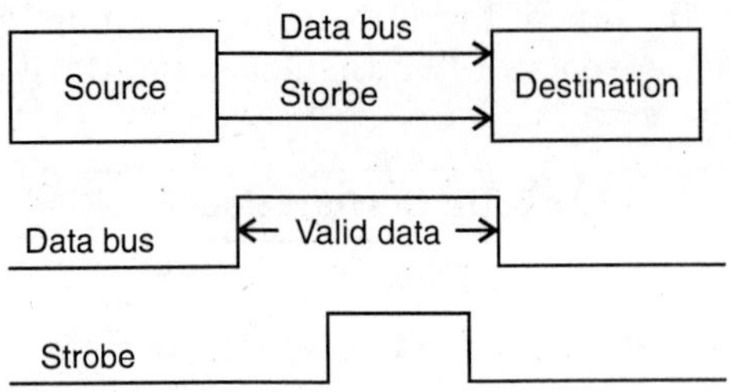

Fig. 8.7.

Source initiated strobe for data transfer

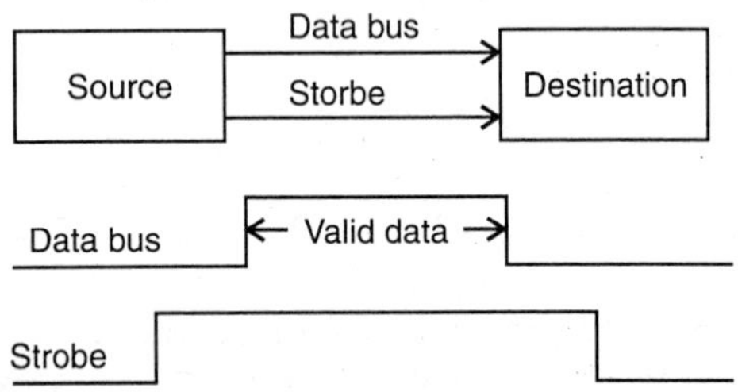

Fig. 8.8.

When activated by the source, strobe informs the destination unit that a valid data is present in the bus. The data bus transfers the information from the source unit to the destination unit this is shown in Fig. 8.7. The source unit first places the data on the data bus and then asserts the strobe pulse after a short delay. This delay is there because the data takes sometime to settle down to a steady value. The data in the data bus and the strobe pulse are held in active state for a sufficient amount of time to allow the destination unit to receive the data. The source unit removes the data from the data bus after disabling the strobe signal. Actual, there is no need for the source to remove the data on the bus is no longer valid. New valid data is available only when the strobe pulse goes high (enabled) again. This entire sequence of operations will be clear from the timing diagram. The strobe here may be a memory–write signal issued by CPU to memory. The CPU acting as a source places the data on the bus and informs the memory (destination unit) about a memory write operation through the strobe.

Destination–initiated strobe for data transfer

In this case, the destination unit first asserts the strobe pulse to inform the source to place the data on the data bus. In response to the strobe sent by the destination, the source places the requested data on the bus. This is known in Fig. 8.8. The data in the data bus and the strobe pulse are held in active state for a sufficient amount of time to allow the destination unit to receive the data. The destinations unit, after receiving the data removes the strobe signal and after a brief delay, the source removes the data from the bus. This entire sequence of operations will be clear from the timing diagram given in Fig. 8.9.

The strobe here can be a memory-read signal issued by the CPU to memory. The CPU acting as a destination generates the strobe to inform the memory (source unit) to place the specified data on bus.

There is no way to inform the source that initiates the transfer that the destination has actually received the data and similarly there is no way to inform the destination that initiates the transfer that the source has actually placed the data on bus.

A set of handshaking lines is required to handle the communication between an interface unit and an input/output device.

8.6.2 Hand Shaking

The problems associated with the strobe method can be solved by using a second control signal to serve as a reply to the unit initiating the transfer. This method of data transfer is known as handshaking. In this method one control line is used by the source that informs the destination about the valid data in the bus. The second control line is used by the destination that informs the source about whether it is ready to accept the data or not. The sequence of control signals generated is dependent on the unit initiating the transfer.

Source initiating Data transfer

This mode of data transfer uses two handshaking signals, data valid, issued by the source and data accepted produced by the destination. This is shown in Fig. 8.9. Following is the sequence of steps followed in the transfer:–

(*a*) The source initiates the transfer by:
   (*i*) Placing the data on the data bus
   (*ii*) Activating the data valid line

(*b*) the destination accepts the data from the bus and then enables the data accepted line.

(*c*) In response, the source disables the data valid line to indicate that the bus does not contain valid data.

(*d*) Finally, the destination disables the data accepted line and the system returns to the initial state.

The entire sequence is listed in Fig. 8.10.

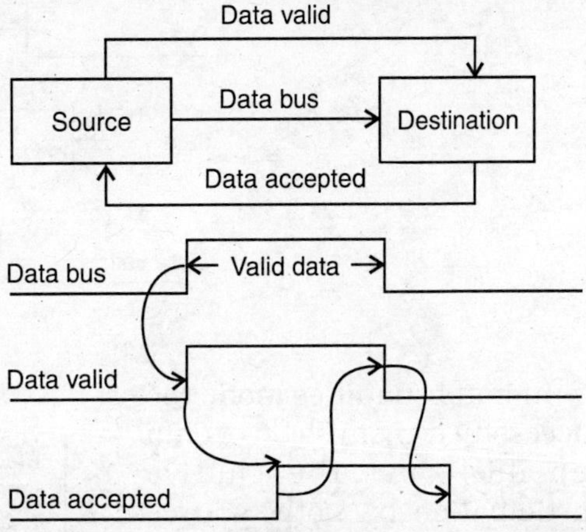

Fig. 8.9.

Destination initiated data transfer

The mode of data transfer, uses two handshaking signals, ready for data, issued by the destination and data valid, produced by the source. This is shown in Fig. 8.9. Following is the sequence of steps:

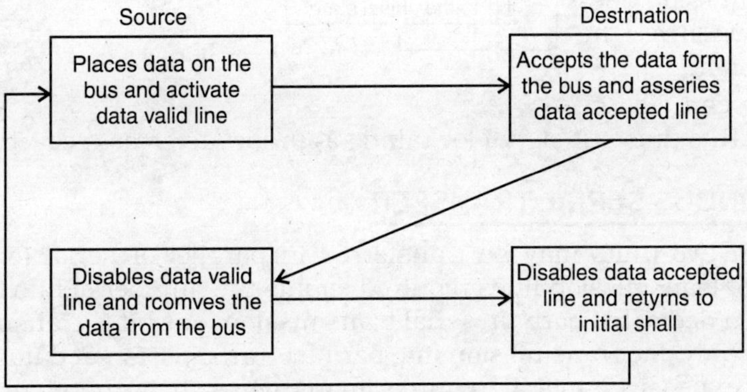

Fig. 8.10.

(*a*) The destination initiates the transfer by generating ready for the data, issued by the destination and data valid. Produced by the source.

(*b*) The source after receiving the ready for data signal places the data on the bus.

(*c*) The destination accepts the data from the bus and then disables the ready for the data line.

(*d*) In response to this, the source disables the data valid line to indicate that the bus does not contain valid data and the system returns to the initial state. The entire sequence is listed in Fig. 8.114.

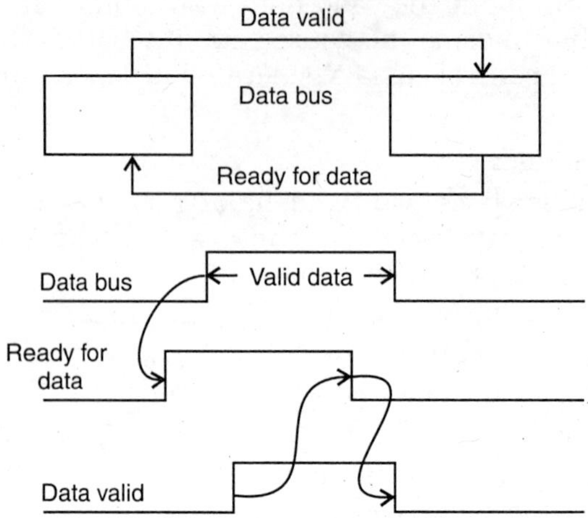

Fig. 8.11.

In hand shadings mode the successful transmission of data depends on the active participation by both, source and the receiver units. The transfer of data is not completed successfully even if any one of the units is faulty. This can be defected by employing time out mechanism that generates an alarm if the transmission does not complete within the specified time interval.

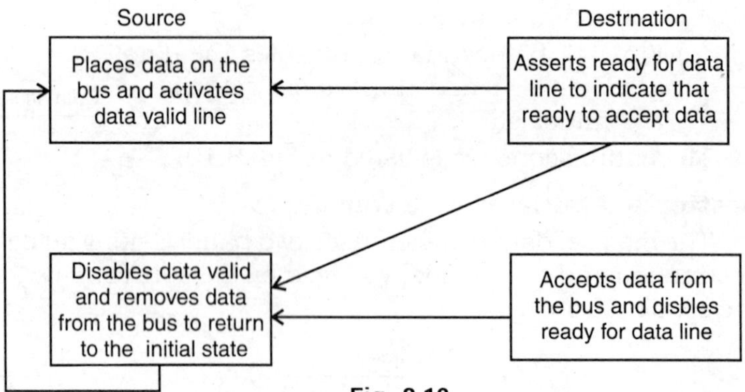

Fig. 8.12.

The Processor can also be interrupted using this time out signal for taking appropriate error recovery action.

8.7 ASYNCHRONOUS SERIAL TRANSFER

This data between two units may be transferred in parallel or serial form. In parallel transmission, all of transmission lines required equals the number of bits in the data, *i.e.* each bit has its own dedicated bath. In serial transmission, one bit is transmitted at a time and thus requires only one transmission line parallel transmission of data is faster but at the same time is expensive since it requires many wires. It is therefore used in short distance transfers where speed is critical. Serial transmission on the other hand is slower but is less expensive as requires only one set of lines. Here we are talking about serial transmission where each bit is sent over a single line, one after the other, the organization looks like one the given in Fig. 8.13. The rate at which serial information is transmitted called as Band rate and is equivalent to the data transfer in bits per second.

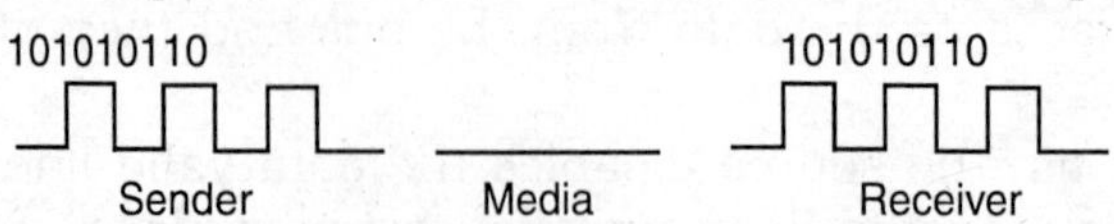

Fig. 8.13.

There are two main methods of serial transmission that are synchronous and asynchronous.

Asynchronous: The term asynchronous means not synchronized not occurring at predetermine or regular intervals and here it is usually used to describe communications in which data are transmitted intermittently rather than in steady. Stream. Heres the time interval between the characters is not fixed. The main difficulty with asynchronous transmission is that the receiver must have a way to distinguish between valid data and noise. In computer communication this is usually achieved using a special start bit and stop bit, at the beginning and end of the data reason asynchronous transmission is sometimes called start stop bit transmission.

Because no signal lines are used to convey clock timing) information, this method groups data to gather in a sequence of five Eight bits, the prefixes them with start bit and stop bit. Each character thus consists of three parts: a start bit, a character bit and a stop bits. The normal convention followed is that the transmission remains at 1-state (high) when idle, *i.e.* when no transmission is transmitted. The start of a character is indicated by transmitting 'θ' as the first bit called the start bit. Similarly to indicate the end of character, '1' is transmitted as the last bit, which is called as stop bit. This is shown in Fig. 8.12. The reviver defects the transmitted character using the following transmission rules.

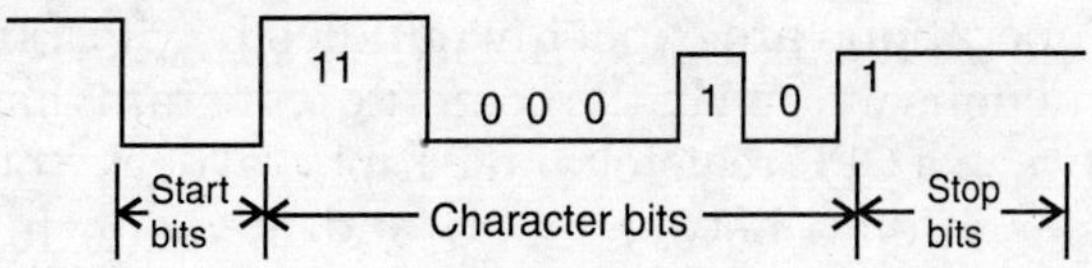

Fig. 8.14.

(*i*) The line is held in 1-state (high) when no characters are transmitted.

(*ii*) The beginning of character is selected from the start bit, which is always 0.

(*iii*) The bits following the start bit are the character bits

(*iv*) The end of a character, is selected from the stop bit.

Synchronous: In synchronous transmission, the trans meter and receiver units share a common clock frequency and data bits are transmitted mutinously. In order to keep the clock frequency in the two units synchronized with each other, it is required that the synchronization signals are transmitted periodically between the two.

In synchronous transmission, the idle state of line is changed to a known character sequence (7 E) used to synchronize the receiver to the transmitter No start/stop bits are needed here and each character is combined with others into a data packet. The data packet is prefixed with a header field and suffixed with a trailer field. The trailer filed includes a checksum value that is used by the receiver to check for errors in the transmission. This shown in Fig. 8.15.

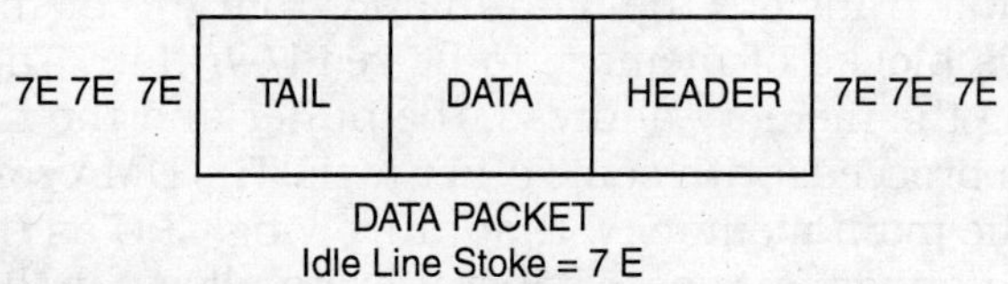

Fig. 8.15.

The header field is used to provide address information (transmitter and receiver), packet type and control data. The data field contains the user data. If the data can not fit into a single packet then multiple packets may be used and are numbered accordingly, Generally it has a fixed size.

The tail filed contains check sum information, which is used by the receiver to check whether, not the packet was corrupted during the transmission.

In asynchronous signal each data element just like a character is prefixed with a start and stop bit. In synchronous signals, each groups or block of characters is prefixed with start and stop codes. This mean using synchronous signals can provide higher speeds (more character per second) than synchronous signals.

8.8 MODES OF TRANSFER

The ultimate source or destination for the information transferred to and from an I/O device is the memory unit, CPU acts as an intermediate path. The problem with I/O is synchronizing a very fast CPU with very slow input/output device. The transfer of data between the central computer and input output devices can be handled in number of ways. In some methods, CPU is intermediate unit while in others data is transferred directly between the memory unit and input/output device. Any of the following methods may be used to handle data transfer to and from the peripherals.

8.8.1 Programmed input/output

All the input/output operations are explicitly handled by CPU. In these case these are actual input/output instructions written in the computer program that initiate data transfer. Usually the transfer is between a CPU register and Peripheral device other set of instructions is required for the transfer of data between CPU and memory. In this method, input/output is directly under program control and requires the CPU to monitor the interface unit until it indicates that it is ready for data transfer, which is time consuming process.

8.8.2 Interrupt Initiated input/output

The Programmed Input–output method keeps the CPU busy needlessly. This can be avoided by using an facility. In this method the interface unit monitors the input/output device to determine if it is ready for data transfer. In the mean time CPU continues with another program. The interface unit issues interrupt request signal to the CPU when the device is ready. In response to the interrupt signal, CPU momentarily suspends the current task and invokes a sub routine to handle input/output and later returns to the task it was originally processing following are the two variants of this type:

(*a*) **Device Polling:** The interrupt handler checks each device to determine the source of interrupt.

(*b*) **Vector Interrupt:** A "numbered" interrupt contains information about which device generated the interrupt request for example. Intel 80 × 86 is like this.

8.8.3 Direct Memory Access (DMA)

Here the interface unit transfers the data to and from the memory unit via memory bus. This method allows blocks of memory to be read/written. The CPU programs the DMA controller by giving the memory address, the buffer and the number of bytes to be read or written and then proceeds with some other task. The DMA controller then reads or writes data directly to the main memory without involving CPU as the intermediate unit. The CPU simply delays its memory access operation to allow for the direct memory I/O transfer. When the transfer is done, the DMA controller interrupts the CPU to indicate that the input/output operation has been completed.

Many computer systems include an input/outputProcessor (IOP) that combines interface logic together with the requirements for DMA. The IOP is capable of handling several peripheral devices using DMA and interrupt facility. Such a system devices the computer into three modules, the main memory, the Central Processing Unit (CPU) and input/output Processor (IOP). CPU may transfer data to or from a number of peripheral devices. The operation treated the input/output subsystem for reading and writing in the same manner as memory, using address bus, data bus and Wr, RD control lines, this requires CPU intervention and is thus costly in time.

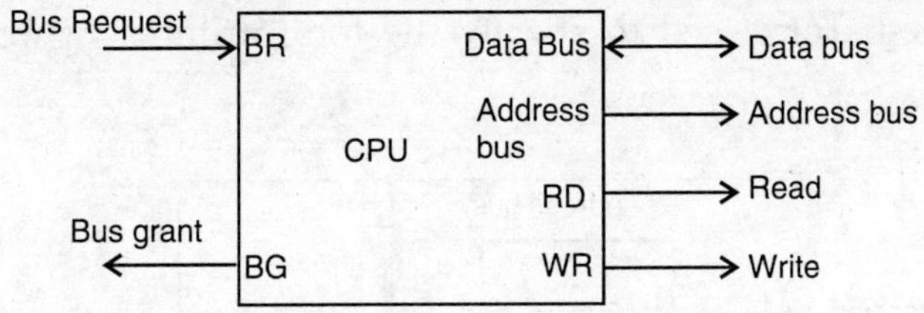

Fig. 8.16.

DMA is the capability provided by some computer bus architectures that allows data to be transferred directly from a peripheral device to the memory. The processor is greed from involvement with the data transfer process and thus it speeds up the overall computer operation.

During DMA transfer, the CPU leaves the control of memory buses and a DMA controller takes over the buses to handle the data transfer directly between the peripheral devices and the memory unit. The two control signals in the CPU, bus request (BR) and bus grant (BG) shown in Fig. 8.16, facilitate the DMA transfer. The DMA controller uses BR input signal to request CPU for the control of buses. When CPU receives this signal, it responds by suspending the execution of the current instruction and placing the system bus (address bus, data bus, and read and write lines) into a high impedance state. In high impedance state, the output is disconnected and thus has no logical significance. The CPU then activates the BG line, which informs the DMA controller that the buses are freed by the Processor. The DMA controller then takes the control of buses to mange the data transfer without the Processor controlling every aspect of the transfer. The DMA controller disables the BR line to terminate the transfer. In response to this, the CPU disables the BG line takes back the control of buses and resumes its operation.

Direct memory Access Channels

Direct memory Access (DMA) channels are system pathways that are used by many devices for transforming information directly to and from memory. DMAs are used most commonly today by Floppy disk drives, tape drives and sound cards. As you already know, the processor is considered as the brain of the machine. In older machines the processor really did almost everything. In addition to executing programs it was also responsible for handling transfer of data to and from peripherals.

Unfortunately, having the processor involved in these transfers is very inefficient, because it is a time consuming process and the processor is tied up, unable to do anything else. The Problem lies in synchronizing a very fast CPU with a very slow input/output device, which we have already discussed previous topics.

The invention of DMA enabled the devices to remove the middle man, the CPU from the path allowing the transfer under DMA can be made in jumps of ways:

(*i*) **Burst transfer:** In this technique, a bock sequence of number of memory words is transferred in a continuous burst. This mode in suitable for high speed input/output like magnetic disks, where it is not possible to stop or slow down the data transmission until an entire bock is transferred.

(*ii*) **Cycle staling:** In this method, one data word is transferred at a time after which the control of the buses must be returned to the CPU.

DMA controller needs the interface circuits to communicate with the CPU and input/ output devices. Fig. 8.17 gives the block diagram of DMA controller. In addition to the usual circuits of an interface, it also includes the following:

(*a*) **Address lines and Address register:** Used for direct communication with the memory unit.

(*b*) **Word count register:** Used to specify the total number of words that must be transferred.

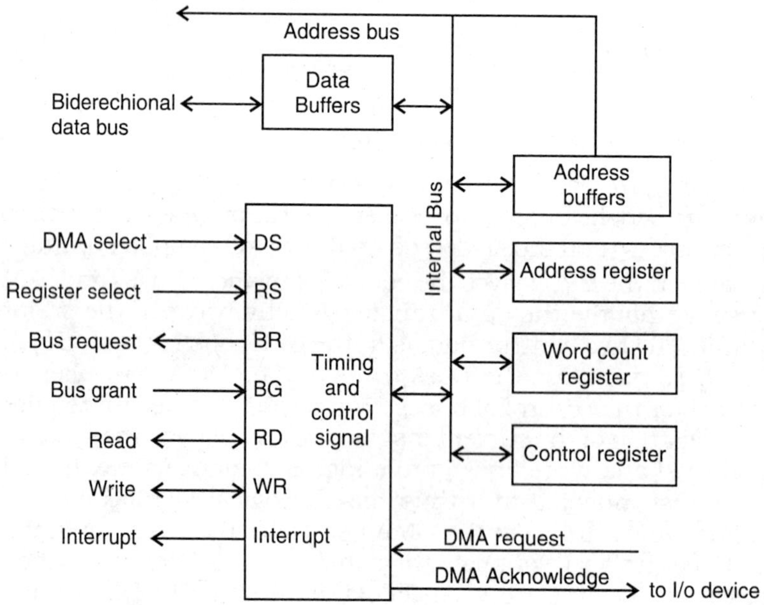

Fig. 8.17.

The processor to do other work and peripheral device to manage the transfers directly, leading to improve performance. Special channels were created, together with the circuiting to control their operation that allowed the data transfer without processor intervention. This circuitry is generally part of the system cheapest on the mother board.

Standard DMA is sometimes called as third partly DMA. This is because the DMA controller is actually doing the transfer, the first two parties being the gender and the receiver of the transfer operation. There is also a type of DMA called as first party DMA. In this case, the peripheral device doing the transfer actually takes control of the system bus and performs the transfer. This is known as bus mastering.

DMA controller: DMA controller is a device that is payable of controlling data transfer between and input/output subsystem and memory unit in the same manner that a processor can control such transfers. The DMA controller issues commands to the memory that behave exactly as the commands issued by the processor (CPU). The DMA controller is thus a second processor in the system but is dedicated to input/output ports directly to memory. The input/output data passes through a DMA controller much faster and efficiently than through the Processor as the DMA channel is specialized to the data transfer job.

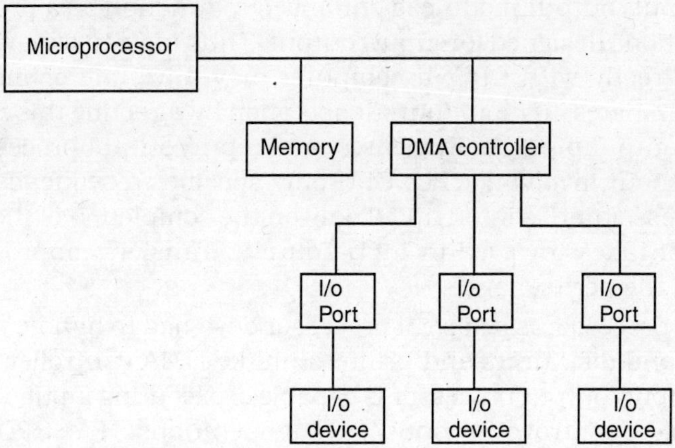

Fig. 8.18.

The DMA controller used data bus and control bus and control lines to communicate with the CPU. The CPU selects the registers in DMA by using the address lines and activating DMA select (DS) and register select (RS) input lines. The DMA communicate with the peripheral device using the DMA request and DMA acknowledges lines through handshaking procedure.

Note that the RD and WR control lines are bidirectional for the vacations given below:

(*i*) When BG = 0, CPU can communicate with the DMA via data bus and using RD and WR controls to read or Write to the DMA registers.

(*ii*) When BG = 1, DMA can communicate directly with the memory unit by placing an address in the address bus to specify the desired location in memory and enabling RD or WR control lines to read or write into the specified locations.

The DMA controller consists of three registers as shown in block diagram of Fig. 8.18.

(*i*) **Address register:** It is used to store an address that specified the desired location in memory is used for direct communication with the memory. The address bits from the address buffers are transferred into the address bus and the contents of the address register are incremented by one after each word transfer.

(*ii*) **Word count register:** It specified the total number of words that are to be transferred and is decremented by one after each transfer the contents of this register are internally tested for zero to determine the end of transfer.

(*iii*) **Control register:** It is used to specify the mode of transfer.

All the DMA registers appear as input/output interface register to the CPU and thus the CPU can communities with any of the DMA register under program control through the data bus.

8.9 INPUT/OUTPUT PROCESSOR AND INPUT/OUTPUT CHANNELS

With the evolution of computer systems, has been an increase in the complexity and sophistication of the components. Earlier, CPU to control a peripheral device directly. Later, in input/output module was added to care of input/output operation. The input/output module is then given direct access to memory via DMA to transfer a block of data between memory and input output devices, without CPU intervention. With the advancement in technology, the input/output module is enhanced to function as a processor with a get of specialized instructions designed for input/output. Thus instead of having each interface unit communicate directly with CPU, a computer may have one or more input/output processors. The CPU initiates an input/output operation by directing this external processor to execute an input/output program in memory and input/output processor and executes tense instruction without involving CPU. CPU only specifies a sequence of input/output activities for the processor and is interrupted only on the completion of the entire sequence. CPU is thus relied of takes involved in input/output transfers, improving the speed of operation and hence the performance.

An input/output processor is a special processor designed to handle peripheral devices like terminals, tape and disk units and printers unlike DMA controller that is initialized entirely by CPU, an input/output processor is capable of executing input/output instruction thus executing complete control over input/output operations. Here CPU does not execute input/output instructions but these instructions are stored in memory and are executed by IOP in the input/output channel. The CPU directs the input/output processor to execute an I/O program in memory. This program specialties the device and actions to be performed Fig. 8.19 gives the block diagram of system with one input/output process. The memory unit as shown can communicate with both CPU and IOP. The CPU handles the data processing task while IOP provides a data transfer path between input/output devices and the memory.

The communication between memory and IOP is via direct memory access method. IOP and the peripheral devices connected to it communicate in a way similar to the program control method. In most of the computer systems, CPU acts as the master processor while IOP is a slave processor. The CPU initiates the input/output program and IOP fetches and executes all input/output instructions. POP operates independent of the CPU and controls the transfer of data between peripheral devices and memory. The CPU instructs the IOP to start an input/output transfer and checks the status conditions reflected in the status register needed to make input/output related decisions. In response to this, the IOP places the status word in the specified location in memory.

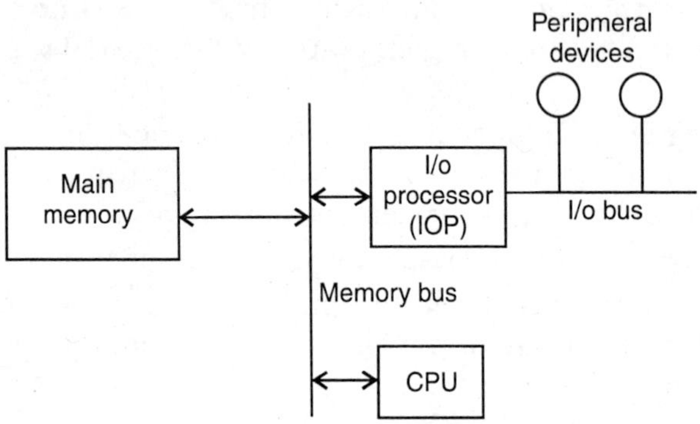

Fig. 8.19.

SOLVED PROBLEMS

Q.1. How do you mean by isolated ? Define memory mapped I/O.

Solution: Many computers use one common bus to transfer information between memory and input/output and the CPU. The distinction between a memory transfer and input/

output transfer "is made through separate read and write lines. The CPU specifies whether-the address on the address line is for a memory word or for an interface register by enabling one of two possible read or write lines. The input/output read and input/output write control lines are enabled during an input/output transfer.

The memory read and memory write control lines are enabled during a memory transfer. This configuration isolates all input/output interface addresses from the addresses assigned to memory and is referred to as the isolated input/output method for assigning addresses in a common bus.

In the isolated input/output configuration, the CPU has distinct input and output instructions, and each of these instructions is associated with the address of an interface register.

The isolated input/output method isolates memory and input/output addresses so that memory address values are not affected by interface address assignment since each has its own address space.

In memory mapped input/output, it use the same address space for both memory and input/output. This is the case in computers that employ only one set of read and write signals and do not distinguish between memory and input/output address.

Computers with memory mapped input/output can use memory type instructions to access input/output data with memory mapped input/output. All instructions that refer to memory are also available for input/output.

Q.2. What is the difference between serial and parallel data transmission ?

Solution: The transfer of data between two units may be done in parallel or serial. In parallel data transmission, each bit of the message has its own path and the total messages is transmitted through *n* separate conductor paths.

In serial data transmission, each bit in the message is sent in sequence one at a time. This method requires the use of one pair of conductors or one conductor and a common ground. Parallel transmission is faster but requires many wires. It is used for short distances and where speed is important. Serial transmission is slower but is less expensive, since it requires only one pair of conductors. Serial transmission can be synchronous or asynchronous.

Q.3. What is the difference between synchronous and asynchronous data transfer in case of serial/transfer?

Solution: In synchronous transmission, the two units share a common clock frequency and bits are transmitted continuously at the rate dictated by the clock pulses. In long distant serial transmission, each unit is driven by a sperate clock of the same frequency. Synchronization signals are transmitted periodically between two units to keep their clocks in step with each other.

In asynchronous transmission, binary information is sent only when it is available and the line remains idle when there is no information to be transmitted. This is contrast to synchronous transmission, where bits must be transmitted continuously to keep clock frequency in both units synchronized with each other.

A serial asynchronous data transmission technique used in many interactive terminals employs special bits that are inserted at both ends of the character code with this technique.

Each character consists of three parts:

a start bit,

a character bits, and

stop bits,

the convention is that the transmitter resets at the 1 state when no characters are transmitted. The first bit, called the *start bit, is* always a 0 and is used to indicate the beginning of a character. The last bit called the stop bit is always a 1.

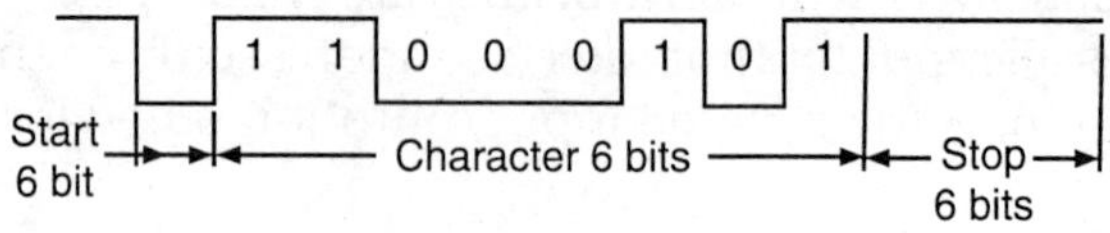

Fig. 8.20. Asynchronous serial transmission

Q.4. Describe different types of modes of transfer?

Solution: Data transfer between the central computer and input/output devices may be handled in a variety of modes. Some modes use the CPU as an intermediate path, others transfer the data directly to and from the memory unit. Data transfer to and from peripherals may be handled in one of the possible modes.

(*i*) Programmed input/output

(*ii*) Interrupt initiated input/output

(*iii*) Direct memory access (DMA).

(*i*) **Programmed input/output:** Programmed input/output operations are the result of input/output instructions written in the computer program. Each data item transfer is initiated by an instruction in the program usually, the transfer is to and from a CPU register and peripheral. Other instructions are needed to transfer the data to and from CPU and memory. In the programmed input/output method, the CPU stage in program loop until the input/output unit indicates that it is ready for data transfer. Transfer of data under programed input/output between CPU and peripheral.

(*ii*) **Interrupted initiated input/output:** When an interface determine that the device is ready, for data transfer, it generates an interrupt request to the computer upon detecting the external interrupt signal. The CPU momentarily stops the task, it is processing, branches to a service program to process the input/output transfer, and then returns to the task it was originally performing.

(*iii*) **Direct Memory Access:** In direct memory access (DMA), the interface transfers data into and out of the memory unit through the memory bus. The CPU initiates the transfer by supplying the interface with the starting address and number of words needed to be transferred and then proceeds to execute other takes when the transfer is made. The DNA requests memory cycles through the memory bus when the request is granted by the memory controller, the DMA transfers the data directly into memory.

Q.5. Define the following:

(*a*) **Priority interrupt**

(*b*) **Polling**

(*c*) **Vector interrupt,**

Solution:

(*a*) **Priority interrupt:** A priority interrupt is a system that establishes a priority over the various sources to determine which condition is to be serviced first when two or more requests arrive simultaneously. The system may also determine which conditions are permitted to interrupt the computer while another interrupt is being serviced. Higher priority interrupt levels are assigned to requests which, if delayed or interrupted, could have serious consequences.

(b) Polling: Establishing the priority of simultaneous interrupts can be done by software or hardware. A polling procedure is used to identify the highest priority sources by software means. In this method there is one common branch address for all interrupts. The program that takes care of interrupts begin at the branch address and polls the interrupt sources in sequences. The order in which they are tested determines the priority of each interrupt. The highest priority sources is tested first and if its interrupt signal is on, control branches to a service routine for this source. Otherwise, the next-lower-priority sources is tested, and so on.

(c) Vector interrupt: The way that the processor chooses the branch address of the service routine varies from one unit to another. In principle there are two methods for accomplishing this:

(i) **Non-vector interrupt:** In a non vector interrupt, the branch address is assigned to a fixed location in memory.

(ii) **Vector interrupt:** In a vector interrupt, the source that interrupts supplies the branch information to the computer. This information is called the interrupt vector. In some computers the interrupt vector is the first address of the I/O service routine.

Q.6. The addresses assigned to the four registers of the interfere of figure given below are equal to the binary equivalent of 12, 13, 14 and 1-5. Show the external circuit that must be connected between an 8-bit input/output address from the CPU and the CS, RSI, and RSO inputs of the interface.

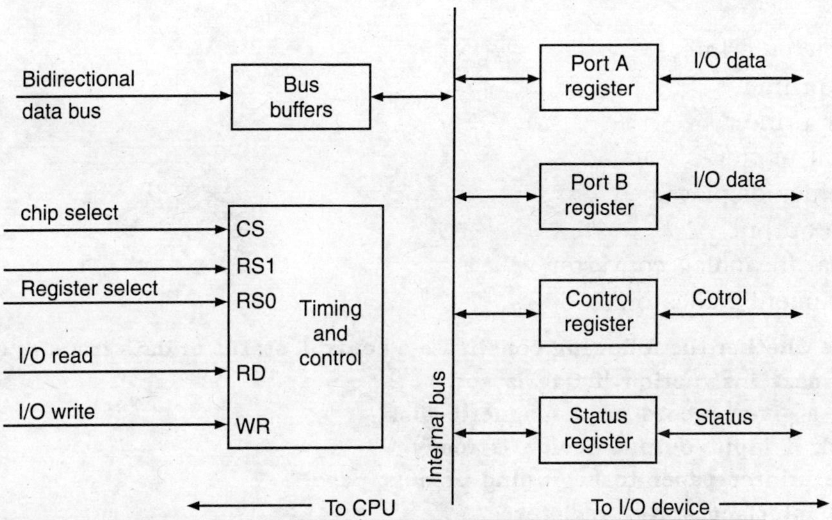

CS	RSI	RSO	Register selected
0	x	x	None: data bus in high - impedance
1	0	0	Port A register
1	0	1	Port B register
1	1	0	Control register
1	1	1	Status register

Fig. 8.21.

Solution:

$A_7 - A_2\ A_1\ A_0$									
12 =	0	0	0	0	1	1	0	0	$CS = A_2\ A_3\ A'_4\ A'_5\ A'_6\ A'_7$
13 =	0	0	0	0	1	1	0	1	$RS1 = A_1$
14 =	0	0	0	0	1	1	1	0	$RS0 = A_0$
15 =	0	0	0	0	1	1	1	1	

To CS RS1 RS0

Q. 7. Six interface units of the type shown in figure given in Q.6 are connected to CPU that uses an I/O address of eight bits. Each one of the six chip select (CS) inputs is connected to a different address lines. Thus the high-order address line is connected to the CS input of the first interface unit and the sixth address line is connected to the CS input of the sixth interface unit. The two low-order address lines are connected to the RSI and RSO of all six interface units. Determine the 8-bit address of each register in each interface.

Solution:

Interface	*Port A*	*Port B*	*Control Reg*	*Status Reg*
1	10000000	10000001	10000010	10000011
2	01000000	01000001	01000010	01000011
3	00100000	00100001	00100010	00100011
4	00010000	00010001	00010010	00010011
5	00001000	00001001	00001010	00001011
6	00000100	00000101	00000110	00000111

Q.8. Give peripheral devices that produce an acceptable output for a person to understand.

Solution:

(*i*) Character printer
(*ii*) Line printer
(*iii*) Laser printer
(*iv*) Digital plotter
(*v*) Graphics display
(*vi*) Voice output
(*vii*) Digital to analog converter
(*viii*) Instrument indicator.

Q.9. Indicate whether the following constitute a control, status or data transfer commands.

(*a*) Skip next instruction if flag is set.
(*b*) Seek a given record on a magnetic disk.
(*c*) Check if input/output device is ready.
(*d*) Move printer paper to beginning of next page.
(*e*) Read interface status register.

Solution:

(*a*) Status command – Checks status of flag bit.
(*b*) Control command – Moves magnetic head in disk.
(*c*) Status command – Checks if device power is on.
(*d*) Control command – Moves paper position.
(*e*) Data input command – Reads value of a register.

Q.10. A CPU with a 20-MHz clock is connected to a memory unit whose access time is 40 ns. Formulate a read and write timing diagrams using a READ strobe and a WRITE strobe. Include the address in the timing diagram.

Solution: 20 MHz = 20×10^6 Hz

$$T = \frac{10^{-6}}{20} = 50 \text{ ns}$$

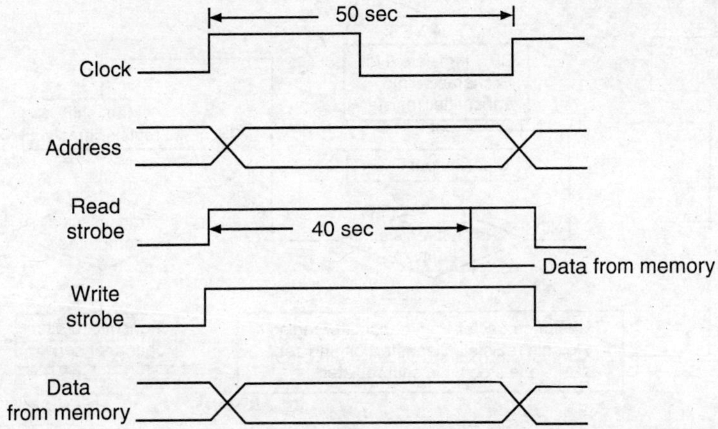

Fig. 8.22.

Q.11. The asynchronous communication interface shown in the figure given below is connected between a CPU and a printer. Draw a flowchart that describes the sequence of operations in the transmitter portion of the interface when the CPU sends characters to be printed.

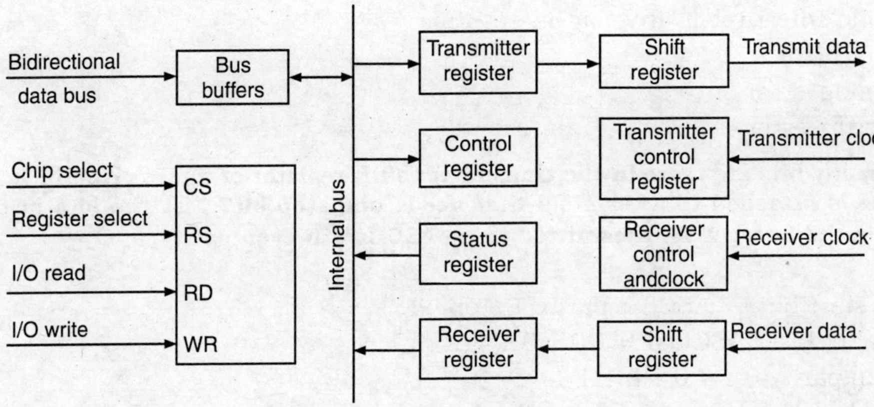

CS	RS	Operation	Register selected
0	x	x	None: data bus in high-impedance
1	0	WR	Transmitter register
1	1	WR	Control register
1	0	RD	Receiver register
1	1	RE	Status register

Fig. 8.23. Block diagram of a typical asynchronous communication interface

Solution: Output flag is a bit in status register.

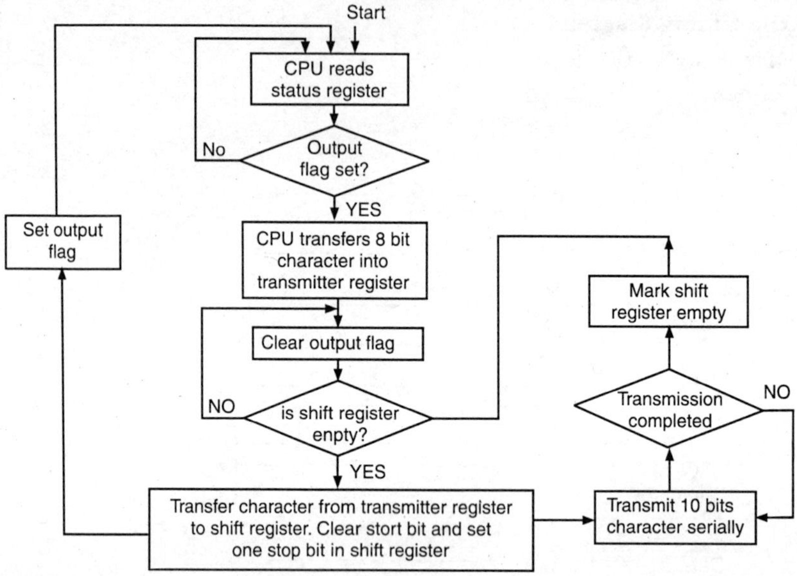

Fig. 8.24.

Q.12. Give six status conditions for the setting of individual bits in the status register of an asynchronous communication interface.

Solution: Status conditions

(*i*) Output flag to indicate when transmitter register is empty.

(*ii*) Input flag to indicate when receiver register is full.

(*iii*) Enable interrupt if any flag is set.

(*iv*) Parity error.

(*v*) Framing error

(*vi*) Overrun error.

Q.13. How many bits are there in the transmitter shift register of figure given in Q.11 when the interface is attached to a terminal that needs one stop bit? List the bits in the shift register when the letter W is transmitted using ASCII with even parity.

Solution:

10 bits: start bit + 7 ASCII + parity + stop bit.

From Table 11 - 1 ASCII W = 1 0 1 0 1 1 1

with even parity = 1 1 0 1 0 1 1 1

with start and stop bits =1 1 1 0 1 0 1 1 1 0

Q.14. How many characters per second can be transmitted over a 1200-band line in each of the following modes? (Assume a character code of eight bits).

(*a*) Synchronous serial transmission.

(*b*) Asynchronous serial transmission with two stop bits.

(*c*) Asynchronous serial transmission with one stop bit.

Solution:

(*a*) $\frac{1200}{8} = 150$ characters per second (cps)

(*b*) $\frac{1200}{11} = 109$ cps

(*c*) $\frac{1200}{10} = 120$ cps

Q.15. Information is inserted into a FIFO buffer at a rate of m bytes per second. The information is deleted at a rate of n byte per second. The maximum capacity of the buffer is k bytes.

(*a*) How long does it take for an empty buffer to fill up when $m > n$?

(*b*) How long does it take for a full buffer to empty when $m < n$?

(*c*) Is the FIFO buffer needed if $m = n$?

Solution:

(*a*) $\frac{k \text{ bytes}}{(m-n) \text{ bytes/sec}} = \frac{k}{(m-n)}$ sec

(*b*) $\frac{k}{m-n}$ sec.

(*c*) No need for FIFO.

Q.16. The bits in the control register of the FIFO shown in figure given below are $F_1\ F_2\ F_3\ F_4$ = 0011. Give the sequence of internal operations when an item is deleted from the FIFO and then a new item is inserted.

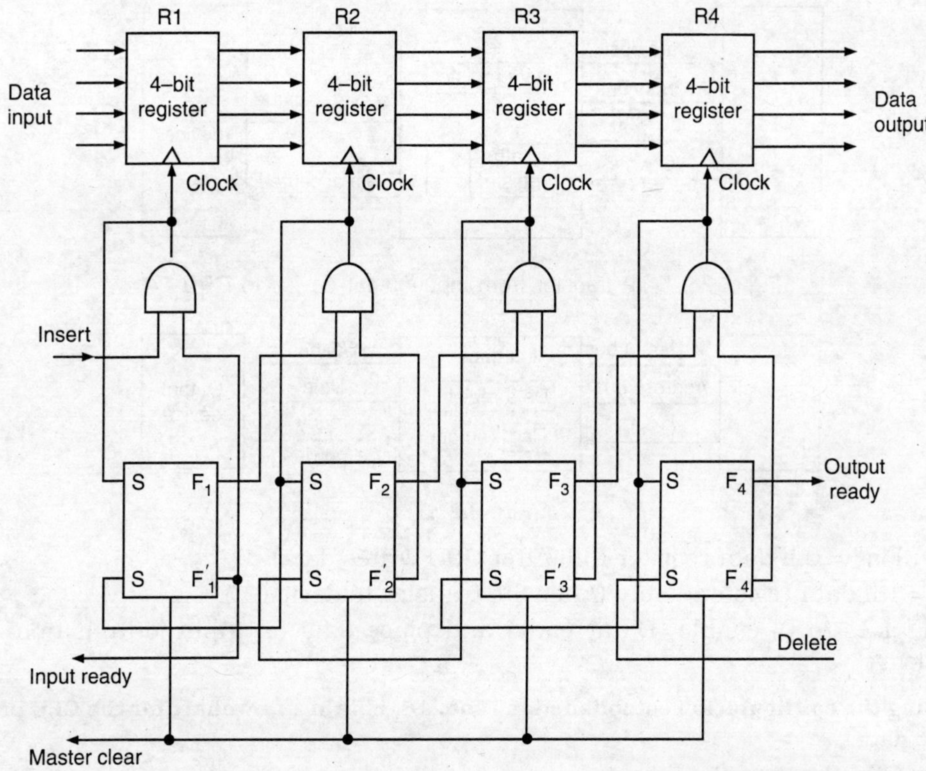

Fig. 8.25.

Solution:

Initial	$F = 0\,0\,1\,1$	Output $\leftarrow R_4$
After delete = 1	$F = 0\,0\,1\,0$	
After delete = 0	$F = 0\,0\,0\,1$	$R_4 \leftarrow R_3$
After insert = 1	$F = 1\,0\,0\,1$	$R_1 \leftarrow$ Input
(Insert goes to 0)	F = 0 1 0 1	$R_2 \leftarrow R_1$
	F = 0 0 1 1	$R_3 \leftarrow R_2$

Q.17. What are the values of input ready and output ready and control bits F_1 through F_4 in figure given in Q.I6 when

(*a*) the buffer is empty ?

(*b*) the buffer is full ?

(*c*) the buffer contains two data items ?

Solution:

	Input ready	Output ready	$F_1 - F_4$
(*a*) Empty buffer	1	0	0 0 0 0
(*b*) Full buffer	0	1	1 1 1 1
(*c*) Two items	1	1	0 0 1 1

Q.18. Show a block diagram similar to figure given below for the data transfer from a CPU to an interface and then to an input/output device. Determine a procedure for setting and clearing the flag bit.

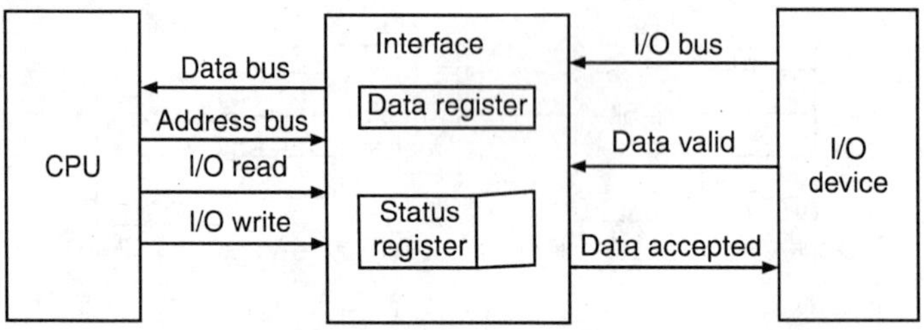

Fig. 8.26. Data transfer from input/output device to CPU

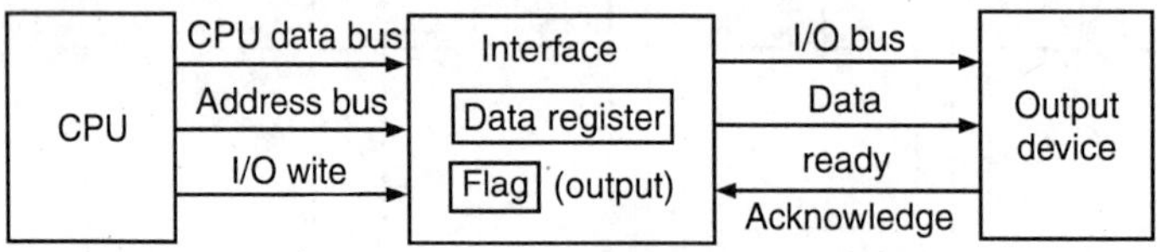

Fig. 8.27.

Solution: Flag = 0 if data register full (After CPU writes data)

Flag = 1 if data register empty (After the transfer to device)

when flag goes to 0, enable "Data ready" and place data on input/output bus. When "Acknowledge"

Q.19. Using the configuration established in Prob. 18, obtain a flowchart for the CPU program to output data.

Solution: CPU Program flow chart.

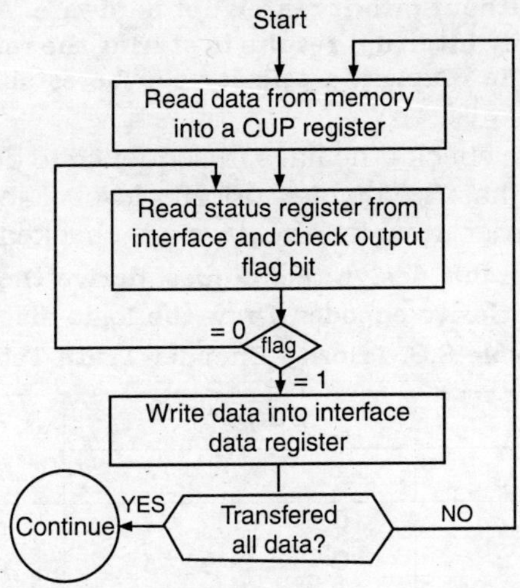

Fig. 8.28.

Q.20. In most computers an interrupt is recognized only after the execution of the instruction. Consider the possibility of acknowledging the interrupt of any time during the execution of the instruction. Discuss the difficulty that may angels.

Solution: If an interrupt is recognized in the middle of an instruction execution, it is necessary to save all the information from control register in addition to processor register. The state of the CPU to be saved is more complex.

Q.21. What happen in the daisy-chain priority interrupt shown in figure given below when device 1 requests an interrupt after device 2 has sent an interrupt request to the CPU but before the CPU responds with the interrupt acknowledge?

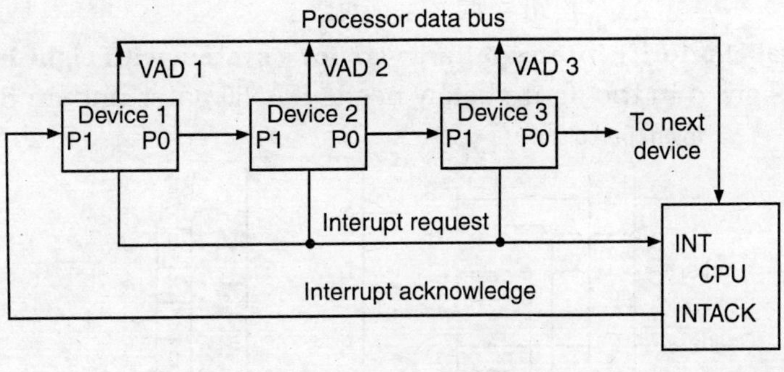

Fig. 8.29. Daisy-chain priority interrupt

Solution:

	Device 1	**Device 2**
1. Initially, device 2 sends an interrupt request	PI = 0; PO = 0; RF = 0	PI = 0; PO = 0; RF = 1
2. Before CPU responds with a acknowledge device 1 sends interrupt request:	PI = 0; PO = 0; RF = 1	PI = 0; PO = 0; RF = 1
3. After CPU sends an acknowledges device 1 has priority:	PI = 1; PO = 0; RF = 1 VAD enable = 1	PI = 0; PO = 0; RF = 1 VAD enable = 0

Q.22. Consider a computer without priority interrupt hardware. Any one of many source can interrupt the computer, and any interrupt results in staring the return address and branching to a common interrupt routine. Explain a priority can be established in interrupt service program.

Solution: The service routine checks the flags in sequence to determine which one is set. The firs flag that is checked has the highest priority level. The priority level of the other source-corresponds to the order in which the flags are checked.

Q.23. Using combinational circuit design techniques, derive the Boolean expression listed in Table given below for the priority encoder. Draw the logic diagram of the circuit.

Table 8.6. Priority Encoder Truth Table.

Inputs				*Outputs*			
I_0	I_1	I_2	I_3	x	y	1ST	*Boolean functions*
1	x	x	x	0	0	1	
0	1	x	x	0	1	1	$x = I'_0 I'_1$
0	0	1	x	1	0	1	$y = I'_0 I'_1 + I'_0 I'_2$
0	0	0	1	1	1	1	(IST) $= I_0 + I_1 + I_2 + I_2$
0	0	0	0	x	x	0	

Solution:

I_0	I_1	I_2	I_3	x	y	1ST
1	x	x	x	0	0	1
0	1	x	x	0	1	1
0	0	1	x	1	0	1
0	0	0	1	1	1	1
0	0	0	0	x	x	0

Map Simplification

$$x = I'_0 I'_1 \quad y = I'_0 I'_1 + I'_0 I'_2$$

Q.24. Design a parallel priority interrupt hardware for a system with eight interrupt sources.

Solution: Same as given in the figure below needs 8 AND gates and an 8 × 3 decoder.

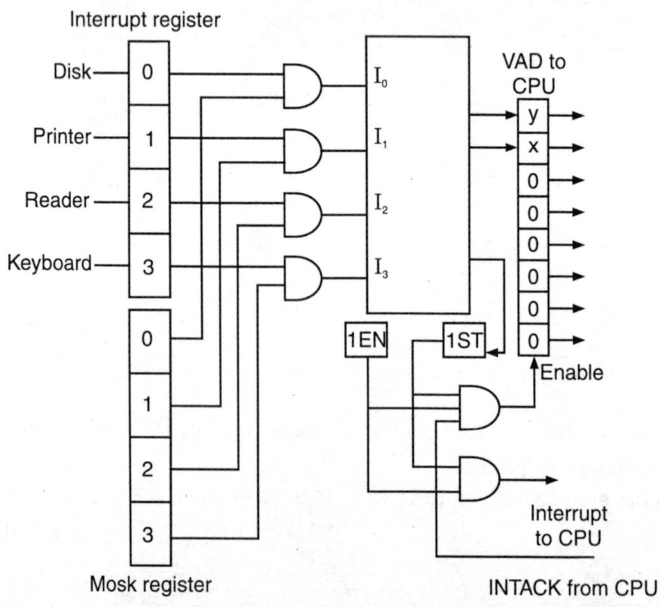

Fig. 8.30. Priority interrupt hardware

Q.25. Obtain the truth table of an 8 × 3 priority encoder. Assume that the three outputs *xyz* from the priority encoder are used to providing a vector address of the form 101*xyz*00. List the eight vector addresses starting from the one with the highest priority.

Solution:

I_0	I_1	I_2	I_3	I_4	I_5	I_6	I_7	*xyz*	*IST*
1	x	x	x	x	x	x	x	000	1
0	1	x	x	x	x	x	x	001	1
0	0	1	x	x	x	x	x	010	1
0	0	0	1	x	x	x	x	011	1
0	0	0	0	1	x	x	x	100	1
0	0	0	0	0	1	x	x	101	1
0	0	0	0	0	0	1	x	110	1
0	0	0	0	0	0	0	1	111	1
0	0	0	0	0	0	0	0	xxx	0

(b)

Binary	hexadecimal
10100000	A0
10100100	A4
10101000	A8
10101100	AC
10110000	B0
10110100	B4
10111000	B8
10111100	BC

Q.26. What should be done in figure given in Q.24 to make the four VAD values equal to the binary equivalent of 76, 77, 78 and 79 ?

Solution: $76 = (01001100)_2$

Replace the six 0's by 0 1 0 0 1 1 *xy*

Q.27. What programming steps are required to check when a source interrupts the computer while it is still being serviced by a previous interrupt request from the same source?

Solution: Set the mask bit belonging to the interrupt source so it can in terrupt again. At the beginning of the service routine check the value of the return address within the source service program, then the same source has interrupted again while being serviced.

Q.28. Why are read and write control lines in a DMA controller bidirectional ? Under what condition and for what purpose are they used as inputs ? Under what condition and for what purpose are they used as outputs ?

Solution: *When* the. CPU communicates with the DMA controller the read and write lines are used as inputs from the CPU to the DMA controller.

When the DMA controller communicate with memory, the read and write lines are used as outputs from the DMA to memory.

Q.29. It is necessary to transfer 256 words from a magnetic disk to a memory section starting from address 1230. The transfer is by means of DMA as shown in figure given below.

(*a*) Give the initial values that the CPU must transfer to the DMA controller.

(*b*) Give the step-by-step account of the actions taken during the input of the first two words.

Solution: (*a*) CPU inititties DMA by transferring:

256 to the words count register.

1230 to the DMA address register.

Bits to the control register to specify a write operations.

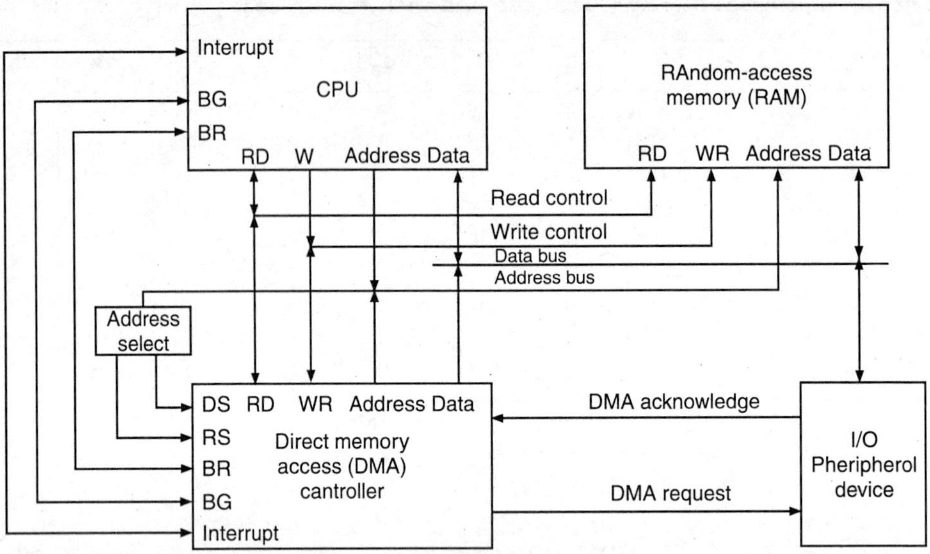

Fig. 8.31. DMA transfer in a computer system

(b) (*i*) input/output device sends a "DMA" request."

(*ii*) DMA sends BR (bus request) to CPU.

(*iii*) CPU responds with *a* BG (bus grant).

(*iv*) Contents of DMA address register are placed in address bus.

(*v*) DMA sends "DMA acknowledge" to input/output device and enables the write control line to memory.

(*vi*) Data word is placed on data bus by input/output device.

(*vii*) Increment DMA address register by 1 and decrement DMA word count register by 1.

(*viii*) Repeat steps 4-7 for each data word transferred.

Q.30. A DMA controller transfers 16-bit words to memory using cycle stealing. The words are assembled from a device that transmits characters at a rate of 2400 characters per second. The CPU is fetching and executing instructions at an average rate of 1 million instructions per second. By how much will the CPU be slowed down because of the DMA transfer?

Solution: CPU refers to memory on the average once (or more) every 2 × 416.6 = 833.3 μ sec.

Characters arrive one every 1/2400 = 416.6 μ sec.

Two characters of 8 bits each are packed into a 16-bit word every 2 × 416.6

= 833.3 μ sec

The CPU is slowed down by no more than (1/833.3) × 100 = O12%.

Q.31. Why does DMA have priority over the CPU when both request a memory transfer?

Solution: The CPU can wait to fetch instructions and data from memory without any damage occurring except loss of time. DMA usually transfer data from a device that cannot be stopped since information continues to flow so loss of data may occur.

Q.32. Draw a flowchart similar to the one is figure given below that describes the CPU-input/output channel communication in the IBM 370.

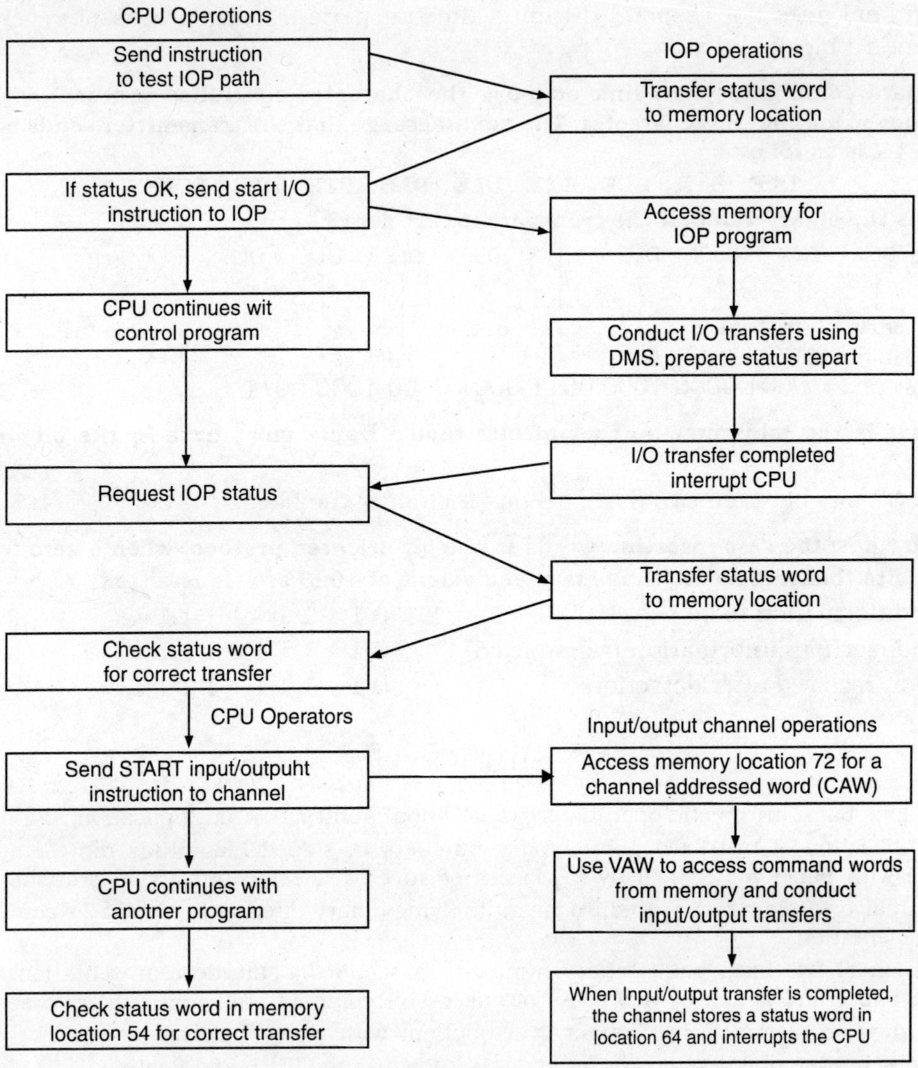

Fig. 8.32.

Q.33. The address of a terminal connected to a data communication processor consists of two letters of the alphabet or a letter followed by one of the 10 numerals. How many different addresses can be formulated.

Solution: There are 26 letters and 10 numerals.

$$26 \times 26 + 26 \times 10 = 936 \text{ possible addresses}$$

Q.34. List a possible line procedure and the character sequence for the communication between a data communication processor and a remote terminal. The processor inquires if the terminal is operative. The terminal responds witrh yes or no. If the response is yes, the processor sends a block of text.

Solution: The processor transmits the address of the terminal followed by ENQ (enquiry) code 00000101.

The terminal responds with either ACK (acknowledge) or NAK (negative Acknowledge) or the terminal does not respond during a timeout period. If the processor receives an ACk, it sends a block of text.

Q.35. A data communications link employs the character controlled protocol with data transparency using the DLE character. The next message that the transmitter sends between STX and ETX is as follows:

DLE STX DLE ETX DLE DLE ETX DLE ETX

What is the binary value of the transparent text data?

Solution:

DLE	STX	DLE	DLE	ETX	DLE	DLE	ETX	DLE	ETX
delete	STX	delete	DLE	ETX	delete	DLE	ETX	delete	ETX

32 – bit text = 0001 0000 1000 0011 0001 0000 1000 0011

Q.36. What is the minimum number of bits that a frame must have in the bit oriented protocol?

Solution: 32 bits between two flags; 48 bits including the flags.

Q.37. Show how the zero insertion words in the bit-oriented protocol when a zero followed by the 10 bits that represents the binary equivalent of 1023 are transmitted.

Solution: Information to be sent (1023): 0 1 1 1 1 1 1 1 1 1 1

After zero insertion, information transmitted: 0 1 1 1 1 1 0 1 1 1 1 1 0

Information received after detection: 0 1 1 1 1 1 1 1 1 1 1

SUMMARY

1. The four basic arithmetic operations are addition, subtraction multiplication and division.
2. Multiplication of two fixed point binary numbers in signed magnitude representation is done with paper and pencil by a process of successive shift and add operations.
3. Booth algorithm gives a procedure for multiplying binary integers in signed 2's complement representation.
4. Division of two fixed point binary numbers in signed magnitude representation is done with paper and pencil by a process of successive compare shift and sub tract operation.
5. The division operation may result in a quotient with an over flow.
6. Representation of floating point numbers is known as the IEEE 754 floating point standard.
7. The IEEE 754 standard has two main representation
   → Single precision
   → Double precision
8. Devices that are under control of the computer are said to be connected on line.
9. Input-output interface provides a method for transferring information between internal storage and external input/output devices.
10. The input/output bus consists of data lines, address lines and control lines.
11. The are three that computer buses can be used to communicate with memory and input/output
    → Use two sperate buses, one for memory and other for input/output
    → Use one common bus for both memory and input/output but have sperate control lines for each

→ Use one common bus for memory and input/output with common control lines.

12. Every input/output activity is initiated by input/output instruction written in the code.
13. An interrupt generated internally by CPU is sometimes termed as trap.
14. IOP is the input output processor
15. Computer components can be broadly divided into two categories - Hardware and Software.
16. Hardware refers to any physical component of a computer. For example, CPU, Monitor, Keyboard, Hard Disk, Floppy Disk, etc.
17. Software refers to the programs which are required to operate the computer
18. All hardware devices except Mother Board are called peripheral devices
19. Motherboard, also called as system board, is the most important hardware component of a microcomputer.
20. Input devices are used to input data, information and instructions into the RAM.
21. Coming to the display devices which is one of the most important peripherals in computer is the graphic display device for display of graphs, diagrams and pictures to make the visual presentation of information more effective for user interaction and decision making.
22. Graphic display is made up of a series of dots called 'pixels' whose pattern produces the image. Each dot on the screen is defined as a separate unit which can be directly addressed. Since each dot can be controlled individually there is much greater flexibility in drawing pictures. There are three categories of display screen technology.
23. CRT Displays: The main components of a CRT terminal are the electron gun, the electron beam controlled by an electromagnetic field and a phosphor coated display screen.
24. The electron gun emits an electron beam which is directed towards the phosphor coated display by the electromagnetic field in order to create an image. The quality of display is indicated by the resolution of the display device. Resolution is determined by the number of pixels horizontally and vertically.

EXERCISE

(I) Fill in the Blanks (Solved)

1. The .... runs against a server, a multiport computer that contains large amounts of memory, allowing multiple terminals to share the server's resources.
2. .... refers to any enhancement to existing IBM VGA standard.
3. The function of .... chip is to support the CPU chip in processing of mathematical calculations.
4. The .... were used as an input medium in electro-mechanical calculating equipment.
5. The layout of the keyboard is like that of the traditional .... typewriter.

(II) True or False (Solved)

1. Input devices are used to input data, information and instructions into the Hard Disk
2. Digitiser is used to create drawings and pictures using digitiser tablet by a process called digitising.
3. Joystick is a pointing device, which is used to select a displayed menu item or draw pictures on the monitor screen
4. OBR (Optical Bar code Reader) devices can sense marks on computer readable paper.
5. A Drum Printer is a line printer in which a revolving cylinder acts as the printing element.

(III) Fill in the Blanks (Unsolved)

1. .... is also called as system board.
2. The function of math co-processor chip is to support the .... in processing of mathematical calculations.

3. Most keyboards have .... sections of keys.
4. .... is a handy device which can be moved on a smooth surface to a simulate the movement of cursor that is desired on the display screen.
5. .... is a pointing device, which is used to select a displayed menu item or draw pictures on the monitor screen.

(IVI) True or false (Unsolved)

1. Both hardware and software are dependent on each other.
2. Motherboard is also called as peripheral.
3. The punched cards were used as an output medium.
4. The keyboard is one of the most common output device for computers.
5. Resolution is determined by the number of pixels horizontally and vertically

(V) Detailed Questions

1. Explain the term Super VGA.
2. What is the difference between Impact Printing and Non-Impact Printing?

Answers

(I) Fill in the Blanks (Solved)

1. Terminal.
2. Super VGA.
3. Math co-processor.
4. Punched card.
5. QWERTY

(II) True or False (Solved)

1. False.
2. True.
3. False.
4. False.
5. True

Glossary

Glossary

ACCESS TIME: The length of time required for a binary WORD in the memory section of a computer to be read by the CENTRAL PROCESSING UNIT (CPU), or the time to read data from a peripheral data storage area.

ACCUMULATOR: An interface REGISTER (memory) in the ARITHMETIC LOGIC UNIT (ALU), that stores interim arithmetic information for future processing. The accumulator is interface between the ALU and other sections of the computer.

ADDER: The digital logic circuits in the ALU section of a computer which implements the adding process (sum and carry) of two or more binary numbers.

ADDRESS: The location of digital information in the MEMORY UNIT of a computer, or a digital code that designates this location.

ALGORITHM: The sequence of operations that defines a solution to a problem in logic.

ALPHANUMERICS: The term that defines the letters of the alphabet (A to Z) and the ten numerals (0 to 9). The term is generally used to mean any text data.

AMERICAN STANDARD CODE FOR INFORMATION INTERCHANGE (ASCII): A 7-bit binary code, providing 128 different binary combinations for standard American keyboards. ASCII is used to encode all 26 letters of the alphabet (upper and lower case), all ten decimal digits (0 to 9), punctuation marks, standard graphics, and special control codes into machine language. Although ASCII has 128 different codes, only 7 bits are needed for each different code. ASCII characters are generally stored inside 8-bit bytes, providing room for the 128 ASCII codes plus another 128 codes, totaling 256 characters. This 8-bit code is referred to as EXTENDED ASCII.

ANALOG COMPUTER: A non-digital system that manipulates linear, or continually changing data.

AND GATE: See LOGIC GATES

ARITHMETIC LOGIC UNIT (ALU) ; The section of a computer that contains a large amount of logic circuitry and performs the four basic arithmetic functions (addition, subtraction, multiplication, and division). Larger units contain circuitry for higher mathematical functions, such as: quadratic equations, etc.

ASCII DATA (ASCII FILE): Data, or the file containing the data, that is constructed with letters of the alphabet, numerals, punctuation marks, and the standard ASCII formatting commands, such as: "carriage return" or "line feed". These terms refer to text information.

ASSEMBLER: A software program that converts (translates) each symbolic instruction written in ASSEMBLY LANGUAGE into the MACHINE LANGUAGE (binary code) of a computer.

ASSEMBLY LANGUAGE: A programming language (source code) that consists of a group of coded letters or labels, called mnemonics. A mnemonic is a memory assist to help recall data. Each mnemonic represents a single instruction that is translated into the binary code of machine language. Mnemonics are easier to use than machine language instructions. For example, the mnemonic "MUL" tells the computer to "MULTIPLY".

ASYNCHRONOUS: A term that describes a non-clocked, or free-running digital signal that triggers successive computer instructions; the completion of one instruction triggers the next. The speed of operation depends only on the speed of the signal generated through the circuit or network. In contrast with SYNCHRONOUS operation, the computer clock controls the speed of the signals in the system.

BACKPLANE: A part of a computer into which the system's PC boards are plugged to provide a common voltage supply, reference, and SYSTEM BUS for all computer sections. A backplane is often called a MOTHERBOARD.

BASIC: The acronym for Beginner's All-purpose Symbolic Instruction Code. BASIC is an easy to use programming language, originally intended for educational purposes; it is available for personal computers in varying degrees of complexity.

BAUD RATE: A measure of speed (up to 2400 bits per second) of transferring information between two or more sections of a computer system, or between two or more computers.

BAUDOT CODE - A 5-bit digital code having 32 possible combinations of binary "0s" and "1s"; for Teletype (Telex) communications systems. BENCHMARK - The specifications for measuring the characteristics of a computer system, or parts of the system, under clearly-defined conditions.

BINARY CODED DECIMAL (BCD): A coding system in which each decimal system numeral (0 to 9) is represented by a 4-digit (4-bit) binary code.

BINARY SYSTEM (BASE 2): A numbering system consisting of only two digits (0 and 1), as contrasted with a DECIMAL SYSTEM that uses ten digits (0 to 9). In electronics, "binary" and "two-state" are synonymous.

BINARY DIGIT (BIT): The term "BIT" is the contraction of Binary DigiT and is part of a binary WORD that consists of combinations of "0s" and/or "1s". There are only two numerals in binary arithmetic (base 2) and is the basis for binary code (Machine Language), the language of the computer. A "bit" has the same significance in binary arithmetic that a decimal digit has in the more familiar decimal (base 10) arithmetic system.

BISTABLE MULTIVIBRATOR: A digital circuit having two stable electrical states. The output signal of this circuit will remain in either state (0 or 1) indefinitely until an external input signal causes the circuit to switch to its other output state. This circuit is generally called a FLIP-FLOP.

BIT: See BINARY DIGIT.

BLACK BOX: A term that refers to an electronic circuit or system within an enclosure without necessarily providing details of its internal elements. The black box concept often offers a useful approach in the design of a system or in the interconnect between two circuits or systems.

BOOLEAN ALGEBRA: Named after George Boole, a 19th century English Mathematician, who first formulated theorems that included a mathematical analysis of the laws of human logic. It uses algebraic-like notation to describe the interaction of variables having only two states - "true" and "false". In electronics, the states are often referred to as "1" and "0" or, "high" and "low". The terms "high" and "low" refer to the voltage levels of the input and output signals of a logic gate (logic circuit). Boolean algebra is used in the design of logic circuits in computers and similar digital systems. The logic gates called AND, OR, and NOT are the three basic logic gates in Boolean algebra. NAND, NOR, XOR and others are combinations of the three basic logic gate operations.

BOOTSTRAP (BOOT): A software program for initiating the operation of a computer. The function of the program is to set up the input and output (input/output) devices and load the OPERATING SYSTEM from a disk, cassette, or built-in READ ONLY MEMORY (ROM).

BREAKPOINT: Location of a place in a program where program execution can be stopped to permit a visual test, printing, or other performance analyses.

BUBBLE MEMORY: A high-density memory medium upon which a magnetic film is grown on a gadolinium-gallium garnet substrate. A small permanent magnet is mounted inside its package, perpendicular to the surface of the substrate. When an external magnetic field is created with an external coil, magnetic "bubble" domains are formed on the internal magnetic film which represent patterns of "1s". The absence of magnetic bubbles will represent patterns of "0s".

BUFFER: 1. A digital logic circuit inserted between other digital circuits to reduce circuit interaction and/or to provide amplification of a digital signal. 2. An intermediate storage circuit used to compensate for a difference in baud rate or, to compensate for different times of occurrence of different events or instructions. 3. A circuit for converting input or output voltages for signal level compatibility when transmitting data from one device to another.

BUNDLED: A term describing a computer system that includes all necessary hardware and software that will allow the system to operate as advertised.

BUS (BUSS): A series of transmission lines connecting the various elements of a computer for distribution of data, control signals, addresses, and/or voltage supply(s) within a computer. See SYSTEM BUS.

BUS CONTROLLER: A circuit that generates commands and control signals for sequencing and timing of the data transmitted on a bus.

BYTE: A set (group) of consecutive binary digits (bits) that forms a unit (word) or sub-unit of digital information in a computer. Depending on the size of the computer, a digital word may contain one or more bytes. Through common usage, a byte is generally accepted as containing eight bits.

CACHE: A high-speed buffer memory that is similar to a SCRATCH PAD MEMORY, but has a larger capacity; it is located in the CENTRAL PROCESSING UNIT. It is filled from the main memory at low speed with instructions and programs and is operated at a higher speed. A cache can also serve as a storage section between the CPU and a hard disk to speed up access of data on the hard disk.

CD-ROM (COMPACT DISK READ-ONLY MEMORY): An aluminized disk, 4.72" in diameter that provides an optical storage medium for digital data up to 650 Megabytes. A laser beam permanently burns data into its surface which cannot be over-written, altered, or erased, providing read-only memory capability. Larger size disks are generally called "laser disks".

CD-R or WORM (COMPACT DISK, WRITE ONCE, READ MANY MEMORY): A blank disk that can be programmed once, permanently storing up to 650 Megabytes of digital data. Once programmed, it has essentially the same features as a CD-ROM.

CD-E or CD R/W (COMPACT DISK, ERASABLE, READ/WRITE MEMORY): Capable of storing up to 650 Megabytes of digital data. It uses a magneto-optical technology that combines the storage capacity and reliability of optical disks and the erasability of magnetic media.

CENTRAL PROCESSING UNIT (CPU): The section(s) in which all the processing circuits of a computer are located. It incorporates a CONTROL UNIT and an ALU.

CHARACTER GENERATOR: A circuit that accepts digital data and forms the corresponding letters and numerals for a monitor or printer.

CLOCK: A digital pulse generator that controls the timing of a computer and, to a great extent, determines the speed (number of instructions per second) capability of the computer. Generally, it is located in the CPU.

CMOS LOGIC: A CMOS technology monolithic IC "logic family" characterized by low power dissipation per gate, high chip density, and relatively high propagation delay per gate compared with bipolar IC logic families.

COMPILER: A software program that converts (translates) a complete software program written in high-level language SOURCE CODE (such as PASCAL or FORTRAN) into machine language. The entire source code is edited, compiled, and run at one time as compared to an INTERPRETER that is run one line at a time.

COMPLEMENT: Reversal of bit values: "1s" become "0s" and "0s" become "1s".

CONCURRENCY: The independent execution of two or more sequences of events that are either occurring, or appearing to occur simultaneously.

CONSOLE: The term referring to a combination of a DISPLAY and KEYBOARD.

CONTROL BUS: A set of transmission lines whose function is to carry synchronization signals and control data as part of the SYSTEM BUS.

CONTROL UNIT (CU): Part of the CPU containing the CLOCK, PROGRAM COUNTER and INSTRUCTION REGISTER. The CONTROL UNIT also generates control signals and manages the CONTROL BUS.

COUNTER: A circuit whose output(s) change state in a specified sequence on receiving appropriate input signals. The circuit can provide a required output pulse after receiving a specified number of input pulses.

CRASH: The term that describes a situation when a part of, or the complete computer, stops working because of a hardware and/or software malfunction. A head-crash in a disk system refers to the accidental impact of the read/write head on the surface of the disk.

DAISY-WHEEL PRINTER: An impact printer that uses a print element shaped like a flat disk or large thimble to form the alphanumerics and punctuation marks that are part of its print element. Unlike dot-matrix, ink-jet, and laser printers, daisy-wheel printers can not print graphics. See LETTER-QUALITY PRINTER.

DATA BASE: A collection of structured information for creating a specific format out of some, or all, of the collected data.

DATA-BASE MANAGEMENT SYSTEM (DBMS): A software program that controls and supervises the updating, editing, and execution of items from multiple files in a data base environment.

DEBUG: The process of detecting, locating, and correcting a problem in a software program or hardware.

DECREMENT: The reduction of the numerical contents of a counter. A decrement of 1 is usually assumed, unless otherwise specified. It is the complementary operation of INCREMENT.

DEVICE: In a computer system, this term refers to a unit of processing equipment external to the CPU. It is sometimes synonymous with the term PERIPHERAL.

DIODE-TRANSISTOR LOGIC (DTL): One of the first bipolar monolithic IC families of logic gates. A diode in an IC logic gate performs the required logic with a transistor amplifying and inverting the output. The DTL family has been made obsolete by the Transistor-Transistor Logic (TTL) family.

DIRECT ADDRESS: A memory accessing mode in which the contents of the accessed location is called the OPERAND.

DIRECT MEMORY ACCESS (DMA): A method of transferring blocks of data directly between an external device and the computer system memory without the need for intervention by the CPU. This method significantly speeds up the data transfer rate, improving system efficiency.

DISC or DISK (DISKETTE): A thin, flexible, plastic, circular memory medium, coated on both sides with metal oxide and enclosed in a protective jacket. Data is stored magnetically in binary digital form on its surface(s). See FLOPPY DISK.

DISK DRIVE: The mechanical/electronic section that can accept and operate a compatible floppy or hard disk. It may include several motors (for disk rotation and reading/writing head positioning), position sensors, and control circuits.

DISK EMULATOR: See RAM Disk.

DISPLAY: A peripheral device serving as a computer readout, such as a cathode-ray tube (CRT), flat-panel (generally for laptop computers), or other readout devices. The screen can be amber, green, or full-color display. Also called a MONITOR.

DISK OPERATING SYSTEM (DOS): A software program on a disk which coordinates the operation, transfer of data, supervision, and control of a computer. This software program must first be booted into the working memory of the computer from the disk before it can operate.

DISK STORAGE: A method of storing software programs and data on a rotating circular disk (either a floppy or hard disk) coated with magnetic material, such as iron oxide. Data is written (stored) and read (retrieved) by movable read/write heads positioned over data tracks on the surface of the disk. Addressable portions of the disk can be selected for read or write operations.

DOCUMENTATION: Information that explains how to use computer hardware or software. It is usually provided as a manual or stored on a disk.

DOT-MATRIX PRINTER: An impact printer with a computer-driven, multi-pin print element (print-head). It create images by imprinting a series of tiny dots on a paper to print a wide variety of character styles and/or finely detailed graphics. Generally, these printers are extremely fast and are used for draft-quality documents and precise graphics.

DOTS PER INCH (DPI): The measurement of density on dot-matrix printers or other dot-matrix devices. As the DPI increases, image clarity increases.

DOWNLOAD: The transfer of files or data from a source of data to a remote computer. See UPLOAD.

DOWN TIME: A period of time during which a computer is not functioning. It is inoperable because of temporary or permanent failure of hardware or software, or when routine hardware or software maintenance procedures are indicated.

DYNAMIC RANDOM ACCESS MEMORY (DRAM): A type of semiconductor memory in which the presence or absence of a capacitive charge in each element of the memory represents the state of the bit (1 or 0). This charge must be periodically recharged (refreshed) to maintain the desired binary state of the element.

EDITOR: A program for preparing and/or modifying a SOURCE PROGRAM or other file by addition, deletion, or change.

EMITTER-COUPLED LOGIC (ECL): A bipolar monolithic IC logic gate family characterized by very high speed operation and relatively high power dissipation compared with other monolithic IC logic families.

EMULATION: The process of imitation (simulation) of one computer system by another. The imitating program, or device (emulator), accepts the same data, executes the same programs, and achieves the same results.

EPROM (ERASABLE PROGRAMMABLE READ-ONLY MEMORY): A general term for a non-volatile, semiconductor memory that can be programmed, erased, and reprogrammed many times without damage to the device. More specific types of EPROMs include: EAPROM (ELECTRICALLY ALTERABLE PROM), EEPROM (ELECTRICALLY ERASABLE PROM) and UV-EPROM (ULTRAVIOLET ERASABLE PROM). Either memory is suitable for prototype and computer development, or to change memory data when new conditions dictate the change.

EVENT: An occurrence during the execution of a task, such as the completion of an input/output operation.

EXECUTE: The command to run a specified instruction or software program.

EXTENDED BINARY CODED DECIMAL INTERCHANGE CODE (EBCDIC): An 8-bit code developed by IBM for their mainframe computers, providing 256 bit-pattern equivalents of standard keyboard symbols.

FAN-IN: The maximum number of output terminals from other logic gates that can be connected to an input terminal of a specified logic gate.

FAN-OUT: The maximum number of input terminals that can be connected to the output terminal of a specified logic gate.

FETCH: The command to obtain an instruction from a stored program.

FILE: A collection of related data treated as a single unit. In a computer, a file can exist on a disk, magnetic tape, or as an accumulation of information in memory. FIRMWARE - A combination of a software program in hardware, such as a READ-ONLY MEMORY (ROM), or a disk that has files or software programs written on its surface.

FIRST IN, FIRST OUT (FIFO): The term refers to the sequence of entering and then retrieving data from a data storage section of a computer. The first data entered is the first data retrieved.

FLAG: An indicator of a specific condition that informs a section of a program that this condition has already occurred and is identified by the presence or absence of the flag. A flag can be implemented in software and/or hardware.

FLASH MEMORY (FLEXIBLE ARCHITECTURE for SHARED MEMORY): Non-volatile semiconductor memory with access time approximately that of EEPROM or EAPROM and density similar to that of an EPROM. FLASH memory is generally limited to about 100,000 erase cycles. Usually packaged in PCMCIA for plugging into a laptop computer card slot.

FLIP-FLOP CIRCUIT: A logic circuit having two stable output states. It has the ability to change from one state to the other when an input pulse is applied in a specified manner. It is also called a BISTABLE MULTIVIBRATOR.

FLOATING-POINT ARITHMETIC: A method in which the decimal point location of a number in an arithmetic operation is determined by the number's exponent value in base 10. All exponents are equalized prior to the operation to set a decimal point in its proper location in the final computation. Floating-point arithmetic extends a computer's mathematical capability beyond the limit imposed by a fixed word length and contributes to easier programming.

FLOPPY DISK (DISKETTE): A relatively inexpensive, flexible, plastic disk for storing digital data on both sides of its magnetic surface. The amount of data depends on the diameter of the disk and the technology for recording the data. See DISK STORAGE.

FLOW CHART: A symbolic representation of the processing steps performed by a software program or a graphic sequence of logic operations implemented in hardware. A flow chart helps to visualize the procedure(s) necessary to design a software program or a final hardware system.

FONT: A style of print typeface having a specific form and size.

FORMAT: An orderly, structured arrangement of data elements (bits, bytes, and/or fields) that is necessary to produce a larger entity, such as: a list, record, table, file, or dictionary. Also, it is the term that describes the preparation of a magnetic disk to allow it to accept digital data.

FORTRAN: A science-oriented high-level software language, derived from the contraction of the terms FORmula TRANslator.

FREQUENCY SHIFT KEYING (FSK): A technique of digitally keying (modulating) an audio tone (carrier wave) so that a commercial telephone line can carry digital data.

FULL DUPLEX: A data transmission mode that provides simultaneous and independent transmission and reception. A conventional telephone communication is an example of this technique. See HALF DUPLEX.

GATE: See LOGIC GATE.

GATE ARRAY: A group of standard logic gates that can be interconnected into a complete circuit or system. Also called LOGIC ARRAY.

GENERAL PURPOSE INTERFACE BUS (GPIB): A BUS specification standard (IEEE 488) for controlling peripheral devices.

GIGO (GARBAGE IN, GARBAGE OUT): The term describing the output of a computer whose operation or accuracy is faulty.

GRAPHICS: Schematic drawings, pictures, line drawings, and/or diagrams generated by data entered into a computer via a keyboard or a data base.

HALF-DUPLEX: A communications mode that allows transmission and reception of digital data between computers, but not simultaneously.

HANDSHAKING: A communications synchronizing technique carried out before and after any transfer of digital data. It consists of a sequence of signals for non-clocked (asynchronous) systems in which a reply is needed to complete a data transfer operation.

HARD COPY: A printed copy of a file, message, or graphic of the visual display on the screen of a computer monitor.

HARD-DISK DRIVE: A sealed unit containing high-density, high-speed, rigid metal disks, and recording heads to store digital data. It reads and writes data faster than floppy disks.

HARDWARE: The physical equipment of a computer system consisting of mechanical and electrical/electronic components.

HEXADECIMAL: The base 16 number system using 16 symbols (0 to 9 and A to F) to represent 16 decimal numerals (0 to 15).

HIGH-LEVEL LANGUAGE (HLL): A programming language (source code) consisting of a unique group of symbols and command statements representing a series of machine operations. A COMPILER or INTERPRETER translates (converts) a HLL into MACHINE LANGUAGE. BASIC, FORTRAN, PASCAL, ALGOL, and ADA are some examples of high-level languages.

IEEE 488: See GENERAL PURPOSE INTERFACE BUS

IMPACT PRINTER: A computer-driven mechanical imprinting device where the characters are formed by the printer-head key striking a ribbon to imprint the character's image onto a paper.

INCREMENT: The increase in the numerical contents of a counter. An increment of one is usually assumed, unless otherwise specified. See DECREMENT.

INK-JET PRINTER: A printer that forms characters by electrostatically aiming and depositing a tiny patterned drop of ink onto the paper to be printed.

INPUT/OUTPUT (input/output) SECTION: The section that interfaces between the computer's SYSTEM BUS and the peripherals feeding data into and taking data out of the computer. Depending on the number of peripherals in a system, the input/output sections can have a single PORT or multiple ports.

INSTRUCTION: A software statement that specifies a machine operation. Also called a COMMAND.

INSTRUCTION SET (INSTRUCTION REPERTOIRE): A description of the total operational capabilities of a computer provided by the computer or the CPU (MPU) manufacturer. It consists of a listing of binary words for each executable commands. It is sometimes called the computer's MICROCODE.

INTERPRETER: A high-level language translator that converts individual high-level computer language program instructions (source code) into machine instructions. It translates and executes each statement line-by-line during the running of the program.

INTERRUPT: The suspension of normal program execution to perform a higher priority service routine as requested by a peripheral device. After completion of the service routine operation, the interrupted program routine is resumed at the point where it was interrupted.

JUMP: An instruction that causes the computer to fetch the next instruction to be executed from a location other than the next sequential location in memory.

KEYBOARD: A peripheral device consisting of alphanumerics, punctuation marks, and other special function keys that are mechanically arranged to allow the entry of data, commands, and other information into the system.

LAN: See LOCAL-AREA NETWORK.

LANDSCAPE: A printer feature, generally controlled by software, which rotates the output image by 90° to print across the length rather than the width of the paper.

LASER (LIGHT AMPLIFICATION by STIMULATED EMISSION of RADIATION): A system that generates high-intensity, highly-focused light for many purposes, including printers, high-density memory media, and a light generator for fiber optic transmission systems.

LASER DISK: See CD-ROMLASER PRINTER: A computer-driven photocopier that creates an original image of the text or graphics from the output of the computer. A computer-controlled laser beam "paints" the desired image inside the photocopier and then prints the image on a sheet of paper.

LAST-IN, FIRST-OUT (LIFO): A method of storing and retrieving data in a stack, table, or list.

LIGHT PEN: A light-sensitive stylus for forming graphics by touching coordinates on a display screen, thereby seeming to draw directly on the screen.

LIBRARY: A collection of standard software instructions, programs, routines, and subroutines in a computer's memory.

LINKAGE: Instructions that connect one program to another, providing continuity of executions between the programs LOCAL AREA NETWORK (LAN) – A combined hardware/ software technique for interconnecting company related multiple computers or computer terminals through a high-speed networking system.

LOGIC GATE: A digital circuit resulting in an output whose state (0 or 1) depends on the specific combination of the states of input signals. Definitions of the more commonly used logic gates are listed below:

AND: All inputs must be in a "1" state to produce a "1" state output.

NAND (NOT AND): All inputs must be in a "1" state to produce a "0" state output.

OR (NOT OR): Any one input, or more, in a "1" state will produce a "0" state output.

NOT (INVERTER): A logic gate having only one input and one output. If the input is in a "1" state, the output is in a "0" state and vice–versa.

OR: Any one input, or more, in a "1" state will produce a "1" state at the output.

XOR (EXCLUSIVE OR): If any of the inputs are in a "1" state, but not if two or more inputs are "1", the output is in a "1" state.

LOOK AHEAD: 1. A feature of a CPU which allows the masking of an interrupt request until the current sequential instruction has been completed. 2. A feature of an adder circuit in the ALU section which allows the circuit to look ahead to see that all the generated arithmetic carrys are available for addition.

LOOPING: The repetition of program instructions until a conditional exit situation is encountered.

MACHINE LANGUAGE: Sets of numeric binary code instructions in a computer which execute its operations. All other programming languages (SOURCE PROGRAMS) must be translated into machine language (OBJECT PROGRAM) before entering the CPU.

MACRO: A combination of commands, instructions, or keystrokes which may be stored in a computer's memory to be executed as a single command by a single keystroke or a simultaneous combination of keystrokes.

MACROASSEMBLER: An assembly language translator that converts macro expressions into several machine language instructions. Although macros simplify program coding and speed up execution of a program, a code for each macro must also be generated.

MAINFRAME COMPUTER: Second largest of the computer family, in capability and, generally, in size, having a WORD-WIDTH of 32 bits and higher. The largest computer is the SUPER COMPUTER with a WORD WIDTH of 64 to 128 bits.

MEMORY: A data storage structure in a computer that accepts binary information for storage in electrical, mechanical, or magnetic form, and retains the information for as long as needed. Selected data can be written, moved, displayed, copied, or erased. The variety of memory media includes: paper, magnetic, bubble, optical laser, and semiconductor memory. Each one has a different method of operation and a specific criterion for being selected.

MICROCOMPUTER: A microprocessor-based computer, consisting of an MPU, internal semiconductor memory, input and output sections, and a system bus, all on one, or several monolithic IC chips inserted into one or several PC boards. The addition of a power supply and connecting cables, appropriate peripherals (keyboard, monitor, printer, disk drives, etc.), an operating system and other software programs can provide a complete microcomputer system. The microcomputer is generally the smallest of the computer family, however, the improvement in performance capability of newer microcomputer systems can make the microcomputer as powerful as larger systems.

MICROPROCESSOR UNIT (MPU): The Central Processor Unit (CPU) implemented in monolithic IC technology, usually, but not necessarily, on one VLSI chip. In many cases, the SYSTEM BUS is also included on the MPU chip.

MINICOMPUTER: Considered to be more capable than a microcomputer but less powerful than a mainframe. Generally, the WORD-WIDTH of the minicomputer is between 12 to 32 bits. MNEMONIC - A symbolic label or code reminder that assists the user in remembering a specific operation or command.

MODEL: A representation of a process or system that can be controlled to demonstrate the effects that various actions will have on the process or system.

MODEM: An acronym for MOdulator/DEModulator that refers to specific equipment that provides a means of communication between two computer systems over conventional telephone lines. Each remote computer requires its own MODEM and a compatible communications software program for proper interfacing.

MONITOR: The visual readout device of a computer system. A monitor can be in several forms:, a Cathode Ray Tube (CRT), a liquid crystal display (LCD), or a flat-panel, full-color display. See DISPLAY.

MOUSE: A manually operated input device for moving or entering positional information and other data or commands by accessing (pointing to) images on a monitor.

MULTITASKING: The technique of using several applications programs (tasks) in a computer system or on several terminals in a network at the same time. Multitasking can simultaneously work with several programs or interrelated tasks that share memories, codes, buffers, and files.

MULTIUSER: The term describing the capability of a computer system to be operated at more than one terminal at the same time.

NEGATIVE LOGIC: This term refers to logic in which the negative voltage represents the "1" state and the zero voltage represents the "0" state.

NESTING: Embedding commands or data in levels of other data so that specific routines or instructions can be executed or accessed continuously in loops, without returning to the main program.

NIBBLE: A sequence of four adjacent bits, or a half-byte. A hexadecimal or BCD coded digit can be represented by a nibble.

NODE: The endpoint of a network branch or the junction of two or more branches.

NON-VOLATILE MEMORY: A memory where stored data remains undisturbed by the removal of electrical power.

OBJECT CODE: Machine language code produced by a translator program, such as an assembler, interpreter, or compiler. Instructions in object code can be executed by a Central Processing Unit (CPU).

OPERATING SYSTEM (OS): A structured software program (set of programmed routines) that manages, controls, coordinates, and sequences the hardware and software resources of a computer system. See BOOTSTRAP. The operating system is sometimes called SYSTEMS SOFTWARE.

OPERATION CODE (OP-CODE): Part of a computer instruction word that designates the function performed by a specific instruction. For example, op-codes for arithmetic instructions include: "ADD", "SUB", "MUL" and "DIV".

OVERFLOW: An error condition occurring in a computer when a mathematical operation produces a result having a magnitude that exceeds the capacity of the computer's arithmetic register.

PASCAL: A high-level programming language that is structured to encourage efficient programming habits (documentation) and is used extensively in educational institutions and engineering environments.

PARALLEL OPERATION: A method of data transmission in which all bits of a digital word are handled simultaneously with each bit on a separate line. Although faster and simpler to install and operate than SERIAL OPERATION, this method requires more transmission lines (real estate).

PARITY: A method of verifying the accuracy of binary data after it has been transferred to or from a storage area.

PARTITIONING: The logical grouping of electronic functions within a given set of hardware components.

PCMCIA (PERSONAL COMPUTER MEMORY CARD INTERNATIONAL ASSOCIATION): A package in a plastic card form containing semiconductor memory, particularly FLASH. These cards are plugged into the card slot in laptop computers.

PERIPHERAL: A term designating the various kinds of machines and devices that work in conjunction with a computer but are not necessarily part of the computer structure. Typically, peripherals refer to: printers, keyboards, monitors, scanners, CD ROM drives, and plotters. A hard drive, floppy disk drive, and a MODEM are considered to be peripheral devices even though they may be physically located inside a computer.

PLUG-COMPATIBLE: A term that describes the ability of peripherals to be interchanged without modification.

POLLING: A process in which a number of peripheral devices, remote stations, or nodes in a computer network are interrogated, one at a time, to determine if service is required.

POP: The instruction that removes a word from the top of a stack.

PORT: An input/output channel (either parallel or serial), terminated at a connector on the computer. It interconnects the computer's input and/or output terminals to an appropriate source and/or destination.

PORTRAIT: A term that designates the position of conventional printing across the width of a page. POSITIVE LOGIC – This logic represents the reverse of NEGATIVE LOGIC. It is the more commonly used form of logic. A positive voltage represents a "1" state and a negative (or zero) voltage represents a "0" state.

PRINT SPOOLER: A device for temporarily storing data to be printed when the printer is functioning. It provides uninterrupted data entry and editing while the printer is active and while other data awaits transmission to the printer.

PROGRAM: A complete sequence of computer software instructions necessary to provide an application, solve a specific problem, perform an action, or respond to external stimuli in a prescribed manner. As a verb, it means to develop a program.

PROGRAM COUNTER (PC): A special-purpose register in the CPU which contains the address of the next instruction to be fetched and executed.

PROGRAMMABLE LOGIC ARRAY (PLA): An unprogrammed, general-purpose logic structure in monolithic IC form consisting of an array of similar, and/or compatible logic gates. Also called PROGRAMMABLE ARRAY LOGIC (PAL).

PROGRAMMABLE READ-ONLY MEMORY (PROM): A blank read-only memory (ROM) that is programmed with external programming equipment after manufacture. Once programmed, it is not re-programmable and is considered to be a ROM.

PROPAGATION DELAY: The time required for the output of a logic gate to respond to a combination of input pulses.

PUSH: The instruction used to deposit a word on top of a stack.

PUSH-DOWN STACK: A dedicated temporary storage register in a computer, sometimes part of a system memory, structured so that data (words) in the stack are retrieved in reverse order of entry. See LIFO.

RAM DISK (DISK EMULATOR): A portion of a hard drive configured to emulate a RAM. It accesses information quickly, but its data must be saved in a non-volatile memory for future use. Otherwise, the information is lost when power is removed.

RANDOM ACCESS: A technique of accessing (reading) a word of data from a memory structure by the CPU. Since a word in the memory can be accessed directly, the time required is independent of its location (address) in the memory structure. It is sometimes called a "direct access" method.

RANDOM ACCESS MEMORY (RAM): A volatile, semiconductor storage structure that accesses temporary data with a random or direct accessing method. It is more accurately referred to as "erasable read/write" memory. Data in this memory can be read by the CPU, edited, altered, erased, or new information written over existing data by computer commands. Its data must be saved for future access by writing it into a non-volatile memory. See NON-VOLATILE MEMORY and VOLATILE MEMORY.

READ-ONLY MEMORY (ROM): A semiconductor memory whose data cannot be erased, or overwritten; it can only be accessed (read) for use by the CPU. The data in a ROM is of a permanent nature and is programmed by the ROM manufacturer. In many cases, its programmed information identifies the dedicated function of a computer. A ROM can also be in the physical form of a module (tape or disk) that is plugged into a computer to change its operation from one program to another.

REAL-TIME DATA: Time-dependent data processed by a computer whose output data is capable of controlling other time-related events, such as traffic control. Real time is the actual time it takes for an event to occur.

REGISTER: A temporary storage unit for quick, direct accessibility of a small amount of data for processing. Most computers include a set of internal registers that can be accessed more quickly than the system's main memory.

RESIDENT SOFTWARE: The program(s) residing in the main memory of a computer system. For convenience of operation, several software programs can reside in RAM after the computer is turned on and booted, and they can be accessed quickly from within another program.

RS-232-C: Identifies an interconnection standard for serial operation. It specifies the configuration and type of connectors in the computer's serial I/O port(s) and peripheral serial port(s).

SCRATCH PAD MEMORY: A group of internal registers used for the temporary storage of data being collected and sorted for immediate processing. It is analogous to a pad of paper for quickly jotting down notes. SECTOR - A section of a recording track on a magnetic disk.

SEMICONDUCTOR MEMORY: Data storage devices formed with semiconductor components (generally monolithic ICs). See RAM, ROM, PROM, EPROM, EAPROM, EEPROM, and FLASH.

SERIAL OPERATION: A method of data transmission where the data is handled in sequence, one bit at a time.

SERVICE ROUTINE: A set of instructions for performing a programmed operation, typically, in response to an interrupt command.

SHIFT: A computer operation consisting of moving a group of adjacent data bits either to the left or to the right by a prescribed number of positions. The move is done in a SHIFT REGISTER for a carry-over operation.

SIMPLEX: A data transmission mode that provides transmission in one direction only.

SIMULATION: The imitation of a logical operation of one computer by another to measure and evaluate the operation of the computer being designed. Simulation is primarily intended to provide an analysis of program logic, independent of hardware environment, and is extremely useful for debugging a new software program prior to committing it to ROM.

SOFTWARE: Programs, languages, procedures, and documentation for a computer system. Software includes: operating systems (system software), language translators (assemblers, interpreters and compilers), subroutine libraries, application programs, and the information in instruction manuals.

SOURCE CODE (SOURCE PROGRAM): A set of computer instructions in hard-copy or stored form. When written in a language other than machine language, the source code requires translation by an assembler (or macroassembler), interpreter, or compiler into object code.

STACK: A dynamic, sequential data list usually contained in the computer system's main memory. It has special provisions to access data from either end. Storage and retrieval of data from the stack is performed automatically by the CPU.

STATIC MEMORY: A type of semiconductor read/write memory (RAM) that does not require periodic refresh cycles. As long as electrical power is ON, the data in a static memory is maintained.

STORAGE: This term is interchangeable with the term MEMORY.

SUBROUTINE: A short program segment that performs a specific function and is available for general use by other programs and routines.

SUPERCOMPUTER: The largest mainframe computer featuring exceptionally high speed operation while manipulating huge amounts of information.

SUPPORT CHIPS: Computer-related circuits other than the CPU. Examples are: main memory (RAM and ROM), input/output ports, and system bus.

STATE: The logic input or output condition of a binary digital circuit - the state is either a "0" or "1".

STORAGE CAPACITY: The amount of data that can be retained in a memory unit. It is expressed either by a number of bits or bytes (8-bit words).

STREAMER/STREAMING TAPES: A small tape recorder, usually in cassette form, is used to store data from a hard disk for the purpose of backing up this data.

SYNCHRONOUS COMMUNICATION: A method of transferring binary data, in serial form, between computers or between a computer and its peripherals. Transmission of data is at a rate set by the computer's clock with synchronization bits located at the beginning of each message or block of data.

SYNTAX: A set of grammatical rules defining valid use of specific commands or instructions in a computer language.

SYSTEMS SOFTWARE: A general term for software that supervises, sequences, and coordinates programs. Systems software may include programs, such as: operating systems, assemblers, interpreters, compilers, software debugging programs, text editors, utilities and peripheral drivers.

TEXT EDITOR: See EDITOR.

THIRD-PARTY SOFTWARE: Software developed by a software company rather than by a computer manufacturer or user.

THROUGHPUT: The number of instructions executed per second, measured in millions of instructions per second (MIPS) or billions of instructions per second (BIPS). TOGGLE - To switch back and forth between two states or conditions of operation, as in a toggle switch.

TOP-DOWN HIERARCHICAL DESIGN: A hardware and/or software design approach that starts at the most general level of a machine or software program. It proceeds, step-by-step, to lower levels, adding detail as the design progresses.

TRACK: A ring on the surface of a magnetic disk. TRACTOR-FEED - A pin-fed device for advancing continuous form paper through a computer printer. TRANSISTOR-TRANSISTOR LOGIC (T2L) - A logic gate family that provides higher-speed and higher-power than the obsolete DTL logic family. The first transistor in the circuit performs the required logic. Another transistor amplifies and inverts the output. Improved pin-compatible versions of this logic family are called TTL-Schottky (T2L-S) and Low Power TTL-S (LPT2L-S).

TRANSLATOR: See ASSEMBLER, MACROASSEMBLER, INTERPRETER, and COMPILER.

TRI-STATE LOGIC: The term that designates the possible conditions of a specific logic gate output: "0", "1" or "undefined".

TRUNCATE: The dropping of digits or characters from one end of a data item causing loss of accuracy or information.

TRUTH TABLE: A tabulation of all possible combinations of states at the inputs of a logic gate which will result in a specific logic state at the output of the gate.

TURNKEY SYSTEM: A complete computer system ready to operate without any hardware or software modification or addition.

UNFORMATTED (UNINITIALIZED) DISK: A blank magnetic disk with no track/sector identification recorded on it that allows users to implement their own track/sector identifications.

USART: UNIVERSAL SYNCHRONOUS/ASYNCHRONOUS RECEIVER TRANSMITTER - A circuit that converts serial-to-parallel or parallel-to-serial operation for communication between two computers or between a computer and its peripherals.

USER-FRIENDLY PROGRAM: A software program that has been designed to easily direct the user through the operation or application of a program. A menu-driven program is considered to be "user-friendly".

UTILITY: A software program designed to perform a computer system's routine housekeeping functions, like copying, deleting files, and/or providing techniques to simplify the execution of a program.

VOCABULARY: A list of operating codes or instructions available to the software programmer for writing a program in a specific language.

VOLATILE MEMORY: A memory whose contents are irretrievably lost when power is removed. If data in RAM must be saved after power shut-down, back-up in nonvolatile memory (magnetic disk, tape, or CD-R) is essential.

WAIT STATE: An internal condition of delay in processing time executed by the CPU when a synchronizing control signal is not present. Wait states synchronize the timing of a CPU with the relatively slower access time of the computer's main memory.

WINCHESTER DRIVE: See HARD-DISK DRIVEWINDOWING - The ability of a program to divide a display screen into smaller sub-units that permit portions of different sections of a program, or different programs, to be displayed on the screen, edited, and copied independently.

WORD PROCESSING (WP): The term refers to a program, that allows creating, editing, formatting, displaying, printing, and storage of text with great flexibility and ease. Different WP programs provide different, and sometimes, more desirable capabilities than others.

WORD: The set of binary bits handled by a computer as a primary unit of data. The width (number of bits) of a computer word depends on the hardware design. Wider words imply higher levels of precision, higher speed, and more intricate instructions. Typically, each location in memory contains one word.

WORM (WRITE-ONCE, READ-MANY): A high-density optical disk memory available in a variety of formats from 5.25" to 14". The WORM can be programmed once, permanently saving a user's data. It then becomes an optical disk read-only memory having essentially the same features as a CD-ROM. Also called CD-R (CD-RECORDABLE).

WORKSTATION: The work area and/or equipment used for computer operations, including computer-aided design (CAD). The equipment generally consists of a monitor, keyboard, printer and/or plotter, and other output devices.

WRITE: The process of storing data into a memory.

X-Y PLOTTER: A computer-driven printing mechanism that draws coordinate points in graph form.

ZERO-WAIT STATE: A condition that results when no delays are inserted into the system software to deliberately detain operation of the CPU. This can be done with high-speed memory and proper design of computer architecture. Zero-wait state is desirable for the high-speed operation of a computer.

ZOOM: The process of proportionately enlarging or reducing an image displayed on a computer monitor.

❑❑❑

Appendices

Appendices

A.1. COMPUTER ORGANIZATION

CO is how operational attributes are linked together and contribute to realize the architectural specification. Computer architecture is the architectural attributes like physical address memory, CPU and how they should be made to coordinate with each other keeping the future demands and goal in mind.

Computer architecture comes before *CO*. It is like building the design and architecture of house takes maximum time and organization is the building house by bricks or by latest technique keeping the basic layout and architecture of house in mind.

Architecture describes what the computer does and organization describes how it does.

Architecture of organization are independent. You can change the organization of computer without changing its architecture.

A.2. ARITHMETIC ALGORITHMS

The solution of any problem that is stated by a finite number of well-defined procedural steps is called an algorithm.

(1) Addition and subtraction of signed-magnitude numbers:

Operation	Add magnitudes	Subtract Magnitudes When A > B	When A < B	When A = B
$(+A) + (+B)$	$+(A + B)$			
$(+A) + (-B)$		$+(A - B)$	$-(B - A)$	$+(A - B)$
$(-A) + (+B)$		$-(A - B)$	$+(B - A)$	$+(A - B)$
$(-A) + (-B)$	$-(A + B)$			
$(+A) - (+B)$		$+(A - B)$	$-(B - A)$	$+(A - B)$
$(+A) - (-B)$	$+(A + B)$			
$(-A) - (+B)$	$-(A + B)$			
$(-A) - (-B)$		$-(A - B)$	$+(B - A)$	$+(A - B)$

Hardware for signed magnitude addition and subtraction:

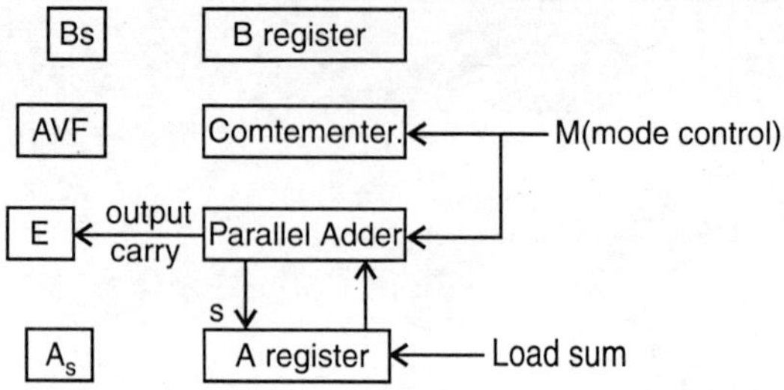

Fig. A.1.

Flowchart for add and subtract operations:

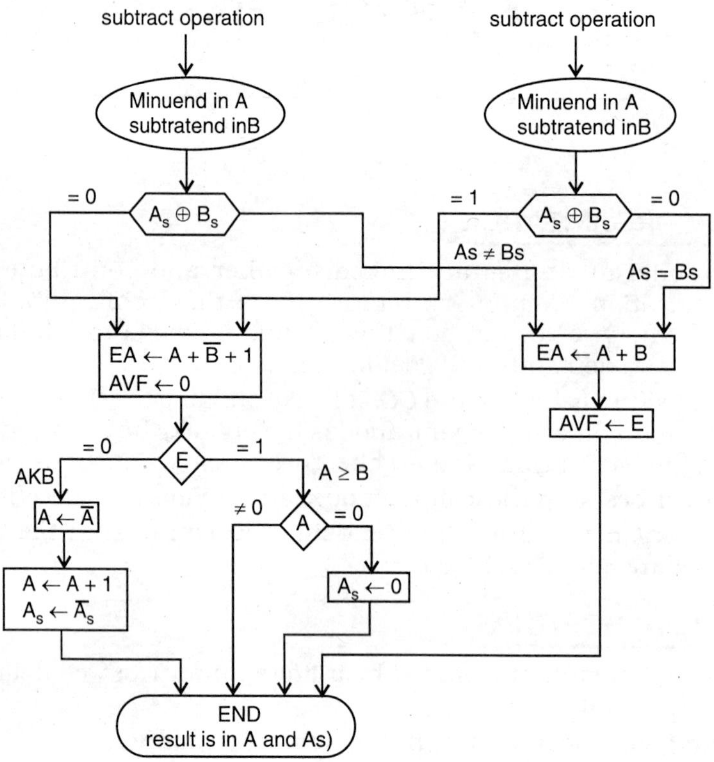

Fig. A.2.

A.3. ADDITION AND SUBTRACTION WITH SIGNED-2'S COMPLEMENT DATA

Hardware for signed-2's complement addition and subtraction.

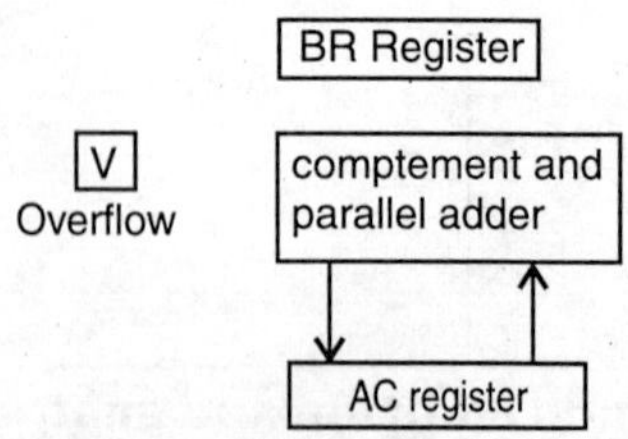

Fig. A.3.

Algorithm for adding and subtracting number is signed-2's complement representation.

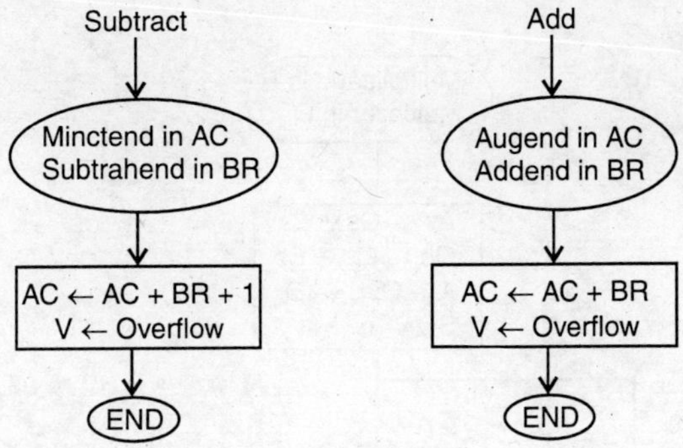

Fig. A.4.

A.4. MULTIPLICATION ALGORITHMS

15	Multiplicand
13	Multiplier
$45 \rightarrow P_1$	(First partial product)
$15 \times \rightarrow P2$	(Second partial product)
$195 \rightarrow P$	(Final product)

101	= equal bits
110	= equal bits
000	→ (First partial product)
1010	→ (Second partial product)
10100	→ (Third product)
11110	→ (Final product)

Hardware for multiply operation:

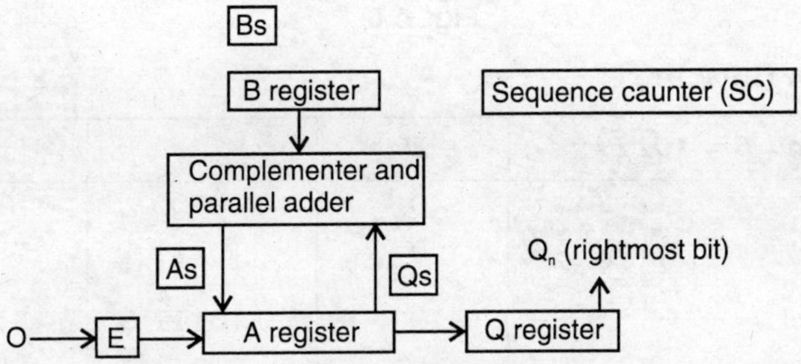

Fig. A.5.

Hardware Algorithm: Flowchart for multiply operation

η = no. of bits in multiplier

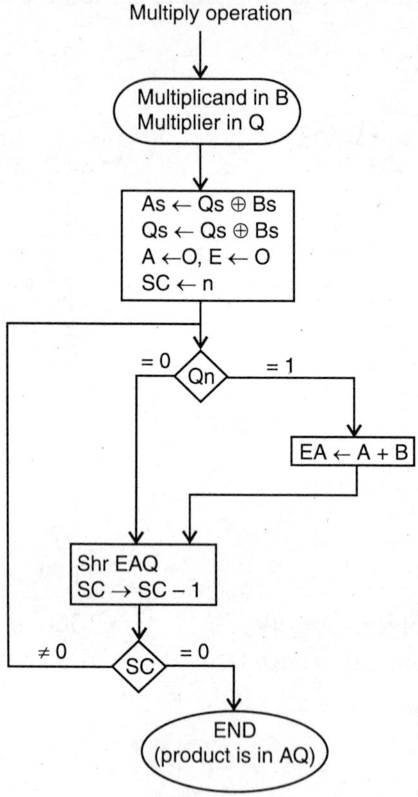

Fig. A.6.

Example of given flowchart:

Multiplicand B = 10111	**E**	**A**	**Q**	**SC**
Multiplier in Q	0	00000	10011	101
Q_n = 1; add $B + A$	0	10111		
First partial product	0	10111		
Shift right EAQ	0	01011	11001	100
Q_n = 1; add B in A		10111		
Second partial product	1	00010		
Shift right EAQ	0	10001	01100	011
Q_n = 0, shift right EAQ	0	01000	10110	010
Q_n = 0, shr EAQ	0	00100	01010	001
Q_n = 1, add B in A		10111		
Fifth partial product	0	11011		
Shr EAQ	0	01101	10101	000
Find product in AQ = 0110110101				

A.5. BOOTH MULTIPLICATION

Booth algorithm for multiplication of signed-2's complement numbers

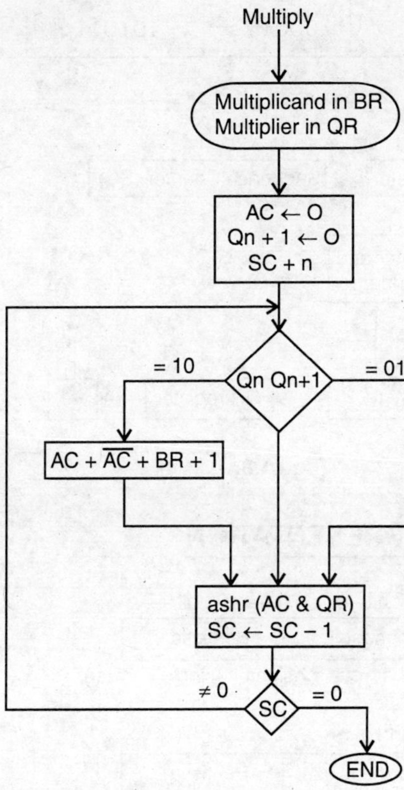

Fig. A.7.

Example:

$$(-9) \times (-13) = +117$$

Multiplicand $9 = 1001$

$\Rightarrow$ $1001\ 0110 + 1 = 0111$

$-9 = 10111 = BR$

$-13 = 10011 = QR$

$BR = 10111$

Q_n	Q_{n+1}	$\overline{BR} + 1 = 01001$	*AC*	*QR*	Q_{n+1}	SC
		Initial	00000	10011	0	101
1	0		01001			
			01001			
		ashr	00100	11001	1	100
1	1	ashr	00010	01100	1	011
			10111			
0	1	Add *BR* in *AC*	11001			
		ashr	11100	10110	0	010

0	0	ashr	11110	01011	0	001
			01001			
1	0		00111			
		ashr	00011	10101	1	000

Ans. is 0001110101.

Hardware for Booth Algorithm:

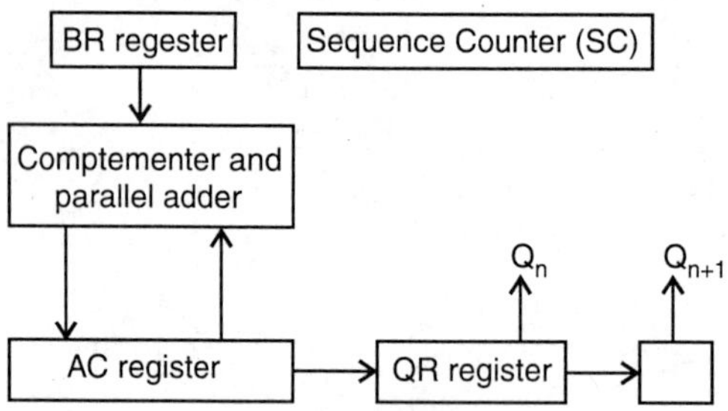

Fig. A.8.

A.6. IEEE FLOATING POINT REPRESENTATION

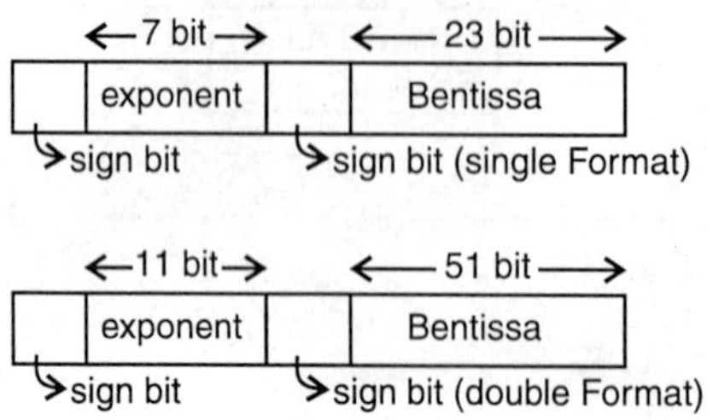

Fig. A.9.

All points are represented in $m \times r^e$.

IEEE: Initiate of electronics and electrical engineering.

A.7. DESIGN OF FAST ADDERS

Faster algorithms and their implementation:

(1) Fast Adder: In the parallel binary adder the carry generated by the ith adder is fed as carry input to the $i + 1$ adder. In this adder the output (C_{out} $S_3S_2S_1S_0$) is available only after the carry is propagated through each of the adders that means from *LSB* to *MSB* adder to intermediate adders.

Hence, the addition process can be considered to be completed only after the above carry propagation delay through adders, which is proportional to the number of stages in bit.

One of the methods of making this faster look ahead carry addition which eliminates the ripple carry delay. This method is based on the carry generating and the carry propagating functions of the full adder.

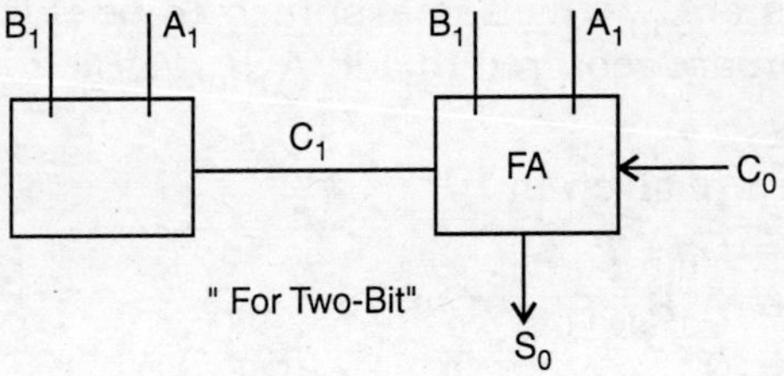

Fig. A.10.

4-Bit Carry Look Ahead Adder: The carry look ahead adder is based on the principle of looking at the lower order bits of the Augend and Addend. If high order carry is generated. This adder reduces the carry delay by reducing the number of gates or number of stages through which carry signal must propagate. To explain its operation consider the truth table of full adder.

ROW	*A*	*B*	C_{in}	*S*	C_{out}	
0	0	0	0	0	0	No carry generation
1	0	0	1	1	0	*i.e.*, $C_{out} = 0$
2	0	1	0	1	0	
3	0	1	1	0	1	Carry propagation
4	1	0	0	1	0	*i.e.*, $C_{out} = C_{in}$
5	1	0	1	0	1	
6	1	1	0	0	1	Carry generation
7	1	1	1	1	1	*i.e.*, $C_{out} = 1$

In row 0 and 1 the carry output C_{out} is always zero and independent of carry input (C_{in}) while in row six and seven the C_{out} is always 1 independent of C_{in}. These are known as carry generate combination.

In row 2, 3, 4 and 5 the carry output is equal to the carry input, *i.e.* $C_{out} = 1$ only when $C_{in} = 1$. These are carry propagate combinations.

Let G_i represents the unity carry ($C_{out} = 1$). Generate condition and P_i represent the propagate condition of *i*th stage of parallel adder.

From truth table, G_i is obtained by summing up the combinations corresponding to sixth and seventh rows as follows:

$$G_i = A_iB_i C_{in} + A_iB_i\overline{C_{in}} = A_iB_i (C_{in} + \overline{C_{in}})$$

$$G_i = A_iB_i \qquad ...(1)$$

Similarly, the carry propagation condition (P_i) occurs when either $A_i = 1$ and $B_i = 0$ or vice-versa. Now P_i is given by

$$P_i = A_i\overline{B_i} + \overline{A_i} + B_i$$

$$P_i = A_i \oplus B_i$$

Consider the addition of 2, 4 bit binary no's $A\ (A_3A_2A_1A_0)$ and $B\ (B_3B_2B_1B_0)$. The unity carry O/P of the $(i-1)$ stage as follows:

$$C_i\,(C_{out}) = G_i + P_i\,C_{i-1}$$

$$C_i = 0, \text{ then } C_{i-1} = \text{initially carry}$$

If $\quad C_i = 1, \text{ then } \overline{C_{i-1}} = 0$

Where C_i for LSB stages is C_{in} which is assumed to be 0 (zero). In a four bit binary adder four stages of addition are required to add A_0B_0, A_1B_1, A_2B_2, A_3B_3.

Therefore,

for $i = 0, 1, 2, 3$ and C_i's are given by

$$C_0 = G_0 + P_0 C_{in}$$

where $G_0 = A_0B_0$, $P_0 = A_0 \oplus B_0$

and $C_{in} = 0$

$$C_0 = A_0B_0$$

$$C_1 = G_1 + P_1C_0 = G_1 + P_1 (G_0 + P_0C_0)$$

$$G = G_1 + P_1G_0 + P_0P_1G_n$$

where $G_1 = A_1B_i$

and $P_1 = A_1 \oplus B_1$

$$G_2 = G_2 + P_2C_1 = G_2 + P_2 (G_1 + P_1G_0 + P_0P_1C_{in})$$

$$G_2 = G_2 + P_2G_1 + P_2P_1G_0 + P_0P_1P_2 C_{in}$$

where $G_2 = A_2B_2$

and $P_2 = A_2 \oplus B_2$

Similarly, $C_3 = G_3 + P_3C_2 = G_3 + P_3 (G_2 + P_2G_1 + P_2P_1G_0 + P_0P_1P_2 C_{in})$

$$C_3 = G_3 + P_3G_2 + P_2P_3G_1 + P_1P_2P_3G_0 + P_0P_1P_2P_3 C_{in}$$

where $G_3 = A_3B_3$

and $P_3 = A_3 \oplus B_3$

Sum of A and B is given by,

$$S = C_3 S_3S_2S_1S_0$$

where, $S_i = A_i \oplus B_i \oplus C_{i-1}$

For $i = 0, 1, 2, 3$

i.e., $S_0 = A_0 \oplus B_0 \oplus C_{in}$

$$S_1 = A_1 \oplus B_1 \oplus C_0$$

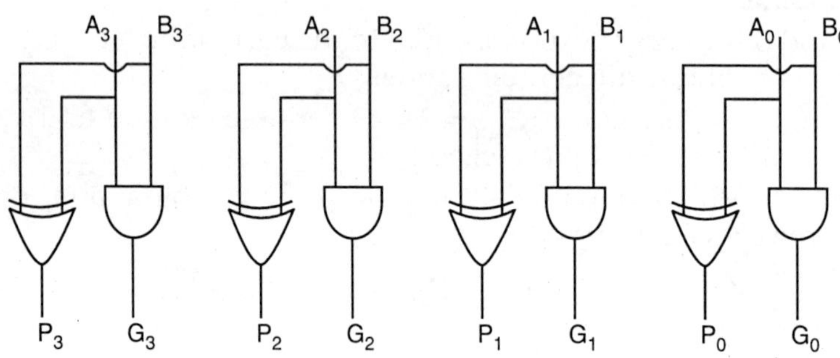

A.11.

Since, $S_0 = A_0 \oplus B_0 \oplus C_{in}$

$\Rightarrow$ $S_0 = P_0 \oplus C_{in}$

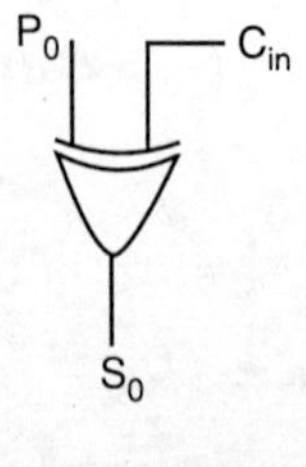

$$S_1 = A_1 \oplus B_1 \oplus C_0$$
$$\Rightarrow \quad S_1 = P_1 \oplus C_0$$
$$\Rightarrow \quad S_1 = P_1 \oplus (G_0 + P_0 C_{in})$$

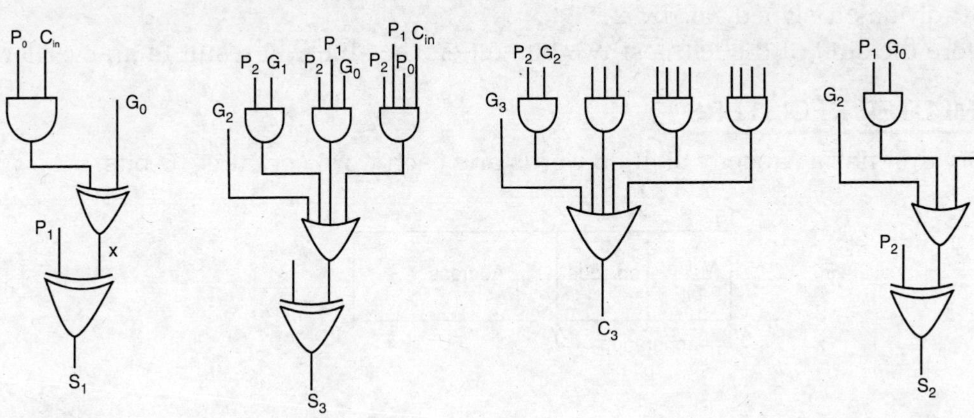

Fig. A.12.

$$S_0 = A_0 \oplus B_0 \oplus C_{in}$$
$$\Rightarrow \quad S_0 = P_0 \oplus C_{in}$$
$$S_1 = A_1 \oplus B_1 \oplus C_0$$
$$\Rightarrow \quad S_1 = P_1 \oplus C_0$$
$$\Rightarrow \quad S_1 = P_1 \oplus (G_0 + P_0\, C_{in})$$

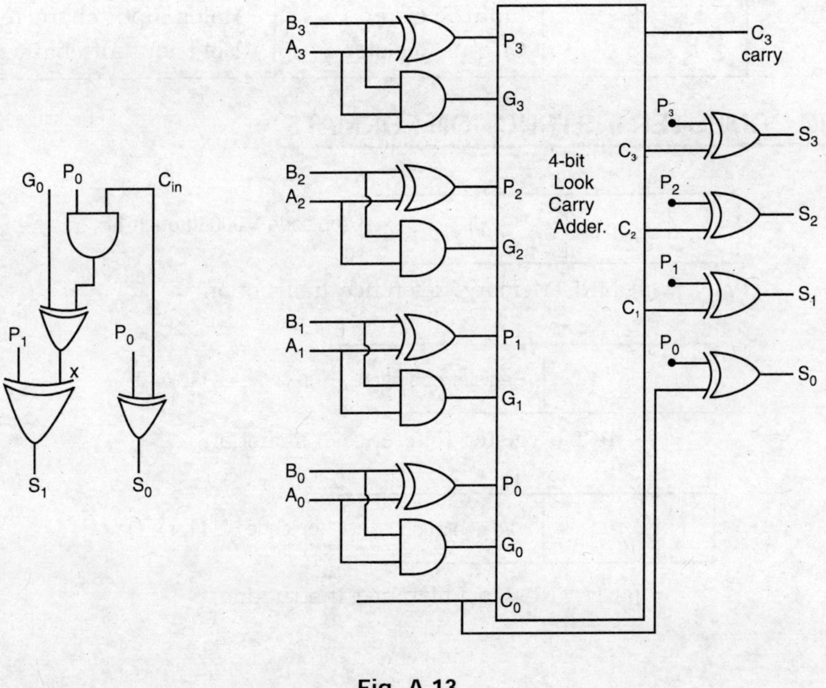

Fig. A.13.

A.8. ARITHMETIC SHIFT

→ An arithmetic shift is meant for signed binary number (integer).
→ *ASL* multiplies a signed no. by 2.
→ *ASR* divides a signed no. by 2.
→ Before the shift, if the leftmost two bits differ, the shift will result in an overflow.

A.9. COMPUTER REGISTERS

The memory unit has a capacity of 4096 words and each word contains 16-bits.

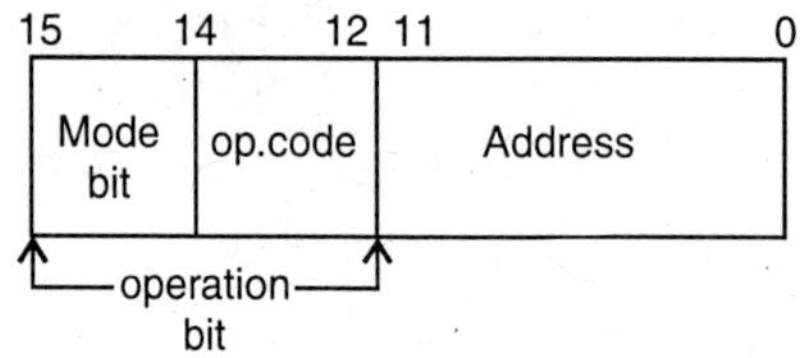

A.10. LIST OF REGISTERS FOR THE BASIC COMPUTERS

Register Symbol	*Number of Bis*	*Register Name*	*Function*
DR	16	Data Register	Holds memory operand.
AR	12	Address Register	Holds address for memory.
AC	16	Accumulator	Processor register.
IR	16	Instruction Register	Holds instruction code.
PC	12	Program Counter	Holds address of instruction.
TR	16	Temporary Register	Holds temporary data.
INPR	8	Input Register	Holds input character.
OUTR	8	Output Register	Holds output character.

A.11. BASIC COMPUTER INSTRUCTION FORMATS

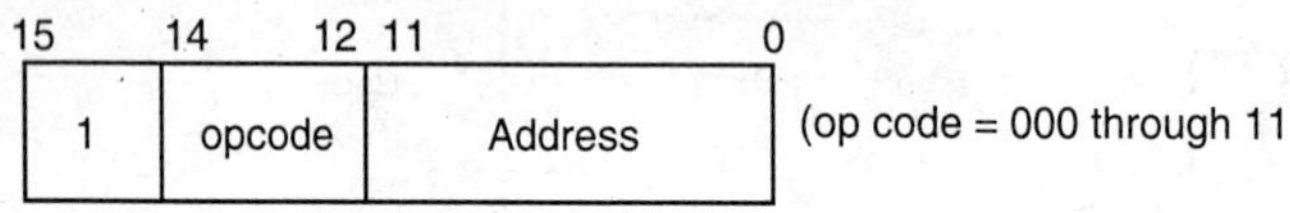

(a) MRI (Memory Reference Instruction)

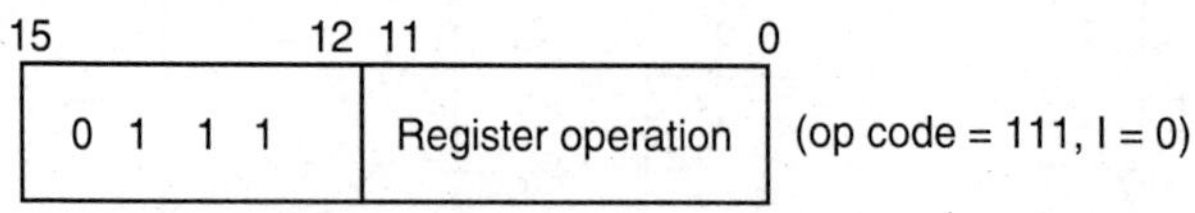

(b) RRI (Register Reference Instruction)

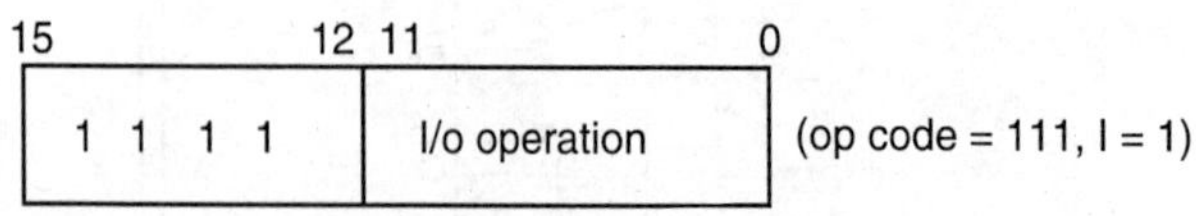

(c) I/o RI (I/o Reference Instruction)

A.12. BASIC COMPUTER REGISTERS AND MEMORY

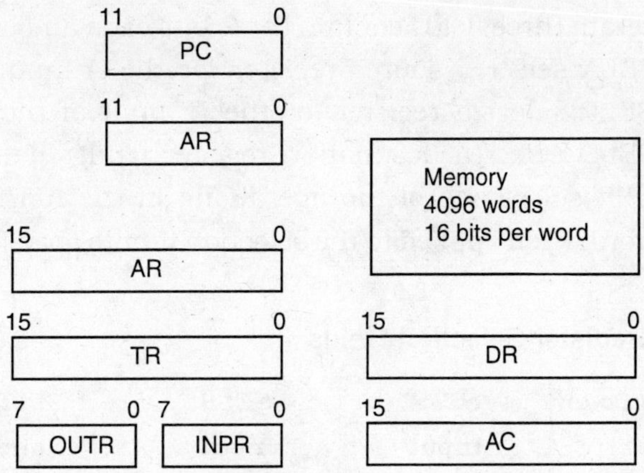

A.13. GENERAL REGISTER ORGANIZATION

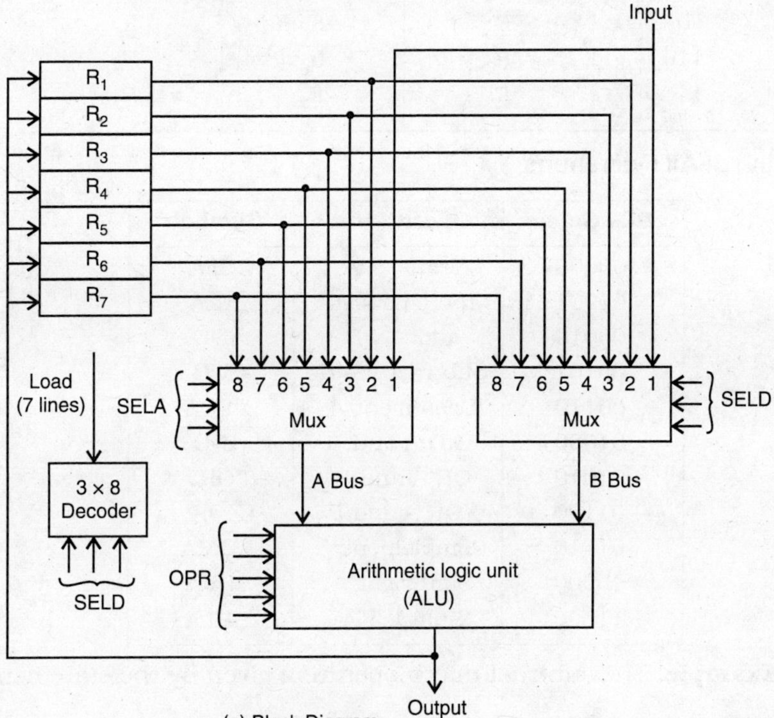

(a) Block Diagram

3	3	3	5
SELA	SELB	SELD	OPR

(b) Control word

Fig. A.14.

A.14. CONTROL WORD

There are 14 binary selection inputs and the unit, and their combined value specifies a control word. In left diagram three fields contain three bits each and one field has five bits.

— The 3 bits of SELA select a source register for the A input of the ALU.

— The 3 bits of SELB select a register for the B input of the ALU.

— The 3 bits of SELD select a destination register using the decoder.

— The 5 bits of OPR select one of the operations in the ALU.

The 14-bit control word when applied to the selection inputs specify a particular micro-operation.

(A) Table: Encoding of Register Selection Fields

Binary Code	***SELA***	***SELB***	***SELD***
000	**Input**	**Input**	**None**
001	R_1	R_1	R_1
010	R_2	R_2	R_2
011	R_3	R_3	R_3
100	R_4	R_4	R_4
101	R_5	R_5	R_5
110	R_6	R_6	R_6
111	R_7	R_7	R_7

B. Table: Encoding of All Operations

OPR Select	***Operations***	***Symbol***
00000	Transfer A	TSFA
00001	Increment A	INCA
00010	Add $A + B$	ADD
00101	Subtract $A - B$	SUB
00110	Decrement A	DECA
01000	AND A and B	AND
01010	OR A and B	OR
01100	XOR A and B	XOR
01110	Complement A	COMA
10000	Shift right A	SHRA
11000	Shift left A	SHLA

Example: The subtract micro-operation given by the statement:

$$R_1 \leftarrow R_2 - R_3$$

Field:	SELA	SELB	SELD	OPR
Symbol:	R_2	R_3	R_1	SUB
Control word:	010	011	001	00101

Note: A memory unit that stores control words is called control memory.

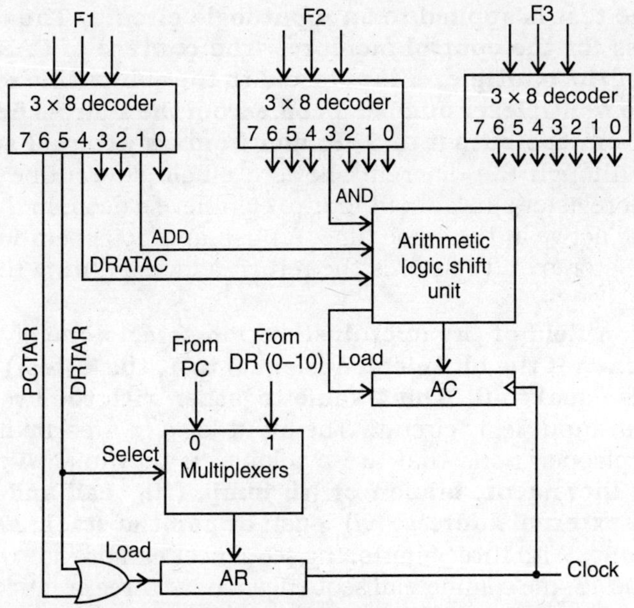

Fig. A.15. Decoding of micro-operation fields

The other outputs of the decoders that are associated with an AC operation must also be connected to the arithmetic logic shift unit in a similar fashion.

A.15. MICROPROGRAM SEQUENCER

The basic components of a micro programmed control unit are the control memory and the circuits that select the next address. The address selection part is called a microprogram sequencer. A microprogram sequencer can be constructed with digital functions to suit a particular application. However, just as there are large ROM units available in integrated circuit packages, so are general-purpose sequencers suits for the construction of microprogram control units. To guarantee a wide range of acceptability, an integrated circuit sequencer must provide an internal organization that can be adapted to a wise range of applications.

The purpose of a microprogram sequencer is to present an address to the control memory so that a microinstruction may be read and executed. The next-address logic of the sequencer determines the specific address source to be loaded into the control address register. The choice of the address source is guided by the next-address information bits that the sequencer receives from the present microinstruction. Commercial sequencers include within the unit an internal register tack used for temporary storage of addresses during microprogram looping and subroutine calls. Come sequencers provide an output register which can function as the address register for the control memory.

The illustrate the internal structure of a typical microprogram sequencer we will show a particular unit that is suitable for use in the microprogram computer example developed in the preceding section. The block diagram of the sequencer is shown in Fig. 7.8. The control memory is included in the diagram to show the interaction between the sequencer and the memory attached to it. **There are two multiplexers in the circuit. The first multiplexer selects an address from one of four sources and routes in into a control address register CAR. The second multiplexer tests the value of a selected status bit**

and the result of the test is applied to an input logic circuit. The output from CAR provides the address for the control memory. The content of CAR is incremented and applied to one of the multiplexer inputs and to the subroutine register SBR. The other three inputs to multiplexer number 1 come from the address field of the present microinstruction, from the output of SBR, and from an external source that maps the instruction. Although the diagram shows a single subroutine register, atypical sequencer will have a register stack about four to eight levels deep. In this way, a number of subroutines can be active at the same time. A push and pop operation, in conjunction with a stack pointer, stores and retrieves the return address during the call and return microinstructions.

The CD (condition field of the microinstruction selects one of the status **bits in the second multiplexer. If the bit selected is equal to 1, the T (test) variable is equal to 1; otherwise, if is equal to 0. The T value together with the** two bits from the BR (branch) field go to an input logic circuit. The input logic in a particular sequencer will determine the type of operations that are available in the unit. **Typical sequencer operations are: (*i*) increment, branch or (*ii*) jump, (*iii*) call and (*iv*) return from subroutine, load an external address, (*vi*) push or pop the stack, and other address sequencing operations.** With three inputs, the sequencer can provide up to eight address sequencing operations. Some commercial sequencers have three or four inputs in addition to the T input and thus provide a wide range of operations.

The input logic circuit in Fig. A.16., **has three inputs, I0, I1 and T, and three outputs, S_0, S_1, and L. Variable S_0 and S_1 select one of the source addresses for CAR. Variable L enables the load input in SBR. The binary values of the two selection variables determine the path in the multiplexer. For example, with S_1S_0 = 10, multiplexer input number 2 is selected and establishes a transfer path from SBR to CAR. Note that each of the four inputs as well as the output of MUX 1 contains a 7-vit address.**

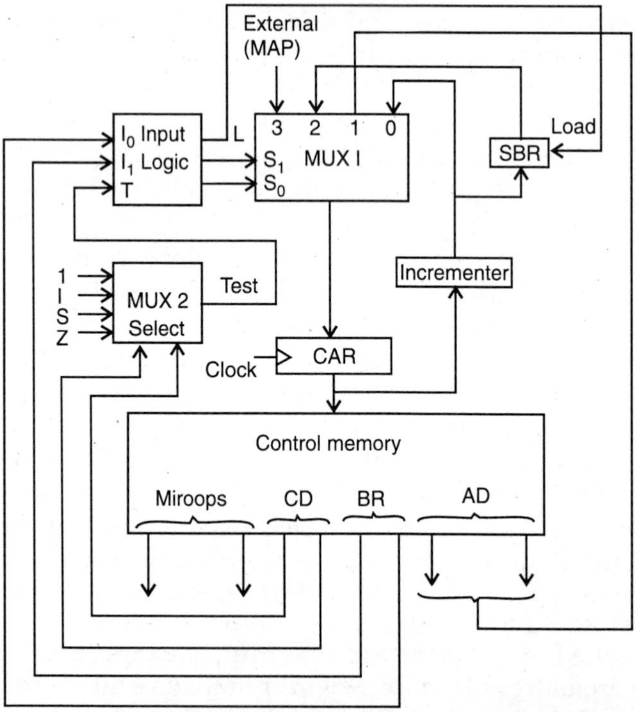

Fig. A.16. Microprogram sequencer for a control memory

The truth table for the input logic circuit is shown in Table 7.4. Inputs I_1 and I_0 are identical to the bit values in the BR field. The function listed in each entry was defined in Table 7.1. The bit values for S_1 and S_0 are determined from the stated function and the path in the multiplexer that establishes the required transfer. The subroutine register is loaded with the incremented value of CAR during a call microinstruction (BR = 01) provided that the status bit condition is satisfied (T = 1). The truth table can be used to obtain the simplified Boolean functions for the input logic circuit:

$$S_1 = I_1$$
$$S_0 = I_1 I_0 + I'_1 T$$
$$L = I'_1 I_0 T$$

Table A.1. Input Logic Truth Table for Microprogram Sequencer

BR Field	*Input* I_1 I_0 T	*MUX 1* S_1 S_0	*Load SBR* L
0 0	0 0 0	0 0	0
0 0	0 0 1	0 1	0
0 1	0 1 0	0 0	0
0 1	0 1 1	0 1	1
1 0	1 0 ×	1 0	0
1 1	1 1 ×	1 1	0

The circuit can be constructed with three AND gates, an OR gate, and inverter.

Note that the incrementer circuit in the sequencer of Fig. A.16. is not a counter constructed with flip-flops but rather a combinational circuit constructed with gates. A combinational circuit incrementer can be designed by cascading a series of half-adder circuits (see Fig. A.16). The output carry from one stage must be applied to the input of the next stage. One input in the first least significant stage must be equal to 1 to provide the increment-by-one operation.

A.16. PERFECTING MICROINSTRUCTION

The disadvantage of microprogram control is that it leads to lower operating speed because of the time it takes to fetch the microinstructions from control memory. This disadvantage can be remove and faster operation can be achieved if the next microinstruction is perfected while other microinstruction is being executed.

Thus the execution time can be overlapped with the patched time but there are same problems. When we prefect next microinstruction some time status plug and the result of currently executed microinstruction are required to determine the address of next microinstruction.

A.17. WIDE BRANCH ADDRESSING

As the no of branches increases generating address for each branch become difficult. To handle this type of situation the best and simple way is used:

Programmable Logic Array (PLA) to generate the required branch address and this way of generating the branch address is called as wide branch addressing. In this method the of code of instruction is translated into the starting address of the corresponding micro routine. The opecode bits of the IR is connected to the inputs to the PLA hence PLA acts as a and outputs the address of the desired micro routine.

A.18. STATIC RAM

The static RAM consists essentially of internal flip-flops that store the binary information. The stored information remains valid as long as power is applied to the unit.

A.19. DYNAMIC RAM

The dynamic RAM stores the binary information in the form of electric charges that are applied to the capacitors. The capacitors are provided in-side the chip by MOS transistors. The stored charge on the capacitors tend to discharge with time and the capacitors must be periodically recharge by refreshing dynamic memory. Refreshing is done by cycling through the words every few milliseconds to restore the decaying charge. The dynamic RAM offers reduced Power consumption and larger storage capacity in a single memory chip.

Most of the memory (main) in general purpose computer is made up of RAM Ics, but a portion of the memory may be constructed with ROM chips. Originally, RAM was used to refer to a random access memory, but now it is used to designate a read/write memory to differentiate if from rad only memory. Although ROM is also a random access. ROM is used to store programs that are permanently resident in the computer.

Among the other things, the ROM portion of main memory is needed for storing an initial programs called a bootstrap loader. The boot strap loader is a program whose function is to start the computer software operating when power is turned on. Since RAM is Volatile, its contents are destroyed when power is turned off. The contents of ROM remains unchanged after power is turned off and on again.

RAM and ROM chips are available in a variety of sizes. If the memory needed for computer is larger than the capacity of one chip, it is necessary to combine a number of chips to form the required memory size.

The next address may also a function of external input conditions. While the micro-operations are being executed, the next address is computed in the next address circuit and then transferred into the control address register to read the next microinstruction.

The next address generator is sometimes called a microprogram sequencer, as it determines the address sequence that is read from control memory.

The control data register holds the present microinstruction while the next address is computed and read from memory. The data register is sometimes called a pipeline register. It allows the execution of the micro-operation specified by the control word simultaneously with the generation of next microinstruction.

The main advantage of the microprogrammed control is the fact that once the hardware configuration is established, there should be no need for further hardware or wiring changes.

A.20. MICROPROGRAMMING

Microprogramming generally involves the use of ROM to store effectively the control signals.

When a computer is microprogrammed, the individual operations between and on registers are called micro-operations. For instance, transferring the program's contents in to the memory address register is a micro-operation. Similarly, incrementing the program counter is a micro-operation. Transferring the contents of accumulator in the another register is also a micro-operation. In each of these cases a micro-operation is initiated by raising a single control signals. Thus sequencing micro-operations involves sequencing the appropriate control signals.

The use of micro program control in a processor to decode and generate various control signals with in a CPU represents a marked and signified and improvement in the system design and operation of the control portion of the CPU.

A.21. IEEE STANDARD 754 FLOATING POINT NUMBER

IEEE Standard 754 floating point is the most common representation today for real number on computers, including Intel-based PC's, McIntoshes, and most Unix platforms. This article gives a brief overview of IEEE floating point and its representation. Discussion of arithmetic implementation may be found in the book mentioned at the bottom of this article.

What are Floating Point Numbers ?

There are several ways to represent real numbers on computers. Fixed point places a radix point somewhere in the middle of the digits, and is equivalent to using integers that represent portions of some unit. For example, one might represent1/100 that of a unit; if you have four decimal digits, you could represent 10.82, or 00.01. Another approach is to use retionals, and represent every number as the ratio of two integers.

Floating-point representation the most common solution-basically represents real in scientific notation. Scientific notation represents numbers as a base number and an exponent. For example, 123.456 could be represented as 1.23456×10^2. In hexadecimal, the number 123. abc might be represented as $1.23abc \times 16^2$.

Floating-point solves a number of representation problems. Fixed-point has a fixed window of representation, which limits it from representing very large or very small numbers. Also, fixed-point is prone to a loss of precision when two large number are divided.

Floating-point, on the other hand, employs a sort of "sliding window" of precision appropriate to the scale of the number. This allows it to represent numbers from 1,000,000,000,000 to 0.0000000000000001 with ease.

Storage Layout

IEEE floating point numbers have three basic components; the sign, the exponent, and the mantissa is composed of the *fraction* and an implicit leading digit (explained below). The exponent base (2) is implicit and need not be stored.

The following figure shows the layout for single (32-bit) and double (64-bit) precision floating-point values. The number of bits for each field are shown (bit ranges are in square brackets):

	Sign	Exponent	Fraction	Bias
Single Precision	1[31]	8[30–23]	23[22–00]	127
Double Precision	1[63]	11[62–52]	52[51–00]	1023

The Sign Bit

The sign bit is as simple as it gets. 0 denotes a positive number; 1 denotes a negative number. Flipping the value of this bit flips the sign of the number.

The Exponent

The exponent field needs to represent both positive and negative exponents. To do this, a *bias* is added to the actual exponent in order to get the stored exponent. For IEEE single-precision floats, this value is 127. Thus, an exponent of zero means that 127 is stored in the exponent field. A stored value of 200 indicates an exponent of (200–127), or 73. For reasons discussed later, exponents of –127 (all 0s) and + 128 (all 1s) are reserved for special numbers.

For double precision, the exponent field is 11 bits, and has a bias of 1023.

The Mantissa

The *mantissa,* also known as the *signific and,* represents the precision bits of the number. It is composed of an implicit leading bit and the fraction bits.

To find out the value of the implicit leading bit, consider that any number can be expressed in scientific notation in many different ways. For example, the number five can be represented as any of these:

$$5.00 \times 10^{6}$$
$$0.50 \times 10^{2}$$
$$5000 \times 10^{-3}$$

In order to maximize the quantity of representable numbers, floating-point numbers are typically stored in *normalized* form. This basically puts the radix point after the first non-zero digit. In normalized form, five is represented as 5.0×10^0.

A nice little optimization is available to us in base two, since the only possible non-zero digit is 1. Thus, we can just assume a leading digit of 1, and don't need to represent it explicitly. As a result, the mantissa has effectively 24 bits of resolution, by way of 23 fraction bits.

Putting it All Together

So, to sum up:

1. The sign bit is 0 for positive, 1 for negative.
2. The exponent's base is two.
3. The exponent field contains 127 plus the true exponent for single-precision, or 1023 plus the true exponent for double precision.
4. The first bit of the mantissa is typically assumed to be 1, *f*, where *f* is the field of fraction bits.

A.22. RANGES OF FLOATING-POINT NUMBERS

Lets consider single-precision floats for a second. Note that we're taking essentially a 32-bit number and re-jiggering the fields to cover a much broader range. Something has to give, and it's precision. For example, regular 32-bit integers, with all precision centered around zero, can precisely store integers with 32-bits of resolution. Single-precision floating-point, on the other hand, is unable to match this resolution with its 24 bits. It does, however, approximate this value by effectively truncating from the lower end. For example:

	11110000	11001100	10101010	00001111	//	32-bit integer
=	+1.1110000	11001100	10101010	x 2^{32}	//	Single-precision Float
=	11110000	11001100	10101010	00000000	//	Corresponding Value

This approximates the 32-bit value, but doesn't yield an exact representation. On the other hand, basis the ability to represent fractional components (which integers lack completely), the floating-point value can represent numbers around 2^{127}, compared to 32-bit integers maximum value around 2^{32}.

The range of positive floating point numbers can be split into normalized numbers (which preserve the full precision of the mantissa), and *denormalized* numbers (discussed later) which use only a portion of the fraction's precision.

	Denormalized	Normalized	Approximate Decimal
Single Precision	$\pm 2^{-149}$ to $(1-2\_^{-23})\times 2^{-126}$	$\pm 2^{-126}$ to $92-2^{-23}) \times 2^{127}$	$\pm \sim 10^{-44.85}$ to $\sim 10^{38.53}$
Double Precision	$\pm 2^{-1074}$ to $1^{-52})\times 2^{-1022}$	$\pm 2^{-1022}$ to $(2-2^{-52})\times 2^{1023}$	$\pm \sim 10^{-323.3}$ to $\sim 10^{308.3}$

Since the sign of floating point number is given by a special leading bit, the range for negative numbers is given by the negation of the above values.

There are five distinct numerical ranges that single-precision floating-point numbers are **not** able to represent:

1. Negative numbers less than $-(2-2^{-23}) \times 2^{127}$ (*negative overflow*)
2. Negative numbers greater than -2^{-149} (*negative underflow*)
3. Zero
4. Positive numbers less than 2^{-149} (*positive underflow*)
5. Positive numbers greater than $(2-2^{-23}) \times 2^{127}$ (*positive overflow*)

Overflow means that values have grown too large for the representation, much in the same way that you can overflow integers. Underflow is a less serious problem because is just denotes a loss of precision, which is guaranteed to be closely approximated by zero.

Here's a table of the effective range (excluding infinite values) of IEEE floating-point numbers:

	Binary	Decimal
Single	$\pm (2-2^{-23}) \times 2^{127}$	$\sim \pm 10^{38.53}$
Double	$\pm(2-2^{-52}) \times 2^{1023}$	$\sim \pm 10^{308.25}$

Note that the extreme values occur (regardless of sign) when the exponent is at the maximum value for finite numbers (2^{127} for single–precision, 2^{1023} for double), and the mantissa is filled with 1s (including the normalizing 1 bit).

Special Values

IEEE reserves exponent field values of all 0s and all 1s to denote special values in the floating-point scheme.

Zero

As mentioned above, zero is not directly representable in the straight format, due to the assumption of a leading 1 (wed need to specify a true zero mantissa to yield a value of zero). Zero is a special value denoted with an exponent field of zero and a fraction field of zero. Note that – 0 and + 0 are distinct values, though they both compare as equal.

Denormalized

If the exponent is all 0s, but the fraction is non-zero (else it would be interpreted as zero), then the value is a *denormalized* number, which does not have an assumed leading 1 before the binary point. Thus, this represents a number $(-1)^5 \times 0.f \times 2^{-126}$, where s is the sign bit and f is the fraction. For double precision, denormalized numbers are of the form $(-1)^s \times 0.f \times 2^{-1022}$. from this you can interpret zero as a special type of denormalized number.

Infinity

The values + infinity are denoted with an exponent of all 1s and a fraction of all 0s. the sign bit distinguishes between negative infinity and positive infinity. Being able to denote infinity as a specific value is useful because it allows operations to continue past overflow situations. *Operations with infinite values are well defined in IEEE floating point.*

Not a Number

The value NaN (*Not a Number) is used to represent a value that does not represent a real number. NaN's are represented by a bit pattern with an exponent of all 1s and a non-zero fraction. There are two categories of NaN; QNaN (QuietI NaN)* and SNaN (signalling NaN).

A QNaN is a NaN with the most significant fraction bit set. QNaN's propagate freely through most arithmetic operations. These values pop out of an operation when the result is not mathematically defined.

An SNaN is a NaN with the most significant fraction bit clear. It is used to signal an exception when used in operations. SNaNs can be handy to assign to uninitialized variables to trap premature usage.

Semantically, QNaNs denote *indeterminate* operations, while SNaN's denote *invalid* operations.

Special Operations

Operations on special numbers are well-defined by IEEE. In the simplest case, any operation with a NaN yields a NaN result. Other operations are as follows:

Operation	Result
n ÷ ± Infinity	0
± Infinity × ± Infinity	± Infinity
± nopnzero ÷ 0	± Infinity
Infinity + Infinity	Infinity
± 0 ¸ ± 0	*NaN*
Infinity – Infinity	*NaN*
± Infinity ¸ ± Infinity	*NaN*
± Infinity × 0	*NaN*

To sum up, the following are the corresponding values for a given representation;

Float Values (*b* = bias)

Sign	Exponent (*e*)	Fraction (*f*)	Value
0	00..0	00..00	+ 0
0	00..00	00..01 : 11..11	Positive Denormalized Real $0.f \times 2^{(-b+1)}$
0	00..00 : 11..10	XX..XX	Positive Normalized Real $1.f \times 2^{(e-b)}$
0	11..11	00..00	+ Infinity
0	11..11	00..01 : 0.1..11	SNaN
0	11..11	10..00 : 11..11	QNaN

1	00..00	00..00	– 0
1	00..00	00..01 : 11..00	Negative Denormalized Real $-0.f \times 2^{(-b+1)}$
1	00..01 : 11.10	XX..XX	Negative Normalized Real $-1.f\,2^{(e-b)}$
1	11.11	00..00	– Infinity
1	11..11	00..01 : 01..11	SNaN
1	11..11	10..00 : 11.11	QNaN

A.23. PERIPHERAL DEVICES

The devices which are connected to a computer and controlled by processor are called **Peripheral Devices**.

Example: Keyboards, Mouse, Display unit, Printers, etc.

A.24. INPUT-OUTPUT INTERFACE

"A hardware unit that plays an important role between the CPU and peripheral devices to supervise the input and output transfers is known as interface".

Input-output interface provides a method for transferring information between internal storage and external I/o devices. Peripherals connected to a computer need special communication links for interfacing them with the CPU. The purpose of the communication link is to restive the differences that exist between the central computer and each peripheral. The major differences are:

1. Peripherad are electromechanical and electromagnetic devices and their manner of operation is different from the operation of the CPU and memory, which are electronic devices.
2. The data transfer rate of peripherals in usually slower than the transfer rate of the CPU.
3. Data codes and formats in peripherad. Different from the word format in the CPU and memory
4. The operating modes of peripherals the different from each other and each must be controlled as so into to disturb the operation of other peripherals connected to the CPU.

To resolve these differences computer systems include special hardware components between the CPU and peripherals to supervise and synchronize all input and output transfers. These components are called interface units because they interface between the processor bus and the peripheral device.

There are fear types of communication that an interface may receive. They are:

1. **Control command:** A control command is issued to activate the peripheral and to inform it what to do. The particular control commands issued depends on the peripheral.

Example: A magnetic tape unit may be instructed to back space the tape by one record, or to start the tape. moving in the forward direction.

2. **Status Command:** A status command is used to text various status conditions in the interface and the peripherals.
3. **Data output command:** A data output command causes the interface to respond by transferring dataform the bus into one of its registers.
4. **Data Input Command:** The data input command is the opposite of the data output. In this call the interface receives written of data from the peripheral and places it in its buffer register.

Input-Output Versus Memory Bus

In addition to communicating with input-output the processor must communicate with the memory unit. Like the input-output bus, the memory bus contains data, address, and read write control lines. The computer buses can communicate to memory and input-output in three different ways.

1. Use two separate buses, one for memory and the other for input-output.
2. Use one common bus for both memory and input-output but have separate control lines for each.
3. Use one common bus for memory and input-output with common control lines.

A.25. SOLATED VERSUS MEMORY MAPPED INPUT-OUTPUT

In isolated input-output configuration the input and output instruction for both interface register and memory are different. When the CPU fetches and decodes the instructions for input-output interfaces, it enables the input-output read or input-output write control line and informs the other components attached to the common bus hot the address in the address bus is for interface register and not for memory words.

Similarly: When the CPU fetches and decodes the instruction from memory, it places the memory address on the address bus and ended the memory areas or memory write control line, which inform the external components that the address is for the memory and input-output addresses are is dated in the method.

Alternative, method is known as memory mapped input-output. In this method the computer use the same address space, for memory and input-output both. In memory mapped input-output there is no specific input or output instructions. The am instruction cache used for manipulating input-output data residing in interfaces and can also be used to manipulate memory words. Computers with memory. Mapped input-output can use memory type instructions to access input-output data and hence, allows the computer to use the same instructions for memory transfers or input output transfers.

Example 1: What is the difference between isolated input-output and memory mapped input-output? What are the advantage and disadvantages of each?

Example 2: What are the three methods of input-output bus structures?

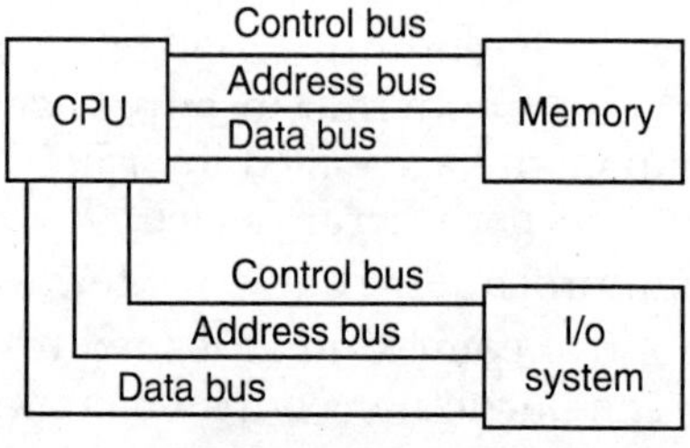

Fig. A.17.

(a) (Isolated input-output) (separate memory and input-output buses)

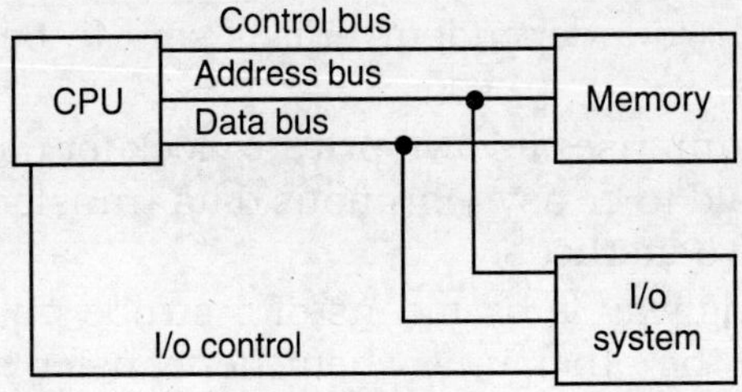

Fig. A.18.

(b) Shared address and data lines.

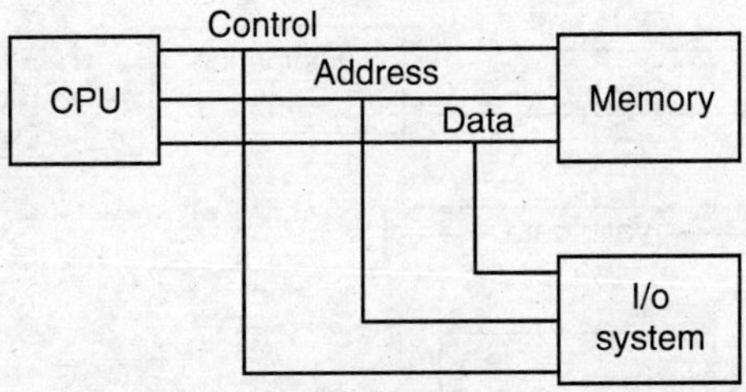

Fig. A.19.

(c) Shared address, data and control lines memory-mapped input-output)
Example of input Interface.

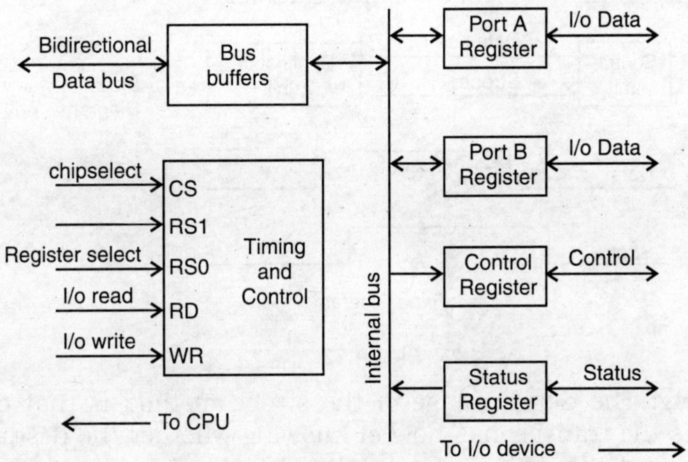

Fig. A.20.

CS	RSI	RSO	Register selected
0	X	X	None: data bus in high impudence
1	0	0	Port A register
1	0	1	Port B register
1	1	0	Control register
1	1	1	Status register

A.26. ASYNCHRONOUS DATA TRANSFER

The data transfer between two independent unit is said to be synchros data transfer if they share a common clock.

But if two independent unit uses its own private clock for data transfers between them when the data transfer is said to be asynchronous data transfers. This asynchronous way of data transfer active by two methods.

1. **Strobe Control:** One way is by means of a strobe pulse supplied by one of the units to indicate to the other unit when the transfer has to occur. This method is known as strobe control. The store control method employs a single control line. The strobe may be activated by either the source or the destination unit.

   (a) Source- Initiated strobe for data transfer

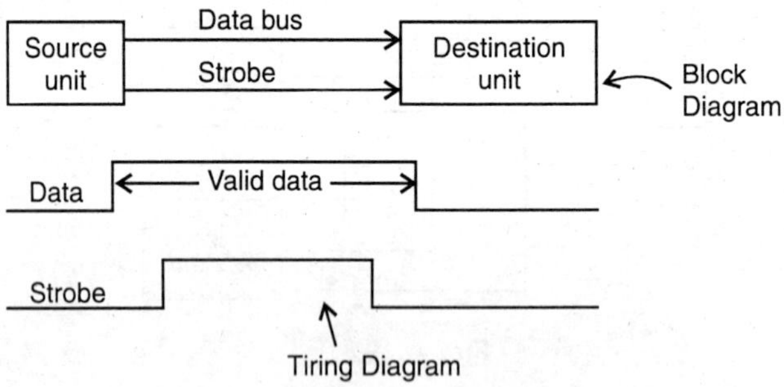

Fig. A.21.

   (b) Destination-Initiated strobe for dat transfer

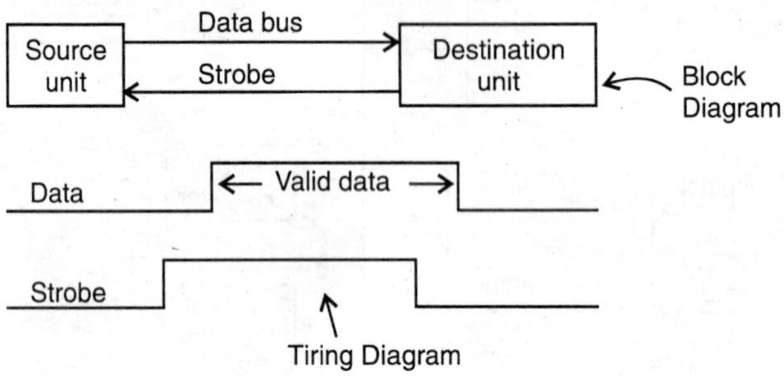

Fig. A.22.

2. **Handshaking:** The disadvantage of the strobe method is that the source unit that initiates the transfer has how of knowing whether the destination unit has actually received the data that was placed in the bus. Similarly, a destination unit that initiates the transfer has no way of knowing whether the source unit has actually placed the data or the bus.

   The handshaking added this problem by introducing a second control signal line that provides a reply to the unit that initiates the transfer.

   One control line is in the same direction as the data flow in the bus from the source to desecration. it is used by the source unit to in form the distortion unit whether there are valid data or not. The other control line is in the other direction from destination to the source.

   (a) **Source:** Initiated Handshaking:

(a) Block Diagram

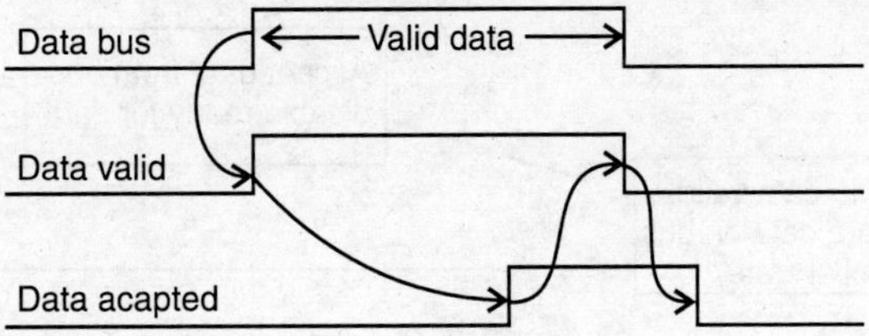

(b) Timing Diagram

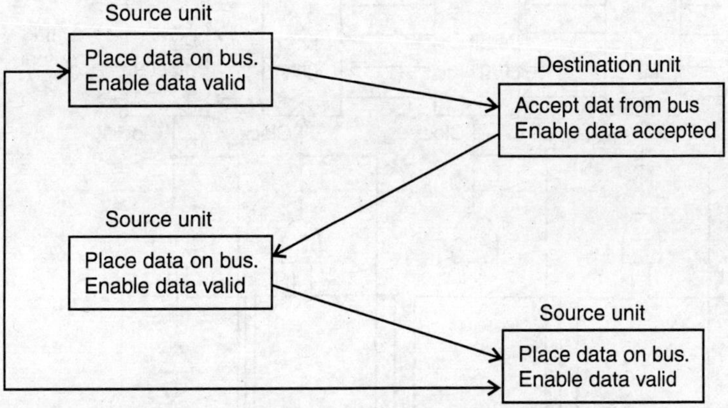

Fig. A.23.

(b) Destination Initiated Handshaking

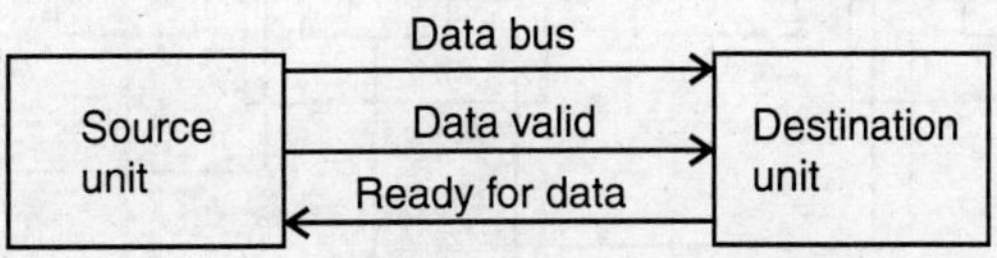

(a) Block Diagram

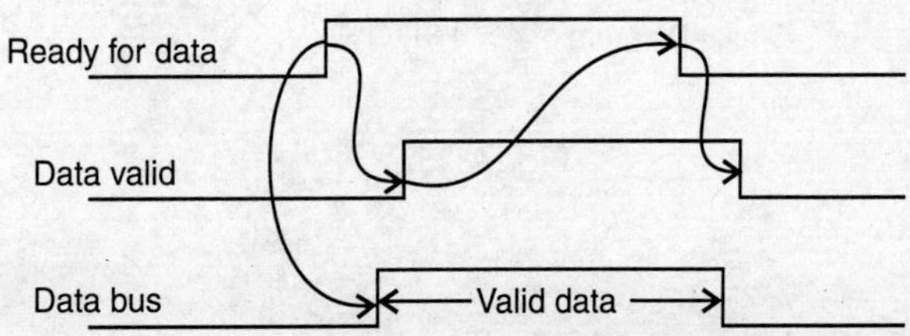

(b) Timing Diagram

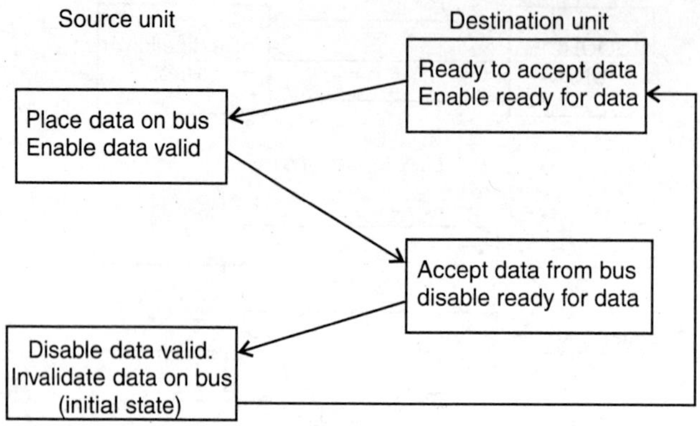

Fig. A.24.

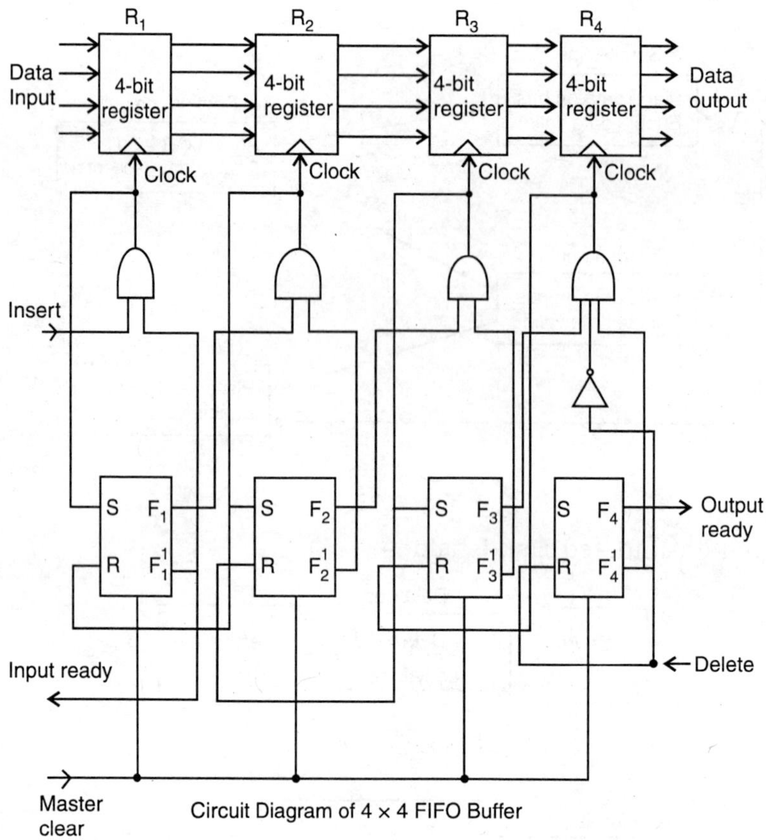

Fig. A.25.

A.27. FIFO BUFFER

The logic diagram of a typical 4 × 4 FIFO buffer side. It consists of four 4-bit register R_I,

I = 1, 2, 3, 4, and a control register with flip-flop

F_i, i = 1, 2, 3, 4, one for each register

The FIFO an store four words of four bits each. The number of bits per word can be increased by increasing the number of bits in each register and the number of words can be increased by increasing the number of register.

A flip-flop F_i in the control register that set to 1 indicates that a 4-bit data word is store in the corresponding register R_1. A 0 to F_i indicate that the corresponding register doesn't contain valid data. The control directs the improvements of data through the registers. Whenever the F_i bit of the control register is set (F_i = 1) the Fi + 1 bit is reset (F'_{i+1} = 1), to accept the data from register R1. The same clock transition sets F_{i+1} to 1 and reset F_i to 0. This causes the control flag to move one position to the right together with the data. Data in the register move down the FIFO toward the output as long as there are empty locations a head of it, This ripple through operation stop when the data reach a register R1 with the next Flip-flop F_{i+1} being set it 01, or bit the last register R4. An overall master cleat is used to initaize all control register flip-flop to 0.

Insertion: Data are inserted into the buffer provided that the input ready signal is enabled. This occurs when the first control flip-flop F1 is reset, indicating that register R1 is empty. Data are loaded from the input lines by enabling the clock, in R1 through the insert control line. The same clock sets F1, which disables the input ready control , indicating that the FIFO is row busy and unable to accept more data. The ripple through process begins provided that R2 and F_1 is cleared. This erables the input ready line, indicating that the input are now available for another data word. If the FIFO is full, F_1 remains set and the input ready line stays in the 0 state.

Detection: The data falling, through the registers stack up at the output end. The output ready control line is enabled when the last control flip-flop F_4 is set, indicating that there are valid data in the output register R_4. The output data form R4 are accepted by a destination unit, which then enables the delete control signal. This resets F_4, causing output ready to disable, indicating that the data on the output are no longer valid. Only after the delete signal goes back to 0 can the data from R_3 move into R_4. If the fIFO is empty, there will be no data in R_3 and F_4 will domain in the reset-states.

A.28. AS SYNCHRONOUS SERIAL TRANSMISSION

A serial asynchronous data transmission technique used special bits that are inserted at both ends of the character code. Each character consists of three parts.

1. a start bit,
2. the character bits,
3. stop bits.

A.29. USART OR UART

USART and Universal synchronous asynchronous receiver transmitter. Intel has developed 8251, a USART which is capable of transmittal data either in synchronous or asynchronous mode.

Example 1: What is the basic advantage of using interrupt initiated data transfer over transfer under program control without an interrupt?

Sol. The problem with programmed input-output is that the CPU has to with a long time for the input-output module of concern to be ready for reception or transmission of data. The CPU, while editing, must repeatedly interrogate the status of the input-output modules. As a result, the level of the system degraded need lessly. An alternative is for the CUP to issue an input-output command to a module and then go on to do some other useful work. The input-output module will then interrupt the CPU to request service when it is ready to exchange data with the CPU.

A.30. DIRECT MEMORY ACCESS (DMA)

"Direct Memory Access is a technique for moving data directly between the main memory and the peripheral devices without the intervention of CPU.

Cycle Stealing: The DMA device take memory bus from the processor and steal one memory cycle required fro read or write.

DMA controller: In DMA the processor realizes the control of the buses to a device called DMA controller.

Bus request (BR): It is the input signal used by the DMA controller to request the CPU to relinquish control if the buses. When this input is active, the CPU terminates the execution of the current instruction and places the address bus, the data bus and control signal into a high impedance state.

Bus grant (BG): It is activated by CPU to inform the external DMA that the buses are in the high impedance state. The DMA that originates the Bus request can now take control of the buses to process the data transfer.

Example 1: Why are read and write control lines in DMA (controller bidirectional? Under what condition and for what purposes are they used as inputs? Under what condition and for what purposes are they used as output?

Solution: When the CPU communicates with the DMA controller the read and write lines are used as inputs from the CPU to the DMA controller when the DMA controller communicate with memory, the read and write lines are used as outputs from the DMA to memory.

1. CPU bus signals for DMA transfer.

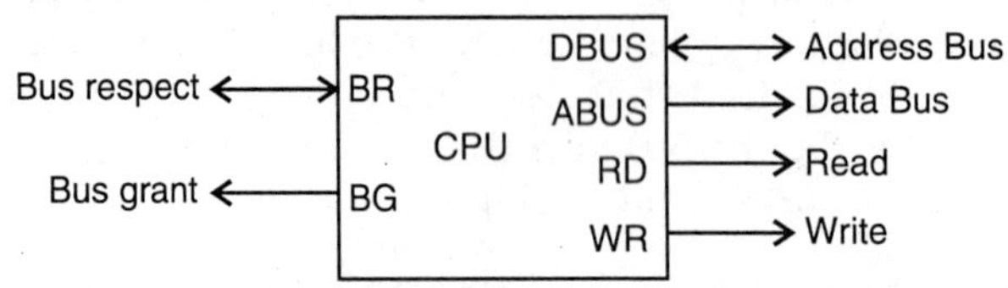

Fig. A.26.

2. Block Diagram of DMA Controller.

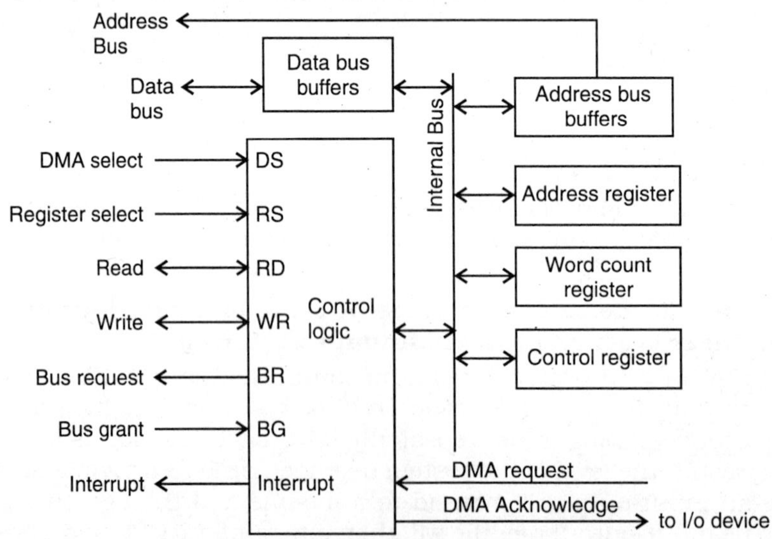

Fig. A.27.

3. DMA transfer in a computer system.

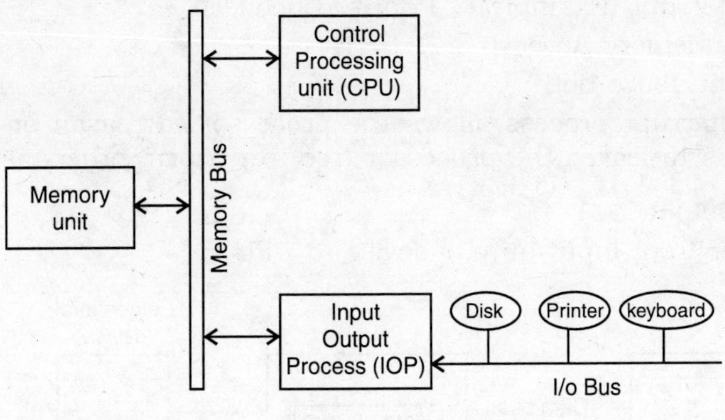

Fig. A.28.

CPU-IOP Communication

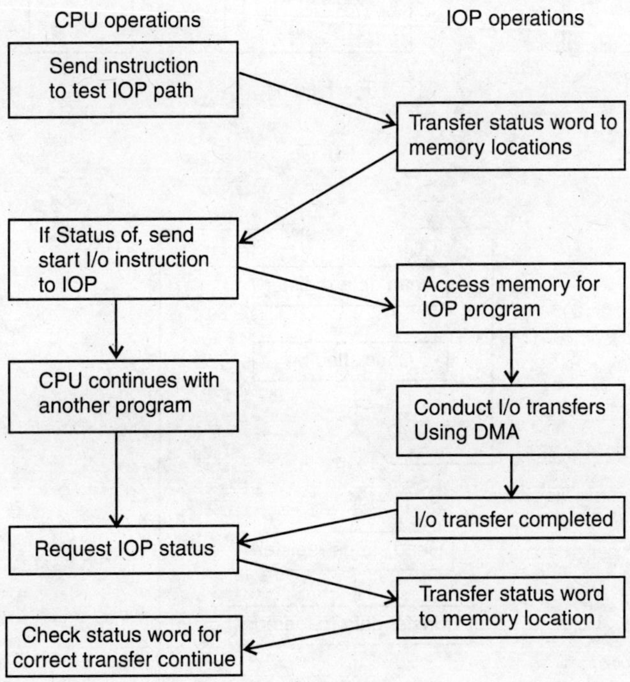

Fig. A.29.

A.31. MODES OF TRANSFER

Data transfer to and from peripherals may be handled in one of the three possible modes.

1. Programmed input-output

   To reduce drawback of (1) we create (2)

2. Interrupt-Imitated input-output

   →Priority Interrupt

→Serial interrupt (Example → Daisy-chin-up)
→parallel interrupt (Example → Priority Encoder)

3. DMA [Direct Memory Access]

It is fast from above both

Interrupt: Interrupt process allows the processor to reason or a service on a demand basis to leaves the processor free to perform other task.

Programmed Input-Output

Data transfer from input-output device to CPU

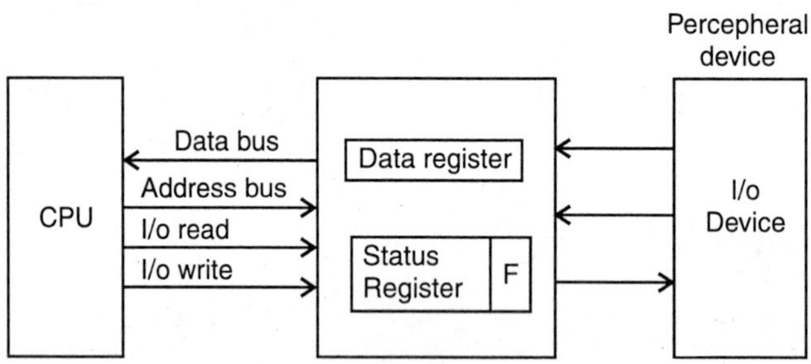

F = Flag bit

(a)

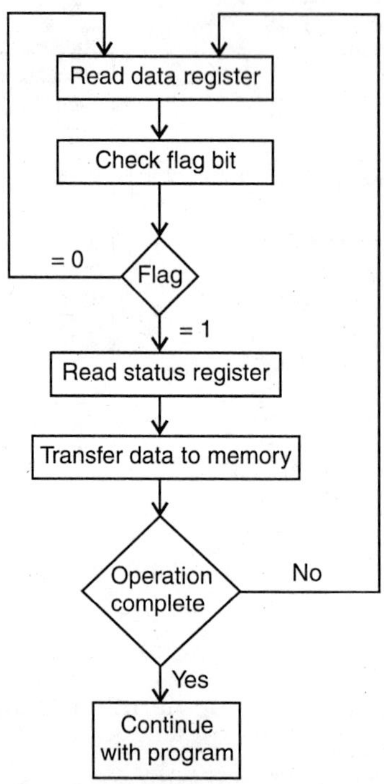

Flowchart for CPU program to input data.

(b)

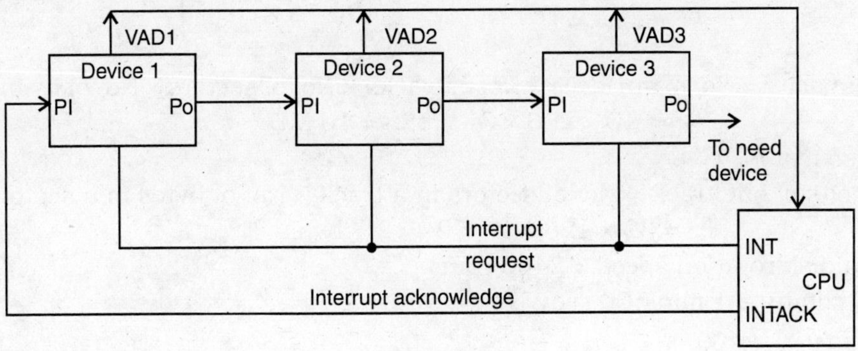

(c)

Fig. A.30.

Daisy-Chain Priority interrupt.

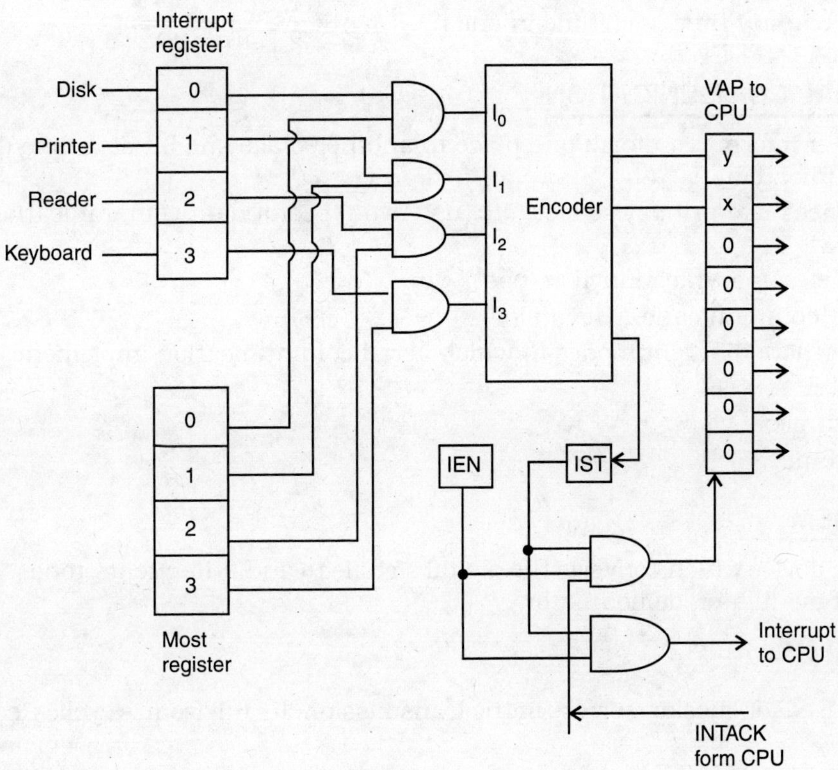

Fig. A.31.

"Priority Encoder or Priority Interrupt Hardware

A certain magnetic disk has following speciates

→

→

→

→

→

(i) Disk capacity = No. of surfaces × No. of tracks × No. of sectors × No. of by in a sector

$$= 10 \times 256 \times 32 \times 128 = 10 \text{ MB.}$$

(ii) Transfer rate

Capacity of one track = No. of sectors in a track × No. of bytes in a sector

$$= 32 \times 128 \times 2^5 \times 2^7 = 2^{12} \text{ bytes.}$$

Given disk rotation speed = 2400 rpm

2400 rotate → 1 min= 60 sec

$$1 \text{ rotation} \rightarrow \frac{60}{2400} = \frac{1}{40}\text{sec}$$

$$\text{in } \frac{1}{40}\text{ sec} = 212 \times 40 \text{ by too}$$

Transferred = 160 k Bytes

$$\text{Latency time} = \frac{1}{2} \text{ (time in one rotating)} = \frac{1}{2} \times \frac{1}{40} = \frac{1}{20}\text{sec}$$

A.32. SERIAL COMMUNICATION

Transferring of information form one place to another place one bit at a time (known as serial communication).

Some devices like printer, sensor, etc. use two types of data communication to send data.

1. Parallel communication II is bast
2. Serial communication. It is slow

There are there different modes in which aerial communication implement

1. Simplex
2. Half-duplex
3. Full-duplex

A.33. MODEM

A modem is a device which converts the digital signals to audio-frequency tones. Modem is short form of modulator-demodulator.

A.34. CRC

It is method used for checking errors in the transmission. Its full from is cyclic Redundancy check.

Binary Number Formats

Name	Size (bits)	Example
Bit	1	1
Nibble	4	0101
Byte	8	0000 0101
Word	16	0000 0000 0000 0101
Double word	32	0000 0000 0000 0000 0000 0000 0000 0101

A.35. ANOTHER METHOD FOR REPRESENTING FLOATING POINT NUMBERS

Floating-Point Numbers

All the numbering systems which we have studied so for fall into category of base r fixed point numbers.

Fixed point number means the fractional point (decimal point) if once finalized will be fixed.

Disadvantage of fixed point no. Vapresertation is that it cannot represent very large number and very small number.

→ The remedy of this disadvantage is Scientific notation.

→ In Scientific Notation, number is in the from of

$$N = M \times 10^E$$

Where M and E are fixed points number called Mantissa and Exponent respectively.

Now with the version of Scientific. Number (notation) is called FLOATING NUMBER POINT.

For into take an example → consider 9 no. $(10000 \times 34)_{10}$

This no. can be represented as,

$$1000\ 234 \times 10^{-3}$$
$$1000\ 23.4 \times 10^{-2}$$
$$1000\ 2.34 \times 10^{-1}$$
$$1000.234 \times 10^{0}$$
$$100.01234 \times 10^{1}$$
$$10.00234 \times 10^{2}$$
$$1.000\ 234 \times 10^{3}$$
$$0.1000234 \times 10^{4}$$

So, in representation of $(1000.234)_{10}$, you must absorved that decimal point is not fixed. It can FLDAT from number to number depending upon Power of 10.

→ In

$$N = \pm M \times ME$$

B = Base of plodding point no. system

It is 10 for decimal system and 2 for Binary system

B = 10 B = 2

FORMAT OF FLOATING POINT NUMBER

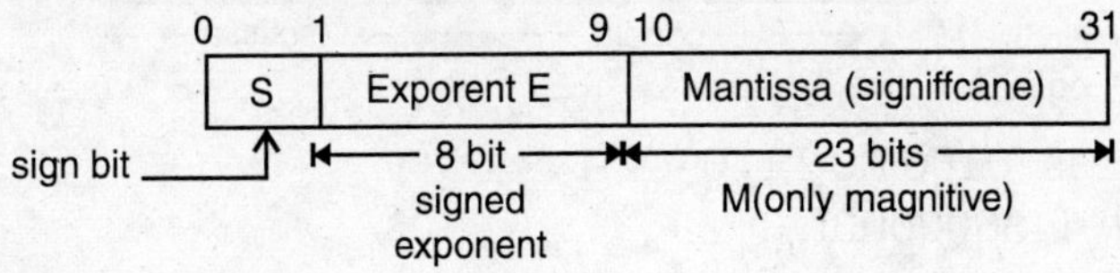

Format of a 32 bit floating point number.

Fight bits are used for exponent E, with remaining 23 bits for mantissa.

E con range from −128 to +127

→ Floating point number are standardized by IEEE. According to it we. represent floating point number is form.

$$\pm 1.M \times 2^{E^1 - 127} \quad \text{(single precision -32 bits)}$$

$$\pm 1.M \times 2^{E^1 - 1023} \quad \text{(Double precision –64 bits)}$$

Here M = Mantissa

$E = E^1 - 127$

$\Rightarrow$ $E^1 = E + 127$ E = exponent of number

The range for E1 is 0 to 255.

1	← 8 →	← 23 →
S	Exponent (E′)	Mantissa

(a) Single precision

1	← 11 →	← 52 →
S	Exponent (E′)	Mantissa

(a) Double Precision

IEEE 754 format

What is 1 M?

Whenever any number is given to you to represent it in floating point no. system, you have to bring down to the format 1.M

Let the no. is 0.0110×2^6

Presently no. has form 0.M where M = 0110

But we want 1.M

So, now benefit no. to left side till you get 1,

$$\left.\begin{array}{l} 0.0110 \times 2^6 \rightarrow \text{Actual number} \\ 0.110 \times 2^5 \rightarrow \text{After } 1^{st} \text{ shift} \\ 1.10 \times 2^4 \rightarrow \text{after } 2^{nd} \text{ shift} \end{array}\right\} \text{shift operation}$$

Now stop here as we have 1.M Now,

$M = 10$

S = plus (positive)

$E' = E + 127 = 4 + 127 = 131 = (10000011)2$

Therefore the floating point no. *i.e.*,

S	E′ = 86 bits	M = 23 bits
0	1 0 0 0 0 0 1 1	1 0 ....
No is positive	← Exponent →	← Msntissa →

$\therefore$ finally number is $(0100\ 00\ 1110\ \text{-----})_2$

Index this no. *i.e.*, $(43800000)_{16}$

A.36. DE-NORMALISED/NORMALISED FLOATING POINT NUMBER

→ While representing 0.0110×2^6 number in floating point, you have shifted the bits and adjusted exponent, till you get 1.M

The number we get after proffering this operation is called "Naimalised number.

→ De-normalised number and 'RAW' number which does not follow 1.M format.

Range →

• Each floating point format (single precision/double precision) has capability to represent the no. within 'Certain' Range.

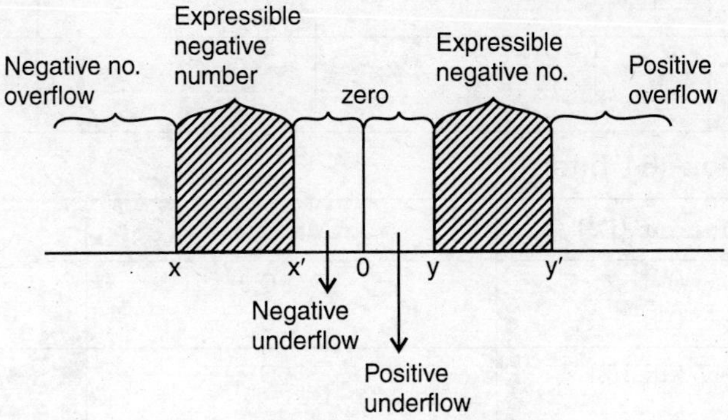

Fig. A.32.

This figure shows the number that can be expressed (+ve or – ve), by shaded part. Below and above the shaded parts, numbers are not representable.

• Fine regions are not included in the groups

1. Negative number, those are under negative overflow.
2. Negative number, those are under negative underflow.
3. Zero.
4. Positive number those are under positive underflow.

→ What is overflow and underflow?

• Overflow → Overflow occur when an arithmetic operation results in magnitude greater than can be expressed with an exponent.

Underflow → Underflow occurs which fractional part is too small. Underflow is less serious problem because the result canbe approximated to zero.

• To represent zero '0' special bit pattern to designated in IEEE format when e = Exponent = 0 and dignificand (Mantissa) = 0, the no. is zero.
• In IEEE format largest exponent field E = $(11111111)_2$. Corresponds to 255 is reserved to denote infinite number that result from some types of overflow. In this standard, certain invalid bit combinations considered to being a special not a number (NAN) category.

Table Interpretation of IEEE 750 Floating-Print Numbers

(a) Single Precision (32 bits)

	Sign	Exponent (E^1)	Mantissa	Value
1. Positive zero	0	0	0	0
2. Negative zero	1	0	0	– 0
3. Plus infinity or min 48 infinity	0 or 1	255 (all 1's)	0	(–1)s ¨
4. NAN	0 or 1	255 (all 1's)	≠ 0	NAN

5. Normalized non-zero number	0 or 1	0 < e < 255	f may be zero	$(-1)^s 2^{E'-127}$ (1.<M) or not
6. Denomalized	0	0	f ≠ 0	$(-1)^{S_2 -126}$(0.M)

(b) Double precision (64-bits)

	Sign	Exponent (E^1)	Mantissa	Value
1. 2.	0 1	0 0	0 0	0 0
3.	0 or 1	2047 (all 1's)	0	(–1)s ∞
4.	0 or 1	2047 (all 1's)	≠ 0	NAN
5.	0 or 1	0 < e < 2047	f may be zero or not	$(-1)^s.2^{E^1-1023}$ (1.M)
6.	0 or 1	0	f ≠ 0	$(-1)^s\ 2^{-1022}$ (0.M)

Represent following in single precision format.
For 'O', signata = 0, M = 0, and E = –127

$$(-1)^s\ 1.M \times 2^{E^1-127}$$

$$\therefore \quad (-1)^0\ 1.0 \times 2^0$$

$$\therefore \quad E = 0$$

S	Exponent (E^1)	Mentis
0	0000 0000	0000 ----0

∴ Flatten point no. is 0000000000011 (hex)

(ii) –1

$$\text{Sin bit} = 1 \rightarrow \text{No. is –Ve}$$

$$(1)10 = 1.0 \times 20 \rightarrow \text{normalized}$$

$$\therefore \quad M = 0$$

$$F = E^1 - 127$$

$$\therefore \quad E^1 = 127 = (01111111)_2$$

S	(E^1)	Mentisa
S	01111111	00------0
1 bit	8 bits	23 bits

∴ Flatting pint nos. is $(10111111100....0)_2$ and (BF800000)16 hex.

A.37. 2D ORGANIZATION

Semiconductor 2D memory (16W, 46/W)

Static memory cells can be organized into a 2D-memory system as shown in fig. The figure illustrates the organization of 156 word, 4 bit per word 200 Memory.

There is an address decoder to decoder the address in 4-bit MAR, which selects one out of 16 word wires. V is the voltage applied to a selected word wire. A number of memory calls equal to the bits/word are connected to each word wire. For i = 1, 2, 3, 4, the bit wires B_{0i} and B_{1i} i of all words are connected together and fed to this sense amplifier S_{Ai} as shown is figure.

To write, 1001 (say), in word 2 the following is done.

• Word address 0010 is stored in MAR. When decoded it selects word W_2 and the voltage line on W_2 is made V.

• The data 1001 is stored in MBR. SA_1 brings voltage on B_{11} to 1, i

SA_2 brings voltage on B_{02} to 0

SA_3 brings voltage on B_{03} to 1

SA_4 brings voltage on B_{14} to 1

Thus, call 1 of word 2 will other a 1, cell 2 of word 2 will store a 0, cell 3 of word 2 will store a 0, cell 4 cell 4 of word 2 will store a 1. To read the contents of award, the address must be needed in MAR and all the memory cells belonging to this word experience voltage V/2. Depending on whether a cell Head a 0 or 1 stored the voltages on the corresponding bit writes do, what is sensed by the respective sense amplifier. If the voltage in B_{0i} dip SA_i will store a 0 in bit 1 of MBR and if the voltage in B1i dip, SA_1 will store 1 in bit in of MBR.

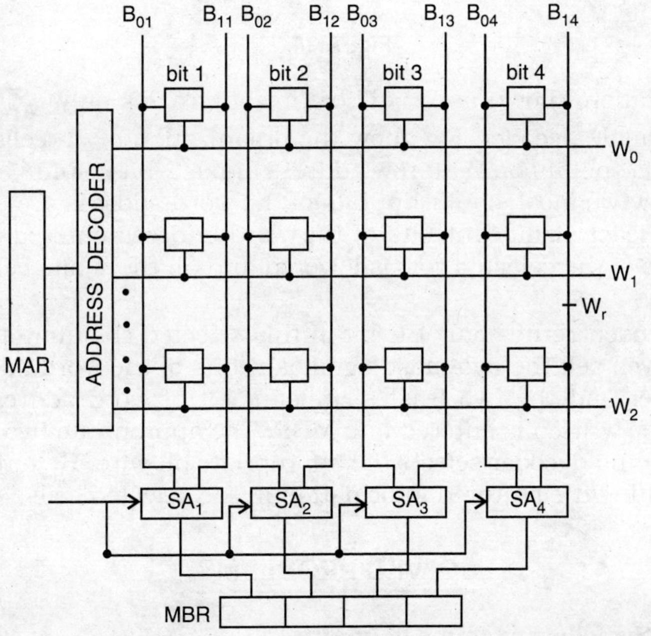

Fig. A.33.

A.38. $2^{1/2}$ D ORGANIZATION

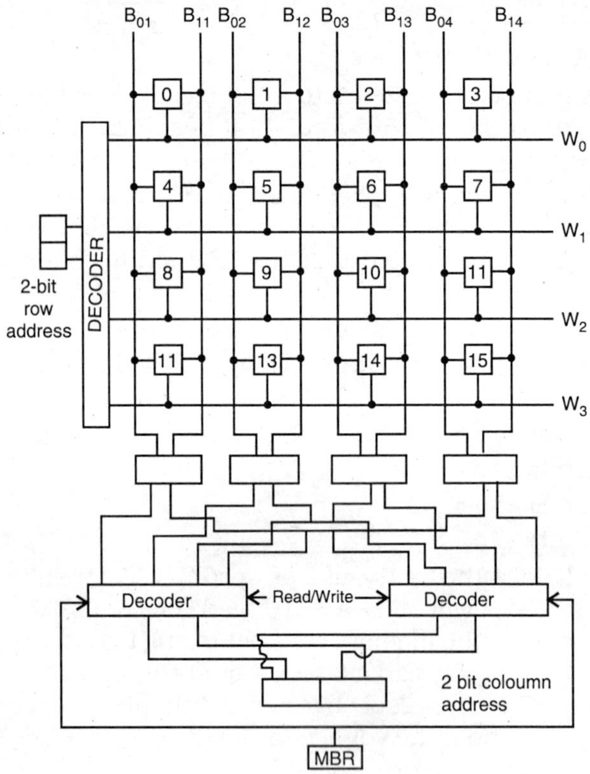

Fig. A.34.

Another organization known as $2^{1/2}$ D organization uses only $\sqrt{W}$ words wires and allows the use of simple decoder. Fig show the organization of 16 cells for 4 wires. The 4-bit word address is split into a 2-bit row address and a 2-bit column address. To read a bit form the memory two most significant bits of the word address are entered bits of the row registered two least significant bits of the word address entered into the coloumn register. The voltage of the selected row is lowered and all the 4 bits cells in this row are read.

The column decoder, routes only the bit in this selected coloumn as data output. To write a bit in the given cell the two most significant bits of the word address are entered into the row register and the two least significant bit of word address in the column register. The bit to be written is entered into MBR. The appropriate two is selected by the decoder and the column decoder selects the appropriate bit wire, B_{1i} or B_{0i} and its voltage is lowered while writing the required information in the selected cells.

SOLVED PROBLEMS

Problem 1. Define the following terms in brief:

(a) Register transfer language

(b) Micro operation

(c) Registers

(d) Control function

Solution: (a) Register transfer language: The symbolic notation used to describe the micro-operation transfers among registers is called a register transfer language. The term register transfer implies the availability of hardware logic circuits that can perform a stated micro-operation and transfer the result of the operation to the same or another registers the word "language" is borrowed from programmes who apply this term to programming language.

(b) Micro Operation: The operations executed on data stored in registers are called micro-operations. A micro-operation is an elementary operation performed on the information stored in one or more registers. The result of operation may replace the previous binary information of a register or may be transferred to another register *e.g.*, shift, count, clear and load.

(c) Registers: The register that gold the address of the memory units is called a memory address register and is designated by the name MAR other designations for registers are:

PC (Program counter)

IR (Instruction register)

R_1 (Processor Register)

(d) Register Transfer: Information transferred from one register to another is designated in symbolic form by means of a replacement operator. The statement $R_2 \leftarrow R_1$, denotes a transfer of the content of transfer R_1 into register R_2. It designates a replacement the contents of R_2 by the content of R_1. By definition, the content of source register R_1 does not change after the transfer.

(e) Control Function: The control function is a Boolean variable that is equal to 1 or 0. The control function is included in the statement as follows:

P, $\quad R_2 \leftarrow R_1$

The control condition is terminated with a colon, it symbolizes the requirement that the transfer operation be executed by hardware only of P = 1.

Problem 2. Describe three state bus buffer. Draw the diagram of three state buffer.

Solution: A bus system can be constructed with three state gates instead of multiplexers. A three state gate is a digital circuit that exhibits three states. Two of these states are signals equivalent to logic 1 and 0 as in a conventional gate the third state is a high independence state behaves like an open circuit, which means that the output is disconnected and does not have a logic significance. Three state gates may perform any conventional logic such as AND or NAND. However, the one most commonly used in the design of a bus system is the buffer gate.

The graphic symbol of a three state buffer gate is shown below:

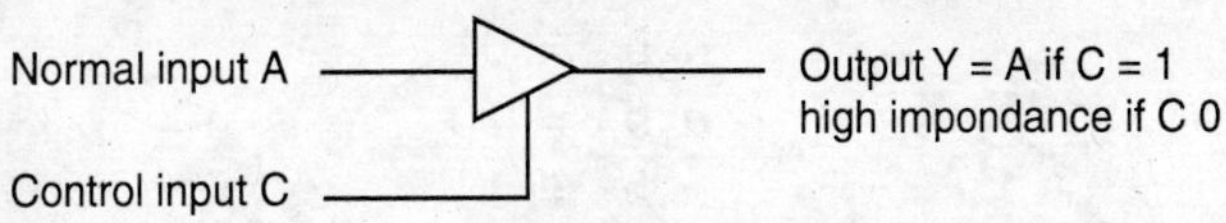

Fig. A.35.

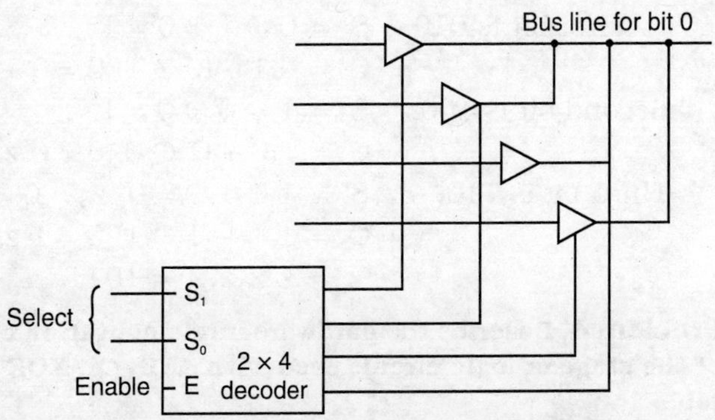

Fig. A.36. Bus line with three state buffers

To construct a common bus for four registers of n-bit using three state buffers, we need n circuit with four buffers. Each group of four buffers receives one significant bit from the four registers. Each common output produces one of the lines for the common bus for a total of n lines only one decoder is necessary to select between the four registers.

Problem 3. Consider the following circuit A = $a_2a_1a_0$ and B = $b_2b_1b_0$ are three bit binary numbers input to the circuit. The output is Z is z_3z_2,z_0. R_0 R_1 and are registers with loading clock shown. The registers are loaded with their input data with the falling edge of the clock pulse. The bits of the input numbers A, B and full address are shown in the circuit.

(a) Find the output 2 for the inputs

A = 110

B = 101

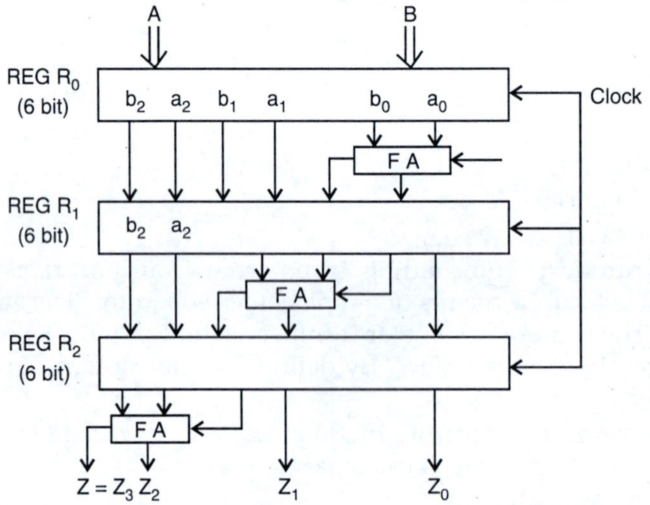

Fig. A.37.

Solution: Given value, A = $110 = a_2a_1a_0$

B = $101 = b_2b_1b_0$

$a_0b_0 = 01$

$a_1b_1 = 10$

$a_2b_2 = 11$

For bit $a_0b_0 = 01$

and third bit = 01

∴ First bit is 010, $S = 0 \oplus 1 \oplus 0 = 1$

$C = 0.1 + (0 \oplus 1)\ 0 = 0 + 0 = 0$

Second bit is 100, $S = 1 \oplus 0 \oplus 0 = 1$

$C = 1.0 + (1 \oplus 0).0 = 0 \quad z_1 = 1$

Third bit is 110, $S = 1 \oplus 1 \oplus = 0 \quad z_2 = 0$

$C = 1.1 + (1 \oplus 1)\ 0 = 1 \quad z_3 = 1$

$z = z_3z_2z_1z_0 = 1011$

Problem 4. Describe the hardware implementation of logic-micro operation. Draw the diagram of the stage of logic circuit used with AND, Or, XOR and NOT GATE with the help of function table.

Solution: The hardware implementation of logic micro-operations requires that logic gates be inserted for each bit or pair of bits in the register to perform the required logic function. Although there are 16 logic micro-operations, most computers use only four AND, OR XOR and complement.

S_1	S_0	Output	Operation
0	0	$E = A \wedge B$	AND
0	1	$E = A \vee B$	OR
1	0	$E = A \oplus B$	XOR
1	1		
1	1	$E = A$	Complement

Function Table

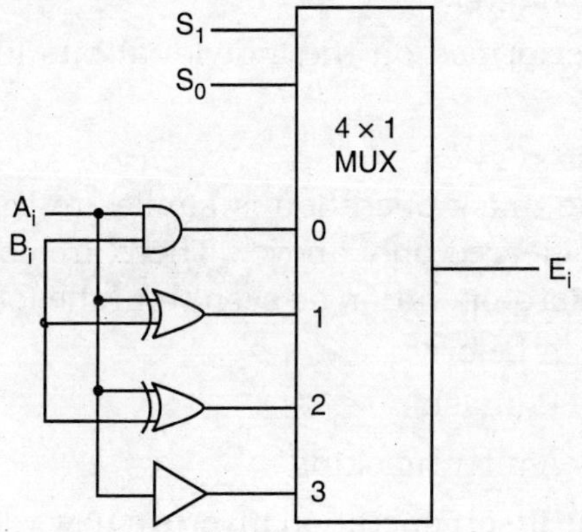

Fig. A.38. Logic diagram of one stage of logic circuit

One stage of circuit that generates the four basic logic micro-operation. It consists of four gates and multiplexer. Each of the four logic operations is generated through a gate that perform the required logic the outputs of gates are applied to the data inputs of the multiplexer. The two selection inputs S_1 and S_0. Choose one of the data inputs of the multiplexer and direct its value to the output. For a logic circuit with *n* bits the diagram mist be repeated *n* times for *i* = 0, 1, 2, 3, *n* the selection variables are applied to all stages.

Problem 5. Define the term and describe its operation:

(*a*) Selective set
(*b*) Selective complement
(*c*) Selective clear
(*d*) Mask operation
(*e*) Ensert operation
(*f*) Clear operation

Solution. (*a*) Selective set: The selective set operation sets to 1. The bits in the register A where there are corresponding is in register B. It does not affect that have O's in the B.

e.g.,	1010	A before
	<u>1100</u>	B(logical operand)
	1110	A (after)

The two left most bits of B are 1's B are 1's, so the corresponding bits of A are set to 1. The two bits of A with corresponding O's in B remain unchanged. Therefore the OR micro-operation can be used to selectively get bits of a register.

(*b*) **Selective complement:** The selective complement operation complements bits in A where there are corresponding 1's in B. It does not affect bit positions that have O's in B.

e.g.,

1010	A before
1100	B (logic operand)
0110	

Therefore exclusive–OR operation can be used to selectively complement bits of register.

(*c*) **Selective clear:** Selective clear operation clears to O. The bits in A only where there are corresponding 1's in B.

e.g.,

1010	A before
1100	B (logic operand)
0010	A after

The boolean operation performed on the individual bits in AB^1. The corresponding logic operation

$$A \leftarrow A \cap B^1$$

(*d*) **Mask operation:** The mask operation is similar to the selective clear operation except that the bits of A are cleared only above. There are corresponding O's in B. The mask operation is an AND micro-operation as seen from the following numerical example

1010	A before
1100	B (mask)
1000	A after masking

(*e*) **Insert operation:** The insert operation inserts a new value into a group of bits this is done by first masking the bits and then OR in them the required value.

e.g.,

0110	1010	A before
0000	1111	B (mask)
0000	1010	After masking
0000	1010	A before
1001	0000	B (insert)
1001	1010	A after insertion

(*f*) **Clear operation:** Clear operation compares the words in A and B and Produces an all O's result if the two numbers are equal. This operation is achieved by an exclusive–OR micro-operation

e.g.,

1010	A
1010	B
0000	$A \leftarrow A \oplus B$

Problem 6. Show the bock diagram of the hardware that implements the following register transfer statements

$$yT_1 \; : \; R_2 \leftarrow R_1, R_1 \leftarrow R_2$$

Solution:

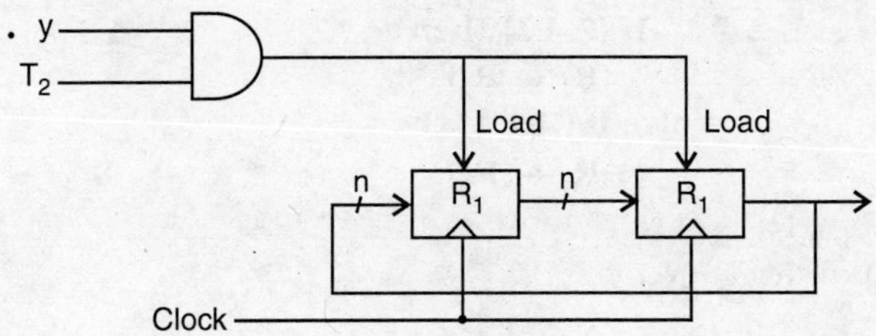

Fig. A.39.

Problem 7. The outputs of four registers R_0, R_1, R_2 and R_3 are connected through 4–to–1 line multiplexers to the inputs of a fifth register, R_5. Each register is eight bits long. The required transfers are dictated by four timing variables to through T_3 as follows:

$$T_0 : R_5 \leftarrow R_0$$
$$T_1 : R_5 \leftarrow R_1$$
$$T_2 : R_5 \leftarrow R_2$$
$$T_3 : R_5 \leftarrow R_3$$

The timing variables are mutually exclusive, which means that only one variable is equal to 1 at any given time, while the other three are equal to 0. Draw a block diagram showing the hardware implementation of register transfers, include the connection S necessary from the four timing variables to the selection inputs of the multiplexers and to load input of register R_5.

Solution:

T_0	T_1	T_2	T_3	S_1	S_0	R_3 load
0	0	0	0	×	×	0
1	0	0	0	0	0	1
0	1	0	0	0	1	1
0	0	1	0	1	0	1
0	0	0	1	1	1	1

$$S_1 = T_2 + T_3$$
$$S_0 = T_1 + T_3$$
$$\text{load} = T_0 + T_1 + T_2 + T_3$$

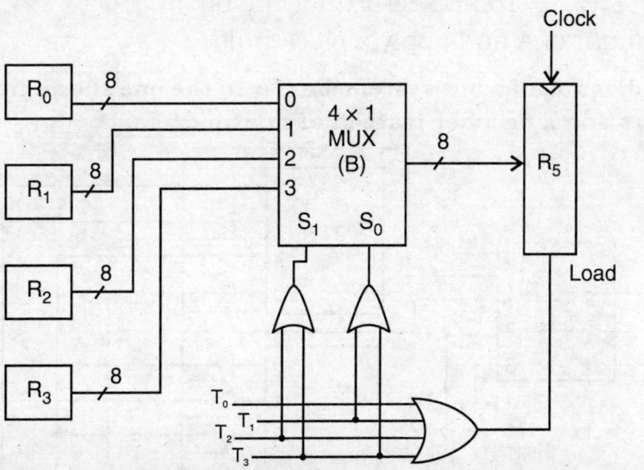

Fig. A.40.

Problem 8. Represent the following conditional, control statement by two register transfer statements with control functions.

If (P = 1) then
($R_1 \leftarrow R_2$)
else if (Q = 1) then
($R_1 \leftarrow R_3$)

Solution: $P : R_1 \leftarrow R_2$
$P^1Q : R_1 \leftarrow R_3$

Problem 9. What has to the bus system of figure given below to be able to transfer information from any register to any other register ? Specifically. Show the connections that must be included to provide a path from the outputs of register C to the inputs of register A.

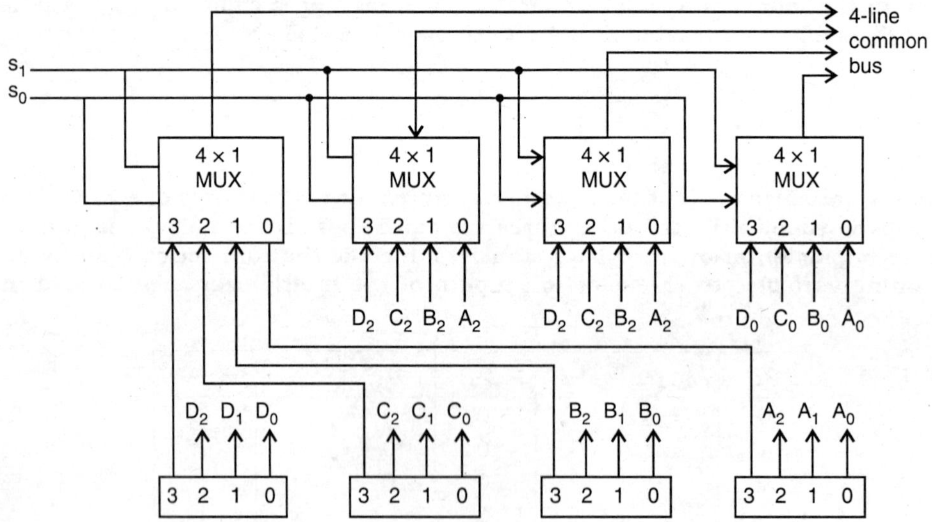

Fig. A.41. Bus system for Four register

Solution: Connect the 4–line common bus to the fair inputs of each register. Provide a "load" control input in each register provide a clock input for each register to transfer from register C to register A.

Apply S_1S_0 = 10 (to select C for the bus)
Enable the load input of A and apply a clock pulse.

Problem 10. Draw a diagram of a bus system similar to the one shown in the figure below but use three-state buffers and a decoder instead of multiplexers.

Solution:

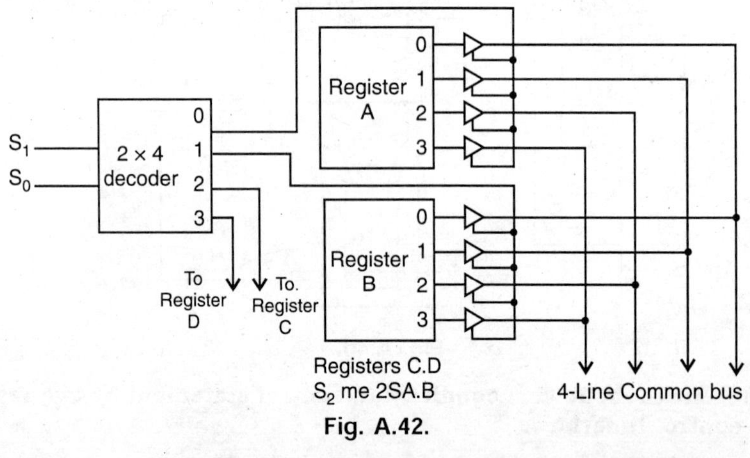

Fig. A.42.

Probelem 11. A digital computer has a common bus system for 16 registers of 32 bits each. The bus is constructed with multiplexers.

(a) How many selection inputs are there in each multiplexers ?

(b) What size of multiplexers are needed ?

(c) How many multiplexers are there in the bus ?

Solution: (a) 4 selection lines to select one of 16 register

(b) 16×1 multiplexers are needed.

(c) 32 multiplexers, one for each bit of the registers.

Problem 12. The following transfer statements specify a memory. Explain the memory operation in each case.

(a) $R_2 \leftarrow M [AR]$

(b) $M [AR] \leftarrow R_3$

(c) $R_5 \leftarrow M [R_5]$

Solution: (a) Read memory word specified by the address in AR into register R_2

(b) Write content of Register R_3 into the memory word specified by the address in AR

(c) Read memory word specified by the address in R_5 and Transfer content to R_5 (destroys previous value)

Problem 13. Draw the block diagram for the hardware that implements the following statements:

$$x + yz \; : \; AR \leftarrow AR + BR$$

Where AR and BR are two n-bit registers and x, y and z are control variables. Include the logic gates for the control function. (Remember that the symbol + designates an OR operation in a control or Boolean function but that it represents an arithmetic plus in a micro-operation)

Solution:

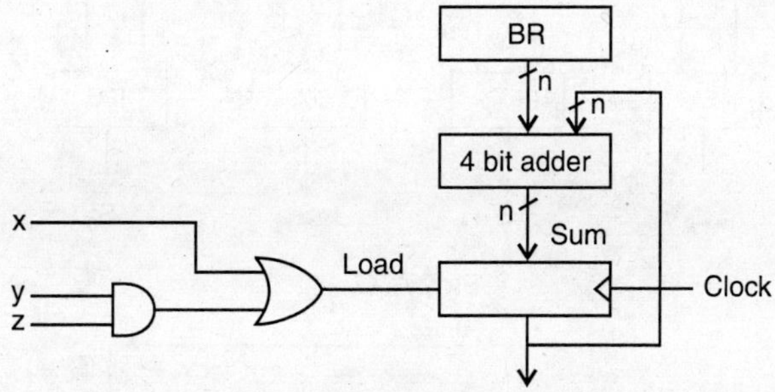

Fig. A.43.

Problem 14. Show the hardware that implements the following statement. Include the logic gates for the control function and a block diagram for the binary counter with a count enable input.

$$xyT_0 + T_1 + y'T_2 : AR \leftarrow AR + 1$$

Solution:

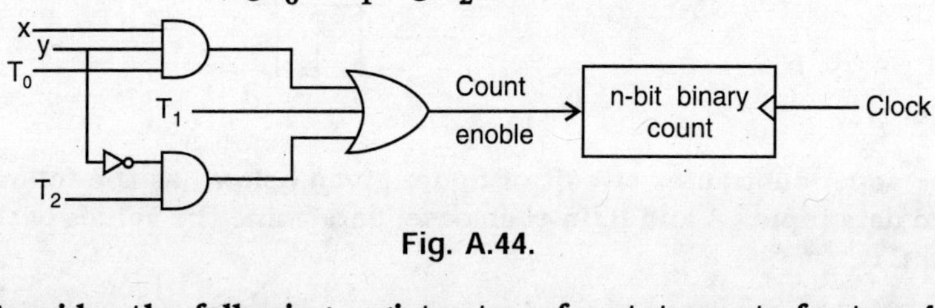

Fig. A.44.

Problem 15. Consider the following register transfer statements for two 4-bit register R_1 and R_2

$$xT : R_1 \leftarrow R_2$$
$$x^1T : R_1 \leftarrow R_2$$

Every time that variable T = 1, either the content of R_2 is added to the content of R_1 if x = 1, or the content of R_2 is transferred to R_1 if x = 0. Draw a diagram showing the hardware implementation of the two statements. Use block diagrams for the two. 4-bit registers, 4-bit adder and a quotable 2-to-1 line multiplexer that selects the inputs to R_1. In the diagram, show how the control variables x and T select the inputs of the multiplexers and the load input of register R_1.

Solution:

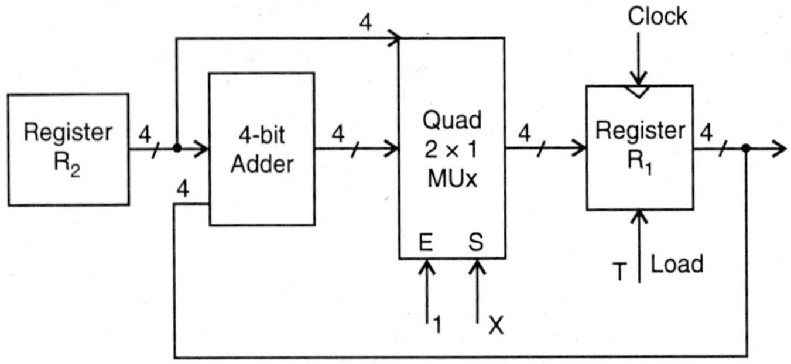

Fig. A.45.

Problem 16. Using a 4-bit counter with parallel load and a 4-bit adder as in figure given below, draw a block diagram that shows how to implement the following statements

$$x : R_1 \leftarrow R_1 + R_2 \qquad \text{Add } R_2 \text{ to } R_1$$
$$x^1y : R_1 \leftarrow R_1 + 1 \qquad \text{Increment } R_1$$

Where R_1 is a counter with parallel load and R_2 is a 4-bit register.

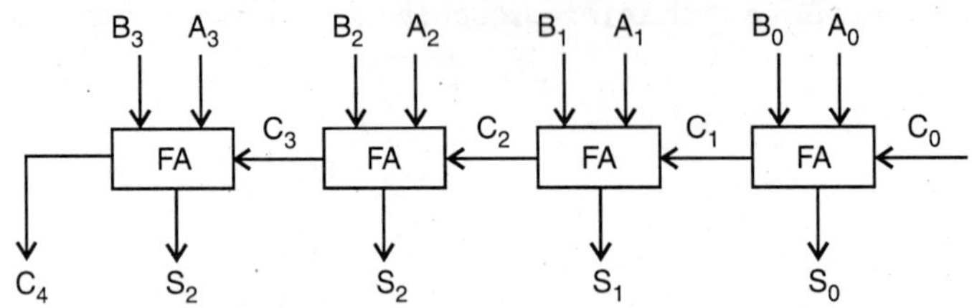

Fig. A.46.

Solution:

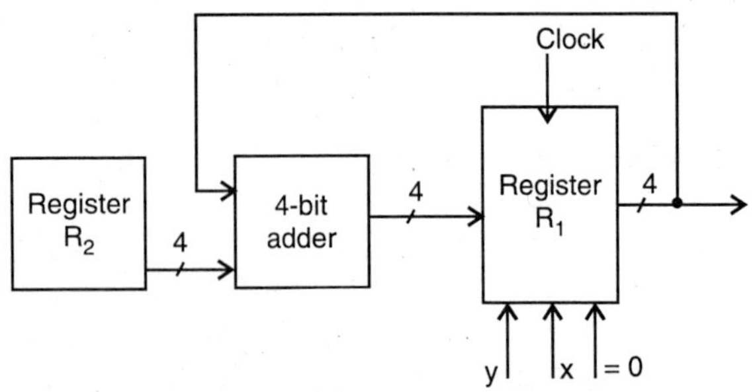

Fig. A.47.

Problem 17. The adder-subtracter circuit of figure given below has the following values for input mode M and data inputs A and B. In each case, determine the values of the outputs: S_3, S_2, S_2, S_1, S_0 and C4

	M	A	B
(*a*)	**0**	**0111**	**0110**
(*b*)	**0**	**1000**	**1001**
(*c*)	**1**	**1100**	**1000**
(*d*)	**1**	**0101**	**1010**
(*e*)	**1**	**0000**	**0001**

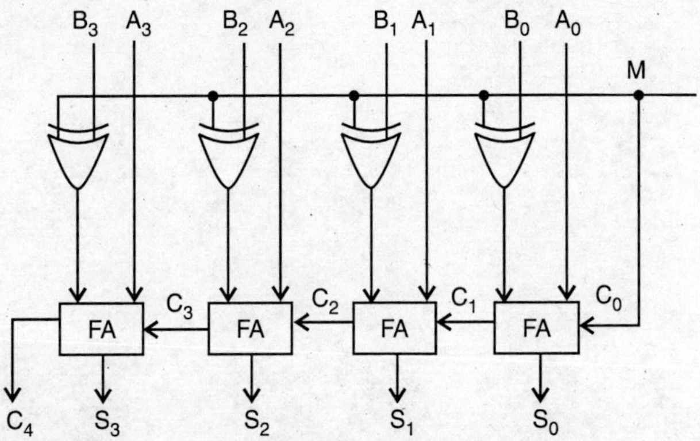

Fig. A.48.

Solution.

M	A	B	Sum	C_4
0	0111	0110	1101	0
0	1000	1001	0001	1
1	1100	1000	0100	1
1	0101	1010	1011	0
1	0000	0001	1111	0

$7 + 6 = 13$
$8 + 9 = 16 + 1$
$12 - 8 = 4$
$5 - 10 = -5$ (in 2's complement)
$0 - 1 = -1$ (in 2's complement)

Problem 18. Design a 4-bit combinational circuit decrements using four full adder circuits.

Solution: A 's complement of 1 = P + 1111

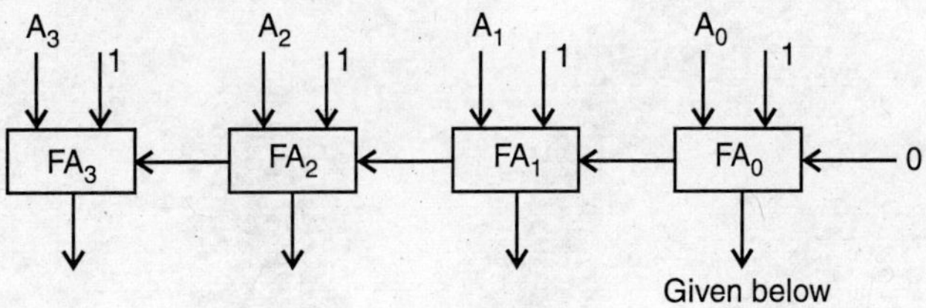

Fig. A.49.

❑❑❑

INDEX

INDEX

We request for your frank assessment regarding various aspects of the book as given below :

COMPUTER ORGANIZATION

By: Pankaj Sharma

Please Fill up the Sheet in Neat and Clean Writing

(i) What topics of your syllabus, important for your examination point of view, are not covered in this book ?

....

....

....

....

(ii) What are the chapters wherein the treatment of the subject matter is not systematic or organised or updated ?

....

....

....

....

(iii) Have you come across misprints/mistakes/factual inaccuracies in the book? Please specify the chapters and the page numbers.

....

....

....

....

(iv) What other books on the same subject have you found/heard better than the present book? Please specify in terms of price and quality

....

....

....

....

(v) Further suggestions and comments for the improvement of this book.

....

....

....

....

(PS: If need be please attach a separate sheet to write your views)

OTHER DETAILS

(i) Who recommended you this book ? (Please tick (✓) in the box near the option relevant to you.

☐ Teacher ☐ Friends ☐ Bookseller

....

....

....

....

(ii) Name of the recommending teacher, his designation and address :

....

....

....

....

(iii) Name and address of the bookseller you purchased the book from :

....

....

....

....

(iv) Name and address of your institution (Please mention the University or Board as the case may be)

....

....

....

....

(v) Your name and complete postal address :

....

....

....

....

(vi) Course you are enrolled in :

....

....

....

....

The best assessment will be awarded each month: The award will be in the form of our publications as decided by the Editorial Board.

Please mail the filled up coupon at your earliest to :

S.K. KATARIA & SONS®

4885/109, Prakash Mahal, Dr. Subhash Bhargav Lane,

Opposite Delhi Medical Association, Daryaganj, New Delhi–110002 (INDIA)

Phone: +91-11-43551243; *Mobile:* +91-9871775858, 7678425209

e-mail: katariabook@gmail.com, katariabooks@yahoo.com

Website: www.skkatariaandsons.com